N J A Sloane
Bell Telephone Labs
Murray Hill NJ 07974
Room 2C-363

17^{00}

Ergebnisse der Mathematik und ihrer Grenzgebiete

Band 44

P. Dembowski

Finite Geometries

Springer-Verlag New York Inc. 1968

Dr. phil. nat. Peter Dembowski
Wissenschaftlicher Rat und Professor
Johann Wolfgang Goethe-Universität
Frankfurt am Main

 · Library of Congress Catalog Card Number 68-15952
Printed in Germany
Title-No. 4588

Introduction

The subject matter of this book is an area of finite mathematics characterized by an interplay of combinatorial, geometric, and algebraic ideas. The best way to explain this further is to give a brief outline of the history of what will be called "finite geometries" here.

The combinatorial parts of the theory date back to the problems of EULER 1782, KIRKMAN 1847, STEINER 1853, on the possibility of arranging a finite number of objects into subsets according to specified rules. Although there are quite a few early contributions to these problems, a systematic treatment was missing until the nineteen-thirties when it was realized that combinatorial configurations are useful in the statistical theory of design of experiments. The notion of "balanced incomplete block design" (YATES 1936) provided a common background for the older investigations, and the constructions of BOSE 1939 and his followers showed the richness of this concept. Designs, as these systems will be called here in short, are the basic combinatorial structures of this book.

The most forceful impulse on the subject originated in projective geometry. This began under the influence of HILBERT's *Grundlagen der Geometrie* (1899), although the first actual computations in finite projective spaces had been performed earlier by v. STAUDT 1856, § 8. Hilbert's coordinatization procedure, later generalized to arbitrary projective planes by HALL 1943, made it possible to translate geometric problems into an algebraic language. For instance, the first examples of non-desarguesian finite projective planes, given by VEBLEN & WEDDERBURN 1907, were actually due to the algebraic constructions of DICKSON 1905. Many of the finite projective planes known today were originally found through similar algebraic ideas.

An alternative approach to the study of projective planes began with a paper by BAER 1942 in which the close relationship between Desargues' theorem and the existence of central collineations was pointed out. Baer's notion of (p,L)-transitivity, corresponding to this relationship, proved to be extremely fruitful. On the one hand, it provided a better understanding of coordinate structures (here SCHWAN 1919 was a forerunner); on the other hand it led eventually to the only coordinate-free, and hence geometrically satisfactory, classification of projective planes existing today, namely that by LENZ 1954a and

Barlotti 1957b. Due to deep discoveries in finite group theory, the analysis of this classification has been particularly penetrating for finite planes in recent years. In fact, finite groups were also applied with great success to problems not connected with (p,L)-transitivity.

Finite projective and affine geometries may be interpreted, in several ways if the dimension is at least 3, as special designs. (The close connection between the two fields was emphasized by Levi 1942.) Hence methods for the analysis of the combinatorial structure of designs may be used to study these geometries, too. One such method is the use of incidence matrices, by which certain parts of linear algebra and number theory become significant for finite geometries. Although these matrices appear, for example, in the book by Levi 1929 (as "incidence tables"), they were not used as genuine mathematical tools until two decades later, when their usefulness was established in three papers by Bruck & Ryser 1949, Bose 1949, and Schützenberger 1949. Since then, incidence matrices and their generalizations have proved indispensable in various investigations on designs and more general structures.

The present state of the theory is one of rapid growth. Finite projective planes still occupy the central position, but the reasons for this may be mainly historical. On the one hand, many results on finite planes have been recognized as special cases of theorems on designs and other configurations; on the other hand, structures which are not projective planes (for example inversive planes, Hjelmslev planes, generalized polygons) are gaining independent interest. The field is influenced increasingly by problems, methods, and results in the theory of finite groups, mainly for the well known reason that the study of automorphisms "has always yielded the most powerful results" (Artin 1957, p. 54). Finite-geometrical arguments can serve to prove group theoretical results, too, and it seems that the fruitful interplay between finite geometries and finite groups will become even closer in the future.

The purpose of the present tract is to give a reasonably complete account of the field described in the preceding paragraphs. There seems to be a need for such a treatment, since the relevant research papers are widely scattered in the literature, and since another comprehensive text does not exist: The booklet by Levi 1942 is largely superseded, and the respective parts dealing with finite geometries in the books by Pickert 1955, Hall 1959, Ryser 1963, Hall 1967b cover only parts of the subject. Moreover, due to the rapid development of the field, most of these treatments are already outmoded. This book, too, is not expected to stay up-to-date for a long time; its aim is to collect the known results, hence to facilitate new research, and thus to keep the subject alive.

Finite geometries have found several areas of application. As mentioned above, these are mainly in statistics, but there are relations to other fields as well (see for example JÄRNEFELT 1951; KUSTAANHEIMO 1950, 1952, 1957a, b). It would have been quite outside the scope of this treatment to cover applications, and so none of them will be given.

The material is divided into seven chapters of which only the first and the last are not exclusively concerned with designs in one form or another. The central part (Chapters 3—5) is occupied by finite projective planes. Each chapter has its own brief introduction describing its contents; thus it will not be necessary to say more about this here.

In order to keep the size of the tract reasonable, the subject had to be presented in a rather condensed fashion. (This is, of course, also in the tradition of the *Ergebnisse* series). Consequently, the successful mastery of the book will require some effort on the part of the reader. He is supposed to be familiar with standard notions of other fields, for example linear algebra and group theory. Many definitions, however, will be repeated for the sake of clarity. Proofs are either briefly outlined or not given at all. In any case, detailed references to the literature will be given whenever called for. These references are of two kinds (not mutually exclusive): some of them identify the original source of the result or definition in question and thus give historical credit, others are meant to guide the reader to a convenient place (usually in a book) where a proof can be found.

The bibliography is intended to serve not only as a source for references, but also as a guide to further investigations not mentioned in the main text, especially on construction techniques for designs and partial designs. There are approximately 1200 entries; in fact, the area covered by the bibliography is wider than that of finite geometries alone. For example, a reader interested in recent results on finite groups (particularly permutation groups) will find quite a few relevant items.

The suggestion to write this book came from Reinhold Baer. Without his continuous encouragements, mainly during the earlier stages, I should never have finished it. I am deeply grateful for his valuable advice concerning the manuscript and, more generally, for his constant interest and support since my student days.

A preliminary manuscript was written in London 1962—63; the present version took shape in Madison 1965—66 and Chicago 1966—67. I spent the interval 1963—65 in Frankfurt, Giessen, and Rome. In this way, I had the privilege of personal contact with a large number of mathematicians while writing this book; this had a very beneficial effect on the quality of the manuscript. I am particularly indebted to W. M. Kantor who contributed many essential improvements, especially to Chapter 2, and to H. Lüneburg who read both the manuscript and the galleys with unsurpassed thoroughness, thus eliminating many errors and obscurities.

Moreover, I owe thanks to all those people and institutions who aided me, by valuable comments on the manuscript, by making available unpublished material of their own, by thorough proofreading, by arranging stimulating personal contacts, by financial support, and in various other ways: J. André, Marianne Baer, A. Barlotti, D. Betten, R. Brualdi, R. H. Bruck, Judita Cofman, H. S. M. Coxeter, L. Dickey, G. Ewald, D. Foulser, M. Hall, Jr., C. Hering, D. G. Higman, K. A. Hirsch, D. R. Hughes, N. Ito, O. Kegel, Marion Kimberley, J. Landin, H. Lenz, L. Lombardo-Radice, J. Malkevitch, T. G. Ostrom, G. Panella, G. Pickert, F. Piper, J. Ryshpan, H. Salzmann, R. Sandler, A. Schleiermacher, F. A. Sherk, L. Solomon, J. Tits, A. Wagner, H. Wielandt, E. Wirsing, Jill Yaqub, the *Mathematisches Forschungsinstitut* in Oberwolfach, and the National Science Foundation.

Last, but certainly not least, I am most thankful to my wife Gertrud for creating and maintaining the peaceful working atmosphere without which I could not possibly have written this book.

Frankfurt am Main, April 1st, 1968

PETER DEMBOWSKI

Contents

Note on exposition

Each of the seven chapters is divided into four sections. As usual, sections are denoted by two integers; for example, 6.3 is the third section of Chapter 6.

The material in each section is numbered in two sequences. Lemmas, theorems, corollaries etc. are listed in bold face numerals without parentheses; some of these are subdivided into parts (**a**), (**b**), (**c**) etc. Equations, definitions, axioms etc. are listed as ordinary numerals in parentheses. A few particularly important conditions (for example the "regularity conditions" in Section 1.1) are denoted by capital letters instead of numerals.

Cross-references are given as in the following typical example: Part (**b**) of theorem **56** in Section 1.4 is referred to as **56b** throughout Section 1.4, and as **1.4.56b** in all other sections. Similarly, equation (34) of Section 1.4 is (34) there and (1.4.34) in other sections.

References to the bibliography are given by stating the name(s) of the author(s) in capitals and the year of appearance of the article in question. When not referring to a specific item in the bibliography, we do not use capitals for proper names.

The set of all x satisfying condition C is denoted by $\{x : C\}$ throughout. The symbol I is used to denote incidence, and we write

$$(Y) = \{x : x \mathrm{I} y \text{ for all } y \in Y\} \quad \text{and} \quad [Y] = |(Y)|,$$

where $|X|$ denotes, as usual, the number of elements in the set X. These are the most important of the less customary notations used here; a detailed list of other special notations appears at the end of the book (cf. p. 369). The reader is also advised to consult the Dictionary, of alternatives for technical terms as used in this book, on p. 367.

Finally, the reader should not be confused by seeming inconsistencies in spelling. For example, the terms "center" and "centre", or "embedding" and "imbedding", are used with different technical meanings.

1. Basic concepts

This chapter is of a preliminary nature. In its first three sections, we introduce the general notions of incidence structure and incidence preserving map, and define incidence matrices, an important tool for the proof of many non-trivial results to be encountered later. We shall develop general theorems on finite incidence structures (many of them originally proved under unnecessarily restrictive hypotheses) and thus set up the framework for the more specialized investigations in later chapters.

In section 1.4, an account of (mostly recent) results in projective and affine geometries over Galois fields will be presented. The reasons that these appear here under the heading of "basic concepts" are (a) that this material appears to be more "classical" (i.e., familiar) to many mathematicians than the rest of the book, (b) that these geometries provide the most obvious, and most homogeneous, examples for various types of more general incidence structures to be encountered later, and (c) that they have often served as incentives for investigations concerned with characterizing finite projective or affine geometries, or certain substructures embedded in them, within wider classes of incidence structures.

1.1 Finite incidence structures

An *incidence structure*[1]) is a triple $\mathbf{S} = (\mathfrak{p}, \mathfrak{B}, \mathrm{I})$, where $\mathfrak{p}, \mathfrak{B}, \mathrm{I}$ are sets with

$$\mathfrak{p} \cap \mathfrak{B} = \emptyset \quad \textit{and} \quad \mathrm{I} \subseteqq \mathfrak{p} \times \mathfrak{B}. \tag{1}$$

The elements of $\mathfrak{p}$ are called *points*, those of $\mathfrak{B}$ *blocks*, and those of I *flags*. Points and blocks will usually be denoted by lower and upper case Latin letters, and sets of points or blocks by lower and upper case

[1]) This should not be confused with the term "Inzidenzstruktur" in PICKERT 1955; Pickert's terminology coincides with what we shall call, with HALL 1943, a "partial plane"; see Section 1.2 below. Other more general concepts which would deserve the name "incidence structure", where there are more than two kinds of elements, have been studied by many authors; we mention here only MOORE 1896 and TITS 1956.

German letters, respectively. Instead of "(p, B) is a flag" we also say "p and B are incident", "p is on B", "B passes through p", or use similar geometric language; instead of $(p, B) \in \mathrm{I}$ we shall usually write $p \, \mathrm{I} \, B$ or $B \, \mathrm{I} \, p$.

The *dual structure* $\overline{\mathbf{S}} = (\overline{\mathfrak{p}}, \overline{\mathfrak{B}}, \overline{\mathrm{I}})$ of the incidence structure $\mathbf{S} = (\mathfrak{p}, \mathfrak{B}, \mathrm{I})$ is defined by

(2) $\overline{\mathfrak{p}} = \mathfrak{B}, \quad \overline{\mathfrak{B}} = \mathfrak{p}, \quad$ *and* $\quad (B, p) \in \overline{\mathrm{I}} \quad$ *if and only if* $\quad (p, B) \in \mathrm{I}.$

It follows from (1) that $\overline{\mathbf{S}}$ is also an incidence structure. If in a statement S on incidence structures the words "point" and "block" are interchanged, we obtain the *dual* statement $\overline{\mathrm{S}}$ of S, and (1) and (2) imply the following *duality principle*:

1. *Let $\mathscr{C}$ be a class of incidence structures such that $\mathbf{S} \in \mathscr{C}$ implies $\overline{\mathbf{S}} \in \mathscr{C}$. Then if the statement S is valid for all $\mathbf{S} \in \mathscr{C}$, so is the dual statement $\overline{\mathrm{S}}$.*

Because of **1**, it will be advantageous to state as many definitions and results as possible in a *self-dual* language, i.e. in the form of statements S with $\overline{\mathrm{S}} = \mathrm{S}$. For this we need a common term for points and blocks: by the *elements* of an incidence structure $\mathbf{S} = (\mathfrak{p}, \mathfrak{B}, \mathrm{I})$ we shall mean the elements of $\mathfrak{p} \cup \mathfrak{B}$.

The *complementary structure* $\mathbf{S}' = (\mathfrak{p}', \mathfrak{B}', \mathrm{I}')$ of the incidence structure $\mathbf{S} = (\mathfrak{p}, \mathfrak{B}, \mathrm{I})$ is defined by

(3) $$\mathfrak{p}' = \mathfrak{p}, \quad \mathfrak{B}' = \mathfrak{B}, \quad \mathrm{I}' = (\mathfrak{p} \times \mathfrak{B}) - \mathrm{I};$$

thus incidence in $\mathbf{S}'$ is defined by non-incidence in $\mathbf{S}$.

If $\mathbf{S} = (\mathfrak{p}, \mathfrak{B}, \mathrm{I})$ is an incidence structure and if $\mathfrak{q}$ and $\mathfrak{C}$ are subsets of $\mathfrak{p}$ and $\mathfrak{B}$, respectively, then the *substructure* of $\mathbf{S}$ defined by $\mathfrak{q}$ and $\mathfrak{C}$ is the incidence structure $\mathbf{T} = (\mathfrak{q}, \mathfrak{C}, \mathrm{J})$, with

(4) $$\mathrm{J} = \mathrm{I} \cap (\mathfrak{q} \times \mathfrak{C});$$

thus a substructure of $\mathbf{S}$ is a set of points and blocks, together with the incidence given in $\mathbf{S}$.

All classes of incidence structures considered later will be defined by certain conditions (incidence axioms) on the set I of flags. Often we shall only consider such substructures which satisfy the same incidence axioms, but here we define four other types of substructures. These consist, intuitively speaking, of the elements incident or nonincident, respectively, with a given fixed element. To be more precise, we introduce the following notation: Let $\mathbf{S} = (\mathfrak{p}, \mathfrak{B}, \mathrm{I})$ be an arbitrary incidence structure, and let $\mathfrak{q} \subseteqq \mathfrak{p}$, $\mathfrak{C} \subseteqq \mathfrak{B}$. We write $\mathfrak{q} \, \mathrm{I} \, \mathfrak{C}$ if $x \, \mathrm{I} \, X$ for all $x \in \mathfrak{q}$ and $X \in \mathfrak{C}$. Identifying X with $\{X\}$ and x with $\{x\}$, we define:

(5) $$(\mathfrak{q}) = \{X \in \mathfrak{B} : X \, \mathrm{I} \, \mathfrak{q}\}, \quad (\mathfrak{C}) = \{x \in \mathfrak{p} : x \, \mathrm{I} \, \mathfrak{C}\}.$$

Also, we write (p) instead of $(\{p\})$ and (B) instead of $(\{B\})$.[1]) Now the *internal structure* $\mathbf{S}_p$ with respect to the point p is the substructure defined by the point set $\mathfrak{p} - \{p\}$ and the block set (p), and the internal structure $\mathbf{S}_B$ with respect to the block B is, dually, the substructure defined by (B) and $\mathfrak{B} - \{B\}$. Similarly, the *external structures* $\mathbf{S}^p$ and $\mathbf{S}^B$ with respect to p and B are the substructures defined by $\mathfrak{p} - \{p\}$, $\mathfrak{B} - (p)$, and $\mathfrak{p} - (B)$, $\mathfrak{B} - \{B\}$, respectively. While incidence structures of the types $\mathbf{S}_B$ and $\mathbf{S}^p$ will occur only occasionally later on, internal structures $\mathbf{S}_p$ and external structures $\mathbf{S}^B$ will play important roles in various parts of Chapters 2—6.[2])

The incidence structures $\mathbf{S} = (\mathfrak{p}, \mathfrak{B}, \mathrm{I})$ considered in this book will usually be *finite*; this means that $\mathfrak{p}$ and $\mathfrak{B}$, and hence also I, are finite sets. It follows that the sets defined in (5) are likewise finite, and we shall need a notation for the numbers of their elements. We define

$$(6) \quad [\mathfrak{q}] = |(\mathfrak{q})| \text{ for every } \mathfrak{q} \subseteq \mathfrak{p}; \quad [\mathfrak{C}] = |(\mathfrak{C})| \text{ for every } \mathfrak{C} \subseteq \mathfrak{B}.$$

Thus $[\mathfrak{q}]$ is the number of blocks incident with every point of $\mathfrak{q}$, and $[\mathfrak{C}]$ is defined dually.[1])

Let $\mathbf{S} = (\mathfrak{p}, \mathfrak{B}, \mathrm{I})$ be a finite incidence structure. We put $|\mathfrak{p}| = v$ and $|\mathfrak{B}| = b$. We define the *parameters*[3]) of $\mathbf{S}$ as the following rational numbers:

$$(7) \quad \begin{aligned} v_0 &= v, \quad v_m = \binom{b}{m}^{-1} \sum_{\substack{\mathfrak{X} \subseteq \mathfrak{B} \\ |\mathfrak{X}| = m}} [\mathfrak{X}], \quad m = 1, \ldots, b; \\ b_0 &= b, \quad b_n = \binom{v}{n}^{-1} \sum_{\substack{\mathfrak{x} \subseteq \mathfrak{p} \\ |\mathfrak{x}| = n}} [\mathfrak{x}], \quad n = 1, \ldots, v. \end{aligned}$$

Thus v_m is the average number of points on m blocks, and b_n is defined dually. By definition, all parameters are non-negative, but clearly

$$v_m = 0 \quad \text{for} \quad m > \max\{[p] : p \in \mathfrak{p}\},$$

and

$$b_n = 0 \quad \text{for} \quad n > \max\{[B] : B \in \mathfrak{B}\}.$$

---

[1]) The notations $(\mathfrak{q})$, $(\mathfrak{C})$, $[\mathfrak{q}]$, $[\mathfrak{C}]$ are not quite satisfactory insofar as $(\emptyset)$ and $[\emptyset]$ are not uniquely defined. However, this will not create confusion later on.

[2]) The ordinary euclidean (affine) plane $\mathbf{E}$, with the lines as blocks, is a classical example both for the type $\mathbf{S}_p$ and for the type $\mathbf{S}^B$: If $\mathbf{S}$ consists of the points and plane sections of an ordinary sphere, then $\mathbf{E}$ is $\mathbf{S}_p$ for any point p of $\mathbf{S}$; but if $\mathbf{S}$ is the real projective plane, with lines as blocks, then $\mathbf{E}$ is $\mathbf{S}^B$ for any line B of $\mathbf{S}$. Both these examples will serve as motivation for later investigations. — One reason that $\mathbf{S}_B$ has not received much interest so far is that in the more familiar situations $\mathbf{S}$ contains two distinct blocks C, D such that $(C) \neq (D)$ in $\mathbf{S}$ but $(C) = (D)$ in some $\mathbf{S}_B$. Almost all incidence structures encountered later will have the property that $(B) = (C)$ implies $B = C$; in these cases the blocks can be interpreted as the sets of points incident with them, and incidence by set theoretic inclusion.

[3]) We use the letters b and v (and not, for example, b and p) because they are quite customary in the statistical literature; the v stands for "varieties".

2. *The parameters v_m, b_n of an arbitrary finite incidence structure satisfy the following inequalities:*

$$v_{m+1} \leqq v_m, \qquad b_{n+1} \leqq b_n \tag{8}$$

$$\left.\begin{array}{l} v_m(b_1 - m) \leqq v_{m+1}(b_0 - m) \\ b_n(v_1 - n) \leqq b_{n+1}(v_0 - n) \end{array}\right\} \quad m = 0, \ldots, b-1; \quad n = 0, \ldots, v-1. \tag{9}$$

In (8), *equality holds precisely for those* m (*resp.* n) *for which there is no point* x *with* $m \leqq [x] < b$ (*no block* X *with* $n \leqq [X] < v$). *In* (9), *equality holds for all* m (*resp.* n) *if and only if* $[p] = b_1$ *for every* $p \in \mathfrak{p}$ ($[B] = v_1$ *for every* $B \in \mathfrak{B}$). *If this condition is not satisfied, then equality in* (9) *holds precisely for* $m = n = 0$ *and* $m > \max\{[p] : p \in \mathfrak{p}\}$, $n > \max\{[B] : B \in \mathfrak{B}\}$.

For the proof of **2**, cf. Dembowski 1961, (3.5) and Satz 1. The main technique of this proof is a simple counting principle: If $\mathbf{T} = (\mathfrak{q}, \mathfrak{C}, \mathrm{J})$ is a substructure of $\mathbf{S} = (\mathfrak{p}, \mathfrak{B}, \mathrm{I})$, then J can be enumerated in two different ways: by first counting blocks and then points on blocks, and in the dual fashion. Hence

$$\sum_{X \in \mathfrak{C}} |(X) \cap \mathfrak{q}| = \sum_{x \in \mathfrak{q}} |(x) \cap \mathfrak{C}| = |\mathrm{J}|. \tag{10}$$

This equation is very useful in many investigations on finite incidence structures. We shall refer to (8) and (9) as the *parameter inequalities* for finite incidence structures.

The parameters of $\mathbf{S}$ are related to those of the dual structure $\bar{\mathbf{S}}$ as follows:

$$\bar{v}_m = b_m, \quad m = 0, \ldots, v; \qquad \bar{b}_n = v_n, \quad n = 0, \ldots, b. \tag{11}$$

The relationship with the parameters v'_m and b'_n of the complementary structure $\mathbf{S}'$ is more complicated:

$$\begin{aligned} v'_m &= \binom{b}{m}^{-1} \sum_{i=0}^{m} (-1)^i \binom{b}{i} \binom{b}{m-i} v_i, \\ b'_n &= \binom{v}{n}^{-1} \sum_{i=0}^{n} (-1)^i \binom{v}{i} \binom{v}{n-i} b_i, \end{aligned} \tag{12}$$

for $m = 0, \ldots, b$ and $n = 0, \ldots, v$. The proof of (12) is not difficult; one uses the "principle of inclusion and exclusion", see Riordan 1958, p. 51, or Ryser 1963, p. 18.

The remarks in **2** on equality in (9) show that incidence structures with $[B] = v_1$ and $[p] = b_1$ for all $B \in \mathfrak{B}$ and $p \in \mathfrak{p}$ (i.e. v_1 and b_1 are not only the average but the exact numbers of points on any block or of blocks through any point) will play a distinguished role. Such incidence structures with equally many points on every block and equally many blocks through every point are called *tactical configura-*

tions.[1]) It is customary to write r and k instead of b_1 and v_1, so that we have $[x] = r$ for every $x \in \mathfrak{p}$ and $[X] = k$ for every $X \in \mathfrak{B}$. For tactical configurations $(\mathfrak{p}, \mathfrak{B}, \mathrm{I})$, the inequalities (8) become

$$(8') \qquad \begin{aligned} b = b_0 = r = b_1 = \cdots = b_v \quad &\text{and} \\ v = v_0 = k = v_1 = \cdots = v_b \quad &\text{if} \quad \mathrm{I} = \mathfrak{p} \times \mathfrak{B} \end{aligned}$$

and

$$(8'') \qquad b_0 > b_1 > \cdots > b_v, \qquad v_0 > v_1 > \cdots > v_b \quad \text{if} \quad \emptyset \neq \mathrm{I} \neq \mathfrak{p} \times \mathfrak{B}.$$

In either case, (9) becomes

$$(9') \qquad b\,k = v\,r, \quad v_m \binom{b}{m} = v \binom{r}{m}, \quad \text{and} \quad b_n \binom{v}{n} = b \binom{k}{n}.$$

We shall now give a combinatorial classification for tactical configurations. We consider the following *regularity conditions* for finite incidence structures:

(R.m) $\quad v_m > 0$, *and* $[\mathfrak{X}] = v_m$ *for all* $\mathfrak{X} \subseteqq \mathfrak{B}$ *with* $|\mathfrak{X}| = m$;

($\bar{\mathrm{R}}$.n) $\quad b_n > 0$, *and* $[\mathfrak{x}] = b_n$ *for all* $\mathfrak{x} \subseteqq \mathfrak{p}$ *with* $|\mathfrak{x}| = n$;

with $m, n = 0, 1, \ldots$

Conditions (R.0) and ($\bar{\mathrm{R}}$.0) merely state that $\mathfrak{p}$ and $\mathfrak{B}$, respectively, are non-empty; the tactical configurations are precisely the finite incidence structures satisfying (R.1) and ($\bar{\mathrm{R}}$.1). We now have the following result:

3. (R.1) *and* ($\bar{\mathrm{R}}$.n) *imply* ($\bar{\mathrm{R}}$.1), ..., ($\bar{\mathrm{R}}$.$n-1$); *and* ($\bar{\mathrm{R}}$.1) *and* (R.m) *imply* (R.1), ..., (R.$m-1$).

This was known to MOORE 1896, and probably to earlier writers. For a short proof, using the counting principle (10), see DEMBOWSKI 1961, Satz 2. It follows from **3** that for any tactical configuration **C** there exist two uniquely determined integers s, t such that (R.1), ..., (R.s) and ($\bar{\mathrm{R}}$.1), ..., ($\bar{\mathrm{R}}$.t) are satisfied, but (R.$s+1$) and ($\bar{\mathrm{R}}$.$t+1$) are violated in **C**. The pair (s, t) is called the *type* of **C**. Clearly, if **C** is of type (s, t), the parameters $v_0, \ldots, v_s$ and $b_0, \ldots, b_t$ are integers.

If $|\mathfrak{p}| = t$ and $|\mathfrak{B}| = s$, then the incidence structure $(\mathfrak{p}, \mathfrak{B}, \mathrm{I})$ with $\mathrm{I} = \mathfrak{p} \times \mathfrak{B}$ is a tactical configuration of type (s, t). The next result shows, among other things, that these trivial structures are the only tactical configurations whose type can be arbitrarily prescribed.

4. *If the tactical configuration* $\mathbf{C} = (\mathfrak{p}, \mathfrak{B}, \mathrm{I})$ *with* $\mathrm{I} \neq \mathfrak{p} \times \mathfrak{B}$ *has type* (s, t) *with* $s, t > 1$ *and at least one of* s *and* t *greater than* 2, *then* $|\mathfrak{B}| = v = |\mathfrak{p}|$, $s = t = v - 1$, *and* $v_i = b_i = v - i$ *for* $i = 0, \ldots, v-1$.

---

[1]) This term seems to be due to MOORE 1896. Note, however, that Moore would have called the present structures "tactical configurations of rank 2".

In this case, the blocks of $\mathbf{C}$ can be identified with the point sets $\mathfrak{x}$ with $|\mathfrak{x}| = v - 1$ (incidence being set theoretic inclusion), so that $\mathbf{C}$ is again of a rather trivial nature. For the proof of **4**, see DEMBOWSKI 1961, Satz 5.[1])

We call the incidence structure $\mathbf{S} = (\mathfrak{p}, \mathfrak{B}, \mathrm{I})$ *nondegenerate* if $\mathfrak{p}$ and $\mathfrak{B}$ are both non-empty and if the following *nondegeneracy conditions* hold:

(N)
(a) *To every block B there are two distinct points non-incident with B,*
(b) *To every point p there are two distinct blocks non-incident with p.*

The following is then an immediate consequence of **4**:

5. *If (s, t) is the type of a nondegenerate tactical configuration, then either $s = 1$ or $t = 1$, or else $s = t = 2$.*

If $\mathbf{C}$ is of type (s, t), then the dual structure $\bar{\mathbf{C}}$ is of type (t, s). It suffices, therefore, to consider tactical configurations of type (s, t) with

$$s \leqq t. \tag{13}$$

Any nondegenerate tactical configuration which is not of type $(1, 1)$ and satisfies (13) is called a *design.*[2]) In other words: a design is a finite incidence structure satisfying conditions (R.1), ($\bar{\mathrm{R}}$.2), and (N), or a nondegenerate tactical configuration of type (s, t) with $t > 1$. The concept of design is fundamental for this book. We shall develop a general theory of designs in Chapter 2, and investigate in detail several special classes of designs in Chapters 3—6.

Few general results are known about tactical configurations which are not designs, i.e. essentially those of type $(1, 1)$. Many configurations of classical geometry are of this kind, for example the configurations of Desargues and Pappus in projective, and those of Miquel and the bundle theorem in inversive geometry (see Sections 1.4, 3.4 and 6.1, respectively). Such "geometric configurations" are the central topic of LEVI 1929; see also EMCH 1929; WEISS 1931; CARMICHAEL 1931a, 1937, Chapters 11 to 14; HARSHBARGER 1931; SHAUB & SCHOONMAKER 1931; KLUG 1932,

[1]) In a slightly different form, the result has also been proved by HSU 1962.

[2]) We use this term instead of the more awkward "balanced incomplete block design" (often abbreviated BIBD) of the statistical literature. Designs are useful for planning and analyzing agricultural and other experiments; this is the reason for the interest of the statisticians in the subject, and they have made important contributions to it (references in Chapter 2). In this book we shall be concerned only with the purely combinatorial aspect of design theory; for the statistical applications, the reader is referred to MANN 1949, COCHRAN & COX 1957. — Note that in the presence of ($\bar{\mathrm{R}}$.2) the nondegeneracy condition (b) is a consequence of (a); see footnote 1) on p. 57.

1934; DE VRIES 1936; JUNG & MELCHIOR 1936, MELCHIOR 1937; FRAME 1938; MAIER 1939; RICKART 1940; KOMMERELL 1941, 1949; BENNETON 1944, 1945a, b; MANDAN 1945; ZACHARIAS 1941, 1948, 1949, 1951a, b, 1952, 1953; COXETER 1950; LAUFFER 1954a, b; BYDZOVSKY 1954; METELKA 1955a, b, 1957; ROSATI 1957a, 1958a; NOVÁK 1959; CROWE 1959, 1961; BURAU 1963. Some special types of tactical configurations of type (1, 1) will be treated in more detail in Chapter 7. In particular, further references to construction techniques for such configurations will be given at the end of Section 7.1.

A finite *regular graph* without loops can be interpreted as a tactical configuration with $v_1 = 2$, and conversely. Also, an arbitrary incidence structure $\mathbf{S} = (\mathfrak{p}, \mathfrak{B}, \mathrm{I})$ gives rise to a graph $G(\mathbf{S})$ whose vertices are the elements of $\mathfrak{p} \cup \mathfrak{B}$ and whose edges are the elements of I; this graph is finite and regular (i.e. each vertex is incident with equally many edges) if and only if $\mathbf{S}$ is a tactical configuration with $v = b$. Thus there is a considerable overlap between the theories of graphs and finite incidence structures.[1] It would, however, lead too far to go into the details here; the reader must be referred to the literature on graph theory, e.g. KÖNIG 1936, BERGE 1958, RINGEL 1959, ORE 1962, and the references collected there.

We conclude this section with the introduction of a concept which will prove useful later on. A *tactical decomposition*[2] of an incidence structure $\mathbf{S} = (\mathfrak{p}, \mathfrak{B}, \mathrm{I})$ is a partition of $\mathfrak{p}$ into disjoint point sets (called the *point classes*) $\mathfrak{x}$, together with a partition of $\mathfrak{B}$ into disjoint block sets (*block classes*) $\mathfrak{Y}$, such that for any such $\mathfrak{x}$, $\mathfrak{Y}$ the substructure $(\mathfrak{x}, \mathfrak{Y}, \mathrm{I} \cap (\mathfrak{x} \times \mathfrak{Y}))$ defined by $\mathfrak{x}$ and $\mathfrak{Y}$ is a tactical configuration (in particular, all the $\mathfrak{x}$, $\mathfrak{Y}$ must be finite, which is of course obvious if $\mathbf{S}$ is finite). In other words: For any point class $\mathfrak{x}$ and any block class $\mathfrak{Y}$, the number of points of $\mathfrak{x}$ on a block $B \in \mathfrak{Y}$ depends only on $\mathfrak{x}$ and $\mathfrak{Y}$, not on B, and can hence be denoted by $(\mathfrak{x}\, \mathfrak{Y})$; dually, the number of blocks of $\mathfrak{Y}$ through $p \in \mathfrak{x}$ depends only on $\mathfrak{Y}$, $\mathfrak{x}$ and can hence be denoted by $(\mathfrak{Y}\, \mathfrak{x})$. The first equation (9') yields, with this notation:

(14) $(\mathfrak{x}\, \mathfrak{Y})\, |\mathfrak{Y}| = (\mathfrak{Y}\, \mathfrak{x})\, |\mathfrak{x}|$ *for any point class* $\mathfrak{x}$ *and block class* $\mathfrak{Y}$.

Furthermore, if $\mathbf{S}$ is finite, we clearly have

(15) $$\sum_{\mathfrak{x}} |\mathfrak{x}| = v \quad \text{and} \quad \sum_{\mathfrak{Y}} |\mathfrak{Y}| = b,$$

[1] As a typical example for the combination of both fields, we mention here ARCHBOLD & SMITH 1962. See also LEVI 1942, Chapter 1, and HOFFMAN 1965. The graph $G(\mathbf{S})$ seems to have been first considered by LEVI 1929 and is therefore sometimes called the "Levi graph" of $\mathbf{S}$. (Other authors call these graphs "bipartite".)

[2] These were introduced, though for more special incidence structures, in DEMBOWSKI 1958.

where $\mathfrak{x}$ ranges over all point classes, $\mathfrak{Y}$ over all block classes. Moreover, if **S** is a tactical configuration, and if $\mathfrak{q}$ is a fixed point class and $\mathfrak{C}$ a fixed block class, we get

$$\sum_{\mathfrak{x}} (\mathfrak{x}\,\mathfrak{C}) = k \quad \text{and} \quad \sum_{\mathfrak{Y}} (\mathfrak{Y}\,\mathfrak{q}) = r, \tag{16}$$

with the same ranges of $\mathfrak{x}$ and $\mathfrak{Y}$ as in (15).

For any point class $\mathfrak{x}$ and block class $\mathfrak{Y}$ of a tactical decomposition, we have $(\mathfrak{x}\,\mathfrak{Y}) = 0$ if and only if $(\mathfrak{Y}\,\mathfrak{x}) = 0$. This observation leads to the definition of the *quotient structure* $\mathbf{S}/\Delta = (\mathfrak{p}/\Delta, \mathfrak{B}/\Delta, \mathrm{I}/\Delta)$ of the incidence structure $\mathbf{S} = (\mathfrak{p}, \mathfrak{B}, \mathrm{I})$ with respect to the tactical decomposition Δ of **S**: The elements of $\mathfrak{p}/\Delta$ are the point classes, those of $\mathfrak{B}/\Delta$ the block classes of Δ, and $(\mathfrak{x}, \mathfrak{Y}) \in \mathrm{I}/\Delta$ if and only if $(\mathfrak{x}\,\mathfrak{Y}) \neq 0$ or, equivalently, $(\mathfrak{Y}\,\mathfrak{x}) \neq 0$. The quotient structures are connected with certain epimorphisms of finite incidence structures (see Section 1.2); they will be used in Section 3.2.

1.2 Incidence preserving maps

Let $\mathbf{S} = (\mathfrak{p}, \mathfrak{B}, \mathrm{I})$ and $\mathbf{T} = (\mathfrak{q}, \mathfrak{C}, \mathrm{J})$ be two incidence structures. An *incidence preserving map* of **S** into **T** is a mapping φ of $\mathfrak{p} \cup \mathfrak{B}$ into $\mathfrak{q} \cup \mathfrak{C}$ such that

$$p \,\mathrm{I}\, B \quad \text{implies} \quad p\varphi \,\mathrm{J}\, B\varphi \quad \text{for all } p \in \mathfrak{p} \text{ and } B \in \mathfrak{B}. \tag{1}$$

We term such a mapping a *homomorphism* if $\mathfrak{p}\varphi \subseteqq \mathfrak{q}$ and $\mathfrak{B}\varphi \subseteqq \mathfrak{C}$, and an *anti-homomorphism* if $\mathfrak{p}\varphi \subseteqq \mathfrak{C}$ and $\mathfrak{B}\varphi \subseteqq \mathfrak{q}$. If **S** is *connected* in the sense that to any two elements $x, y \in \mathfrak{p} \cup \mathfrak{B}$ there exists a sequence $u_1, \ldots, u_n$ of elements in $\mathfrak{p} \cup \mathfrak{B}$ such that $x \,\mathrm{I}\, u_1 \,\mathrm{I} \ldots \mathrm{I}\, u_n \,\mathrm{I}\, y$, then every incidence preserving map of **S** into **T** is either a homomorphism or an anti-homomorphism; but there exist other incidence preserving maps whenever **S** is not connected. Only connected incidence structures will be considered later.

An *epimorphism* of **S** onto **T** is a homomorphism φ with $\mathfrak{p}\varphi = \mathfrak{q}$ and $\mathfrak{B}\varphi = \mathfrak{C}$, and an *isomorphism* is a one-one epimorphism whose inverse is likewise incidence preserving (and hence also an isomorphism).[1] **S** and **T** are *isomorphic* if there exists an isomorphism of **S** onto **T**. The terms *anti-epimorphism* and *anti-isomorphism* are defined in the obvious fashion: here we postulate $\mathfrak{p}\varphi = \mathfrak{C}$ and $\mathfrak{B}\varphi = \mathfrak{q}$, and that the inverse of an anti-isomorphism is also an anti-isomorphism. **S** and **T**

[1]) Trivial examples show that there exist one-one epimorphisms of **S** onto **T** whose inverses are not incidence preserving. Hence if the words "isomorphism" and "isomorphic" are to have their usual meanings, the condition that φ^{-1} be incidence preserving is indispensable. For special classes of incidence structures, however (e.g. for designs), the condition is superfluous.

are *dual* to each other if there exists an anti-isomorphism of $\mathbf{S}$ onto $\mathbf{T}$. *Automorphisms* (and *anti-automorphisms*) are isomorphisms (anti-isomorphisms) of an incidence structure onto itself, and a *polarity* is an anti-automorphism of order 2. $\mathbf{S}$ is *self-dual* if there exists an anti-automorphism, and *self-polar* if there exists a polarity of $\mathbf{S}$. The set of all automorphisms and anti-automorphisms of $\mathbf{S}$ is a group, denoted by $\mathsf{A}(\mathbf{S})$, and the automorphisms form a subgroup $\mathrm{Aut}\,\mathbf{S}$ which is of index 2 if $\mathbf{S}$ is self-dual and coincides with $\mathsf{A}(\mathbf{S})$ if $\mathbf{S}$ is not self-dual.

Let φ be an incidence preserving map of $\mathbf{S}$ into itself. An element (point or block) of $\mathbf{S}$ is called *φ-fixed* if it is mapped by φ onto itself, and *φ-absolute* if it is incident with its φ-image.[1]) Similarly, if Φ is a set of incidence preserving maps of $\mathbf{S}$ into $\mathbf{S}$, an element is called *Φ-fixed* or *Φ-absolute* if it is φ-fixed or φ-absolute for every $\varphi \in \Phi$, respectively. In almost all later applications, Φ will either consist of just one element, namely an automorphism or a polarity, or else be a subgroup of $\mathrm{Aut}\,\mathbf{S}$. These subgroups are the *automorphism groups* of $\mathbf{S}$; their study is one of the more important subjects of this book. The substructure of all Φ-fixed or Φ-absolute elements of $\mathbf{S}$ often gives useful information about $\mathbf{S}$ and Φ, as we shall see frequently later on. Here we mention only a few rather obvious facts.

An incidence structure $\mathbf{S}$ is called a *partial plane* if $[p, q] \leqq 1$ for distinct points p, q or, equivalently, $[B, C] \leqq 1$ for distinct blocks B, C [cf. (1.1.6) for notation]. The following are two simple but useful properties of partial planes.

1. *Let φ be an automorphism of a partial plane. Suppose that φ either fixes or interchanges two distinct points p, q, both incident with the block B. Then φ fixes B. Dually, if φ fixes or interchanges two distinct blocks $B, C \mathrel{\mathrm{I}} p$, then $p\varphi = p$.*

This is an obvious consequence of the definitions.

2. *Let π be a polarity of a partial plane, and p a π-absolute point. Then $p\pi$ is the only π-absolute block incident with p and, dually, p is the only π-absolute point on $p\pi$.*

For, if $p \neq q \mathrel{\mathrm{I}} q\pi$ and $q \mathrel{\mathrm{I}} p\pi$, then $[p, q] > 1$ because both $p\pi$ and $q\pi$ are incident with p and q.

For the remainder of this section, we shall mainly be concerned with an arbitrary automorphism group Γ of a finite incidence structure $\mathbf{S} = (\mathfrak{p}, \mathfrak{B}, \mathrm{I})$; we shall not always repeat these hypotheses. Γ induces

[1]) If there is no danger of confusion, we say simply "fixed" and "absolute" instead of "φ-fixed" and "φ-absolute". This remark applies also to other terms to be defined presently, like "Φ-fixed", "Φ-absolute", "Γ-conjugate", and "Γ-orbit".

permutation groups on each of the sets $\mathfrak{p}$, $\mathfrak{B}$, I, and if P is an arbitrary permutation group theoretic property, such as (semi-) regularity, primitivity, or k-transitivity[1]), we say that Γ is *point*-P, *block*-P, or *flag*-P if the induced group on $\mathfrak{p}$, $\mathfrak{B}$, or I has property P. Since $|\mathfrak{p}|$ is in general different from $|\mathfrak{B}|$, the representations of Γ on $\mathfrak{p}$ and $\mathfrak{B}$ cannot be expected to be similar[2]), so that point-P and block-P are in general non-equivalent properties. We shall see later (cf. also Sections 1.3 and 2.3) that under additional hypotheses the equivalence of point- and block-P may be proved for certain P (e.g., transitivity). Here we note the following easily proved result:

3. *Let Γ be an automorphism group of $\mathbf{S} = (\mathfrak{p}, \mathfrak{B}, I)$. The representations of Γ on $\mathfrak{p}$ and $\mathfrak{B}$ are similar if, and only if, there exists a permutation π of $\mathfrak{p} \cup \mathfrak{B}$ such that*

(**a**) $\mathfrak{p}\,\pi = \mathfrak{B}$ *and* $\mathfrak{B}\,\pi = \mathfrak{p}$, *and*
(**b**) $\pi\Gamma = \Gamma\pi$

Consequently, if $\mathbf{S}$ is self-dual, then $\operatorname{Aut}\mathbf{S}$ has similar representations on $\mathfrak{p}$ and $\mathfrak{B}$. Another special case is:

4. *If Γ is point and block transitive on $\mathbf{S} = (\mathfrak{p}, \mathfrak{B}, I)$ and if $\Gamma_p = \Gamma_B$ for some $p \in \mathfrak{p}$ and $B \in \mathfrak{B}$, then the representations of Γ on $\mathfrak{p}$ and $\mathfrak{B}$ are similar.*

Here Γ_p and Γ_B denote the stabilizers of p and B in Γ. The desired π is given by $(p\gamma)\,\pi = B\gamma$ and $(B\gamma)\,\pi = p\gamma$, for all $\gamma \in \Gamma$.

Two points p and q are called Γ-*conjugate* if there exists $\gamma \in \Gamma$ such that $p\gamma = q$. Among blocks and flags, Γ-conjugacy is defined similarly. Clearly, Γ-conjugacy is an equivalence relation on each of the sets $\mathfrak{p}$, $\mathfrak{B}$, I; the equivalence classes of this relation are the Γ-(point, block, or flag) *orbits* of $\mathbf{S}$. The Γ-fixed elements are, of course, just those points and blocks which constitute full Γ-orbits by themselves. The following is an immediate consequence of the definitions:

5. *The point and block orbits of Γ form a tactical decomposition of* $\mathbf{S}$.

This is true even when Γ is a finite automorphism group of an infinite incidence structure. Result **5** is the main reason for the importance of tactical decompositions. Γ is *point, block,* or *flag transitive* if there is only one point, block, or flag orbit, respectively. Hence **5** implies:

---

[1]) For the definitions of these standard permutation group theoretic terms, the reader is referred to WIELANDT 1964. — Note that in general none of the three representations of Γ on $\mathfrak{p}$, $\mathfrak{B}$, and I need be faithful.

[2]) We say that Γ acts *similarly* on $\mathfrak{p}$ and $\mathfrak{B}$ if there exists a one-one mapping β from $\mathfrak{p}$ onto $\mathfrak{B}$ and an automorphism α of Γ such that $x\beta\gamma = x\gamma^{\alpha}\beta$ for all $x \in \mathfrak{p}$ and all $\gamma \in \Gamma$.

6. *A finite incidence structure which admits a point and block transitive automorphism group is a tactical configuration.*

Also, the following is easily verified:

7. *If* $[x] > 0$ *for all* $x \in \mathfrak{p}$ *and* $[X] > 0$ *for all* $X \in \mathfrak{B}$, *then any flag transitive automorphism group is also point and block transitive.*

In some cases, point transitivity may be deduced from block transitivity and vice versa:

8. *Let* $\mathbf{C}$ *be a tactical configuration with parameters* v, b, k, r, *and* Γ *a block transitive automorphism group of* $\mathbf{C}$. *Then the number of points in any* Γ*-point orbit is a multiple of* $v(v, k)^{-1}$; *hence the number of point orbits is at most* (v, k), *and it is equal to* (v, k) *if and only if there are exactly* $v(v, k)^{-1}$ *points in every point orbit. In particular,* Γ *is point transitive if* v *and* k *are relatively prime.*

The simple proof uses **5**; see LÜNEBURG 1966c, Lemma 1. Clearly, dual conclusions hold if Γ is assumed to be point transitive.

Along similar lines, the following may be proved:

9. *Let* $\mathbf{C}$ *be a tactical configuration and* Γ *an automorphism group such that, for some block* B *of* $\mathbf{C}$,

(**a**) *B is not fixed under* Γ,
(**b**) Γ_B *is transitive on the points incident with* B, *and*
(**c**) Γ_B *is transitive on the points non-incident with* B.

If v *and* k *are relatively prime,* Γ *is doubly point transitive, so that* $\mathbf{C}$ *is a design if nondegenerate.*

Here one uses the familiar fact of permutation group theory that if $|x G|$ and $|y G|$ are relatively prime, then G_x is transitive on $y G$ (applied with $G = \Gamma_B$ and $x \mathrm{I} B \not\mathrm{I} y$). Note that (**a**)—(**c**) are satisfied if Γ is flag transitive on $\mathbf{C}$ and on the complementary structure $\mathbf{C}'$.

10. *Let* $\mathbf{C}$ *be a tactical configuration with* $(v, k) = 1$. *If* $\mathbf{C}$ *admits a sharply point transitive automorphism group, then* $v \leqq b$.

Proof: Assume $v > b$; then $\Gamma_B \neq 1$ for every block. But $\varphi \neq 1$ in Γ_B permutes the k points on B regularly, say in cycles of c each, $c > 1$. It follows that c divides k and $|\Gamma_B|$ which in turn divides $|\Gamma| = v$, so that $(v, k) \geqq c$, a contradiction.

11. *Let* Γ *be a point and block transitive automorphism group of* $\mathbf{C} = (\mathfrak{p}, \mathfrak{B}, \mathrm{I})$, *and let* $p \in \mathfrak{p}$ *and* $B \in \mathfrak{B}$. *Then the number of* Γ_p*-orbits of blocks equals that of* Γ_B*-orbits of points.*

In fact, both these orbit numbers are equal to the number s of orbits of the permutation group Γ^* induced by Γ in $\mathfrak{p} \times \mathfrak{B}$: Every Γ^*-orbit contains pairs (p, X), by point transitivity of Γ, and $Y \in X\Gamma_p$ if and only if (p, X) and (p, Y) are in the same Γ^*-orbit. This proves that the number of Γ_p-orbits of blocks is s, and the rest follows by a dual argument. Note that finiteness is not used essentially in this proof.[1)]

We shall now show how a tactical configuration with a point and block transitive automorphism group may be reconstructed within this group, and after that we give a similar reconstruction within a flag transitive group.

Let G be a finite group, P a subgroup of G, and D a union of cosets Px. We call "points" the cosets Px, "blocks" the sets Dy, where $x, y \in G$, and define "incidence" by set theoretic inclusion:

$$Px \mathrm{I} Dy \quad \text{if and only if} \quad Px \subseteqq Dy. \tag{2}$$

Then the incidence structure $\mathbf{C} = \mathbf{C}(G, P, D) = (\{Px\}, \{Dy\}, \mathrm{I})$ so defined is a tactical configuration, and the mappings

$$\gamma(g): x \to x g, \tag{3}$$

called the *right translations* of G, yield a group Γ of automorphisms of $\mathbf{C}$ which is point and block transitive. Also, the representation Γ of G is faithful if and only if P contains no normal subgroup of G: since blocks are determined by their points, $\gamma(g)$ is the identical automorphism if and only if $Pxg = Px$ for all $x \in G$, and this is equivalent to $g \in \bigcap_{x \in G} x^{-1} Px$.

The importance of this construction lies in the fact that every tactical configuration which satisfies

$$(B) \neq (C) \quad \text{whenever} \quad B \neq C \tag{4}$$

for all blocks B, C [cf. (1.1.5)] and which admits a point and block transitive automorphism group can be represented as $\mathbf{C}(G, P, D)$:

12. *Let* $\mathbf{C}$ *be a tactical configuration satisfying* (4), *and* Γ *a point and block transitive automorphism group of* $\mathbf{C}$. *Then* $\mathbf{C}$ *is isomorphic to* $\mathbf{C}(\Gamma, \Pi, \Delta)$, *where* Π *is the stabilizer* Γ_p, *and where*

$$\Delta = \Delta(p, B) = \{\xi \in \Gamma : p\xi \mathrm{I} B\}, \tag{5}$$

for an arbitrary point-block pair (p, B).

For the proof (which, incidentally, is quite straightforward), the reader is referred to Dembowski 1965b, p. 61–62. The result is, however, much older; for special cases of it see, for example, Bose 1939,

[1)] For another proof of **11**, cf. Lüneburg 1967a, Satz (2.5).

HALL 1947, BRUCK 1955. A subset of the form (5) of Γ will be called here a *quotient set*[1]) of Γ. If B denotes the stabilizer Γ_B of the block B, we have

$$\Delta = \Pi\,\Delta\,\mathsf{B}; \tag{6}$$

hence Δ is not only a union of cosets $\Pi\,\xi$, but also a union of cosets $\eta\,\mathsf{B}$. It follows from (6) that

$$|\Delta| = |\Pi|\,k = |\mathsf{B}|\,r, \tag{7}$$

where k and r are, as in Section 1.1, the number of points per block, or blocks per point, respectively, in **C**. Clearly, the identical automorphism is in Δ if and only if $p\,\mathrm{I}\,B$, and a change of the initial elements p and B affects Δ as follows:

$$\Delta(p\xi, B\eta) = \xi^{-1}\,\Delta(p, B)\,\eta. \tag{8}$$

Two quotient sets Δ and Δ' of Γ are *equivalent* if $\Delta' = \xi^{-1}\,\Delta\eta$ for suitable $\xi, \eta \in \Gamma$. By (8), equivalent quotient sets define the same configuration.

We give two applications of **12**. The first of these shows essentially that **C** is self-dual if Γ is abelian; more precisely:

13. *If* **C** *satisfies* (4) *and admits an abelian automorphism group* Γ *which is faithful and transitive both on points and on blocks, then* **C** *admits a polarity. Furthermore, if* $v = b$ *is odd, there is a polarity with* $k = r$ *absolute points.*

This was proved in a special case by HALL 1947; his argument applies also to yield the present more general result: As transitive abelian permutation groups are regular, we have $\Pi = \mathsf{B} = 1$; hence we can represent **C**, by **12**, as $(\mathbf{C}\Gamma, 1, \Delta)$, and the points may be identified with the elements of Γ. But then the correspondence π between points and blocks given by

$$\xi^{\pi} = \Delta\,\xi^{-1}, \quad (\Delta\eta)^{\pi} = \eta^{-1}, \quad \text{for} \quad \xi, \eta \in \Gamma, \tag{9}$$

is easily seen to be a polarity. The π-absolute points are the ξ with $\xi^2 \in \Delta$; if $v = |\Gamma|$ is odd, then $\gamma \to \gamma^2$ is an automorphism, and hence the number of these ξ is $|\Delta| = k$. (See also DEMBOWSKI 1965b, p. 62—63.)

The second application of **12** is concerned with the normalizer of Γ in the group $\mathrm{Aut}\,\mathbf{C}$ of all automorphisms of **C**.

---

[1]) These are often called "difference sets" (HALL 1947, HALL & RYSER 1951, BRUCK 1955) because they were first considered in abelian (even cyclic) groups only, and for the case of designs, where Δ can indeed be interpreted as a set of elements of the form $\xi^{-1}\eta$ in Γ (see result **2.3.29** below).

14. *Let the tactical configuration* $\mathbf{C}$, *satisfying* (4) *and possessing a point and block transitive automorphism group* Γ, *be represented as* $\mathbf{C}(\Gamma, \Pi, \Delta)$, *as in* **12**. *Then a permutation* φ *of the points of* $\mathbf{C}$ (*i.e. of the cosets* $\Pi\,\xi$) *is contained in the normalizer of* Γ *in* $\mathrm{Aut}\,\mathbf{C}$ *if, and only if, there exists an automorphism* α *of* Γ *and two elements* $\beta, \gamma \in \Gamma$ *such that*

$$\Delta = \beta^{-1}\,\Delta\gamma \tag{10}$$

and

$$(\Pi\,\xi)^{\varphi} = \Pi\,\beta\,\xi^{\alpha}, \quad (\Delta\eta)^{\varphi} = \Delta\gamma\,\eta^{\alpha} \quad \text{for all} \quad \xi, \eta \in \Gamma. \tag{11}$$

This was proved in a special case by BRUCK 1955. The same argument applies, however, to yield the present more general result; see DEMBOWSKI 1965b, p. 63—64.

14 can be further improved by using the concept of a *multiplier*[1]) of a quotient set Δ of Γ. By this is meant an automorphism μ of Γ such that

$$\Pi^{\mu} = \Pi \quad \text{and} \quad \Delta^{\mu} = \Delta\gamma \quad \text{for suitable} \quad \gamma \in \Gamma. \tag{12}$$

Then the permutation φ induced by μ among the cosets $\Pi\xi$ satisfies conditions (10) and (11) of **14**; hence it is an automorphism in the normalizer of Γ. Conversely, it can be shown that any such automorphism φ for which the element β of **14** is the identity is induced by a multiplier. Clearly the multipliers of Δ form a group; let Φ denote the automorphism group of $\mathbf{C}$ induced by this group. Then

$$\Phi \cap \Gamma \leqq \Pi \tag{13}$$

and

15. *The normalizer of* Γ *in* $\mathrm{Aut}\,\mathbf{C}$ *is the product* $\Phi \cdot \Gamma$.

For proofs, the reader is again referred to BRUCK 1955 (for a special case) or DEMBOWSKI 1965b, p. 64—66.

Result **7** above shows that in general flag transitivity implies point and block transitivity. The next result gives necessary and sufficient conditions, on a quotient set Δ of Γ, that Γ be flag transitive on $\mathbf{C} = \mathbf{C}(\Gamma, \Pi, \Delta)$.

16. *Let* Γ *be a point and block transitive automorphism group of a tactical configuration* $\mathbf{C}$ *satisfying* (4), *let* (p, B) *be a flag of* $\mathbf{C}$, Π *and* B

[1]) In its present form, this notion is due to BRUCK 1955. The essential features of multipliers are, however, present in the fundamental paper of HALL 1947 which was intended for a more special situation. Note that in this section we say nothing about the existence of multipliers. The paper of Hall just mentioned contains the first such existence theorem; this will be treated in detail in Section 2.3.

the stabilizers of p *and* B *in* Γ, *and* Δ *the quotient set* $\Delta(p, B)$. *Then the following conditions are equivalent:*

(a) Γ *is flag transitive;*
(b) Γ *contains a subgroup* Λ *with* $\Delta = \Pi\,\Lambda$;
(c) $\Delta = \Pi\,\mathsf{B}$;
(d) $\Pi \cap \mathsf{B}$ *has index* k *in* B.

This is due to HUGHES 1965a, and again the proof can also be found in DEMBOWSKI 1965b, p. 66—67.

We turn now to representations of tactical configurations within flag transitive automorphism groups. As suggested by **16c**, quotient sets are superfluous in this case; we can work instead with the subgroup B. Let G be an arbitrary group, and let P and B be two subgroups of G. Then the incidence structure $\mathbf{K} = \mathbf{K}(G, P, B)$ whose points are the cosets Px, whose blocks are the cosets By, and whose incidence is defined by

(14) $$Px \,\mathrm{I}\, By \quad \text{if and only if} \quad Px \cap By \neq \emptyset,$$

is a tactical configuration, and the right translations (3) again induce an automorphism group Γ on $\mathbf{K}$ which here is not only point and block, but even flag transitive. The following result is analogous to **12**:

17. *Let* $\mathbf{C}$ *be a tactical configuration which admits a flag transitive automorphism group* Γ. *Then* $\mathbf{C}$ *is isomorphic to* $\mathbf{K}(\Gamma, \Pi, \mathsf{B})$, *with* $\Pi = \Gamma_p$ *and* $\mathsf{B} = \Gamma_B$ *for an arbitrary flag* (p, B).

This was first used by TITS 1956, later by HIGMAN & MCLAUGHLIN 1961. For the proof, which is easy, see also DEMBOWSKI 1965b, p. 68—69. In view of later applications, we translate two important combinatorial properties of $\mathbf{K}(G, P, B)$ into group theoretical language:

18. *In* $\mathbf{K}(G, P, B)$, *there exists a block* By *incident with two given points* Pa *and* Pb *if, and only if,* $ab^{-1} \in PBP$, *and the dual is likewise true.*

In particular, any two points are joined by a block if and only if $G = PBP$, and any two blocks have a common point if and only if $G = BPB$.

19. $\mathbf{K}(G, P, B)$ *is a partial plane (i.e.* $[p, q] \leqq 1$ *for* $p \neq q$*) if and only if*

(15) $$PB \cap BP = P \cup B.$$

Results **18** and **19** are due to HIGMAN & MCLAUGHLIN 1961; the proofs can again be found in DEMBOWSKI 1965b, p. 69—71.

We remark that the preceding results are not as special as they may seem: in many situations one has a group which is not transitive, but has not too many orbits (of points, blocks, or flags). One can then consider the tactical configurations formed by selected point and block orbits; on these the given group induces point and block transitive automorphism groups. Simultaneous consideration of the representations $\mathbf{C}(G, P, D)$ induced on these subconfigurations often yields valuable information.[1]) Examples will be encountered in Sections 2.4, 4.3, and 6.4.

Next, we mention a useful result on primitive automorphism groups.

20. *If a finite incidence structure admits a point (block) primitive automorphism group Γ which is soluble, then the number of points (blocks) is a power of a prime.*

For Γ, being soluble, possesses an elementary abelian normal subgroup which, by primitivity, is still transitive. As abelian permutation groups are regular, the number of points is the order of this normal subgroup.[2])

In conclusion of this section, we mention another class of incidence structures admitting point transitive automorphism groups.[3]) Let G be a group and $\mathscr{C}$ a *covering* of G, i.e. a set of subgroups such that G is the set theoretical union of the subgroups in $\mathscr{C}$. Define points as the elements of G, blocks as the cosets Cx, with $C \in \mathscr{C}$, and incidence by set theoretic inclusion. The incidence structures so defined will be denoted by $\mathbf{J} = \mathbf{J}(G, \mathscr{C})$. As before, the right translations (3) are automorphisms of $\mathbf{J}$, hence G operates faithfully as a sharply point transitive automorphism group on $\mathbf{J}$. The covering $\mathscr{C}$ is a *partition* if every non-identity element of G is in only one subgroup of $\mathscr{C}$. The following is easily verified:

21. *If $\mathscr{C}$ is a covering of G, then $\mathbf{J}(G, \mathscr{C})$ satisfies*

(16) $[x, y] \geqq 1$ *for any two distinct points x, y.*

Furthermore, $\mathbf{J}(G, \mathscr{C})$ is a partial plane [i.e. (16) holds with equality] if and only if $\mathscr{C}$ is a partition of G.

In fact, $[x, y]$ is precisely the number of subgroups in $\mathscr{C}$ containing $x y^{-1}$.

Finally, we remark that $\mathbf{J}$ is a tactical configuration if and only if (G is finite and) all $C \in \mathscr{C}$ have the same order k. In this case, the

[1]) In the special case where G is abelian, this approach has been used by Bose 1939 to construct large classes of designs; cf. Section 2.4.

[2]) The permutation group theoretical background for this can be found in Wielandt 1964, 4.4 and 8.8.

[3]) Examples of incidence structures representable in this fashion will appear in Sections 2.4, 3.1, 5.2, and 7.2.

parameters of **J** are given by

$$v = |G|, \quad b = |\mathscr{C}| \cdot |G| k^{-1}, \quad r = |\mathscr{C}|. \tag{17}$$

1.3 Incidence matrices

Let Δ be a tactical decomposition of a finite incidence structure $\mathbf{S} = (\mathfrak{p}, \mathfrak{B}, \mathrm{I})$, and let the (point and block) classes of Δ be numbered in an arbitrary but fixed way: $\mathfrak{p}_1, \ldots, \mathfrak{p}_m$ and $\mathfrak{B}_1, \ldots, \mathfrak{B}_n$. Then we define two matrices $C = (c_{ij})$ and $D = (d_{ij})$ by

$$c_{ij} = (\mathfrak{p}_i \, \mathfrak{B}_j), \quad d_{ij} = (\mathfrak{B}_i \, \mathfrak{p}_j), \tag{1}$$

where $(\mathfrak{p}_i \, \mathfrak{B}_j)$ means, as in Section 1.1, the number of points of $\mathfrak{p}_i$ on a block of $\mathfrak{B}_j$, and where $(\mathfrak{B}_i \, \mathfrak{p}_j)$ is defined dually. C and D are called the *incidence matrices* of Δ, with respect to the chosen numbering of the Δ-classes. Clearly, C and D are integral (m, n)- and (n, m)-matrices, respectively. Also, the incidence matrices C', D' of Δ with respect to another numbering of the Δ-classes are related to C and D by

$$C' = QCR^T, \quad D' = RDQ^T, \tag{2}$$

for suitable permutation matrices Q and R (the superscript ^{T} denotes transposition).

Besides C and D, we consider the diagonal matrices

$$P = \operatorname{diag}(|\mathfrak{p}_1|, \ldots, |\mathfrak{p}_m|), \quad B = \operatorname{diag}(|\mathfrak{B}_1|, \ldots, |\mathfrak{B}_n|). \tag{3}$$

The relation (1.1.14) for tactical decompositions can then be written as

$$BC^T = DP. \tag{4}$$

Furthermore, we clearly have from (1.1.15)

$$\operatorname{tr} P = v \quad \text{and} \quad \operatorname{tr} B = b, \tag{5}$$

with $v = |\mathfrak{p}|$ and $b = |\mathfrak{B}|$ as in Section 1.1; and if $\mathbf{S}$ is a tactical configuration, (1.1.16) becomes

$$J_m C = k J_{mn}, \quad J_n D = r J_{nm}, \tag{6}$$

where we denote by J_{mn} the (m, n)-matrix all of whose entries are 1, and where J_m means the same as J_{mm}.

The equation (4) clearly implies

$$CBC^T = CDP \tag{7}$$

and

$$(CD)^T = P^{-1}(CD) P, \quad (DC)^T = B^{-1}(DC) B, \tag{8}$$

so that CD and DC are transformed into their transposes by P and B, respectively. It follows immediately from (4) that

1. *C is nonsingular if and only if D is nonsingular.*

Clearly, this can happen only when $m = n$. This case where C and D are nonsingular is particularly important (but it seems difficult to determine the exact geometric significance of nonsingularity) because then (7) says that the matrices B and CDP are *congruent.* This observation permits the application of the theory of congruence of quadratic forms (over the field of rationals), due to MINKOWSKI 1890 and HASSE 1923. It would lead too far afield to develop this theory here; we shall give only the definitions and results relevant to the applications of this book. For proofs and further details the reader is referred to JONES 1950, Chapter 2.

For any two non-zero integers a, b, the *p-norm residue symbol* $\left(\frac{a, b}{p}\right)$, for an arbitrary prime number p, is defined to be 1 or -1 according whether the equation $a x^2 + b y^2 = 1$ has a solution in the field of p-adic numbers or not. Alternatively, it can be shown that, if $a = a' p^\alpha$ and $b = b' p^\beta$, with $(a', p) = (b', p) = 1$, then

$$\left(\frac{a, b}{p}\right) = (-1)^{\alpha\beta(p-1)/2} \left(\frac{a'}{p}\right)^\beta \left(\frac{b'}{p}\right)^\alpha \qquad \text{if} \quad p \neq 2$$

where $\left(\frac{x}{p}\right)$ denotes the Legendre symbol, and

$$\left(\frac{a, b}{2}\right) = (-1)^{(a'-1)(b'-1)/4 + \beta(a'^2-1)/8 + \alpha(b'^2-1)/8}.$$

Now for any symmetric integral (n, n)-matrix M, let $D(i)$ denote the determinant of the (i, i)-submatrix in the upper left hand corner of M (principal minor), provided this determinant is $\neq 0$; otherwise put $D(i) = 1$. Then the *Hasse symbols* $H_p(M)$, for all primes p, are defined by

$$H_p(M) = \left(\frac{-1, -D(n)}{p}\right) \prod_{i=1}^{n-1} \left(\frac{D(i), -D(i+1)}{p}\right).$$

The main results needed later are the following:

2. *If the integral matrices L and M are congruent* (*i.e.* $L = CMC^T$ *for a nonsingular C*), *then* $H_p(L) = H_p(M)$ *for all primes* p.

For the proof see JONES 1950, p. 32—35. The converse of **2** is also true (and considerably deeper), but will not be needed here.

3. *If the nonsingular integral diagonal* $(2k, 2k)$*-matrix D is congruent to nD, where n is a positive integer, then* $\left(\frac{n, (-1)^k \det D}{p}\right) = 1$ *for all primes* p.

This can either be derived from **2** or proved more directly, cf. LENZ 1962, Satz 3. As an immediate application of **2**, we state the following *general nonexistence theorem for tactical decompositions*:

4. *Let C, D, P, B be four nonsingular integral matrices, with P and B diagonal. Then C, D, P, B cannot be the matrices of a tactical decomposition of a finite incidence structure, as defined by* (1) *and* (3), *unless $H_p(B) = H_p(CDP)$ for all p.*

We shall encounter various special cases of this theorem later on; see in particular Sections 2.1, 2.3, and 3.2. The first and most celebrated result of this sort was the theorem of BRUCK & RYSER 1949; cf. result **3.2.13** below.

The most important incidence matrices are those of the *trivial tactical decomposition,* every class of which consists of just one element. In this case the numbers c_{ij} and d_{ji} given by (1) are 1 or 0 according as the ith point and the jth block are incident or not; hence C and D consist entirely of zeros and ones. Also, we have $P = I_v$ and $B = I_b$, where I_n denotes the (n, n)-identity matrix, so that (4) reduces to $D = C^T$. Here we speak of C as an *incidence matrix of* **S** (rather than of the trivial tactical decomposition of **S**). Clearly, every matrix of zeros and ones can be interpreted as an incidence matrix of a finite incidence structure.[1]) The following *general nonexistence theorem for finite incidence structures* is a first special case of **4**:

5. *A finite incidence structure cannot have a nonsingular incidence matrix C unless $H_p(CC^T) = H_p(C^TC) = 1$ for all p.*

This result also shows the significance of the (v, v)-matrix $CC^T = (g_{ij})$ and of the (b, b)-matrix $C^TC = (h_{ij})$. These matrices permit valuable insight into the combinatorial structure of **S**: if C is the incidence matrix with respect to the numbering $p_1, \ldots, p_v$ of the points and $B_1, \ldots, B_b$ of the blocks, then, by definition:

$$(9) \qquad g_{ij} = \begin{cases} [p_i] & \text{for } i = j \\ [p_i, p_j] & \text{for } i \neq j \end{cases}, \qquad h_{ij} = \begin{cases} [B_i] & \text{for } i = j \\ [B_i, B_j] & \text{for } i \neq j \end{cases};$$

here we have used the notation (1.1.6) again. As a consequence, we have the following result, with I_b, J_b etc. as defined above [cf. (6)]:

6. *The finite incidence structure* $\mathbf{S} = (\mathfrak{p}, \mathfrak{B}, \mathrm{I})$ *satisfies the regularity conditions* (R. 1) *and* (R. 2) *of Section* 1.1 *if, and only if, an arbitrary*

[1]) In particular, every result on matrices of zeros and ones can be interpreted as a result on finite incidence structures. There is an extensive literature on this subject; we mention here FULKERSON 1960; FULKERSON & RYSER 1961, 1962, 1963; HABER 1960, 1963; TAUSSKY 1960; and RYSER 1951, 1952, 1956, 1957, 1958a, b, 1960a, b, c, 1963. The last reference is Ryser's "Combinatorial Mathematics"; Chapter 6 of this book is exclusively concerned with matrices of zeros and ones, and many further references are given there.

incidence matrix C *of* **S** *satisfies*

(10) $$C^T C = (v_1 - v_2)\, I_b + v_2 J_b.$$

Dually, **S** *satisfies* ($\bar{\mathrm{R}}$.1) *and* ($\bar{\mathrm{R}}$.2) *if and only if*

(11) $$C C^T = (b_1 - b_2)\, I_v + b_2\, J_v.$$

Here the v_i and b_i $(i = 1, 2)$ are the parameters as defined in (1.1.7). As J_b has rank 1, the matrix $M = (v_1 - v_2)\, I_b + v_2\, J_b$ has $v_1 - v_2$ as an eigenvalue of multiplicity $\geqq b - 1$. On the other hand, $(1, \ldots, 1)$ is an eigenvector of M, with corresponding eigenvalue $v_1 + v_2(b - 1)$. Hence:

7. *Let* C *be an incidence matrix of the finite incidence structure* **S**. *If* **S** *satisfies* (R.1) *and* (R.2), *then* $C^T C$ *has one eigenvalue* $v_1 + v_2(b - 1)$ *and* $b - 1$ *eigenvalues* $v_1 - v_2$, *whence*

(12) $$\det C^T C = (v_1 - v_2)^{b-1}\, [v_1 + v_2(b - 1)].$$

Dually, if **S** *satisfies* ($\bar{\mathrm{R}}$.1) *and* ($\bar{\mathrm{R}}$.2), *then* $C C^T$ *has one eigenvalue* $b_1 + b_2(v - 1)$ *and* $v - 1$ *eigenvalues* $b_1 - b_2$, *whence*

(13) $$\det C C^T = (b_1 - b_2)^{v-1}\, [b_1 + b_2(v - 1)].$$

Now suppose that $\emptyset \neq \mathrm{I} \neq \mathfrak{p} \times \mathfrak{B}$. From (1.1.8″) we can then conclude $v_1 > v_2$ if (R.1) and (R.2) hold, so that $C^T C$ is of rank b, by (12). On the other hand, the rank of $C^T C$ is at most that of C and hence $\leqq \min(b, v)$. This shows:

8. *If* $\mathbf{S} = (\mathfrak{p}, \mathfrak{B}, \mathrm{I})$ *satisfies* $\mathrm{I} \neq \mathfrak{p} \times \mathfrak{B}$, *then*

(14) $v \geqq b$ *in case* (R.1) *and* (R.2) *hold, and dually*

(15) $b \geqq v$ *in case* ($\bar{\mathrm{R}}$.1) *and* ($\bar{\mathrm{R}}$.2) *hold.*

The first proof of (15), under the unnecessarily strong hypothesis that **S** is a design, is due to FISHER 1940; (15) is therefore often called *Fisher's inequality.*[1]) The present treatment is that of BOSE 1949. As a consequence of **8**, we have $v = b$ for every design of type (2,2), cf. **1.1.5.** This result has an interesting converse:

9. *If* $\mathbf{S} = (\mathfrak{p}, \mathfrak{B}, \mathrm{I})$ *satisfies* $\emptyset \neq \mathrm{I} \neq \mathfrak{p} \times \mathfrak{B}$ *and* $v = b$, *then* (R.1) *and* (R.2) *together are equivalent to* ($\bar{\mathrm{R}}$.1) *and* ($\bar{\mathrm{R}}$.2).

For the proof, which uses (6) and the fact that C is nonsingular if $v = b$, see RYSER 1950, or RYSER 1963, p. 103–104.

Result **8** has been considerably generalized by BLOCK 1967. If C is an incidence matrix of a tactical decomposition of the finite incidence

[1]) Many proofs of this inequality have been given, as well as of the partial converse **9** below. As further references in this direction, we mention here BOSE 1939; MAJUMDAR 1953; DEMBOWSKI 1961; HSU 1962.

structure $\mathbf{S}$ and if A is an incidence matrix of $\mathbf{S}$ itself, then $c_{ij} = (\mathfrak{p}_i\, \mathfrak{B}_j)$ is the sum of the entries in any column of the submatrix A_{ij} of A which consists of the rows corresponding to the points of $\mathfrak{p}_i$ and the columns corresponding to the blocks of $\mathfrak{B}_j$ $(1 \leqq i \leqq m, 1 \leqq j \leqq n)$. Let $\varrho = \varrho(A)$ denote the rank of A, so that there exist ϱ linearly independent rows in A. The remaining $v - \varrho$ rows represent at most $v - \varrho$ distinct point classes, consequently there are $s \geqq m - (v - \varrho)$ other point classes, whose union is represented by a linearly independent set S of rows of A. Now the s rows of C corresponding to these point classes must be linearly independent also: a dependence relation among them would give rise to a dependence relation among the rows of S. Hence the rank $\varrho(C)$ of C is at least s, so that we now have $\varrho(C) \geqq s \geqq m - [v - \varrho(A)]$, or[1])

$$v + \varrho(C) \geqq m + \varrho(A). \tag{16}$$

A dual argument gives

$$b + \varrho(C) \geqq n + \varrho(A), \tag{17}$$

and as $\varrho(A) \leqq \min(v, b)$ and $\varrho(C) \leqq \min(m, n)$, we can conclude:

10. *Let* $\mathbf{S}$ *be an incidence structure with* v *points and* b *blocks, and suppose that* $\mathbf{S}$ *has an incidence matrix of rank* ϱ. *Then the numbers* m *and* n, *of point and block classes, respectively, of an arbitrary tactical decomposition of* $\mathbf{S}$, *satisfy*

$$v + n \geqq m + \varrho \quad \textit{and} \quad b + m \geqq n + \varrho.$$

This is Theorem 2.1 of Block 1967. As immediate consequences, we note that $m \leqq n$ $[n \leqq m]$ in case $\mathbf{S}$ has an incidence matrix A with linearly independent rows [columns], and $m = n$ if A is nonsingular.

Next, we discuss representations of automorphisms and anti-automorphisms of finite incidence structures by means of their incidence matrices. The following is a simple consequence of (2):

11. *Let* C *be an incidence matrix of* $\mathbf{S} = (\mathfrak{p}, \mathfrak{B}, \mathrm{I})$. *Then the automorphisms of* $\mathbf{S}$ *correspond in a one-one fashion to the pairs* (Q, R) *of* (v, v)- *and* (b, b)-*permutation matrices* Q *and* R, *respectively, for which*

$$QC = CR. \tag{18}$$

Here Q represents the permutation of the points, and R^{-1} that of the blocks, induced by the automorphism represented by the pair (Q, R). Corollary:

---

[1]) Note that for the proof of (16) it was used only that every block of the substructure $\mathbf{S}_{ij}$ defined by the points of $\mathfrak{p}_i$ and the blocks of $\mathfrak{B}_j$ has the same number of points of $\mathbf{S}_{ij}$, and not the dual of this condition. Hence (16) holds for a more general type of decomposition, called "right tactical decomposition" by Block 1967. The inequality (17) and result **10** below can be generalized accordingly.

12. *If* $\mathbf{S} = (\mathfrak{p}, \mathfrak{B}, \mathrm{I})$ *has a nonsingular incidence matrix, then every automorphism of* $\mathbf{S}$ *induces similar permutations on* $\mathfrak{p}$ *and* $\mathfrak{B}$.

For the proof, rewrite (18) as $R = C^{-1}QC$ and observe that conjugate permutation matrices represent similar permutations. Result **12** is due to R. Brauer 1941 who also showed how to derive the following from **12**:

13. *If* $\mathbf{S}$ *has a nonsingular incidence matrix, then any automorphism group of* $\mathbf{S}$ *has equally many point and block orbits.*

These facts were rediscovered for designs of type (2,2) by Parker 1957. In this context, compare also Lüneburg 1967a, (2.5).

For anti-automorphisms, we have the following analogue to **11**:

14. *Let* C *be an incidence matrix of* $\mathbf{S} = (\mathfrak{p}, \mathfrak{B}, \mathrm{I})$. *Then the anti-automorphisms of* $\mathbf{S}$ *correspond in a one-one fashion to the pairs of permutation* (v, v)*-matrices* Q, R *for which*

$$QC^T = CR. \tag{19}$$

Here Q and R^{-1} represent the permutations $\varkappa$ and ϱ defined by the anti-automorphism α corresponding to (Q, R) as follows: $p_i \alpha = B_{i\varkappa}$, $B_j \alpha = p_{j\varrho}$. Note that the numbering of the blocks can be so chosen that $\varkappa = 1$; for this C we then have α represented by (I_v, R), and (19) becomes

$$C^T = CR. \tag{19'}$$

We list two corollaries:

15. $\mathbf{S}$ *is self-polar if and only if* $\mathbf{S}$ *has a symmetric incidence matrix.*

This follows from (19′) with $R = I_v$.

16. *Every self-dual finite incidence structure possesses a normal incidence matrix* C.

For (19′) shows $C = (CR)^T = R^{-1}C^T$ (permutation matrices are orthogonal), whence $C^T = RC$. This and (19′) show not only that $CC^T = C^TC$, i.e. **16**, but also the following analogue to **12**:

17. *The square of an anti-automorphism of* $\mathbf{S} = (\mathfrak{p}, \mathfrak{B}, \mathrm{I})$ *induces similar permutations in* $\mathfrak{p}$ *and* $\mathfrak{B}$.

For now we have $RC = CR$, and **11** shows that the same permutation matrix R represents the point as well as the block permutation induced by the automorphism ϱ^2.

The representation (19) of anti-automorphisms ϱ can be used to determine the number $a(\varrho)$ of ϱ-absolute points, which is of course equal to that of ϱ-absolute blocks. If the points of $\mathbf{S}$ are numbered arbitrarily $p_1, \ldots, p_v$, and the blocks so that $B_i = p_i \varrho$, then the number of absolute points is the sum of the entries in the main diagonal in the corresponding incidence matrix C. Hence

(20) $$a(\varrho) = \operatorname{tr} C.$$

This will be used in Section 2.1 to determine $a(\varrho)$ for anti-automorphisms ϱ of designs.

Finally, we mention a concept of *generalized incidence matrices*, due to HUGHES 1957e. Here an automorphism group Γ of the finite incidence structure $\mathbf{S} = (\mathfrak{p}, \mathfrak{B}, \mathrm{I})$ is given, and the matrices to be defined will be related to those of the tactical decomposition of the Γ-orbits of $\mathbf{S}$ (cf. **1.2.5**). However, the elements of the generalized incidence matrices are not integers, but elements of the group algebra $\mathfrak{F}\Gamma$ of Γ, over any field $\mathfrak{F}$ whose characteristic does not divide $|\Gamma|$.[1]) Let $\mathfrak{p}_1, \ldots, \mathfrak{p}_m$ be the point orbits and $\mathfrak{B}_1, \ldots, \mathfrak{B}_n$ the block orbits of Γ, and choose representative elements $p_i \in \mathfrak{p}_i$ and $B_j \in \mathfrak{B}_j$. Let Δ_{ij} denote the quotient set $\{\gamma \in \Gamma : p_i \gamma \mathrm{I} B_j\}$; cf. (1.2.5). Then the generalized incidence matrix of $\mathbf{S}$ with respect to Γ is the matrix $G = (\gamma_{ij})$, with

(21) $$\gamma_{ij} = \sum_{\xi \in \Delta_{ij}} \xi \in \mathfrak{F}\Gamma.$$

Let Π_i and B_j denote the stabilizers Γ_{p_i} and Γ_{B_j}, respectively, and consider the integral diagonal matrices

(22) $$P^* = \operatorname{diag}(|\Pi_1|, \ldots, |\Pi_m|), \qquad B^* = \operatorname{diag}(|\mathsf{B}_1|, \ldots, |\mathsf{B}_n|).$$

From $|\Gamma| = |\Pi_i|\,|\mathfrak{p}_i| = |\mathsf{B}_j|\,|\mathfrak{B}_j|$ it follows that

(23) $$P P^* = |\Gamma|\, I_m \quad \text{and} \quad B B^* = |\Gamma|\, I_n,$$

where P and B are the matrices (3) again. Furthermore, if g denotes the sum of all elements of Γ in $\mathfrak{F}\Gamma$, and if C, D are the incidence matrices defined by (1), we get from (1.2.7) and the easily proved fact that $\gamma_{ij}\, g = |\Delta_{ij}|\, g$:

(24) $$G(B^*)^{-1} g = D g \quad \text{and} \quad (P^*)^{-1} G g = C g.$$

An application of these facts, in the case $m = n = 1$ where Γ is point and block transitive,[2]) will be given in Section 2.3.

1.4 Geometry of finite vector spaces

We begin this section with a brief introduction to general projective and affine geometry. Later we shall restrict ourselves to finite desarguesian geometries.

---

[1]) The idea of utilizing this group algebra is due to BRUCK 1955.

[2]) For $m, n > 1$, the generalized incidence matrices were originally used by HUGHES 1957e, for the proof of result **4.1.12** below. We shall see there, however, that they are not really necessary for that proof.

Let $\mathbf{S} = (\mathfrak{p}, \mathfrak{L}, \mathrm{I})$ be an incidence structure, whose blocks will be called *lines*, such that the following conditions (1)—(3) are satisfied.

(1) $$[p, q] = 1 \qquad \text{for any two distinct } p, q \in \mathfrak{p},$$

i.e. there exists exactly one line incident with both p and q; cf. (1.1.6). This line will be denoted by $p + q$.

(2) $$[L] \geqq 3 \qquad \text{for every line } L \in \mathfrak{L},$$

i.e. every line is incident with at least three different points. (1) and (2) show that $(L) = (M)$ implies $L = M$, for all $L, M \in \mathfrak{L}$, cf. (1.1.5). We can, therefore, interpret the lines L as the point sets (L); accordingly we shall henceforth write $p \in L$ for $p \,\mathrm{I}\, L$, and use also other set theoretic notations.

(3) *If two distinct lines L, M have a common point p, if q, r are points $\neq p$ of L, and if s, t are points $\neq p$ of M, then the lines $q + s$ and $r + t$ have a common point.*

We define *subspaces* of $\mathbf{S}$ to be those subsets S of $\mathfrak{p}$ for which the following condition holds:

(4) $$\textit{If} \quad p \neq q \quad \textit{and} \quad p, q \in S, \textit{ then } p + q \subseteqq S.$$

Immediate examples of subspaces are the empty set $\emptyset$, the one-point subsets $\{p\}$, where $p \in \mathfrak{p}$, the lines, and the set $\mathfrak{p}$ of all points. A *hyperplane* is a maximal subspace H, i.e. $\mathfrak{p}$ is the only subspace properly containing H.

Now we define: A *projective geometry* is the system of all subspaces of an incidence structure $\mathbf{S}$ satisfying (1)—(3). An *affine geometry* is the system of point sets $S - (H \cap S)$, where S ranges over all subspaces of a projective geometry, and where H is a fixed hyperplane.[1] These point sets are the *subspaces* of the affine geometry. If there is danger of confusion, we shall speak of projective and affine subspaces, respectively, but usually the meaning of the term "subspace" will be clear from the context.

It follows from (4) that intersections of subspaces are again subspaces of the projective or affine geometry under consideration. The

[1] Notice that there is a close connection with the concept of "external structure" of 1.1. Our definition of projective and affine geometries is essentially that of VEBLEN & YOUNG 1916, except that we make no nondegeneracy assumptions at this stage. In particular, the empty set is a projective geometry here. It is possible to define affine geometries without reference to the embedding projective geometry, see for example LENZ 1961, p. 138. However for such a definition of affine geometries it is necessary to have more than two kinds of basic objects, for example points, lines, and planes, except in the two-dimensional case. For definitions of projective geometries with points and hyperplanes as basic objects see WINTERNITZ 1940; ESSER 1951; also WYLER 1953.

subspace S is said to be *spanned* by the point set $\mathfrak{s}$, written $S = \langle \mathfrak{s} \rangle$, if S is the intersection of all subspaces containing $\mathfrak{s}$. We define

$$S + T = \langle S \cup T \rangle \qquad \text{for all subspaces } S, T;$$

this is clearly compatible with the addition of points defined earlier.[1])

The point set $\mathfrak{s}$ is called *independent* provided

$$x \notin \langle \mathfrak{s} - \{x\} \rangle \qquad \text{for every point } x \in \mathfrak{s}.$$

It can be shown that all independent sets spanning a given subspace S have the same cardinality $r = r(S)$, called the *rank* of S. An ordered maximal independent set of S will be called a *basis* of S. Clearly if $r(S)$ is finite, then the bases of S are precisely the $r(S)$-tuples of independent points in S. The *dimension* of S is $d(S) = r(S) - 1$. Rank and dimension of the geometry under consideration are those of the unique largest subspace, i.e. of $\mathfrak{p}$ or $\mathfrak{p} - H$, respectively. We shall, of course, only be interested in finite-dimensional geometries here; for these we have, in the projective case, the *rank formula*

$$(5) \quad r(S) + r(T) = r(S \cap T) + r(S + T) \qquad \text{for all subspaces } S, T.$$

Note that the empty subspace has rank 0 and dimension -1.

Let $\mathcal{G}$ be a projective geometry of rank n and S a distinguished subspace of rank r. If the subspaces of ranks $r+1$ and $r+2$ containing S are considered as new "points" and "lines", with inclusion as incidence, then the incidence structure so defined again satisfies (1)—(3). The corresponding projective geometry is called the *quotient geometry* $\mathcal{G}/S$. A subspace of $\mathcal{G}$ of rank m which contains S has rank $m - r$ as a subspace of $\mathcal{G}/S$.

Let $\mathbf{S} = (\mathfrak{p}, \mathfrak{L}, \in)$ be an incidence structure satisfying (1)—(3), where the lines are point sets, and let $\mathcal{G}$ be the corresponding projective geometry. Suppose that for some subset $\mathfrak{p}'$ of $\mathfrak{p}$, and for some set $\mathfrak{L}'$ of subsets of $\mathfrak{p}'$, satisfying the condition

(4') *Every* $L' \in \mathfrak{L}'$ *is contained in some* $L \in \mathfrak{L}$,

conditions (1)—(3) are again satisfied. Then the incidence structure $\mathbf{S}' = (\mathfrak{p}', \mathfrak{L}', \in)$ determines a projective geometry $\mathcal{G}'$, and any such $\mathcal{G}'$ is called a *subgeometry* of $\mathcal{G}$. As (4) clearly implies (4'), every subspace is also a subgeometry, but the converse is not true in general.

---

[1]) With respect to this addition and intersection, every projective or affine geometry is a lattice, and this lattice is modular in the projective and semimodular in the affine case. For further details about these lattices, and for the possibility of defining projective and affine geometries in terms of lattice theory, see BIRKHOFF 1948 (particularly Chapter VIII), BAER 1952, Chapter VII, and DUBREIL-JACOTIN, LESIEUR & CROISOT 1953, troisième partie.

We proceed now to the definition of the class of desarguesian projective and affine geometries. In an arbitrary geometry, a point set $\mathfrak{s}$ is termed *collinear* if there exists a line containing $\mathfrak{s}$. A *triangle* is an ordered triple of non-collinear points. For the triangle (p_1, p_2, p_3), we put $p_i + p_k = L_j$, where i, j, k are distinct in $\{1, 2, 3\}$. Clearly, the L_j are then three distinct lines. We say that the triangles $\mathfrak{t} = (p_1, p_2, p_3)$ and $\mathfrak{t}' = (p_1', p_2', p_3')$ form a *couple* if

(i) the p_i and p_i' are six different points,
(ii) the lines $M_i = p_i + p_i'$ are all distinct, } for $i = 1, 2, 3$.

It follows from (i) and (ii) that $L_i \neq L_i'$ for $i = 1, 2, 3$.

Furthermore, $\mathfrak{t}$ and $\mathfrak{t}'$ are called *central* to each other if the lines M_i of (ii) have a common point c; this (necessarily unique) point is then

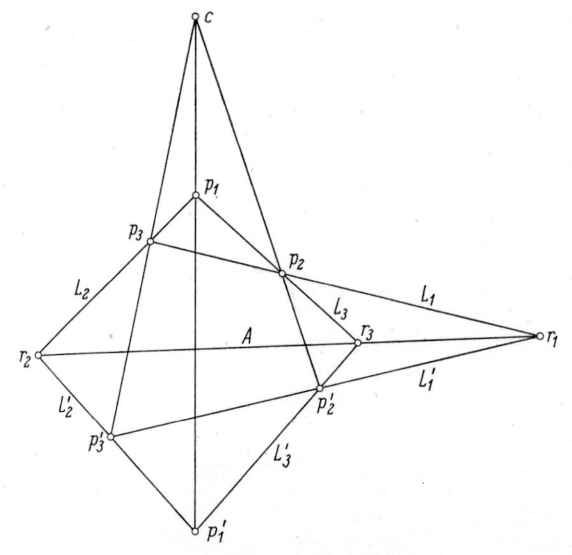

Fig. 1. Theorem of Desargues

called the *center* of $\mathfrak{t}$ and $\mathfrak{t}'$. We say that $\mathfrak{t}$ and $\mathfrak{t}'$ are *axial* to each other if (a) the lines L_i and L_i' have a common point r_i $(i = 1, 2, 3)$ and if (b) the points r_1, r_2, r_3 are collinear; the (unique) line A containing the r_i is then called the *axis* of $\mathfrak{t}$ and $\mathfrak{t}'$. Now in any projective geometry the dual conditions

(D) *Every central couple of triangles is axial*
and
(D') *Every axial couple of triangles is central*

can be shown to be equivalent; see Fig. 1. A projective geometry satisfying one, and hence both, of these conditions is called *desarguesian*,[1]) and an affine geometry is desarguesian if the corresponding projective geometry is desarguesian.

1. *All projective and affine geometries of dimension $\neq 2$ are desarguesian.*

For the proof of this important result, see VEBLEN & YOUNG 1910, p.41. The dimensional restriction is quite essential: 2-dimensional geometries, which will be called *projective* and *affine planes*, need not be desarguesian, as many examples in Chapter 5 will show.[2]) More about the general theory of projective and affine planes will be found in Section 3.1, and finite planes are the principal topic of Chapters 3–5.

Let V be a left vector space over an arbitrary (not necessarily commutative) field $\mathfrak{F}$. A subspace of V is said to be of rank r if its bases consist of r elements each. A *coset* of V is a subset of the form

$$S + a = \{x + a \,\vdots\, x \in S\},$$

where S is a subspace and a an element of V. The rank of a coset is that of the corresponding subspace of V. One verifies that

2. *The non-zero subspaces of V form a desarguesian projective geometry.*

This geometry will be called $\mathcal{P}(V)$; its points are the subspaces of rank 1, its lines the subspaces of rank 2 of V. (As a matter of fact, the terminology has been so chosen that "subspace of rank r" means the same for V and $\mathcal{P}(V)$.)

3. *The cosets of V form a desarguesian affine geometry.*

This geometry is called $\mathcal{A}(V)$; its points are the cosets of rank 0, i.e. essentially the elements of V, and its lines are the cosets of rank 1. It should be noted that the projective geometry corresponding to $\mathcal{A}(V)$ is not $\mathcal{P}(V)$ but rather $\mathcal{P}(V')$, where V' is the left vector space over $\mathfrak{F}$ whose hyperplanes are isomorphic to V.[3])

The significance of the geometries $\mathcal{P}(V)$ and $\mathcal{A}(V)$ lies in the following fundamental theorem:

---

[1]) The classical *Theorem of Desargues* asserts the validity of (**D**) in Euclidean (projective) geometry. There is a close connection between (**D**) [or (**D'**)] and the existence of central collineations to be defined on p. 30 below. A detailed analysis of this connection in arbitrary projective planes will be given in Section 3.1.

[2]) All these will be finite, but there are also large classes of infinite non-desarguesian planes known.

[3]) For a proof of **3**, with the present definition of affine geometry, see GRUENBERG & WEIR 1967, Section 2.6. A certain difficulty lies in showing that $\mathcal{A}(V)$ can be embedded in some $\mathcal{P}(V')$; that $\mathcal{P}(V)$ must then be isomorphic to a hyperplane of $\mathcal{P}(V')$ [the ideal hyperplane of $\mathcal{A}(V)$], is rather easy to see.

4. *Every desarguesian projective or affine geometry is isomorphic to a suitable* $\mathscr{P}(V)$ *or* $\mathscr{A}(V)$, *respectively.*

This result is essentially due to HILBERT 1899. For the proof, see also BAER 1952, Chapter VII, ARTIN 1957, Chapter II, LENZ 1965, Kapitel III, and LÜNEBURG 1965a, 1966d.

Now we restrict ourselves to the finite case. Then $\mathfrak{F}$ is a Galois field $GF(q)$, with $q = p^e$ a power of a prime, and V is of finite rank n over $\mathfrak{F}$. In fact, V is uniquely determined by q and n; hence we can write $V = V(n, q)$. Then, if we put $\mathscr{P}(d, q) = \mathscr{P}(V(d+1, q))$ and $\mathscr{A}(d, q) = \mathscr{A}(V(d, q))$, we get from **4**:

5. *The only finite desarguesian d-dimensional projective and affine geometries are the* $\mathscr{P}(d, q)$ *and* $\mathscr{A}(d, q)$, *for all prime powers* q.

Let $N_r(n, q)$ denote the number of subspaces of rank r of $V(n, q)$. Using the counting principle (1.1.10) and the fact that $|V(n, q)| = q^n$, one derives the recurrence relation

$$N_{r+1}(n, q)\, N_r(r+1, q) = N_r(n, q)\,(q^{n-r} - 1)\,(q-1)^{-1},$$

which implies

$$N_r(n, q) = N_{n-r}(n, q) = \prod_{i=0}^{r-1} \frac{q^{n-i} - 1}{q^{i+1} - 1}. \tag{6}$$

Also, the number $M_r(n, q)$ of cosets of rank r of $V(n, q)$ is related to $N_r(n, q)$ by

$$M_r(n, q) = q^{n-r}\, N_r(n, q). \tag{7}$$

These equations yield the number of subspaces, of any prescribed dimension, of $\mathscr{P}(d, q)$ and $\mathscr{A}(d, q)$. In particular, the number of points of $\mathscr{P}(d, q)$ is

$$N_1(d+1, q) = (q^{d+1} - 1)\,(q-1)^{-1} = \sum_{i=0}^{d} q^i, \tag{8}$$

and this is also the number of hyperplanes in $\mathscr{P}(d, q)$. The numbers of points and hyperplanes in $\mathscr{A}(d, q)$ are, respectively, $M_0(d, q) = q^d$ and $M_{d-1}(d, q) = \sum_{i=1}^{d} q^i$.

The geometries $\mathscr{P}(d, q)$ and $\mathscr{A}(d, q)$ give rise to various series of tactical configurations: Let s and t be integers with $0 \leqq s < t \leqq d - 1$, call "points" the s-dimensional, "blocks" the t-dimensional (projective or affine) subspaces of $\mathscr{P}(d, q)$ or $\mathscr{A}(d, q)$, and define incidence by set theoretic inclusion. The incidence structures so defined are all non-degenerate tactical configurations, and they are of type (1, 1) if and only if $s > 0$. We shall be particularly interested in the case $s = 0$; these tactical configurations will be denoted by $\mathbf{P}_t(d, q)$ and $\mathbf{A}_t(d, q)$, respectively. All $\mathbf{P}_t(d, q)$ with $t < d - 1$ and all $\mathbf{A}_t(d, q)$ with $q > 2$ are designs of type $(1, 2)$; the $\mathbf{P}_{d-1}(d, q)$ are of type $(2, 2)$, and the $\mathbf{A}_t(d, 2)$ of type $(1, 3)$. The desarguesian projective and affine planes $\mathscr{P}(2, q)$ and $\mathscr{A}(2, q)$ are essentially the same as $\mathbf{P}_1(2, q)$ and $\mathbf{A}_1(2, q)$, respectively; these we denote simply by $\mathbf{P}(q)$ and $\mathbf{A}(q)$ later on.

Some of the parameters of the $\mathbf{P}_t(d, q)$ and $\mathbf{A}_t(d, q)$ are, of course, given by (6)—(8). We give a more complete list now. The parameters of $\mathbf{P}_t(d, q)$ are:

$$(9)\qquad \begin{aligned} v &= N_1(d+1, q), & b &= N_{t+1}(d+1, q),\\ k &= N_1(t+1, q), & r &= N_t(d, q), \end{aligned}$$

and those of $\mathbf{A}_t(d, q)$ are

$$(10)\qquad v = q^d, \quad b = N_t(d, q)\, q^{d-t}, \quad k = q^t, \quad r = N_t(d, q).$$

In Section 2.1 we shall also be interested in the parameter b_2 of these designs (which will then be called λ). Similarly to (9) and (10), one verifies that $b_2 = N_{t-1}(d-1, q)$, for both $\mathbf{P}_t(d, q)$ and $\mathbf{A}_t(d, q)$.

We give another application of (6). By a *t-spread* of a projective geometry is meant a set $\mathscr{S}$ of t-dimensional subspaces such that every point is contained in exactly one subspace of $\mathscr{S}$. In other words, the set of all points of the geometry is the disjoint union of the subspaces in $\mathscr{S}$. Spreads play an important role in the construction of certain non-desarguesian affine planes; see Section 5.1. In the finite case, if $\mathscr{P}(d, q)$ possesses a t-spread, the number $N_1(d+1, q)$ of all points must be divisible by the number $N_1(t+1, q)$ of points in any t-dimensional subspace. This observation and (8) yield the following result:

6. *There does not exist a t-spread of $\mathscr{P}(d, q)$ unless $d + 1$ is divisible by $t + 1$.*

Conversely, if $t + 1$ divides $d + 1$, there always exist t-spreads of $\mathscr{P}(d, q)$, but only for a few small values q, d, t have all t-spreads of $\mathscr{P}(d, q)$ been determined. Further details will be found in Section 3.1 and Chapter 5.

A *collineation* of a projective or affine geometry is a permutation of its points which maps lines onto lines. It follows from (4) that every subspace is then mapped onto a subspace. The set of all i-dimensional subspaces fixed by the collineation φ is denoted by $\mathscr{F}_i(\varphi)$, and $\mathscr{F}(\varphi)$ is the union of all $\mathscr{F}_i(\varphi)$. More generally, if Φ is a set of collineations, then $\mathscr{F}_i(\Phi) = \bigcap_{\varphi \in \Phi} \mathscr{F}_i(\varphi)$, and $\mathscr{F}(\Phi)$ is the union of all $\mathscr{F}_i(\Phi)$.

7. *In any projective geometry, if every $L \in \mathscr{F}_1(\Phi)$ contains at least three points of $\mathscr{F}_0(\Phi)$, then $\mathscr{F}_0(\Phi)$, together with the point sets $L \cap \mathscr{F}_0(\Phi)$, where $L \in \mathscr{F}_1(\Phi)$, is a subgeometry.*

We denote this subgeometry again by $\mathscr{F}(\Phi)$; this is not quite precise but will not lead to confusion. The study of $\mathscr{F}(\Phi)$ is of great importance in many situations, especially in non-desarguesian planes; cf. Chapter 4. If S is a fixed subspace of Φ, clearly Φ induces a set of collineations both in the subgeometry and in the quotient geometry with respect to S.

A point p is a *center* of the collineation φ if the collineation induced in the quotient geometry with respect to p is the identity. In other words, all subspaces containing p are fixed (it suffices to ask that the lines or the hyperplanes through p be fixed). A hyperplane H is an *axis* of φ if the collineation induced in the subgeometry H is the identity, i.e. if every point of H is fixed.

8. *A non-trivial collineation has at most one center and at most one axis.*

This is not only true in projective and affine geometries but also in many designs; see Section 2.3.

9. *A collineation of a projective geometry has a center if and only if it has an axis.*

This is not true for affine geometries and, a fortiori, not for designs in general. The proofs of **8** and **9** for dimension 2 can be found in Pickert 1955, p. 62—65, and the general result is derived from this by a simple inductive argument.

A collineation is *central* or *axial* if it has a center or an axis, respectively. By **9**, both concepts are equivalent in a projective geometry; we shall be concerned only with this projective case for some time. A central collineation is called an *elation* or a *homology* if it has an incident or non-incident center-axis pair, respectively; hence the identity is the only central collineation which is both an elation and a homology. The group generated by all elations is called the *little projective group*, and that generated by all central collineations the *full projective group*. Both the little and the full projective groups are normal subgroups in the group of all collineations of the projective geometry under consideration.

We now restrict our attention to desarguesian projective geometries of finite rank >2 (i.e. of dimension $\geqq 2$). In view of **4**, this means that we investigate the geometries $\mathscr{P}(V)$, with a vector space V of rank $d+1>2$ over a field $\mathfrak{F}$; and $\mathfrak{F}$ will later be assumed finite, i.e. $\mathfrak{F}=GF(q)$, with q a prime power p^e. A *semilinear automorphism* of V is a one-one map σ of V onto V such that

$$(x+y)^\sigma = x^\sigma + y^\sigma \quad \text{for all} \quad x, y \in V \tag{11}$$

and, for some automorphism σ' of $\mathfrak{F}$ (called the *companion automorphism* of σ),

(12) $$(fx)^\sigma = f^{\sigma'} x^\sigma \qquad \text{for } f \in \mathfrak{F},\ x \in V.$$

A *linear* automorphism is a semilinear automorphism whose companion automorphism is the identity, and a linear automorphism is called *unimodular* if its determinant (determinant of any representing matrix) is 1. If $\mathfrak{F}$ is non-commutative, the determinant is that of DIEUDONNÉ 1943; see also ARTIN 1957, Section IV. 1. It is easily checked that every semilinear automorphism maps subspaces of V onto subspaces and hence induces a collineation in $\mathscr{P}(V)$. Conversely:

10. *Every collineation of $\mathscr{P}(V)$ is induced by a semilinear automorphism of V. A collineation of $\mathscr{P}(V)$ is in the full projective group of $\mathscr{P}(V)$ if and only if it is induced by a semilinear automorphism σ whose companion automorphism is an inner automorphism of $\mathfrak{F}$; hence σ is linear if $\mathfrak{F}$ is commutative. The collineations in the little projective group of $\mathscr{P}(V)$ are precisely those induced by unimodular automorphisms of V.*

For proofs of these fundamental results, see BAER 1952, Chapter III, and ARTIN 1957, Chapter IV. The groups of all semilinear, linear, and unimodular automorphisms of the vector space V are denoted by $\Gamma L(V)$, $GL(V)$ and $SL(V)$, and the collineation groups of $\mathscr{P}(V)$ induced by them by $P\Gamma L(V)$, $PGL(V)$ and $PSL(V)$, respectively. The kernel of the canonical epimorphism $\Gamma L(V) \to P\Gamma L(V)$, i.e. the set of all semilinear automorphisms of V leaving every subspace of V invariant, consists of the mappings $\eta(f)$ defined by $x^{\eta(f)} = fx$, for $x \in V$ and $0 \neq f \in \mathfrak{F}$, and is therefore isomorphic to the multiplicative group of $\mathfrak{F}$.

In the finite case $V = V(n, q)$, we write $\Gamma L_n(q)$ instead of $\Gamma L(V(n, q))$, and $GL_n(q)$, $SL_n(q)$, $P\Gamma L_n(q)$, $PGL_n(q)$, $PSL_n(q)$ are defined similarly.

11. *The group $P\Gamma L_{d+1}(q)$, of all collineations of the d-dimensional desarguesian projective geometry $\mathscr{P}(d, q)$ over $GF(q)$, where $d \geqq 2$ and $q = p^e$, is of order*

$$|P\Gamma L_{d+1}(q)| = e\, q^{d(d+1)/2} \prod_{i=2}^{d+1} (q^i - 1).$$

The full projective group of $\mathscr{P}(d, q)$ is $PGL_{d+1}(q)$, of order

$$|PGL_{d+1}(q)| = e^{-1}\, |P\Gamma L_{d+1}(q)| = q^{d(d+1)/2} \prod_{i=2}^{d+1} (q^i - 1),$$

and the little projective group of $\mathscr{P}(d, q)$ is the simple group $PSL_{d+1}(q)$, of order

$$|PSL_{d+1}(q)| = (q-1, d+1)^{-1}\, |PGL_{d+1}(q)|,$$

where $(q-1, d+1)$ is the greatest common divisor of $q-1$ and $d+1$.

The proof of this appears essentially in Artin 1957, Chapter IV; in particular see his Theorem 4.11.

The next result shows that central collineations can be recognized by the numbers of their fixed points alone.

12. *A non-trivial collineation* φ *of* $\mathscr{P}(d, q)$ *is an elation if and only if*

$$|\mathscr{F}_0(\varphi)| = (q^d - 1)(q - 1)^{-1};$$

and φ *is a homology if and only if*

$$|\mathscr{F}_0(\varphi)| = 1 + (q^d - 1)(q - 1)^{-1}.$$

A proof[1]) can be found in Norman 1965.

A *translation* of the vector space V is a mapping of the form

$$\tau(a) : x \to x + a.$$

Clearly every translation induces a collineation of the affine geometry $\mathscr{A}(V)$; this collineation will also be called a translation.[2]) The translations form an abelian group $\mathsf{T} = \mathsf{T}(V)$ isomorphic to the additive group of V, and T is a normal subgroup of the group of all collineations of $\mathscr{A}(V)$. Considered as collineations of the projective geometry corresponding to $\mathscr{A}(V)$, the translations are just the elations with the distinguished hyperplane as axis. As an affine analogue of **11**, we have:

13. *The group of all collineations of the d-dimensional desarguesian affine geometry* $\mathscr{A}(d, q)$ *over* $GF(q)$ *is* $\mathsf{T} \cdot \Gamma L_d(q)$, *of order* $q^d |\Gamma L_d(q)|$.

We remark here (see the context of **11**) that

$$|\Gamma L_d(q)| = (q - 1)\,|P\Gamma L_d(q)|.$$

For much of the remainder of this section, we shall be concerned with various collineation groups of $\mathscr{P}(d, q)$, i.e. subgroups of $P\Gamma L_{d+1}(q)$, and their transitivity properties. These groups may be considered as permutation groups on the subspaces of equal dimension t, for $0 \leqq t \leqq d - 1$, and therefore as automorphism groups of the designs $\mathbf{P}_t(d, q)$ whose parameters were given in (9). There exist certain relationships between the transitivity properties of these different permutation representations; this will be discussed in a more general setting in Section 2.3. At this point, we mention only:

14. *A collineation group of* $\mathscr{P}(d, q)$ *is transitive on the points if and only if it is transitive on the hyperplanes.*

---

[1]) An analogous result holds in non-desarguesian projective planes, cf. Section 4.1.

[2]) In Section 2.3 we shall define translations of designs. It will be seen that the present concept is a special case of that definition.

This is a special case of result **2.3.1b** below. Considering the quotient geometry at a point, we get the following immediate corollary:

15. *Let Γ be a collineation group of $\mathscr{P}(d, q)$ or $\mathscr{A}(d, q)$, fixing the point p. Then Γ is transitive on the lines through p if and only if it is transitive on the hyperplanes through p.*

We are now going to discuss more special transitivity properties of collineation groups of $\mathscr{P}(d, q)$ and $\mathscr{A}(d, q)$. We begin with an account of *sharp transitivity on points* in $\mathscr{P}(d, q)$, and for this we need the concept of nearfield. A *nearfield*[1]) is an algebraic structure $\mathfrak{N}$ with two operations called addition and multiplication, such that $\mathfrak{N}$ together with addition is a group $\mathfrak{N}^+$, and $\mathfrak{N} - \{0\}$, where 0 is the neutral element of $\mathfrak{N}^+$, together with multiplication is a group $\mathfrak{N}^\times$. Moreover, we require that $x\,0 = 0$ and

(13) $$(x + y)\,z = x\,z + y\,z \qquad \text{for all } x, y, z \in \mathfrak{N}.$$

It follows from (13) that the mappings

$$\mu(z): x \to x\,z, \qquad \text{with } z \neq 0,$$

form an automorphism group of $\mathfrak{N}^+$ which is isomorphic to $\mathfrak{N}^\times$ and sharply transitive on the non-zero elements of $\mathfrak{N}$. In fact, nearfields and groups with sharply transitive automorphism groups are essentially identical objects; this is the reason for the importance of nearfields in the present context.

16. *The additive group $\mathfrak{N}^+$ of an arbitrary nearfield $\mathfrak{N}$ is commutative.*

In the finite case, $\mathfrak{N}^+$ is easily proved to be elementary abelian (Zassenhaus 1935a). For the proof of **16** in the infinite case, see B. H. Neumann 1940, or Karzel 1965c, p. 66—67.

The finite nearfields have been completely determined by Zassenhaus 1935a. This will be presented in Section 5.2; now we need only an important special class of finite nearfields which will be called here *regular nearfields*.[2]) Let q be a prime power and n an integer all of whose prime divisors divide $q - 1$. Also, suppose that $n \not\equiv 0 \bmod 4$ in case $q \equiv 3 \bmod 4$. Let c be a primitive element (i.e. a generator of the multi-

[1]) Our definition differs inessentially from what Zassenhaus 1935a calls "vollständiger Fastkörper"; instead of (13) he postulates the other distributive law $x(y + z) = x\,y + x\,z$. A more significant variation is found in André 1955, where also the solvability of all equations $-x\,a + x\,b = c$ is postulated. Nearfields with this additional property will be called *planar* here; it is an easy exercise to show that all finite nearfields are planar. More results about finite nearfields will be found in Section 5.2.

[2]) These nearfields were discovered by Dickson 1905. For an intrinsic characterization of them, see Ellers & Karzel 1964.

plicative group) of the Galois field $\mathfrak{F} = GF(q^n)$. Then c^n generates a multiplicative subgroup G whose cosets are represented by the elements $c_i = c^{(q^i-1)/(q-1)}$, $i = 0, \ldots, n-1$. If $f \in \mathfrak{F}$ is in the G-coset represented by c_i, we associate with f the automorphism $\alpha(f): x \to x^{q^i}$ of $\mathfrak{F}$. We define a new multiplication $\cdot$ in $\mathfrak{F}$ by $x \cdot 0 = 0$ and

$$x \cdot y = x^{\alpha(y)} y \qquad \text{for } y \neq 0. \tag{14}$$

Then the elements of $\mathfrak{F}$, together with the original addition in $\mathfrak{F}$ and the multiplication defined by (14), form a nearfield with centre $GF(q)$; cf. DICKSON 1905, ZASSENHAUS 1935a. We denote this nearfield by $N(n, q)$; clearly $N(1, q) = GF(q)$. The nearfields $N(n, q)$ are the *regular* nearfields mentioned above.

Now we return to the projective geometries $\mathscr{P}(V)$ of dimension $\geqq 2$. We are interested in determining all sharply transitive collineation groups of $\mathscr{P}(V)$. The connection with nearfields is that every such collineation group is a quotient group $\mathfrak{N}^\times/\mathfrak{K}^\times$ of the multiplicative group of a nearfield $\mathfrak{N}$, over the multiplicative group of a subfield $\mathfrak{K}$ of $\mathfrak{N}$ (DEMBOWSKI 1963, ELLERS & KARZEL 1963). In the finite case, $\mathfrak{N}$ must be regular. Further analysis shows:

17. *To every collineation group Γ of $\mathscr{P}(d, p^e)$ which is sharply transitive on the points there corresponds a divisor n of $e(d+1)$ such that $n(p^e-1)$ divides $p^{e(d+1)}-1$ and such that there is a regular nearfield $N(n, p^{e(d+1)/n})$. This nearfield $\mathfrak{N}$ then contains a subfield $\mathfrak{K} \cong GF(p^e)$, and $\Gamma \cong \mathfrak{N}^\times/\mathfrak{K}^\times$; the order of Γ is $(p^{e(d+1)}-1)(p^e-1)^{-1}$, and Γ is generated by two collineations γ and δ satisfying the relations*

$$\gamma^s = 1, \quad \delta^n = \gamma^t, \quad \delta\gamma\delta^{-1} = \gamma^{p^{e(d+1)n^{-1}}}, \tag{15}$$

where

$$s = (p^{e(d+1)}-1)\, n^{-1} (p^e-1)^{-1} \text{ and } t = (p^{e(d+1)}-1)\, n^{-1} (p^{e(d+1)/n}-1)^{-1}.$$

Conversely, every group satisfying these relations is of order $(p^{e(d+1)}-1)(p^e-1)^{-1}$ and operates as a collineation group Γ of $\mathscr{P}(d, p^e)$, sharply transitive on the points. Finally, Γ is contained in the projective group of $\mathscr{P}(d, p^e)$ if and only if n divides $d+1$.

This was proved by ELLERS & KARZEL 1964. The simplest special case is that where $n = 1$:

18. *Every $\mathscr{P}(d, q)$ admits a cyclic transitive group of collineations.*

For this see SINGER 1938; a short proof will be given in Section 2.4, see p. 105. The following supplement to **18** is also a consequence of **17**:

19. *Every abelian transitive collineation group of $\mathscr{P}(d, q)$ is cyclic.*

This follows also from a result of KARZEL 1962, 1964, 1965a which says that if a nearfield $\mathfrak{N}$ contains a subfield $\mathfrak{K}$ (with $[\mathfrak{N} : \mathfrak{K}]$ not necessarily finite) such that $\mathfrak{K}^\times$ is normal in $\mathfrak{N}^\times$ and $\mathfrak{N}^\times/\mathfrak{K}^\times$ is commutative, then $\mathfrak{N}$ is a commutative field.[1]) For other special cases of **17**, see ZAPPA 1953 and ROSATI 1957b.

Let Γ be a sharply transitive collineation group of $\mathscr{P}(d, q)$, let S be a subspace of $\mathscr{P}(d, q)$, and let $\mathscr{S}$ be the Γ-orbit containing S, i.e. the set of all images of S under collineations in Γ. (By **14** above, Γ must also be transitive on the hyperplanes of $\mathscr{P}(d, q)$, so that we can take, for example, $\mathscr{S}$ as the set of all hyperplanes.) Then the incidence structure of the points of $\mathscr{P}(d, q)$ and the subspaces in $\mathscr{S}$ is a tactical configuration $\mathbf{C}$, and Γ is an automorphism group of this tactical configuration, transitive on both points and blocks. It follows from **1.2.12** that we can write $\mathbf{C} = \mathbf{C}(\Gamma, 1, \Delta)$, with Δ a suitable quotient set in Γ. The following result then gives all multipliers of Δ [cf. (1.2.12) and **1.2.15**], at least in the linear case.

20. *Let Γ be a sharply transitive subgroup of the full projective group of $\mathscr{P}(d, q)$, with $q = p^e$, and let* N *denote the normalizer of Γ in the group of all collineations of $\mathscr{P}(d, q)$. Then* N *splits over Γ, a complement being isomorphic to the automorphism group of the regular nearfield $N(n, p^{e(d+1)/n})$ determined by Γ as in* **17**. *This automorphism group is essentially the group of multipliers of any quotient set Δ corresponding to a subspace of $\mathscr{P}(d, q)$, and it is cyclic of order a divisor of $e(d+1)$.*

For the proof see DEMBOWSKI 1963, Satz 2; the result is, mutatis mutandis, also valid in the infinite case. The automorphism groups of the $N(n, q)$ were determined by ZASSENHAUS 1935a; with the exception $N(2,3)$ not coming into play here, they can be considered as subgroups of the automorphism groups of the corresponding Galois fields $GF(q^n)$. For the special case $n = 1$ of **20**, see HIGMAN & MCLAUGHLIN 1961.

Result **20** shows that a collineation group of $\mathscr{P}(d, q)$ with a sharply transitive normal subgroup (for example, a Frobenius group) cannot be very large. This can be used to prove the nonexistence of collineation groups with certain prescribed transitivity behaviour, for example:

21. *The only $\mathscr{P}(d, q)$ which admit a collineation group sharply transitive on incident point-line or point-hyperplane pairs are the projective planes $\mathscr{P}(2, 2)$ and $\mathscr{P}(2, 8)$.*

---

[1]) This has the consequence, among other things, that an abelian transitive collineation group of an infinite desarguesian projective geometry is non-cyclic (KARZEL 1965b). In particular, infinite projective planes with transitive cyclic groups are non-desarguesian (examples of such planes exist, see HALL 1943 who conjectured Karzel's result). It has been conjectured that just the opposite is true in the finite case, i.e. that every finite plane with a cyclic transitive group must be desarguesian. More about this problem in Section 4.4.

That such a group must be a Frobenius group is true under weaker hypotheses (HIGMAN & MCLAUGHLIN 1961). The proof of **21** for point-line pairs is in DEMBOWSKI 1963a; in view of **15** the result also holds for point-hyperplane pairs. That $\mathcal{P}(2, 2)$ and $\mathcal{P}(2, 8)$ actually admit sharply flag transitive groups was shown by HIGMAN & MCLAUGHLIN 1961; the case $d = 2$ of **21** was proved there also.

22. *The only $\mathcal{P}(d, q)$ with $d \geqq 3$ which admits a collineation group sharply transitive on lines is $\mathcal{P}(4, 2)$.*

Such a group must again be a Frobenius group on the points. Result **22**, due to LÜNEBURG 1965b, has the following corollary:

23. *A collineation group of $\mathcal{P}(d, q)$, where $d \geqq 3$, is never sharply transitive on incident point-plane pairs.*

Here we mean by a *plane* a two-dimensional subspace. For $d = 3$, the result follows from **21**. Let $d > 3$ and assume Γ sharply transitive on incident point-plane pairs of $\mathcal{P}(d, q)$. For any point p, the stabilizer Γ_p induces a sharply line transitive group in the quotient geometry with respect to p, whence $d = 5$ and $q = 2$ by **22**. On the other hand, Γ would be primitive on points (see **2.3.7a** below) and of odd order $3^2 \cdot 5 \cdot 7 \cdot 31$, hence soluble (FEIT & THOMPSON 1963). But then **1.2.20** would imply that the number of points is a power of a prime, whereas $\mathcal{P}(5, 2)$ has 63 points.[1]

The word "sharply" in the hypotheses of **20**—**23** is quite essential. In particular, there are many transitive collineation groups of $\mathcal{P}(d, q)$; these have by no means been classified.[2] If, however, one assumes more than transitivity on points, it can sometimes be shown that such a collineation group must be fairly large. We shall now discuss some results of this kind.

As a preliminary, we remark:

24. *The little projective group $PSL_{d+1}(q)$ of $\mathcal{P}(d, q)$ is transitive on the bases of $\mathcal{P}(d, q)$. The stabilizer of a basis $\mathfrak{b}$, in the full projective group $PGL_{d+1}(q)$, is sharply transitive on the points which are in none of the hyperplanes $\langle \mathfrak{b} - \{x\} \rangle$, $x \in \mathfrak{b}$.*

This is easily verified,[3] and in fact one can prove a similar statement also for the stabilizer of a basis in the little projective group. **24** ob-

[1] This proof is due to W. M. Kantor.

[2] However, in the plane case $d = 2$, a complete enumeration of the subgroups of $PGL_3(q)$ was given by MITCHELL 1911 for q odd, and by HARTLEY 1926 for q even. For higher dimensions, HUPPERT 1957a has determined all soluble transitive collineation groups. More recently, HERING 1968a has shown that, except possibly for $\mathcal{P}(5, 2)$, such a group has at most one non-cyclic composition factor.

[3] The result also holds in the infinite case, provided the dimension is finite and the ground field commutative.

viously implies that the little projective group is doubly transitive on points; this implies in turn transitivity on incident point-line pairs. In fact, $PSL_{d+1}(q)$ is even transitive on the *nests*, i.e. on the sequences $S_0, \ldots, S_{d-1}$, where S_i is an i-dimensional subspace contained in S_{i+1} $(i = 0, \ldots, d-2)$.

The following results will give criteria under which, conversely, a collineation group of $\mathcal{P}(d, q)$ contains $PSL_{d+1}(q)$.

25. *Suppose that the collineation group Γ of $\mathcal{P}(d, q)$ satisfies the following conditions:*

(**a**) *No proper subspace of $\mathcal{P}(d, q)$ is left invariant (as a whole) by Γ.*
(**b**) *There exist nontrivial central collineations $\alpha, \beta \in \Gamma$ which either have distinct centers but the same axis, or, dually, the same center but distinct axes.*
(**c**) *Unless α, β of* (**b**) *are both elations, Γ is generated by homologies.*

Then either $d = 2$, $q = 4$, and $\Gamma \cong A_6$ or S_6, or else Γ contains the little projective group of $\mathcal{P}(d, q)$.

(Here A_n and S_n denote the alternating and symmetric groups of degree n, respectively.) These results are due to Piper 1966b, 1967, 1968b. The first of these papers deals with the case where q is odd and α, β are elations, the third with the case q even and α, β elations (this is where the exceptions $\Gamma \cong A_6$ or S_6 arise, see also **4.3.23a**), and the second with the case where at least one of α, β is a homology.

A similar result, with slightly stronger hypotheses, is:

26. *Let Γ be a collineation group of $\mathcal{P}(d, q)$ such that either every incident point-hyperplane pair is center-axis for a nontrivial elation in Γ, or every non-incident point-hyperplane pair is center-axis for a nontrivial homology in Γ. Then Γ contains the little projective group of $\mathcal{P}(d, q)$.*

For the proof see Higman 1962, propositions 3 and 5. The arguments involved in this proof (as well as those of the following corollary) will be discussed in detail under more general hypotheses in Sections 2.1 and 2.3.

27. *If Γ is transitive on incident point-hyperplane pairs and contains a central collineation $\neq 1$, or if Γ is transitive (on points or on hyperplanes, cf.* **14**) *and contains a homology $\neq 1$, then Γ contains the little projective group of $\mathcal{P}(d, q)$.*

This follows from **25**; direct proofs, of slightly weaker results, are in Wagner 1961 and Higman 1962. Now the existence of central collineations can sometimes be proved. For example, if for some hyperplane H the stabilizer Γ_H is not faithful on the points of H, then obviously

H is the axis of a central collineation in Γ. In view of this remark, the following result is of particular value:

28. *Suppose that* $d > 2$, $(d, q) \neq (3, 2)$, *and let* S *be a subspace of dimension* $\geqq d/2$ *of* $\mathcal{P}(d, q)$. *Suppose that* Γ *is a collineation group of* $\mathcal{P}(d, q)$ *such that the stabilizer* Γ_S *is faithful on the points of* S *and, considered as collineation group of the subgeometry* S, *contains the little projective group of* S. *Then every collineation of* Γ *which induces an elation in* S *is itself an elation of* $\mathcal{P}(d, q)$.

This is Proposition 6 of HIGMAN 1962; the proof involves complicated matrix computations. The hypothesis $(d, q) \neq (3, 2)$ is essential: there exists a counterexample, isomorphic to the alternating group A_7, which is transitive on the incident point-line-plane triples of $\mathcal{P}(3, 2)$. For a discussion of this (unique) counterexample (which was discovered by C. Jordan; cf. DICKSON 1901, p. 290), see WAGNER 1961, Appendix 1.[1]) The most important application of **28** is:

29. *If* $(d, q) \neq (3, 2)$ *and* Γ *is a collineation group of* $\mathcal{P}(d, q)$ *such that for every hyperplane* H *the stabilizer* Γ_H *induces the little projective group of the subgeometry* H, *then* Γ *contains the little projective group of* $\mathcal{P}(d, q)$.

This was also proved by WAGNER 1961, Theorem 3.[2])

Concerning transitivity on nests, as defined above, the following has been proved:

30. *If* $(d, q) \neq (2, 2), (2, 8), (3, 2)$, *then every nest transitive collineation group of* $\mathcal{P}(d, q)$ *contains the little projective group.*

(HIGMAN 1962.) Moreover, if $(d, q) = (2, 2), (2, 8)$, or $(3, 2)$, the examples mentioned in **21** and after **28**, respectively, are the only nest transitive groups which do not contain the little projective group.

The proof of **30** is by induction on the dimension d. The main difficulty lies in the case $d = 2$, settled by HIGMAN & MCLAUGHLIN 1961. Some of the results for $d = 2$ also hold in the non-desarguesian case; these will be discussed in Sections 4.3 and 4.4.

It is an unsettled problem whether or not the conclusion of **30** still holds when the hypothesis of nest transitivity is replaced by that of double transitivity (on points or on hyperplanes; we shall see later that these conditions are equivalent). The following are some results concerning this question.

---

[1]) See also MOORE 1899, §§ 2—4, 6.

[2]) The result may also be derived from **28** and a recent theorem of ITO 1967a from which it follows that if Γ is doubly transitive and contains a central collineation $\neq 1$, then Γ contains all elations; cf. **2.3.35b** below.

31. *A doubly transitive collineation group of $\mathcal{P}(d, q)$ is transitive on the triples (p, H, K), where p is a point and H, K are hyperplanes, with $p \in H$, $p \notin K$.*

(WAGNER 1961, Lemma 2.) Such a group will then also be transitive on the dual figures (p, q, H) with $p \in H$, $q \notin H$, and in particular transitive on incident as well as on non-incident point-hyperplane pairs. Also, the collineation group induced by the stabilizer of a hyperplane H is transitive on non-incident point-hyperplane pairs of the subgeometry H.

32. *Let Γ be a doubly transitive collineation group of $\mathcal{P}(d, q)$, and let P be a plane of $\mathcal{P}(d, q)$. Then the stabilizer Γ_P, considered as a collineation group of the subgeometry P, contains the little projective group of P.*

(WAGNER 1961, Lemma 3.) For the proof it is shown, with the help of **31**, that Γ_P is transitive on non-incident point-line pairs of P. That this suffices to prove **32**, even under weaker hypotheses, will be seen in Section 4.4. Together with **29**, **32** implies:

33. *If $d \leqq 4$ and $(d, q) \neq (3, 2)$, then every doubly transitive collineation group of $\mathcal{P}(d, q)$ contains the little projective group.*

This is Theorem 4 of WAGNER 1961.[1]) All known doubly transitive collineation groups of $\mathcal{P}(d, q)$, where $(d, q) \neq (3, 2)$, contain the little projective group, so that possibly no restrictions on the dimension are necessary at all. The following result shows that if there are other counterexamples besides A_7 on $\mathcal{P}(3, 2)$, then these must involve some so far unknown finite simple groups:

34. *Every normal subgroup $\neq 1$ of a doubly transitive collineation group of $\mathcal{P}(d, q)$ is also doubly transitive.*

This is Theorem 2 of WAGNER 1961; for a generalization see **2.3.35f** below.

At this point, it is interesting to pause and consider the question as to what extent the results **17**—**34** have analogues in the affine case; this problem seems to have received attention only in special cases so far. The quotient geometry with respect to a point of a d-dimensional affine geometry is a $(d-1)$-dimensional projective geometry; hence **17**—**27** carry over, with obvious changes, if $\mathcal{P}(d, q)$ is replaced by $\mathcal{A}(d+1, q)$ and Γ is supposed to have a fixed point. In particular, to every point p of $\mathcal{A}(d, q)$, $d \geqq 2$, there exists a cyclic collineation

---

[1]) Using **28**, D. G. Higman has shown that the restriction $d \leqq 4$ of **33** may be relaxed to $d \leqq 6$. More recently, W. M. Kantor proved that the conclusion of **33** holds whenever d is a power of 2, provided q is odd. These results are unpublished.

group fixing p and transitive on the remaining points (analogue to **18**, BOSE 1942b). For $d = 2$, the non-cyclic groups of this type were determined by ROSATI 1957c. The collineation groups of the affine geometries $\mathcal{A}(d, q)$ which are sharply transitive on points seem to have not yet been completely determined.[1] Clearly, the translation group T of $\mathcal{A}(d, q)$ is of this kind, and this example shows that, unlike **20**, the normalizer of such a group may be the group of all collineations. **21** has the following affine analogue:

35. *$\mathcal{A}(d, q)$ admits a collineation group sharply transitive on incident point-line or point-hyperplane pairs if, and only if, d and $q - 1$ are relatively prime.*

The proof for point-line pairs is in LÜNEBURG 1966a; nearfields again play an important role here. As in **21**, the result for point-hyperplane pairs follows from **15**.

As in the projective case, the word "sharply" in **35** is essential, and in dimensions $d > 2$ the groups which are transitive on incident point-line pairs have not yet been determined. For dimension 2, however, such a determination has been made by FOULSER 1964a; his results may be summarized as follows.[2] As $\mathfrak{K} = GF(q^2)$ is a vector space of rank 2 over $\mathfrak{F} = GF(q)$, the points of $\mathcal{A}(2, q)$ may be identified with the elements of $\mathfrak{K}$, and the group Φ of the mappings

(16) $\quad x \to a\, x^\alpha + b \quad$ with $a, b \in \mathfrak{K}$ and $\alpha \in \operatorname{Aut} \mathfrak{K}$

is a collineation group of $\mathcal{A}(2, q)$. It is easy to see that Φ is doubly transitive on the points, and clearly $\mathsf{T} \subset \Phi$. Let Λ be the collineation group of $\mathcal{A}(2, q)$ which, considered as a collineation group of the corresponding projective geometry $\mathcal{P}(2, q)$, is generated by all elations fixing the ideal line; clearly $\mathsf{T} \subset \Lambda$, and Λ is also doubly transitive.

36. *Let Γ be a collineation group of $\mathcal{A}(2, q)$ which is transitive on incident point-line pairs. Then $\mathsf{T} \subset \Gamma$, and apart from finitely many exceptions Γ either contains Λ or is conjugate to a subgroup of Φ. Furthermore, if Γ is isomorphic to another collineation group Δ which is also transitive on incident point-line pairs, then with one exception Γ is conjugate to Δ.*

For the proof, as well as for the description of the exceptional cases, the reader is referred to FOULSER 1964a. The class of collineation groups which are transitive on incident point-line pairs is rather large for $d = 2$, as **36** shows, and one might expect similar situations for higher

---

[1] For the plane case $d = 2$, see DEMBOWSKI 1965c and DEMBOWSKI & OSTROM 1968.

[2] These results are independent of those of MITCHELL 1911 and HARTLEY 1926 mentioned in footnote [2] of p. 36.

dimensional affine groups. However, if transitivity on incident point-line pairs is replaced by the stronger assumption of nest transitivity, we get a situation similar to **30** in the affine case as well:

37. *Let Γ be a nest transitive collineation group of $\mathscr{A}(d, q)$, where $d > 2$ and $(d, q) \neq (3, 2), (3, 8), (4, 2)$. Then Γ, considered as collineation group of the corresponding projective geometry $\mathscr{P}(d, q)$, contains all elations fixing the ideal hyperplane.*

This is Theorem 6 of FOULSER 1964a; the proof uses **30**.

We return to projective geometries. A *correlation* of such a geometry $\mathscr{G}$ (which at present need be neither desarguesian nor finite) is a permutation δ of the subspaces which inverts inclusion, i.e.

(17) $\quad S \subseteqq T$ *implies* $S^\delta \supseteqq T^\delta$, *for all subspaces* S, T *of* $\mathscr{G}$.

It can be shown (cf. BAER 1952, p. 96) that $\mathscr{G}$ admits a correlation only if the dimension of $\mathscr{G}$ is finite. In a d-dimensional geometry, every correlation interchanges the set of i-dimensional with that of $(d - 1 - i)$-dimensional subspaces; in particular, points are mapped onto hyperplanes, and vice versa. Also, every correlation of $\mathscr{P}(d, q)$ induces an anti-automorphism (in the sense of Section 1.2) of the tactical configuration consisting of i- and $(d - 1 - i)$-dimensional subspaces of $\mathscr{P}(d, q)$, where $0 \leqq i \leqq d - 1$. A subspace S of $\mathscr{G}$ will be termed *totally isotropic, isotropic,* or *nonisotropic,* with respect to the correlation δ, according as $S \cap S^\delta$ is S, nonempty, or empty, respectively. Clearly a totally isotropic subspace must have dimension $\leqq (d - 1)/2$, and therefore rank $\leqq (d + 1)/2$. Points are either nonisotropic or totally isotropic; in the latter case we speak also of *absolute* points, which is consistent with the terminology of Section 1.2. Dually, a hyperplane H is *absolute* if it contains the point H^δ.

There exist projective planes which admit no correlations; finite examples will be given in Section 5.2. Every finite desarguesian projective geometry, however, does admit correlations, as the following discussion will show. Let V be a left vector space of finite rank over a field $\mathfrak{F}$ which, at present, need not be commutative. A *sesquilinear form*[1]) on V is a mapping s from $V \times V$ into $\mathfrak{F}$ such that

(18) $\quad s(x + x', y + y') = s(x, y) + s(x, y') + s(x', y) + s(x', y')$

for all $x, x', y, y' \in V$

[1]) These are also called "generalized hermitians" or "semibilinear forms". Our terminology is that of DIEUDONNÉ 1955.

and, with some anti-automorphism α of $\mathfrak{F}$ (called the *companion anti-automorphism* of the sesquilinear form s),

(19) $\quad s(fx, gy) = fs(x, y)\, g^{\alpha} \qquad$ for all $f, g \in \mathfrak{F}$ and $x, y \in V$.

Given any sesquilinear form s on V, we associate with every subspace S of V the set

(20) $$S^{\delta} = \{x \in V : s(x, S) = 0\};$$

here $s(x, S) = 0$ means $s(x, y) = 0$ for all $y \in S$. It is easily seen that S^{δ} is again a subspace of V, and that (17) is satisfied. Hence the mapping δ defined by (20) will be a correlation of the projective geometry $\mathscr{P}(V)$ if and only if it is a permutation of the subspaces of V; but it can be shown (BAER 1952, p. 103) that this is the case if and only if the sesquilinear form s is *nondegenerate* in the sense that $V^{\delta} = 0$, i.e. $s(x, y) = 0$ for all $y \in V$ (if and) only if $x = 0$. Thus for every nondegenerate sesquilinear form on V, (20) defines a correlation of $\mathscr{P}(V)$. Conversely:

38. *If δ is a correlation of the projective geometry $\mathscr{P}(V)$, then there exists a nondegenerate sesquilinear form s on V such that δ is given by* (20).

For the proof of this important result, see BIRKHOFF & v. NEUMANN 1936, or BAER 1952, p. 102. In view of **4** and **38**, we can now say that every correlation of a desarguesian projective geometry can be represented by a nondegenerate sesquilinear form.

A *polarity* is a correlation of order 2; this is again consistent with the terminology of Section 1.2. It is easily seen that a correlation π of a projective geometry $\mathscr{G}$ is a polarity if and only if, in addition to (17),

(21) $\quad S \subseteq T^{\pi}$ *implies* $T \subseteq S^{\pi} \qquad$ *for all subspaces S, T of $\mathscr{G}$.*

For the desarguesian case, it now follows that a nondegenerate sesquilinear form s on V represents a polarity of $\mathscr{P}(V)$ if and only if

(22) $\quad s(x, y) = 0$ *implies* $s(y, x) = 0 \qquad$ *for all $x, y \in V$.*

Now the following can be proved:

39. *Let V be a left vector space over $\mathfrak{F}$ and s a nondegenerate sesquilinear form with companion anti-automorphism α. Then if s satisfies* (22), *there are exactly the following three possibilities for s:*

(**a**) $\alpha = 1$ *(hence $\mathfrak{F}$ is commutative), $s(x, y) = s(y, x)$ for all $x, y \in V$, and if F has characteristic 2, then $s(z, z) \neq 0$ for some $z \in V$.*

(**b**) $\alpha = 1$ *(hence $\mathfrak{F}$ is commutative), and $s(x, x) = 0$ for all $x \in V$.*

(**c**) $\alpha^2 = 1 \neq \alpha$ *(here $\mathfrak{F}$ may be noncommutative) and $s(x, y) = s(y, x)^{\alpha}$ for all $x, y \in V$.*

Proof: BIRKHOFF & v. NEUMANN 1936, BAER 1952, p. 111; ARTIN 1957, p. 112—114. In case (**b**) we clearly have $s(x, y) = -s(y, x)$ for all $x, y \in V$; this reduces to $s(x, y) = s(y, x)$ if $\mathfrak{F}$ has characteristic 2, but even then (**a**) and (**b**) are clearly inequivalent.

We call a polarity *orthogonal, symplectic* or *unitary*[1]) if it is represented by a sesquilinear form of type (**a**), (**b**), or (**c**) above, respectively. The following results, whose proofs appear in Chapter IV of BAER 1952, show how these different types of polarities may be distinguished geometrically. A polarity is symplectic if and only if every point is absolute. A polarity is unitary if (and in the finite case only if, see **47** below) there exists a line containing at least one non-absolute and at least three absolute points. Such a line cannot be totally isotropic, for totally isotropic subspaces clearly consist of absolute points only. The converse of this statement is not true in general, but it can be shown that if π is not symplectic and the subspace S is not totally isotropic, then S can consist entirely of absolute points only if π is orthogonal and $\mathfrak{F}$ has characteristic 2 (cf. **41** below).

The polarities of the finite geometries $\mathscr{P}(d, q)$ give rise to several interesting classes of finite groups as well as of finite incidence structures. Before defining these, it is convenient to introduce the concept of "quadric" which, as we shall see, is closely related to that of the set of absolute points of an orthogonal polarity. A *quadratic form* on the vector space V over the commutative field F is a mapping Q of V into $\mathfrak{F}$, satisfying

$$Q(fx) = f^2 Q(x) \qquad \text{for all } f \in \mathfrak{F} \text{ and } x \in V \tag{23}$$

and the following condition: The mapping S from $V \times V$ into $\mathfrak{F}$ given by

$$S(x, y) = Q(x + y) - Q(x) - Q(y) \qquad \text{for all } x, y \in V \tag{24}$$

is a symmetric bilinear form.[2]) A *quadric* of the projective geometry $\mathscr{P}(V)$ is then defined to be any set of points which is represented by the vectors $x \in V$ with $Q(x) = 0$, for some quadratic form Q. If the characteristic of $\mathfrak{F}$ is not 2, then a quadric, as well as the representing quadratic form, is called *nondegenerate* if the bilinear form S given by (24) is nondegenerate. If, on the other hand, $\mathfrak{F}$ has characteristic 2,

[1]) We choose these terms because the groups connected with these polarities (see p. 44—45 below) are well known under these names. The more customary terminology for "symplectic polarity" is "null system" or "null polarity".

[2]) That is, a sesquilinear form with $S(x, y) = S(y, x)$ and, therefore, identity companion anti-automorphism. It is easily seen that this coordinate-free definition of a quadratic form (ARF 1941, DIEUDONNÉ 1955, § 16, ARTIN 1957, p. 108) reduces, after choosing a basis of V, to the more familiar form $Q(x_1, \ldots, x_n) = \sum_{i,j-1}^{n} a_{ij} x_i x_j$.

then $S(x, x) = 0$ for all $x \in V$, and S cannot be nondegenerate unless d is odd (cf. **39b** and **46** below). In this case, Q is called nondegenerate if $Q(v) \neq 0$ whenever $v \neq 0$ and $S(v, x) = 0$ for all $x \in V$ (cf. DIEUDONNÉ 1955, p. 32—33).

40. *If $\mathfrak{F}$ is of characteristic $\neq 2$, then the nondegenerate quadrics of $\mathscr{P}(V)$ are precisely the sets of absolute points with respect to the orthogonal polarities of $\mathscr{P}(V)$.*

In fact, if π is represented by the symmetric bilinear form s (see **39a**), then the absolute points of π are represented by the vectors $x \in V$ with $Q(x) = s(x, x) = 0$. It follows that $S(x, y) = 2s(x, y)$ and if the characteristic of $\mathfrak{F}$ is $\neq 2$, then S and s represent the same polarity of V. On the other hand, orthogonal polarities are easily described if the characteristic is 2, at least in the finite case:

41. *The absolute points with respect to an orthogonal polarity π of $\mathscr{P}(d, 2^e)$ are precisely those of a distinguished hyperplane $H(\pi)$ which is itself absolute if and only if d is odd.*

For in this case the equation $s(x, x) = 0$ describing the absolute points reduces to a non-trivial linear equation over $GF(2^e)$. The assertion about $H(\pi)$ is a consequence of **1.2.2** for $d = 2$; in higher dimensions it follows from this by induction, using the fact that a d-dimensional projective space cannot admit a symplectic polarity unless d is odd; see **46** below.

We shall now define the "classical" groups connected with polarities and quadrics of the desarguesian geometries.[1] Let s be a nondegenerate sesquilinear form on the vector space V over $\mathfrak{F}$, suppose that s satisfies (22), and call π the polarity of $\mathscr{P}(V)$ represented by s. A collineation of $\mathscr{P}(V)$, represented by the semilinear automorphism σ with companion automorphism σ', can be shown to commute with π if and only if there exists a non-zero element $f = f(\sigma) \in \mathfrak{F}$ such that

$$s(x^\sigma, y^\sigma) = s(x, y)^{\sigma'} f(\sigma) \qquad \text{for all } x, y \in V. \tag{25}$$

(Proof: BAER 1952, p. 144.) The collineation group of $\mathscr{P}(V)$ induced by all σ with (25), i.e. the centralizer of π, will be denoted by $\Gamma(\pi)$. If $\mathfrak{q}$ is the quadric given by the nondegenerate quadratic form Q, we define $\Gamma(\mathfrak{q})$ as the collineation group induced by the σ for which there exists $g = g(\sigma) \neq 0$ in $\mathfrak{F}$ with

$$Q(x^\sigma) = Q(x)^{\sigma'} g(\sigma) \qquad \text{for all } x \in V. \tag{25'}$$

It is easily seen that if π is orthogonal, $\mathfrak{F}$ of characteristic $\neq 2$, and if $\mathfrak{q}$ is the set of absolute points of π, then $\Gamma(\pi) = \Gamma(\mathfrak{q})$. Finally, we

---

[1]) For a more complete theory of these groups, in particular for their abstract structure, see DIEUDONNÉ 1955, also DICKSON 1901, VAN DER WAERDEN 1935.

define $\Gamma_0(\pi)$ and $\Gamma_0(\mathfrak{q})$ as the (normal) subgroups of $\Gamma(\pi)$ and $\Gamma(\mathfrak{q})$ induced by those σ for which (25) reduces to, respectively,

(26) $\quad s(x^\sigma, y^\sigma) = s(x, y)$ and $Q(x^\sigma) = Q(x) \qquad$ for all $x, y \in V$.

Clearly, both $\Gamma_0(\pi)$ and $\Gamma_0(\mathfrak{q})$ are contained in the full projective group $PGL(V)$ of $\mathscr{P}(V)$; in fact it can be shown that Γ_0 is generated by the central collineations in Γ.

From now on we suppose that $\mathfrak{F}$ is commutative. The following fact is fundamental:

42. *If S and T are subspaces of equal rank which are both totally isotropic with respect to the polarity π of $\mathscr{P}(V)$, then there exists a collineation $\gamma \in \Gamma_0(\pi)$ such that $S\gamma = T$.*

In other words, $\Gamma_0(\pi)$ [and a fortiori $\Gamma(\pi)$] is transitive on the subspaces of constant rank which are totally isotropic with respect to π. Result **42** is a special case of "Witt's theorem" (Witt 1937); for the proof see Dieudonné 1955, § 11; also Artin 1957, Theorem 3.9, and Lenz 1961, pp. 335—340. In view of **40**, result **42** remains true, provided $\mathfrak{F}$ is of characteristic $\neq 2$, if S and T are supposed to be contained in a nondegenerate quadric $\mathfrak{q}$ and if $\Gamma_0(\pi)$ is replaced by $\Gamma_0(\mathfrak{q})$. In fact, an analogue of Witt's theorem for quadrics can be proved also in certain cases where the characteristic is 2 (cf. Arf 1941, Dieudonné 1955, § 16), and this suffices for the following result:

43. *All maximal totally isotropic subspaces with respect to the polarity π, or all maximal subspaces contained in a nondegenerate quadric $\mathfrak{q}$, have the same rank.*

This invariant is called the *index* of π or $\mathfrak{q}$, respectively, and will be denoted here by $m(\pi)$ or $m(\mathfrak{q})$. Clearly, if $\mathscr{P}(V)$ has dimension d, i.e. if V has rank $d+1$, then $m \leqq (d+1)/2$.

We return now to the finite case. The following results give a complete classification of polarities (non-orthogonal if $p = 2$) and quadrics in $\mathscr{P}(d, p^e)$. We say that two polarities π, π', or two nondegenerate quadrics $\mathfrak{q}, \mathfrak{q}'$, are *equivalent* if there exists a collineation φ such that $\pi' = \varphi^{-1}\pi\varphi$ or $\mathfrak{q}' = \mathfrak{q}\varphi$, respectively. If this is the case, then such a φ can even be found in $PGL_{d+1}(p^e)$.

44. *Let $t = (q-1, 2)$. If d is even, then $\mathscr{P}(d, q)$ contains exactly*

$$Q(d, q) = t\, q^{d(d+2)/4} \prod_{i=1}^{(d-1)/2} (q^{2i+1} - 1)$$

nondegenerate quadrics. These are all equivalent of index $d/2$, and for each of them the group $\Gamma_0(\mathfrak{q})$ is the projective orthogonal group $PO_{d+1}(q)$, of *order* (cf. **11**)

$$|PO_{d+1}(q)| = |PGL_{d+1}(q)| \big/ Q(d, q) = t^{-1} q^{d^2/4} \prod_{i=1}^{d/2} (q^{2i} - 1).$$

If d is odd, then there exist exactly two equivalence classes of nondegenerate quadrics, of indices $(d + \varepsilon)/2$, where $\varepsilon = \pm 1$. The number of such quadrics is

$$Q(d, q; \varepsilon) = t\, q^{(d+1)^2/4} (q^{(d+1)/2} + \varepsilon) \prod_{i=1}^{(d-1)/2} (q^{2i+1} - 1),$$

and for each of these, $\Gamma_0(\mathfrak{q})$ is the projective orthogonal group $PO_{d+1}(q, \varepsilon)$, of order

$$\begin{aligned} |PO_{d+1}(q, \varepsilon)| &= |PGL_{d+1}(q)| \big/ Q(d, q; \varepsilon) \\ &= t^{-1} q^{(d^2-1)/4} (q^{(d+1)/2} - \varepsilon) \prod_{i=1}^{(d-1)/2} (q^{2i} - 1). \end{aligned}$$

Using **42**—**44**, we can now find the number of subspaces of given rank in a nondegenerate quadric:

45. *Let $\mathfrak{q}$ be a nondegenerate quadric of $\mathscr{P}(d, q)$. If a line contains more than two points of $\mathfrak{q}$, then it is contained in $\mathfrak{q}$. The number of subspaces of rank r contained in $\mathfrak{q}$, where $1 \leqq r \leqq m(\mathfrak{q})$, is*

$$T_r(\mathfrak{q}) = N_r(m, q) \prod_{i \in I} (q^i + 1),$$

with $m = m(\mathfrak{q})$, $N_r(m, q)$ as in (6), and

$$I = \begin{cases} \{m - r, \ldots, m - 1\} & \text{if } m = (d+1)/2 \quad (d \text{ odd}) \\ \{m - r + 1, \ldots, m\} & \text{if } m = d/2 \quad (d \text{ even}) \\ \{m - r + 2, \ldots, m + 1\} & \text{if } m = (d-1)/2 \quad (d \text{ odd}). \end{cases}$$

Clearly **44** and **45**, when restricted to the case $p \neq 2$, give ample information about the system of totally isotropic subspaces of orthogonal polarities of $\mathscr{P}(d, q)$. Similar results hold for the symplectic and unitary polarities:

46. *$\mathscr{P}(d, q)$ admits symplectic polarities if and only if d is odd. If this is the case, then the number of such polarities is*

$$S(d, q) = t\, q^{(d^2-1)/4} \prod_{i=1}^{(d-1)/2} (q^{2i+1} - 1),$$

where t is the g.c.d. of 2 and $q - 1$; these polarities are all equivalent of index $(d+1)/2$, and for any one of them the group $\Gamma_0(\pi)$ is the projective symplectic group $PSp_{d+1}(q)$, of order

$$|PSp_{d+1}(q)| = |PGL_{d+1}(q)| \big/ S(d, q) = t^{-1} q^{(d+1)^2/4} \prod_{i=1}^{(d+1)/2} (q^{2i} - 1).$$

Also, the number of totally isotropic subspaces of rank r with respect to π *is given by*

$$T_r(\pi) = \prod_{i=0}^{r-1} (q^{d+1-2i} - 1) \Big/ (q^{i+1} - 1).$$

47. $\mathcal{P}(d, q)$ *admits unitary polarities if and only if q is a square:* $q = s^2$. *If this is the case, then the number of such polarities is*

$$U(d, q) = U(d, s^2) = s^{d(d+1)/2} \prod_{i=2}^{d+1} (s^i + (-1)^i)$$

These are all equivalent of index $[(d+1)/2]$, *and their representing sesquilinear forms have companion automorphism* $x \to x^s$. *For each of these polarities* π, *the group* $\Gamma_0(\pi)$ *is the projective unitary group* $PGU_{d+1}(q)$, *of order*

$$|PGU_{d+1}(q)| = |PGL_{d+1}(q)| \Big/ U(d, q) = s^{d(d+1)/2} \prod_{i=2}^{d+1} (s^i - (-1)^i).$$

A line of $\mathcal{P}(d, q)$ *consists entirely of absolute points with respect to* π *if, and only if, it is totally isotropic. Every other line contains either at most one or exactly* $s + 1$ *absolute points. Finally, the number of totally isotropic subspaces of rank r with respect to* π *is given by*

$$T'_r(\pi) = \prod_{i=d+2-2r}^{d+1} (s^i - (-1)^i) \Big/ \prod_{j=1}^{r} (s^{2j} - 1).$$

The proofs of **44**—**47** are all implicitly contained in DICKSON 1901; for more direct treatments, see also ARTIN 1957, pp. 122, 143—148, 177, 210; DIEUDONNÉ 1955, p. 16; PRIMROSE 1951; SEGRE 1959a, p. 4; RAY-CHAUDHURI 1962a; DAI & FENG 1964a, b; WAN 1964; WAN & YANG 1964; PLESS 1967.[1]) We shall see in Sections 2.4 and 7.1 how these results can be used for the construction of certain tactical configurations, some of which are designs.

Note that the numbers $T_1(\mathfrak{q})$, $T_1(\pi)$, $T'_1(\pi)$ in **45**—**47** are different from zero if $d \geqq 2$, so that all nondegenerate quadrics in $\mathcal{P}(d, q)$ are non-empty and all polarities have absolute points (for this see also **41**). The second of these statements is true for more general incidence structures as well; cf. **2.1.17** and **3.3.2** below.

Let $\mathfrak{q}$ be a nondegenerate quadric in $\mathcal{P}(d, q)$. Then $\Gamma_0(\mathfrak{q})$ acts transitively on the points of $\mathfrak{q}$. Also, $\mathfrak{q}$ always contains two points x, y such that the line $x + y$ has only x and y with $\mathfrak{q}$ in common. Now if $\mathfrak{q}$ contains a line L, it is clear that a collineation mapping $x + y$ onto L cannot preserve $\mathfrak{q}$ and hence is not in $\Gamma(\mathfrak{q})$. This shows that $\Gamma(\mathfrak{q})$ cannot

[1]) In these papers, results about degenerate quadrics can also be found (these are defined by quadratic forms which are not nondegenerate), as well as results on the number of isotropic subspaces whose intersections with their π-images are of prescribed rank.

be doubly transitive on the points of $\mathfrak{q}$ if $m(\mathfrak{q}) > 1$. Conversely, suppose that $m(\mathfrak{q}) = 1$. Then it follows from **44** that $d = 2$ or 3, and **45** implies that $\mathfrak{q}$ has the following properties:

(27) $$|\mathfrak{q} \cap L| \leqq 2 \quad \text{for any line } L, \text{ and}$$

(28) $$|\mathfrak{q}| = q^{d-1} + 1.$$

In these cases the group $\Gamma_0(\mathfrak{q})$ turns out to be triply transitive on the points of $\mathfrak{q}$; also, there is a canonical isomorphism of $\Gamma_0(\mathfrak{q})$ onto $PGL_2(q)$ if $d = 2$, and onto $PGL_2(q^2)$ if $d = 3$ (cf. ARTIN 1957, Theorems 5.20 and 5.21). The following results will characterize this situation in a more general setting.

In an arbitrary projective geometry $\mathcal{G}$ (which at this point need be neither desarguesian nor finite), we define an *ovoid*[1]) as a set $\mathfrak{q}$ of points satisfying (27) and

(28′) *For any $x \in \mathfrak{q}$, the union of all lines L with $\mathfrak{q} \cap L = \{x\}$ is a hyperplane.*

This hyperplane will be called the *tangent hyperplane* of $\mathfrak{q}$ at x, and will be denoted by $T(x)$. It is easily seen that (28) holds for all ovoids if $\mathcal{G} = \mathcal{P}(d, q)$. Clearly, if H is a hyperplane of $\mathcal{G}$ such that $|\mathfrak{q} \cap H| > 1$, then $\mathfrak{q} \cap H$ is an ovoid of the subgeometry H of $\mathcal{G}$. Combining this fact with (28) for $\mathcal{G} = \mathcal{P}(d, q)$, we see that the tactical configuration of the points and nontrivial hyperplane intersections of a hypothetical ovoid in $\mathcal{P}(4, q)$ would have parameters $v = q^3 + 1$, $k = q^2 + 1$, and $r = q(q^2 + q + 1)$. This is impossible because, according to (1.1.9′), k is a divisor of $v\,r$. Hence there is no such ovoid, and:

48. *If $\mathcal{P}(d, q)$ contains an ovoid, then $d \leqq 3$.*

We have already remarked that, in the finite case, (27) and (28′) imply (28). The converse is almost always true also:

49. *Let $\mathfrak{q}$ be a point set in $\mathcal{P}(d, q)$ which satisfies condition* (27).

(**a**) *If $d = 2$ and $q \equiv 1$ mod 2, then $|\mathfrak{q}| \leqq q + 1$, and equality holds if and only if $\mathfrak{q}$ is an ovoid.*
(**b**) *If $d = 2$ and $q \equiv 0$ mod 2, then $|\mathfrak{q}| \leqq q + 2$, and $|\mathfrak{q}| = q + 1$ if and only if $\mathfrak{q}$ is an ovoid.*
(**c**) *If $d = 3$ and $q > 2$, then $|\mathfrak{q}| \leqq q^2 + 1$, and equality holds if and only if $\mathfrak{q}$ is an ovoid.*

[1]) This term is due to TITS 1962a. For projective planes, we shall later use the term "oval", see Section 3.2. If a set of k points satisfies (27), i.e. if no three of its points are collinear, then SEGRE 1959a—d calls it a *k-arc* if $d = 2$ and a *k-cap* if $d > 2$. We shall see below (cf. **49**) that maximal arcs and caps are usually ovoids, provided that $d \leqq 3$.

Results (**a**) and (**b**) are also true in nondesarguesian finite projective planes; cf. QVIST 1952, Theorems 2 and 5. For the proofs of these and related results, see also Section 3.2. The proof of (**c**) is more difficult; see QVIST 1952, Theorem 6, and BARLOTTI 1955, p. 7—9. Simple examples show that the restriction $q > 2$ in (**c**) is essential. For the dimensions $d > 3$, little seems to be known about point sets $\mathfrak{q}$ satisfying (27); in particular no sharp upper bound for $|\mathfrak{q}|$ is known.[1])

It follows from **49** that, except for the case of $\mathcal{P}(3, 2)$, the ovoids of $\mathcal{P}(d, q)$ are precisely the point sets $\mathfrak{q}$ satisfying (27) and (28), and in particular all nondegenerate quadrics of index 1 are ovoids. Surprisingly, the converse of this result is also true if the characteristic is not 2:

50. *If q is odd, then every ovoid in $\mathcal{P}(d, q)$, where $d = 2, 3$, is a nondegenerate quadric, and hence (by* **40***) the set of absolute points of an orthogonal polarity.*

This was proved by SEGRE 1954, 1955a for $d = 2$, and by BARLOTTI 1955 and PANELLA 1955 for $d = 3$. The hypothesis that q be odd is essential: we shall see below that for $d = 2$ as well as for $d = 3$ there exist ovoids which are not quadrics if the characteristic is 2. Hence there arises the problem of characterizing the quadrics among the ovoids in $\mathcal{P}(d, 2^e)$.

51. *Let $\mathfrak{q}$ be an ovoid in $\mathcal{P}(d, 2^e)$, where $d = 2$ or 3 and $e > 1$, and denote by $\Gamma(\mathfrak{q})$ the group of all collineations of $\mathcal{P}(d, q)$ leaving $\mathfrak{q}$ invariant.*[2]) *Then the following conditions are equivalent:*

(**a**) *$\mathfrak{q}$ is a nondegenerate quadric of index 1.*
(**b**) *$\Gamma(\mathfrak{q})$ is d-fold transitive on the points of $\mathfrak{q}$.*
(**c**) *For any plane P with $|P \cap \mathfrak{q}| > 1$, the intersection $P \cap \mathfrak{q}$ is a quadric of index 1 in the subgeometry P.*
(**d**) *For any plane P as in* (**c**), *the stabilizer $\Gamma(\mathfrak{q})_P$ is doubly transitive on the points of $P \cap \mathfrak{q}$.*
(**e**) *$\Gamma(\mathfrak{q})$ has a subgroup isomorphic to $PO_3(2^e) \cong PGL_2(2^e)$ if $d = 2$, or isomorphic to $PO_4(2^e, -1) \cong PGL_2(2^{2e})$ if $d = 3$.*[3])
(**f**) *The intersection of $\Gamma(\mathfrak{q})$ with the projective group of $\mathcal{P}(d, 2^e)$ contains a subgroup of order $2^{e(d-1)} + 1$.*

---

[1]) The following papers are concerned with k-arcs and k-caps [cf. footnote [1]) on p. 48], as well as with the problem of embedding these into larger arcs and caps: BARLOTTI 1956, 1957a, 1965; BOSE 1947; BOSE & SRIVASTA 1964; DICOMITE 1962; FELLEGARA 1962; LOMBARDO-RADICE 1956, 1962; LUNELLI & SCE 1958; SCAFATI 1957; SCE 1960; SEGRE 1955b, 1957, 1958, 1959a—d; TALLINI 1956a—c, 1957, 1964. Further references are in the excellent bibliographies of SEGRE 1959b and BARLOTTI 1965.

[2]) This is clearly consistent with our earlier notation in (25′) and context.

[3]) These isomorphies hold also when the characteristic is not 2; cf. ARTIN 1957, Theorems 5.20, 5.21.

That (**a**) implies the other conditions is straightforward. For the converses concerning (**b**)—(**d**), see TITS 1962a, Théorèmes 3.1.1 and 4.1.2, where a more general situation (including the infinite case) is treated.[1]) Clearly, (**c**) and (**d**) reduce to (**a**) and (**b**), respectively, if $d = 2$. For (**e**), see LÜNEBURG 1964c, Korollar 2 (for the case $d = 2$) and LÜNEBURG 1964b, Theorem 2 (for $d = 3$; here a result on finite inversive planes is used. The connection between ovoids and certain inversive planes will be discussed in detail in Chapter 6). Finally, that (**f**) implies (**a**) is a consequence of a result of WIELANDT 1954; in fact, all subgroups of order $2^{e(d-1)} + 1$ of $PGL_{d+1}(2^e)$, $d = 2$ or 3, are cyclic and transitive on the points of some quadric of index 1 in $\mathscr{P}(d, 2^e)$. For a proof of this in case $d = 3$, see LÜNEBURG 1966c, lemma 2; the argument for $d = 2$ is quite similar.

The following considerations will show that there actually exist ovoids which are not quadrics, in projective geometries $\mathscr{P}(d, 2^e)$, $d = 2, 3$. We call p a *special* point of the ovoid $\mathfrak{q}$ if there exists a group Γ of elations such that

(29) $p\gamma = p$ *for all* $\gamma \in \Gamma$,

(30) $\mathfrak{q}\gamma = \mathfrak{q}$ *for all* $\gamma \in \Gamma$, *and*

(31) Γ *is transitive on the points* $\neq p$ *of* $\mathfrak{q}$.

It is easy to prove that, in a finite projective geometry $\mathscr{P}(d, p^e)$, an ovoid can contain a special point only if $p = d = 2$ (TITS 1962a). If a special point p exists, then Γ is an elementary abelian 2-group, and all the elations in Γ have the tangent $T(p)$ as axis, but their centers are distinct. Suppose that p is a special point of the ovoid $\mathfrak{q}$ in $\mathscr{P}(2, 2^e)$; distinguish $T(p)$ as ideal line and introduce coordinates in the corresponding affine plane such that p is the ideal point of the y-axis, that $(0, 0)$ and $(1, 1)$ are in $\mathfrak{q}$, and such that $y = 0$ is the equation of the tangent at $(0, 0)$.[2]) Then the affine points of $\mathfrak{q}$ are of the form (x, x^α), where α is a permutation of $GF(2^e)$, fixing 0 and 1, and such that the mapping

$$x \to x^\alpha x^{-1}, \qquad 0 \neq x \in GF(2^e),$$

is a permutation of the multiplicative group of $GF(2^e)$. (For this fact, one does not need that p is special.) The hypothesis that p be special yields furthermore that α is an automorphism of the additive group of $GF(2^e)$, and conversely every additive automorphism α of $GF(2^e)$ satisfying the above conditions defines an ovoid $\mathfrak{q}(\alpha)$ with the ideal

---

[1]) An improvement of (**c**) is SEGRE 1959c, Theorem V: for the conclusion that $\mathfrak{q}$ is a quadric it suffices to know that $q(q^2 - q + 2)/2$ of the $q(q^2 + 1)$ non-trivial plane sections are conics (here $q = 2^e$).

[2]) This is a special case of the general procedure of introducing coordinates in an affine plane, which will be discussed in detail in Section 3.1.

point of the y-axis special. In particular, the field automorphisms $\alpha : x \to x^{2^k}$ with $(k, e) = 1$ satisfy all these conditions. Thus,

$$y = x^{2^k}, \quad \text{where } (k, e) = 1, \tag{32}$$

always represents an ovoid in $\mathscr{P}(2, 2^e)$. It is clear that this is a quadric only if $k = 1$; hence (32) yields ovoids with special points which are not quadrics. In fact:

52. *If $2 \leqq k \leqq e - 2$, $e = 5$ or $e \geqq 7$, and $(k, e) = 1$, then the set of points in $\mathscr{P}(2, 2^e)$ consisting of the ovoid defined by (32) and the ideal point of the x-axis (note that no three of these $2^e + 2$ points are collinear) contains no quadric.*

This was shown by SEGRE 1957. — The problem of finding all ovoids in $\mathscr{P}(2, 2^e)$ is in general unsolved; for partial results see SEGRE 1957, 1959a, b, 1962a. We conclude the discussion of the two-dimensional case with another characterization of the quadrics in $\mathscr{P}(2, 2^e)$:

53. *An ovoid in $\mathscr{P}(2, 2^e)$ is a quadric if and only if all its points are special.*

Clearly, all points of an ovoid are special if more than one point of it is special. For the proof of **53**, see again TITS 1962a, especially 2.4.1.

We turn now to the discussion of ovoids in $\mathscr{P}(3, 2^e)$.

54. *Let $\mathfrak{q}$ be an ovoid in $\mathscr{P}(3, q)$, $q > 2$. Then there exists a polarity of $\mathscr{P}(3, q)$ which interchanges every point of $\mathfrak{q}$ with its tangent plane. This polarity is orthogonal if q is odd and symplectic if q is even.*

For odd q, this is of course a consequence of **50**. For the proof in case $q = 2^e > 2$, see SEGRE 1959c, Theorem III.[1]) This shows that there is an intimate connection between ovoids and symplectic polarities in $\mathscr{P}(3, 2^e)$; in fact the only known ovoids in $\mathscr{P}(3, 2^e)$ which are not quadrics can be most naturally described by means of symplectic polarities. We proceed to develop this now.

The points of $\mathscr{P}(3, q)$, together with the totally isotropic lines with respect to a symplectic polarity π, form a tactical configuration $\mathbf{W} = \mathbf{W}(q)$, with parameters

$$v = N_1(4, q) = (q + 1)(q^2 + 1) = T_2(\pi) = b, \quad \text{and} \quad k = q + 1 = r.$$

This symmetry suggests that $\mathbf{W}$ may be self-dual. It is not known whether this is always the case, but the following result decides the question whether or not $\mathbf{W}$ is self-polar, for the case where q is even:

---

[1]) Another proof of this result will be indicated in Section 6.2.

55. *Suppose that $q = 2^e$. Then the configuration* $\mathbf{W}(q)$ *in* $\mathcal{P}(3, q)$ *possesses a polarity if and only if e is odd, and if this is the case, all polarities of* $\mathbf{W}(q)$ *are equivalent.*

Proof: TITS 1962b, Théorème 3.6.[1]) The following result exhibits a close connection between polarities of $\mathbf{W}(q)$ and certain ovoids in $\mathcal{P}(3, q)$:

56. *If ψ is a polarity of the configuration* $\mathbf{W}(q)$ *in* $\mathcal{P}(3, q)$, *then*

(**a**) *the ψ-absolute points form an ovoid* $\mathfrak{t} = \mathfrak{t}(\psi)$ *in* $\mathcal{P}(3, q)$, *and*
(**b**) *the ψ-absolute lines form a 1-spread of* $\mathcal{P}(3, q)$.

Proof: TITS 1962b, Proposition 3.3 and Théorème 5.1; the definition of "spread" was given in the context of **6** above. Results **55** and **56** both hold, mutatis mutandis, also in the infinite case.

Now let $\mathfrak{F} = GF(2^e)$, with e odd. Then $\mathfrak{F}$ possesses a unique automorphism σ such that

$$x^{\sigma^2} = x^2 \qquad \text{for all } x \in \mathfrak{F}, \tag{33}$$

namely $\sigma: x \to x^{2^{(e+1)/2}}$. Also, let ψ be a polarity of $\mathbf{W}(2^e)$ in $\mathcal{P}(3, 2^e)$. Such a ψ exists by **55**. TITS 1962, proposition 5.3, has shown that then the ovoid $\mathfrak{t}(\psi)$ of the ψ-absolute points in $\mathcal{P}(3, 2^e)$ can be represented as follows: Let $\mathcal{A}(3, 2^e)$ be the affine geometry obtained by deleting one of the tangent planes of $\mathfrak{t}(\psi)$, and introduce affine coordinates (x, y, z) in $\mathcal{A}(3, 2^e)$ such that $z = 0$ is a tangent plane with $(0, 0, 0)$ as point of contact, and such that $(0, 1, 1)$, $(1, 0, 1)$, and $(1, 1, 1)$ are on $\mathfrak{t}(\psi)$. Then $\mathfrak{t}(\psi)$ consists of the ideal point of the z-axis and the points of $\mathcal{A}(3, 2^e)$ satisfying the equation

$$z = x\,y + x^{\sigma+2} + y^{\sigma}. \tag{34}$$

If $e = 1$, then σ is the identical automorphism, and (34) becomes $z = x\,y + x + y$, which represents a quadric. For $e > 1$, however, the $\mathfrak{t}(\psi)$ are obviously not quadrics because (34) is then of degree >2. These are the only known ovoids in three-dimensional finite projective geometries which are not quadrics; they were discovered by TITS 1960.[2]) The subgroup of the full projective group $PGL_4(q)$, $q = 2^{2m+1} > 2$, which leaves $\mathfrak{t}(\psi)$ invariant is the group $Sz(q)$ of SUZUKI [1960; 1962, Section 13], of order $q^2(q^2+1)(q-1)$; it is doubly transitive on the points of $\mathfrak{t}(\psi)$, and only the identity of $Sz(q)$ fixes more than two

[1]) Clearly, the polarities of $\mathbf{W}(q)$, as they interchange the points with certain lines of $\mathcal{P}(3, q)$, are not polarities of a projective geometry. TITS 1962b calls these polarities "σ-polarities", where σ refers to the automorphism connected with them, to be defined by (33) below.

[2]) The smallest $\mathfrak{t}(\psi)$ which is not a quadric, namely that in $\mathcal{P}(3, 8)$, had been previously found by a different method by SEGRE 1959c.

points of $\mathfrak{t}(\psi)$. The following results provide more intrinsic characterizations of these ovoids.

57. *Let $\mathfrak{q}$ be an ovoid in $\mathscr{P}(3, 2^e)$ which is not a quadric, and let $\Gamma(\mathfrak{q})$ be defined as in* **51**. *Then the following conditions are equivalent:*

(**a**) *$\mathfrak{q}$ is of the type $\mathfrak{t}(\psi)$, i.e. representable in the form* (34).
(**b**) *$\Gamma(\mathfrak{q})$ is doubly, but not triply, transitive on the points of $\mathfrak{q}$.*
(**c**) *$\Gamma(\mathfrak{q})$ contains a subgroup isomorphic to $Sz(2^e)$.*

It follows that in either one of these cases, e must be odd and >1. For the proof concerning (**a**) and (**b**), see Tits 1966, Théorème 2. For (**c**), see Lüneburg 1965c, Satz 8.

There is another more complicated characterization of the $\mathfrak{t}(\psi)$ which will be useful in Chapter 6: Let L be a tangent line of the ovoid $\mathfrak{q}$ in $\mathscr{P}(3, q)$, i.e. L contains a point $p \in \mathfrak{q}$ and is contained in the plane $T(p)$. We say that L is a *special* tangent of $\mathfrak{q}$ if there exists a subgroup Δ of $\Gamma(\mathfrak{q})_p$ such that

(35) *Every $\delta \in \Delta$ induces an elation, with center L and axis $T(p)$, in the quotient geometry $\mathscr{P}(3, q)/p$ with respect to p; and*

(36) *For every plane $P \neq T(p)$ containing L, Δ is transitive on the points $\neq p$ of $\mathfrak{q} \cap P$.*

Then we have the following result:

58. *An ovoid in $\mathscr{P}(3, 2^e)$ is a nondegenerate quadric of index 1 if and only if all its tangents are special, and it is of the type $\mathfrak{t}(\psi)$ described by* (34) *if and only if there is exactly one special tangent in each of its points.*

Proof: Tits 1966, Théorème 1. Again, a more general situation is discussed there, and the equivalence of (**a**) and (**b**) in **57** is in fact derived from **58**. In the case of the ovoids $\mathfrak{t}(\psi)$, the special tangents are precisely the ψ-absolute lines of the configuration $\mathbf{W}(q)$ defined above; they form a 1-spread of $\mathscr{P}(3, 2^e)$, cf. **56b**.

As we have seen, the concept of ovoid provides a satisfactory characterization of the nondegenerate quadrics of index 1, and it is natural to ask for similar characterizations of the quadrics with index >1. As was pointed out above, these have a less homogeneous structure, and accordingly there do not seem to exist similarly satisfying results, except for the following special case:

59. *The point set $\mathfrak{q}$ in $\mathscr{P}(3, q)$, q odd, is a nondegenerate quadric of index 2 if and only if the following conditions are satisfied:*

(**a**) $|\mathfrak{q}| = (q+1)^2$,
(**b**) *$|\mathfrak{q} \cap L| > 2$ implies $L \subset \mathfrak{q}$, for any line L, and*
(**c**) *$\mathfrak{q}$ does not contain a plane.*

Proof: BARLOTTI 1955, Teorema 2, PANELLA 1955, § 2. Note that the union of a plane P and a line not contained in P satisfies (**a**) and (**b**); hence (**c**) cannot be avoided. In fact, Barlotti shows that this example is the only non-quadric satisfying (**a**) and (**b**). The proof uses again the case $d = 2$ of **50**.

It is natural to ask for geometric characterizations of the system of totally isotropic subspaces of a polarity π, not only in the orthogonal but also in the symplectic and unitary cases.[1]) Such a polarity has index 1 only if $d = 2$ and π is unitary. In this case $q = s^2$ (see **47**), and the π-absolute points together with the π-nonisotropic lines form a design with parameters $v = s^3 + 1$, $k = s + 1$, $r = s^2$ (see Section 1.1 for definitions). Some of these results carry over to the nondesarguesian case; they will be discussed in more detail in Section 3.3. In the present (desarguesian) situation, these designs have an automorphism group which is doubly point transitive; this will follow from the next result. For any polarity π of $\mathcal{P}(d, q)$, let $\Phi(\pi)$ denote the intersection of the group $\Gamma(\pi)$ with the little projective group of $\mathcal{P}(d, q)$.[2])

60. *Let π be a unitary polarity of $\mathcal{P}(2, q)$. Suppose that the subgroup Γ of $\Gamma(\pi)$ is transitive on incident pairs of π-absolute points and π-non-isotropic lines. Then Γ contains $\Phi(\pi)$, which is doubly transitive on the π-absolute points.*

Proof: HIGMAN & MCLAUGHLIN 1965, Theorem 3.

As in the case of quadrics, it follows from **42** that if a polarity π (non-orthogonal if the characteristic is 2) has index >1, then $\Gamma(\pi)$ and a fortiori $\Phi(\pi)$ cannot be doubly transitive on the π-absolute points. In particular, this applies to all symplectic and unitary polarities in dimensions $d \geqq 3$. For the remainder of our discussion, let π always be such a symplectic or unitary polarity of $\mathcal{P}(d, q)$, $d \geqq 3$. It is not difficult to verify (cf. HIGMAN & MCLAUGHLIN 1965) that $\Phi(\pi)$ is then of *rank* 3 when considered as permutation group of the π-absolute points, i.e. it permutes these points transitively, and for any π-absolute point x, the stabilizer $\Phi(\pi)_x$ permutes the remaining π-absolute points in exactly two orbits. (One of these orbits consists of the π-absolute points $\neq x$ on the hyperplane $x\pi$, the other of the remaining π-absolute points.)

[1]) There is a certain connection between this question and that of the existence of (k, n)-*arcs* and (k, n)-*caps*, i.e. sets of k points in $\mathcal{P}(d, q)$, $d = 2$ or $d > 2$, respectively, such that some n but no $n + 1$ of these are collinear. For more information see BARLOTTI 1956b, 1965; COSSU 1961; LUNELLI & SCE 1964; MAISANO 1960; D'ORGEVAL 1960; TALLINI 1957, 1960, 1961.

[2]) The group $\Phi(\pi)$ is usually called $PSp_{d+1}(q)$ or $PSU_{d+1}(q)$, according as π is symplectic or unitary. In the symplectic case, $\Phi(\pi) = \Gamma_0(\pi)$ (cf. **46**), while in the unitary case the index of $\Phi(\pi)$ in $\Gamma_0(\pi)$ is the g.c.d. $(d + 1, q + 1)$.

Conversely, HIGMAN & McLAUGHLIN (1965, Theorem 5) have shown that if $d \leqq 7$, and if π is non-symplectic when q is even, any rank 3 subgroup of $\Gamma(\pi)$ contains $\Phi(\pi)$. Whether or not the restriction $d \leqq 7$ is essential is not known. The result rests on the following theorem, proved in the same paper:

61. *If π is symplectic or unitary in $\mathscr{P}(d, q)$, $d \geqq 3$, and if Σ is a subset of $\Gamma(\pi)$ which contains either*

(i) *an elation $\neq 1$ with center x and axis $x\pi$, for every π-absolute point x, or else*

(ii) *a homology $\neq 1$ with center y and axis $y\pi$, for every point y which is not π-absolute,*

then Σ generates a subgroup of $\Gamma(\pi)$ which contains $\Phi(\pi)$.

Proof: HIGMAN & McLAUGHLIN 1965, Theorems 1 and 2. Actually, the hypothesis can be relaxed for the alternative (ii). In the same paper it is also shown (Theorem 4) that any nontrivial normal subgroup of a rank 3 subgroup of $\Gamma(\pi)$ is again of rank 3, except perhaps when $q \leqq 4$ in the symplectic and $q \leqq 16$ in the unitary case. Clearly, we have here a situation similar to that of **33** and **34** above.

We conclude with the following geometric characterization of the 3-dimensional symplectic groups:

62. *Suppose that the collineation group Γ of $\mathscr{P}(3, q)$, $q = p^e > 2$, is of rank 3 when considered as permutation group of the points, and that the orbit lengths of a stabilizer Γ_x are 1, $q(q+1)$, and q^3. Also, assume that either*

(i) *the subgroup of Γ_x fixing the orbit of length $q(q+1)$ pointwise has order $\geqq q$, or else that*

(ii) *e is a power of 2.*

Then the Γ_x-orbit of length $q(q+1)$ consists of the points $\neq x$ of a plane $P(x)$ containing x, the correspondence $\pi: x \longleftrightarrow P(x)$ is a symplectic polarity, and Γ contains $\Phi(\pi)$.

This is contained in HIGMAN 1964, Theorem 2. As a matter of fact, it is not necessary to assume that the objects permuted by Γ are the points of $\mathscr{P}(3, q)$; that this must be the case (and in particular that q must be a prime power) can be proved (HIGMAN 1964, Theorem 1) by appealing to a result to be discussed in the next chapter (cf. **2.1.22**). The conditions (i) and (ii) may be not essential.

2. Designs

In this chapter we shall be concerned with the general theory of designs (these were defined in Section 1.1, and the definition will be repeated in 2.1). The discussion will by no means be complete: we shall concentrate mostly on results and problems significant for projective, affine, and inversive planes. These three types of designs are the subject matter of Chapters 3—6 below; only a few facts about them will be mentioned here.

Our discussion falls into three main parts. The first of these consists of Sections 2.1 and 2.2; it covers the combinatorial part of the subject, viz., results dealing with the parameters of designs and subdesigns, numbers of points and blocks in certain subsets, and existence problems for given parameters. We pay particular attention to the question as to when a design may be interpreted as a substructure, of a specified type, of a larger design; these results are collected under the heading of "embeddings and extensions" in Section 2.2.

The second part is Section 2.3, dealing with various aspects of automorphisms of designs, and in particular with results characterizing the "geometric" designs $\mathbf{P}_{d-1}(d, q)$ and $\mathbf{A}_{d-1}(d, q)$, of points and hyperplanes in a finite projective resp. affine geometry (cf. Section 1.4), by properties of their automorphism groups.

The third part, Section 2.4, gives various construction methods for designs, and thus furnishes examples other than the $\mathbf{P}_t(d, q)$ and $\mathbf{A}_t(d, q)$. We rely heavily on group theoretical methods, which yield particularly homogeneous designs. Other construction techniques will be mentioned rather more briefly; it would lead too far afield to attempt a complete coverage of all known designs here.

2.1 Combinatorial properties

Designs were defined in Section 1.1 as nondegenerate tactical configurations of type (s, t) with $s \leqq t > 1$, and some basic properties of designs are obvious consequences of **1.1.5** and **1.3.8**. For the convenience of the reader, these results will be restated in the present section.

A *design* is a finite incidence structure $\mathbf{D} = (\mathfrak{p}, \mathfrak{B}, \mathrm{I})$ satisfying the regularity conditions (R. 1) and ($\bar{\mathrm{R}}$. 2) of Section 1.1, as well as part (a) of the nondegeneracy condition (N),[1] i.e.

(1) $[B] = v_1 > 0$ for every block $B \in \mathfrak{B}$,

(2) $[p, q] = b_2 > 0$ for any two distinct points $p, q \in \mathfrak{p}$, and

(3) $|\mathfrak{p}| \geqq v_1 + 2$.

We use here the notation (1.1.6) again: the bracket symbol denotes the number of elements incident with every element of the set within the brackets. (1) and (2) imply

(4) $[p] = b_1 > 0$ for every point $p \in \mathfrak{p}$;

see **1.1.3**. Hence designs are tactical configurations. It is customary to write $b_0 = |\mathfrak{B}| = b$, $b_1 = r$, $b_2 = \lambda$, $v_0 = |\mathfrak{p}| = v$, $v_1 = k$ for designs. With this notation, (1)—(4) become

$$[B] = k, \ [p, q] = \lambda, \ v \geqq k + 2, \quad \text{and} \quad [p] = r,$$

for any $p, q \in \mathfrak{p}$ and $B \in \mathfrak{B}$. Furthermore, (1.1.9′) yields the equations

(5) $r v = b k$ and $\lambda(v-1) = r(k-1)$.

These are easily proved directly by counting in two ways either all flags of **D** or all those flags (x, X) for which $x \neq p \text{ I } X$, where p is some fixed point.

Result **1.3.8** shows that for any design we have FISHER's (1940) inequality

$$v \leqq b,$$

and from **1.3.9** it follows that equality holds if and only if the dual of (2) is satisfied: $[B, C] = \bar{\lambda} =$ const. for distinct blocks B, C. But then (5) and $b = v$ show that $\bar{\lambda} = \lambda$; hence:

1. *A design has equally many points and blocks if and only if*

(6) $[B, C] = \lambda$ *for any two distinct blocks* $B, C \in \mathfrak{B}$.

Note that this is also true if (3) is not satisfied.

---

[1]) For designs, part (a) of (N) implies part (b), for if (a) holds, there is a block B and two points, $p, q \text{ I\!\!\!/ } B$, and if (b) were false, all blocks $\neq B$ would pass through p and q, so that $b_1 = b_2$, which is incompatible with (a). Conversely, (b) does not imply (a), even when (R.1) and ($\overline{\text{R}}$.2) hold; the simplest example for this (which the author owes to J. Malkevitch) is given by the incidence matrix $\begin{pmatrix} 1 & 1 & 1 & 1 & 0 & 0 \\ 1 & 1 & 0 & 0 & 1 & 1 \\ 0 & 0 & 1 & 1 & 1 & 1 \end{pmatrix}$. Note that (a) is not essential for all of the following results.

Any design with $b = v$, or, equivalently, a design satisfying condition (6), will be called here a *projective design*[1]). Immediate examples of projective designs are the tactical configurations $\mathbf{P}_{d-1}(d, q)$ of points and hyperplanes of the finite projective geometries $\mathscr{P}(d, q)$; the parameters of these were given in (1.4.9). Other examples are the non-desarguesian finite projective planes. These are precisely the projective designs with $\lambda = 1$ which are not of the form $\mathbf{P}_1(2, q)$; many examples will be given in Chapter 5. We shall see, mainly in Section 2.4, that besides these there exist many other projective designs. At this point, we observe that

$$(5') \quad v = b, \quad r = k, \quad \textit{and} \quad \lambda(v-1) = k(k-1), \quad \textit{for projective designs.}$$

Also, the defining conditions for projective designs, namely (1)—(4) and (6), form a self-dual system of statements, whereas this is not the case for nonprojective designs. Hence:

2. *The dual structure of a design* **D** *is again a design if and only if* **D** *is projective.*[2])

Note that this does not imply that projective designs are self-dual. It is true that the $\mathbf{P}_{d-1}(d, q)$ are all self-dual, but there exist non-desarguesian finite projective planes which are not self-dual; see Chapter 5.

In view of **1** and **2**, the question arises what can be said about the numbers $[B, C]$, for distinct blocks B, C, in a non-projective design. The (b, b)-matrix formed by these integers was investigated by CONNOR 1952a; among other things he proved the following result:

3. *Let* B, C *be distinct blocks in an arbitrary design. Then*

$$(7) \qquad [2\lambda k + r(r - k - \lambda)]\, r^{-1} \geqq [B, C] \geqq k + \lambda - r.$$

Note that in the projective case this reduces to (6) again. For other results concerning the $[B, C]$, cf. MAJUMDAR 1953; PARKER 1963b; SEIDEN 1963a; STANTON & SPROTT 1964. Some related questions will be discussed in Section 2.2.

---

[1]) Other names for the same concept: "symmetric balanced incomplete block design" (BOSE 1939; this is used by the statisticians), "(v, k, λ)-configuration" (RYSER 1963), "λ-plane" (BRUCK 1955), "λ-space" (DEMBOWSKI & WAGNER 1960). The fact that a projective design has equally many points and blocks is, at least from a geometric point of view, less significant than (6). Also, projective designs are related to finite projective geometries (of arbitrary dimension) in much the same fashion as are finite projective planes to the desarguesian ones. These observations have prompted us to choose the present terminology.

[2]) The dual structure of non-projective designs has been investigated by SHRIKHANDE 1952; ROY 1954a—c; BOSE & CLATWORTHY 1955; RAMAKRISHNAN 1956; ROY & LAHA 1956a, 1957; AGRAVAL 1963. Some of these results will be discussed in Section 7.1.

The complementary structure $\mathbf{D}'$ [see (1.1.3)] of a design $\mathbf{D}$ is always a design, with parameters

(8) $$v' = v,\ b' = b,\ k' = v - k,\ r' = b - r,\ \lambda' = b - 2r + \lambda.$$

Clearly, $\mathbf{D}'$ is projective if and only if $\mathbf{D}$ is projective. The *order* of any design is defined to be the integer

(9) $$n = r - \lambda;$$

it follows immediately from (8) that any design and its complementary design have the same order. Also, (5) implies that

(10) $$\lambda \lambda' = n(n-1) \qquad \textit{for projective designs,}$$

so that for the classification problem of projective designs it suffices to consider those with $\lambda \leqq n - 1$ or $r = k > 2\lambda$. The order n of a design $\mathbf{D}$ plays an important role in certain nonexistence theorems for projective designs; see in particular results **11** – **15** below. At this point we only remark that, for arbitrary designs,

(11) $$\lambda v + n = r k;$$

this is a simple consequence of (5) and (9).

We present now some results concerning Fisher's inequality $b \geqq v$ for designs. The result is, because of (5), equivalent to $r \geqq k$; combining both these inequalities one easily derives

(12) $$b \geqq v + r - k.$$

This can be improved if $k \mid v$. Note that (5) implies, without restrictions on the parameters:

$$b = r + (r - v \lambda k^{-1})(v - 1).$$

In particular, $r - v \lambda k^{-1} > 0$. This proves half of the following result:

4. *For any design, k is a divisor of v if and only if*

(13) $$b = m(v-1) + r$$

for some integer $m > 0$. Consequently, $k \mid v$ implies $b \geqq v + r - 1$.

For the complete proof, which is equally easy, see MAJUMDAR 1954. Another variation is

(14) $$b \geqq \frac{r k (r-1)}{r - k + \lambda(k-1)} \geqq v;$$

this was proved, using matrix arguments, by NAIR 1943.

5. *If the design* $\mathbf{D}$ *is of type (s, t), $t \geqq 2$ (see Section 1.1 for definition), then*

(15) $$(v - i)\binom{v}{i} \leqq b \binom{k}{i} \qquad \textit{for } i = 0, \ldots, t-2.$$

Equality holds if and only if $i = t - 2$ and all substructures formed by $v - t + 2$ points and the blocks incident with each of the $t - 2$ remaining points are projective designs.

This is due to HUGHES 1965a; Fisher's inequality is the case $i = 0$, and for $t = 2$ we have **1** again. The proof involves internal substructures of designs with respect to points, as defined in Section 1.1; these will be discussed in more detail in Section 2.2, where a proof of **5** will be outlined (cf. p. 76).

The last generalization of Fisher's inequality to be discussed here involves tactical decompositions of designs, as defined in Section 1.1. If Δ is a tactical decomposition of the design $\mathbf{D}$, then besides (1.1.14)—(1.1.16) we have the relation

$$\sum_{\mathfrak{X}} (\mathfrak{q}\ \mathfrak{X})\ (\mathfrak{X}\ \mathfrak{r}) = \lambda\ |\mathfrak{q}| + n\ \delta(\mathfrak{q}, \mathfrak{r}), \tag{16}$$

for any two point classes $\mathfrak{q}, \mathfrak{r}$ of Δ. Here $\mathfrak{q}$ and $\mathfrak{r}$ need not be distinct, and $\delta(\mathfrak{q}, \mathfrak{r}) = 0$ or 1 according as $\mathfrak{q} \neq \mathfrak{r}$ or $\mathfrak{q} = \mathfrak{r}$. Also, n in (16) is the order of $\mathbf{D}$ defined by (9), and $\mathfrak{X}$ ranges over all block classes of Δ. For the proof of (16), see DEMBOWSKI 1958, p. 64—65.[1]) Now let $C = (c_{ij}) = ((\mathfrak{p}_i\ \mathfrak{B}_j))$ and $D = (d_{ij}) = ((\mathfrak{B}_i\ p_j))$ be the incidence matrices of Δ, as defined in 1.3, with respect to an arbitrary numbering of the point and block classes of Δ. Then, if $P = \operatorname{diag}(|\mathfrak{p}_1|, \ldots, |\mathfrak{p}_m|)$ as in 1.3 (m is the number of point classes of Δ), the matrix equation

$$CD = n\ I_m + \lambda\ P J_m \tag{16'}$$

is equivalent to (16); here I_m and J_m are, as in Section 1.3, the (m, m) identity matrix and the (m, m) matrix (s_{ij}) with $s_{ij} = 1$ for all i, j. Using (11), one now derives easily from (16′) that

$$\det CD = r\ k\ n^{m-1} \neq 0, \tag{17}$$

where n is again the order of $\mathbf{D}$. Hence CD is always nonsingular, and similarly to **1.3.8** we obtain the following result (KANTOR 1968b, Theorem 4.1):

6. *If a tactical decomposition of a design has m point classes and m' block classes, then*

$$m \leqq m'. \tag{18}$$

Equality holds if $\mathbf{D}$ *is projective.*

---

[1]) The restriction imposed there that the design be projective is unnecessarily strong; this is essential only for the equation $\sum_{\mathfrak{x}} (\mathfrak{C}\ \mathfrak{x})\ (\mathfrak{x}\ \mathfrak{D}) = \lambda\ |\mathfrak{C}| + n\ \delta(\mathfrak{C}, \mathfrak{D})$ dual to (12), which holds in any projective design.

The assertion about the projective case follows by duality.[1] Note that we cannot say "and only if" as in the corresponding results **1** and **5**; a counterexample is, for instance, the decomposition with only one point class and one block class. The trivial decomposition with $|\mathfrak{p}_i| = |\mathfrak{B}_i| = 1$ for all i gives Fisher's inequality again. In Section 2.2, we shall encounter a class of non-projective designs for which (18) can be further improved.

In the projective case, C and D are both square matrices, and (17) becomes $\det CD = k^2 n^{m-1}$. On the other hand, (1.3.4) shows that $(DP)^2 = BC^T DP$, with $B = \mathrm{diag}(|\mathfrak{B}_1|, \ldots, |\mathfrak{B}_m|)$, as in 1.3. Consequently, $\det CD \cdot \det PB = (\det DP)^2$ is a square. Now if $\det PB$ is also a square (this is true in particular if Δ is *symmetric* in the sense that $P = B$), it follows that n^{m-1} must be a square. Hence:

7. *A projective design of non-square order n does not admit a symmetric tactical decomposition whose point (and block) class number m is even.*

(DEMBOWSKI 1958, p. 82) The most important special case of this result is that of the trivial decomposition for which $P = B = I_v$:

8. *If the number v of points (and blocks) of a projective design* **D** *is even, then the order $n = r - \lambda$ of* **D** *is a square.*

(SCHÜTZENBERGER 1949, CHOWLA & RYSER 1950.) This result is a first non-trivial restriction of the possible parameters for projective designs beyond (5). A purely arithmetic corollary of **8** is:

9. *If $v = b$ is a power of 2, then so is $n = k - \lambda$, and $v = 4n$.*

For the proof, see MANN 1965a, p. 213. Note that v must then also be a square. Result **9** characterizes the prime 2:

10. *If $v = b = p^m$ for some prime $p > 2$, then $n = k - \lambda$ is not a power of p.*

The proof is similar to that of **9**; cf. MANN 1965a, p. 215.

11. *Suppose that the projective design* **D** *of order n has a tactical decomposition whose point and block classes have lengths $|\mathfrak{p}_i|$ and $|\mathfrak{B}_i|$, respectively $(i = 1, \ldots, m)$. Define*

$$\pi_i = \prod_{j=1}^{i} |\mathfrak{p}_j| \quad \textit{and} \quad \beta_i = \prod_{j=1}^{i} |\mathfrak{B}_j|, \qquad i = 1, \ldots, m.$$

[1] A short matrix-free proof for $m = m'$ in the projective case appears in DEMBOWSKI 1958, p. 65. Clearly, **6** is a special case of **1.3.10**; cf. BLOCK 1967, Corollary 2.2.

Then the following norm residue symbol relations must hold for every prime p:

$$(19)\qquad \begin{aligned} &\prod_{i=1}^{m-1}\left(\frac{\beta_i,\beta_{i+1}}{p}\right)\left(\frac{n^i\pi_i,\ -n^{i+1}\pi_{i+1}}{p}\right)=\left(\frac{n,\lambda\beta_m}{p}\right)\\ &\prod_{i=1}^{m-1}\left(\frac{\pi_i,\pi_{i+1}}{p}\right)\left(\frac{n^i\beta_i,\ -n^{i+1}\beta_{i+1}}{p}\right)=\left(\frac{n,\lambda\pi_m}{p}\right). \end{aligned}$$

For the proof, one observes first that B is congruent to $\lambda PJP + nP$, by (1.3.7) and (16′). As J is of rank 1 and congruent to the matrix (n_{ij}) with $n_{11} = 1$ and $n_{ij} = 0$ otherwise, it follows (cf. LENZ 1962, p. 117) that the $(m+1, m+1)$-matrix $B' = \operatorname{diag}(|\mathfrak{B}_1|, \ldots, |\mathfrak{B}_m|, \lambda n)$ is congruent to $n P'$, with $P' = \operatorname{diag}(|\mathfrak{p}_1|, \ldots, |\mathfrak{p}_m|, \lambda n)$. Application of **1.3.2** and elementary simplifications then yield the first relation (19), and the second follows dually.[1]

Result **11** is quite complicated and not very useful in its full generality. In special cases, however, the equations (19) become much more manageable; we proceed now to list some of these.

12. *If a projective design of non-square order n admits a symmetric tactical decomposition with m point (and block) classes, then, with $D = \det P = \det B$:*

$$(20)\qquad \left(\frac{n,\ (-1)^{(m-1)/2}\lambda D}{p}\right) = 1 \quad \textit{for every prime } p.$$

Equivalently, the diophantine equation

$$(20')\qquad n\,x^2 + (-1)^{(m-1)/2}\lambda D\, y^2 = z^2$$

has a nontrivial integral solution.

This is essentially due to LENZ 1962; Lenz carries out the case $\lambda = 1$ only, but the present result can be obtained by the same method. The result is a fairly direct consequence of **1.3.3**. Note that $m-1$ must be even, because of **7**.

We call a tactical decomposition *standard* if, for some integer f with $0 \leq f \leq m$, the point and block classes can be so numbered that

$$|\mathfrak{p}_i| = |\mathfrak{B}_i| = \begin{cases} 1 & \text{for } i = 1, \ldots, f\\ s & \text{for } i = f+1, \ldots, m, \end{cases}$$

where $s > 1$. If $f = 0$, then all $|\mathfrak{p}_i| = |\mathfrak{B}_i| = s$, and if $f = m = v$, then all $|\mathfrak{p}_i| = |\mathfrak{B}_i| = 1$, which is the trivial decomposition. In any

[1]) Formulas equivalent to (19) appear in BLOCK 1967, Theorem 6.1. The reader is warned that still another formula of this kind (DEMBOWSKI 1958, Satz 11) is false, due to computational errors. A correction is found in DEMBOWSKI 1962b, but (19) is simpler. — For further relations between the $|\mathfrak{p}_i|$ and $|\mathfrak{B}_i|$, see DEMBOWSKI 1958, Satz 9, and BLOCK 1967, Section 5.

case, a standard decomposition is symmetric, and f, s, v and m are related by

$$v = f + (m - f)\, s.$$

Now **12** yields the following result:

13. *If a projective design of non-square order n admits a standard tactical decomposition, then*

$$n\, x^2 + (-1)^{(m-1)/2}\, s^{f+1}\, \lambda\, y^2 = z^2$$

has a nontrivial integral solution.

This is due to HUGHES 1957d; the assumption made there that the decomposition be the orbit decomposition of some automorphism group is superfluous.

Finally, we consider the two extreme cases $f = 0, m$ of **13**. For $f = 0$, we get

14. *If a projective design of non-square order n admits a tactical decomposition all of whose classes have the same number s of elements, then*

$$n\, x^2 + (-1)^{(m-1)/2}\, s\, \lambda\, y^2 = z^2$$

has a nontrivial integral solution.

For this, see HUGHES 1957a (again under unnecessary restrictions), and DEMBOWSKI 1958, p. 87. It turns out that the assumption $s > 1$ made in the definition of a standard decomposition is inessential for **14**, so that **14** is also correct if $s = 1$. This is, of course, the special case $f = m$ of **13**:

15. *If the number v of points (and blocks) of a projective design is odd, then*

$$n\, x^2 + (-1)^{(v-1)/2}\, \lambda\, y^2 = z^2$$

has a nontrivial integral solution.

This result, complementary to **8**, is due to CHOWLA & RYSER 1950 and SHRIKHANDE 1950. (See also SHRIKHANDE & RAGHAVARAO 1964.) While Shrikhande's proof uses essentially the approach outlined here, i.e. the Minkowski-Hasse result **1.3.2**, the method of Chowla and Ryser is more direct and avoids **1.3.2**. The special case $\lambda = 1$ of **15** was established by BRUCK & RYSER 1949. In fact, this paper of Bruck and Ryser initiated most of the research resulting in **11**—**15**.

The next part of our discussion will be concerned with the number of absolute elements of a design **D** with respect to an anti-automorphism ϱ. Clearly, such a design must be self-dual and hence projective, by **2**. We shall use several results of Section 1.3 again. First, if C is an incidence matrix of **D**, then

(21) $$C C^T = C^T C = n\, I_v + \lambda\, J_v;$$

see **1.3.6**. In particular, C is normal and can hence be written $C = U^{-1}EU$, where U is unitary, i.e. $U^{-1} = U^*$, the conjugate transpose of U, and $E = \operatorname{diag}(e_1, \ldots, e_v)$, where the e_j are the eigenvalues of C. This implies that the eigenvalues of $CC^T = CC^*$ are the $e_j \bar{e}_j = |e_j|^2$, $j = 1, \ldots, v$, with the bar denoting complex conjugation. But the eigenvalues of CC^T were given in **1.3.7**. Observing that $C(1, \ldots, 1)^T = k(1, \ldots, 1)^T$, which shows that $(1, \ldots, 1)$ is an eigenvector with eigenvalue k, we arrive at the following result:

16. *The eigenvalues of any incidence matrix of a projective design of order $n = k - \lambda$ are of the form*

$$k, \quad \text{and} \quad e^{i a_j}\sqrt{n}, \quad \text{with} \quad 0 \leqq a_j < 2\pi, \qquad j = 1, \ldots, v-1.$$

The number $a(\varrho)$ of absolute points of the anti-automorphism ϱ equals the trace of a suitable incidence matrix C [see (1.3.20)], and the permutation ϱ^2 induced among the points is represented by the permutation matrix $R = C^{-1}C^T$ [cf. (1.3.19)]. Now **16** and the normality of C imply that the eigenvalues of R are 1, e^{-2ia_j} $(j = 1, \ldots, v-1)$. On the other hand, the eigenvalues of permutation matrices are roots of unity, and it follows that

$$g = \sum_{j=1}^{v-1} e^{i a_j}$$

is an algebraic integer. From (1.3.20) and **16** we can now infer that

$$[a(\varrho) - k]^2 = n\,g^2, \tag{22}$$

whence g^2 is a rational integer, and $[a(\varrho) - k]^2 \equiv 0 \bmod n$. This implies:

17. *Let ϱ be an anti-automorphism of the projective design* **D** *of order $n = k - \lambda = n^* s^2$, where n^* is square-free. Then*

(a) $$a(\varrho) \equiv \lambda \bmod n^*s,$$

and

(b) *If ϱ is a polarity, then $a(\varrho) = k + g\sqrt{n}$ for some integer $g \equiv v - 1$ mod 2. In particular, if n is not a square, then $g = 0$ and $a(\varrho) = k$.*

For if ϱ is a polarity, then $R = I_v$ and $e^{ia_j} = \pm 1$; in this case the algebraic integer g is of the form $\Sigma(\pm 1) \equiv v - 1 \bmod 2$. Result **17** is due to Hoffman, Newman, Straus, Taussky 1956. These authors carry out the proof only for $\lambda = 1$ but point out that their method also applies to designs and even more general tactical configurations.[1]) They also give some sufficient conditions, other than that stated in (**b**), for $a(\varrho) = k$; for this compare also result **1.2.13**. Also, for $\lambda = 1$, some of these results had been previously obtained by Baer 1946a and Ball 1948 who used a matrix-free approach (which, incidentally,

[1]) Some of these will be discussed in Section 7.1.

also carries over to the case of arbitrary λ). It is easily derived from (1.4.9) and **17a** that every anti-automorphism of $\mathbf{P}_{d-1}(d, q)$ has absolute elements (for polarities, we know more than that from the results of Section 1.4). Other results on absolute points of polarities in finite projective planes will be given in Section 3.3.

Concluding our discussion of results on incidence matrices of projective designs, we mention the following existence theorem:

18. *A projective design with parameters v, k, λ exists if, and only if, there exist permutations $\pi_1, \ldots, \pi_k$ of a set $\mathfrak{p}$ of size v such that*

(**a**) *if $x^{\pi_i} = x^{\pi_j}$ for some $x \in \mathfrak{p}$, then $i = j$;*
(**b**) *if $x \neq y$, then exactly λ among the permutations $\pi_i \pi_j^{-1}$ map x onto y.*

Proof (MANN 1964): Represent the π_i by permutation matrices P_i $(i = 1, \ldots, k)$ and put $C = \sum_{i=1}^{k} P_i$. Then (**a**) implies that C consists of zeros and ones, and (**b**) shows that C satisfies (21); hence C is an incidence matrix of a projective design. Conversely, any such incidence matrix can be written as a sum of k permutation matrices (by a theorem of KÖNIG 1936, p. 239); the corresponding permutations must satisfy (**a**) and (**b**).

Most of the remainder of this section will be occupied by an investigation which will yield, among other things, a first intrinsic characterization of the designs $\mathbf{P}_{d-1}(d, q)$. Using notation (1.1.5), we define the *lines* of an arbitrary design $\mathbf{D} = (\mathfrak{p}, \mathfrak{B}, \mathrm{I})$ as the point sets of the form $((x, y))$, for distinct $x, y \in \mathfrak{p}$. In other words, a line consists of all points incident with every block through two distinct points. Clearly, if $\lambda = 1$, the lines are just the sets (B), for $B \in \mathfrak{B}$; hence they can be identified with the blocks in this case. Also, the lines of a projective or affine geometry, as defined in the beginning of 1.4, are the same as the lines in the present sense, for each of the designs $\mathbf{P}_t(d, q)$ and $\mathbf{A}_t(d, q)$.

The following statements are easy consequences of the definition of lines:

19. *Let* $\mathbf{D}$ *be an arbitrary design, with parameters v, b, k, r, λ.*

(**a**) *Given two distinct points, there is a unique line containing them both.*
(**b**) *If $\mathfrak{l}$ is any line, then $2 \leqq |\mathfrak{l}| \leqq (b - \lambda)/(r - \lambda)$.*
(**c**) *If two distinct points of $\mathfrak{l}$ are on the block B, then $x \mathrm{I} B$ for all $x \in \mathfrak{l}$.*
(**d**) *If all lines have equally many, say h, points, then $h \leqq k$ (equality holds if and only if $\lambda = 1$), and the points and lines form a design $\tilde{\mathbf{D}}$ with parameters*

$$(23) \quad \tilde{v} = v,\ \tilde{b} = v(v-1)\,h^{-1}(h-1)^{-1},\ \tilde{k} = h,\ \tilde{r} = (v-1)(h-1)^{-1},\ \tilde{\lambda} = 1.$$

Next, we define the *planes* of $\mathbf{D}$ as the point sets $((x, y, z))$, where x, y, z are three non-collinear points. Hence a plane consists of all points which are incident with every block through three points not contained in the same line. Again, this terminology is a generalization of that for projective and affine geometries. In general it may happen, for example when $\lambda = 1$, that $(x, y, z) = \emptyset$, in which case the plane $((x, y, z))$ consists of all points of $\mathbf{D}$. More generally, one plane may be a proper subset of another.[1]) In order to exclude this undesirable possibility, we define a design to be *smooth* if any three non-collinear points are contained in equally many blocks. Clearly, every design with $\lambda = 1$ or of type $(1, t)$ with $t > 2$ is smooth, but also the designs $\mathbf{P}_t(d, q)$ and $\mathbf{A}_t(d, q)$ of type $(s, 2)$, with $s = 1$ or 2, are smooth.

20. *Let* $\mathbf{D}$ *be a smooth design, with parameters* v, b, k, r, λ.

(**a**) *Given three non-collinear points, there is a unique plane containing them all.*
(**b**) *If* $\mathfrak{l}$ *is a line and* p *a point not in* $\mathfrak{l}$, *then there is a unique plane containing both* p *and* $\mathfrak{l}$.
(**c**) *If* B *is a block and* $\mathfrak{e}$ *a plane containing two distinct points* $x, y \mathrel{I} B$ *and a point not on* B, *then* $((x, y)) = \mathfrak{e} \cap (B)$, *i.e. the common points of* $\mathfrak{e}$ *and* B *are precisely those of the line joining* x *and* y.
(**d**) *Let* t *denote the number of blocks containing three non-collinear points. Then every line consists of equally many, namely*

$$h = (\lambda k - t v)(\lambda - t)^{-1}$$

points.
(**e**) *If* $\lambda > 1$, *then for any point* p *of* $\mathbf{D}$ *the lines and blocks through* p *form a design* $\mathbf{D}^* = \mathbf{D}^*(p)$ *with parameters*

$$v^* = (v-1)(h-1)^{-1}, \quad k^* = (k-1)(h-1)^{-1}, \quad b^* = r, \quad r^* = \lambda, \quad \lambda^* = t.$$

The proofs of (**a**)—(**d**) are in Dembowski & Wagner 1960; the assumption made there that $\mathbf{D}$ be projective is superfluous. Result (**e**) is easily proved[2]) from (**d**) and **19d**, and has the following consequence:

21. *Let* $\mathbf{D}$ *be smooth and* t *the number of blocks containing three non-collinear points. Then*

$$t \leqq \lambda(\lambda - 1)/(r - 1),$$

and equality holds if and only if $\lambda = 1$ *or the design* $\mathbf{D}^*$ *of* **20e** *is projective.*[2])

This is shown by applying **1** to $\mathbf{D}^*$. Also, it is not difficult to see that $\mathbf{D}^*$ is projective if $\mathbf{D}$ is,[2]) but the converse of this is not true. These

---

[1]) The reason for this is, of course, that the analogue to (2) is missing: the number of blocks containing three non-collinear points will in general depend on these points.

[2]) For these results, the author is indebted to J. Ryshpan.

facts are important tools in the proof of the following theorem which gives the promised intrinsic characterization of the $\mathbf{P}_{d-1}(d, q)$:

22. *Let* **D** *be a design with parameters* v, b, k, r, λ, *such that* $(X) = (Y)$ *implies* $X = Y$, *for arbitrary blocks* X, Y. *The following properties of* **D** *are equivalent:*

(**a**) **D** *is isomorphic to the system of points and hyperplanes of a (not necessarily desarguesian) finite projective geometry.*

(**b**) *Every line meets every block.*[1]

(**c**) *Every line consists of* $(b - \lambda)(r - \lambda)^{-1}$ *points.*

(**d**) *Every plane is contained in exactly* $\lambda(\lambda - 1)(k - 1)^{-1}$ *blocks.*

(**e**) **D** *is smooth and projective.*

That (**a**) implies (**b**)—(**e**) is clear, and the equivalence of (**d**) and (**e**) follows from **21** and its context, because of $r \geqq k$. The proof of the remaining implications is essentially[2] contained in DEMBOWSKI & WAGNER 1960.

The three-dimensional case will be of special interest in Chapter 6; hence we state this separately:

23. *A finite incidence structure* $\mathbf{S} = (\mathfrak{p}, \mathfrak{B}, \mathrm{I})$ *is isomorphic to the system of points and planes of some* $\mathcal{P}(3, q)$ *if and only if there exists an integer* $s > 1$ *such that*

(**a**) $|\mathfrak{p}| = |\mathfrak{B}| = (s + 1)(s^2 + 1)$,

(**b**) $[B] = s^2 + s + 1$ *for every block* $B \in \mathfrak{B}$,

and

(**c**) $[p, q] = [(p, q)] = s + 1$ *for any two distinct points* $p, q \in \mathfrak{p}$.

The notation in (**b**) and (**c**) is again that of (1.1.5) and (1.1.6). If conditions (**a**)—(**c**) hold, then $s = q$ is clearly the order of the Galois field underlying **S**. Result **23** can be derived fairly easily from **22**; a direct proof appears in DEMBOWSKI 1964b, Satz 1.

In conclusion of this section, we prove a rather isolated result which will have a useful application in Section 2.3.

---

[1] This condition is stronger than necessary. Actually, it suffices to assume that there exists just one block meeting every line. For the proof of this, see KANTOR 1969a, Theorem 4.

[2] The hypothesis made there that **D** be projective is not necessary except for (**e**). [That smoothness alone does not suffice to prove (**a**) is seen by considering the $\mathbf{A}_{d-1}(d, q)$.] The arguments in DEMBOWSKI & WAGNER 1960, p. 468 carry over with only minor changes to the present case: (**b**) and (**c**) are equivalent and imply **D** to be smooth with $t = [\lambda(b - \lambda) - r(r - \lambda)]/(b - r)$. But this condition, together with (5) and **20d**, implies (**c**) and is therefore equivalent to (**b**) and (**c**). For the proof that (**b**), (**c**) and (**d**) together imply (**a**), only smoothness is needed, not the value of t. Thus (**b**) and (**c**) imply in particular that **D** is projective, whence (5) shows that the value for t just given is the same as that in (**d**). Hence conditions (**a**)—(**d**) are equivalent.

24. *Let* $\mathbf{D}^{(1)}$ *and* $\mathbf{D}^{(2)}$ *be two designs, with parameters* $v^{(i)}, \ldots, \lambda^{(i)}$ $(i = 1, 2)$ *such that*

(a) $v^{(2)} = s\, v^{(1)}$ *for some integer* $s > 1$, *and*

(b) $r^{(1)} = r^{(2)} = r$ *and* $\lambda^{(1)} = \lambda^{(2)} = \lambda$.

Then

$$0 < s\, k^{(1)} - k^{(2)} < s - 1 \quad \textit{and} \quad b^{(1)} < b^{(2)}. \tag{24}$$

Also, if d *denotes the g. c. d.* (r, λ), *then* $r\, d^{-1}$ *divides* $s - 1$*; furthermore*

$$d \geqq k^{(1)} \quad \textit{and} \quad d\,(d-1) \geqq \lambda \tag{25}$$

and

$$(k^{(1)} - 1)\,(s-1)\,(d-1) \leqq r \leqq \lambda\,(k^{(2)} - 3). \tag{26}$$

Proof: As $\lambda(s-1) = \lambda(v^{(2)} - 1) - \lambda(v^{(2)} - s)$, it follows from **(a)**, **(b)** and the second equation (5) that

$$\lambda\,(s-1) = r\,[k^{(2)} - 1 - s\,(k^{(1)} - 1)] = r\,(s - 1 - D), \tag{27}$$

where D denotes the difference $sk^{(1)} - k^{(2)}$. As $\lambda < r$ and $\lambda(s-1) > 0$, the first assertion (24) follows immediately. Also, the first equation (5) now gives

$$[b^{(2)} - b^{(1)}]sk^{(1)} = b^{(2)}D > 0,$$

proving the second part of (24). Next, it is clear from (27) that $rd^{-1} | (s-1)$; we put $s - 1 = h\, r\, d^{-1}$, where h is some positive integer. Fisher's inequality for $\mathbf{D}^{(2)}$ yields $r \geqq k^{(2)}$; hence

$$d = r\,h/(s-1) \geqq k^{(2)}\,h/(s-1) = k^{(1)}\,h \cdot s/(s-1) - h \cdot D/(s-1)$$
$$> h\,[k^{(1)} - 1] \geqq k^{(1)} - 1,$$

where $D < s - 1$ is used again. This proves the first claim of (25); moreover,

$$d \geqq 1 + h\,[k^{(1)} - 1] = 1 + [k^{(1)} - 1]\,(s-1)\,d\,r^{-1},$$

which is equivalent to the first inequality (26). As $r\,d^{-1}$ divides $v^{(1)} - 1$ by (5), we have $(v^{(1)} - 1)\,dr^{-1} \geqq 1$, whence $\lambda \leqq dr^{-1}(v^{(1)} - 1)\,\lambda = d\,(k^{(1)} - 1) \leqq d\,(d-1)$, since $d \geqq k^{(1)}$. This proves the second claim of (25). Finally, (27) implies that $\lambda\,(s-1) \geqq r > 0$ and

$$k^{(2)} - 1 \geqq s\,(k^{(1)} - 1) + 1 = s + 1 + s\,(k^{(1)} - 2) \geqq s - 1 + 2 \geqq r\,\lambda^{-1} + 2$$

or $k^{(2)} - 3 \geqq r\,\lambda^{-1}$, which is equivalent to the second inequality (26).

From $d = (r, \lambda) > 1$ it follows in particular that $\lambda > 1$ under the hypotheses of **18**. This was first shown by Higman & McLaughlin 1961, p. 386—387, and in fact the present proof is just a refinement of their argument. There exist $\mathbf{D}^{(1)}$ and $\mathbf{D}^{(2)}$ with $v^{(1)} = 5$, $v^{(2)} = 35$, $r = 18$, $\lambda = 9$; this shows that the conditions **(a)** and **(b)** of **24** are compatible.

2.2 Embeddings and extensions

Let $\mathbf{D}$ be a design and B a block of $\mathbf{D}$. The internal and external structures $\mathbf{D}_B$ and $\mathbf{D}^B$ of $\mathbf{D}$ with respect to B were defined in Section 1.1 as the substructures of the points incident or non-incident, respectively, with B, together with the blocks different from B.

1. *$\mathbf{D}_B$ is a design for every block B if, and only if, $\mathbf{D}$ is projective with $\lambda > 1$. $\mathbf{D}^B$ is a design for every B if, and only if, $\mathbf{D}$ is projective.*

This is easily proved; see for example DEMBOWSKI 1965 b, p. 127—128. If $\mathbf{D}$ is projective and has parameters $v = b$, $k = r$, and λ, then the parameters of $\mathbf{D}_B$ and $\mathbf{D}^B$ are given by

$$(1) \qquad v_{(B)} = k,\ b_{(B)} = b - 1,\ r_{(B)} = r - 1,\ k_{(B)} = \lambda,\ \lambda_{(B)} = \lambda - 1 \qquad (\text{here } \lambda > 1),$$

$$(2) \qquad v^{(B)} = v - k,\ b^{(B)} = b - 1,\ r^{(B)} = r,\ k^{(B)} = k - \lambda,\ \lambda^{(B)} = \lambda.$$

The question arises under what circumstances a given design may be interpreted as $\mathbf{D}_B$ or $\mathbf{D}^B$, for a suitable projective design $\mathbf{D}$ and one of its blocks B. This, and the corresponding question about internal and external structures with respect to points, is the main topic of the present section; but we shall also discuss some related questions.

As (1) implies $\lambda_{(B)} + 1 = k_{(B)}$, we have the following result:

2. *A necessary condition for a design to be isomorphic to $\mathbf{D}_B$, with $\mathbf{D}$ projective, is that its parameters satisfy the equation $\lambda + 1 = k$.*

This may be used to show that among the designs $\mathbf{P}_t(d, q)$ and $\mathbf{A}_t(d, q)$, whose parameters were given in (1.4.9) and (1.4.10), only $\mathbf{A}_1(2, 2)$ is of the kind in question: $\mathbf{A}_1(2, 2) = (\mathbf{P}_1(2, 2)')_B$ for any block B of the design complementary to $\mathbf{P}_1(2, 2)$.

The jth *multiple* of an incidence structure $\mathbf{S} = (\mathfrak{p}, \mathfrak{B}, \mathrm{I})$ is obtained from $\mathbf{S}$ by considering every block of $\mathbf{S}$ in j distinct copies, with incidences as in $\mathbf{S}$. More precisely, we define $j\,\mathbf{S} = (\mathfrak{p}, j\,\mathfrak{B}, \mathrm{I}^{(j)})$, where $j\,\mathfrak{B}$ is the union of j disjoint copies of $\mathfrak{B}$, and $p\ \mathrm{I}^{(j)}\ X$, for $p \in \mathfrak{p}$ and $X \in j\,\mathfrak{B}$, if and only if $p\ \mathrm{I}\ B$ for the block $B \in \mathfrak{B}$ corresponding to X. Clearly, $j\,\mathbf{S}$ is a design if and only if $\mathbf{S}$ is. Now, while $\mathbf{P}_{d-1}(d, q)$ is never isomorphic to $\mathbf{P}_d(d+1, q)_B$, a suitable multiple of it always is. Hence representability in the form $\mathbf{D}_B$ is, at least in this case, not a very useful concept. The main reason for this is, of course, that in $\mathbf{D}_B$ blocks can in general not be distinguished by the points with which they are incident.

The situation is very different when we consider representations in the form $\mathbf{D}^B$. We shall call a design $\mathbf{E}$ *embeddable* if there exists a projective design $\mathbf{D}$ and a block B of $\mathbf{D}$ such that $\mathbf{E} \cong \mathbf{D}^B$; we shall also say that $\mathbf{E}$ is *embedded* in $\mathbf{D}$. The following is analogous to **2**:

3. *A necessary condition for a design to be embeddable is that its parameters satisfy the equation*

$$n = r - \lambda = k. \tag{3}$$

Here n is the order of the design, as defined in (2.1.9). It follows easily from (2.1.5) that (3) is equivalent to either one of the following conditions:

$$\lambda v = k(r-1), \quad or \quad b = v + r - 1. \tag{4}$$

Hence these conditions are also necessary for embeddability. However, (3) and (4) are not sufficient, as we shall see.

4. *Let* $\mathbf{E} = (\mathfrak{p}, \mathfrak{B}, \mathrm{I})$ *be a design satisfying condition* (3). *Then* $\mathbf{E}$ *is embeddable if and only if there exists a system* $\mathscr{S}$ *of sets of blocks in* $\mathbf{E}$ *such that*

(a) $|(p) \cap \mathfrak{X}| = \lambda$ *for every* $p \in \mathfrak{p}$ *and* $\mathfrak{X} \in \mathscr{S}$ [*for notation see* (1.1.5)]
(b) $|\mathfrak{X} \cap \mathfrak{Y}| = \lambda - 1$ *for any two distinct* $\mathfrak{X}, \mathfrak{Y} \in \mathscr{S}$, *and*
(c) *Every block is contained in exactly* λ *sets* $\mathfrak{X} \in \mathscr{S}$.

This is essentially[1]) Theorem 2.1 of Hall & Connor 1954; the sets in $\mathscr{S}$ correspond in a one-one fashion to the points of the block B for which $\mathbf{E} \cong \mathbf{D}^B$. If $\lambda > 1$, then $(\mathscr{S}, \mathfrak{B}, \mathrm{J})$, with $\mathfrak{X}\,\mathrm{J}\,B$ defined by $B \in \mathfrak{X}$ (for $\mathfrak{X} \in \mathscr{S}$ and $B \in \mathfrak{B}$) is a design isomorphic to $\mathbf{D}_B$. Hall & Connor 1954 have derived the following theorem from **4**:

5. *If* E *satisfies* (3) *and* $\lambda \leqq 2$, *then* $\mathbf{E}$ *is embeddable.*[2])

While this is rather difficult to prove for $\lambda = 2$, the case $\lambda = 1$ is easily established. In fact, if we define a *parallelism* of an arbitrary design $\mathbf{D}$ as an equivalence relation $\|$ among the blocks, satisfying[3])

(5) *To any* $p \in \mathfrak{p}$ *and* $B \in \mathfrak{B}$ *there exists a unique* $C \in \mathfrak{B}$ *such that* $p\ \mathrm{I}\ C \| B$,

then:

---

[1]) These authors include a fourth condition: $|\mathfrak{X}| = r - 1$ for every $\mathfrak{X} \in \mathscr{S}$. But, by **1.1.3**, this is a consequence of **(a)**—**(c)**.

[2]) The hypothesis $\lambda \leqq 2$ is essential here: Bhattacharya 1944a, b has exhibited an example with $v = 16$, $b = 24$, $k = 6$, $r = 9$, $\lambda = 3$ [hence satisfying (3)], in which there exist two blocks with four common points. Clearly such a design cannot be embedded into a projective design with $\lambda = 3$.

[3]) This condition is usually known as the "Euclidean axiom" or the "Axiom of parallels". A generalization of it, and its dual, will be considered in Section 7.3. In the statistical literature, designs with a parallelism are generally called "resolvable balanced incomplete block designs". It should be pointed out that a design may admit more than one parallelism; see the next footnote. In the most important special case, however [see (7) and (8) below], the parallelism is necessarily unique.

6. *The following properties of a design* **E** *with* $\lambda = 1$ *are equivalent:*

(**a**) $n = k = r - 1$, *i.e.* (3) *holds,*
(**b**) **E** *possesses a parallelism,*
(**c**) **E** *is embeddable,*
(**d**) **E** *is an affine plane.*

This follows readily from **4** and the definitions. The system $\mathcal{S}$ of **4** consists here of the equivalence classes of the parallelism in **E**. This parallelism, incidentally, is necessarily unique.

Parallelisms in arbitrary designs (not necessarily embeddable, or with $\lambda = 1$), are of independent interest. It is easily seen that if **D** possesses a parallelism, then every *parallel class* of blocks, i.e. every equivalence class of the relation $\|$, consists of equally many blocks. If this number is called m, then $m > 1$ because of (2.1.3), and

(6) $\quad v = k\,m, \quad b = r\,m \quad$ (*for designs with parallelism*).

Hence if $v\,k^{-1} = b\,r^{-1}$ is not an integer, there cannot exist a parallelism. It follows from this observation that the designs $\mathbf{P}_t(d, q)$, with parameters (1.4.9), cannot admit a parallelism unless $t + 1$ divides $d + 1$; see also **1.4.6**. It is not known whether, conversely, $d + 1 \equiv 0 \mod t + 1$ always implies the existence of a parallelism in $\mathbf{P}_t(d, q)$.[1])

The designs $\mathbf{A}_t(d, q)$, on the other hand, have an obvious parallelism: the cosets $S + a$ and $T + b$ are parallel if and only if $S = T$ (cf. **1.4.3**). Also, the $\mathbf{A}_t(d, q)$ are in a natural way substructures of the corresponding $\mathbf{P}_t(d, q)$, but they are not in general "embedded" in them in the present sense: if $t < d - 1$, then more than one block of $\mathbf{P}_t(d, q)$ must be deleted in order to obtain $\mathbf{A}_t(d, q)$.[2]) However, $\mathbf{A}_{d-1}(d, q)$ is clearly

---

[1]) Little seems to be known about this question, except for small q and $t = 1$, $d = 3$. For $(t, d, q) = (1, 3, 2)$, any parallelism gives a solution of Kirkman's (1847, 1850) "schoolgirl problem": *Can 15 schoolgirls walk in five rows of three each, every day of the week, so that every girl is together in one row with every other girl at least once?* The schoolgirl problem is a special case of a much more general problem posed by Steiner 1853 (who was apparently not aware of Kirkman 1847). While Steiner's problem is not yet completely solved (some solutions will be discussed in Section 2.4), the solutions of Kirkman's problem are all known, see Heffter 1897; Davis 1911; Eckenstein 1911; Cole 1922; White, Cole, Cummings 1925; Emch 1929; Hall & Swift 1955. Not all of these give rise to parallelisms of $\mathbf{P}_1(3, 2)$, but several of them do, and it turns out that $\mathbf{P}_1(3, 2)$ possesses more than one parallelism. There exist parallelisms also in $\mathbf{P}_1(3, 3)$; this is an unpublished result of D. Mesner. See also Mesner 1967.

[2]) This gives rise to a concept considerably wider than embeddability: under what circumstances can an incidence structure be completed, by adjoining new points and blocks, to a design with prescribed parameters? Contributions to this general problem, mostly attacked via incidence matrices, are Connor 1952a, 1953; Hall & Connor 1954; Hall & Ryser 1954; Hall & Newman 1963; Parker 1965; Johnsen 1965.

embeddable into $\mathbf{P}_{d-1}(d, q)$. These designs $\mathbf{A}_{d-1}(d, q)$ are special cases of the following concept: An *affine design*[1]) is a design with a (necessarily unique) parallelism satisfying the following condition: there exists an integer $\mu > 0$ such that

(7) $$\text{if} \quad B \nparallel C, \quad \text{then} \quad [B, C] = \mu.$$

In other words: in affine designs, any two non-parallel blocks have equally many points in common. The parallelism of an affine design is *natural* in the sense that it may be defined by incidence properties alone, namely:

(8) $$B \| C \quad \text{if and only if} \quad B = C \quad \text{or} \quad [B, C] = 0.$$

This holds for affine designs, but is a weaker condition than (7). Note that the "only if" part of (8) holds also for other parallelisms.

The following result is supplementary to **2.1.6**.

7. *A tactical decomposition of an affine design* **A** *induces a natural equivalence relation among the parallel classes of* **A**, *two such classes being equivalent if and only if they contain blocks of the same block class. Let t and t' denote the numbers of point and block classes, respectively, and s that of equivalence classes of parallel classes of blocks. Then*

$$s + t = t' + 1.$$

Since $1 \leqq s \leqq r$, it follows that $0 \leqq t' - t \leqq r - 1$.

For the proof, see Norman 1968, Theorem 1. Only the case of a group orbit decomposition is treated there, but his argument (the essential part of which is also in Dembowski 1965b, pp. 164—165) proves the present claim. See also Block 1967, Corollary 2.3. It is not difficult to show, by means of examples in $\mathbf{A}_{d-1}(d, q)$, that $t' - t$ can actually assume all values between 0 and $r - 1$. The extreme case $t' = t + r - 1$ always holds for the trivial decomposition; this is a consequence of the next result.

8. *Let* **D** *be a design with a parallelism. Then*

(9) $$r \geqq k + \lambda \quad \text{and equivalently} \quad b \geqq v + r - 1.$$ [2])

Furthermore, equality holds in one (or, equivalently, in both) of these relations if, and only if, **D** *is affine. If this is the case, then the parameters of* **D** *can be expressed by means of the integers m and μ defined in (6) and*

[1]) "Affine resolvable balanced incomplete block designs" in the statistical literature. The connection between projective and affine designs may not be as intimate as between projective and affine geometries: it is not known whether affine designs can always be embedded into projective designs. Concerning this question, see result **9** below. Examples of affine designs which are neither affine planes nor isomorphic to $\mathbf{A}_{d-1}(d, q)$ will be given in Section 2.4.

[2]) Note that (9) follows from (6) alone, because of **2.1.4**.

(7) *as follows:*

$$(10)\quad \begin{aligned} v &= \mu m^2, \; k = \mu m, \; \lambda = (\mu m - 1)(m-1)^{-1}, \\ r &= (\mu m^2 - 1)(m-1)^{-1}, \; b = m(\mu m^2 - 1)(m-1)^{-1}. \end{aligned}$$

In particular, $\lambda - \mu = (\mu - 1)(m-1)^{-1}$, *so that* $m - 1$ *is a divisor of* $\mu - 1$.

This is due to BOSE 1942a.

It follows from **8** that affine designs satisfy the condition (3) necessary for embeddability. However, it is not known whether affine designs must always be embeddable. The best result in this direction is the following:

9. *Let* **A** *be an affine design, with parameters as in* (10), *and* $\lambda > 1$. *Then* **A** *is embeddable if and only if there exists a projective design* $\mathbf{D}_0$ *with parameters*

$$(11)\quad v_0 = b_0 = (\mu m^2 - 1)(m-1)^{-1}, \; k_0 = r_0 = (\mu m - 1)(m-1)^{-1}, \; \lambda_0 = (\mu - 1)(m-1)^{-1}.$$

More specifically, for any such $\mathbf{D}_0$ *and any one-one correspondence* ω *between the parallel classes* $\mathfrak{X}$ *of* **A** *and the blocks* $\mathfrak{X}^\omega$ *of* $\mathbf{D}_0$, *there exists a projective design* $\mathbf{P} = \mathbf{P}(\mathbf{D}_0, \omega)$ *such that* $\mathbf{P}^W \cong \mathbf{A}$ *and* $\mathbf{P}_W \cong m\,\mathbf{D}_0$, *for some block* W *of* **P**, *and* $(\mathfrak{X}^\omega) = (\mathfrak{X})$ *in* **P**, *for every parallel class* $\mathfrak{X}$ *of* **A**.

Here $(\mathfrak{X}^\omega) = (\mathfrak{X})$ means, intuitively, that the blocks of $\mathfrak{X}$ intersect in the block $\mathfrak{X}^\omega$ of $\mathbf{D}_0$, considered as a set of points on W. Result **9** is due to SHRIKHANDE 1951b. We note a corollary: If **A** is embeddable, then $\mathbf{A} \cong \mathbf{P}^W$, and the projective design **P** has parameters

$$(12)\quad v^* = b^* = b + 1, \quad k^* = r^* = r, \quad \lambda^* = \lambda,$$

with b, r, λ as in (10). Now **9** implies the first of the following results:

10. *Suppose that there exists an affine design with parameters* (10).

(**a**) *If there is a projective design with parameters* (11), *then there also exists a projective design with parameters* (12).
(**b**) *If* v *and* r *are odd, then* $k = \mu m$ *is a square.*
(**c**) *If* v *is odd and* r *is even, then* $k^2/v = \mu$ *is a square.*
(**d**) *If* $r \equiv 2$ *or* $3 \bmod 4$, *then the square-free factor of* $v/k = m$ *has no prime divisor* $\equiv 3 \bmod 4$.

Result (**a**) is due to SHRIKHANDE 1951b, and (**b**)—(**d**) to MAJUMDAR 1953, Theorems 6, 7. Majumdar's proofs, by matrix arguments, resemble those of CHOWLA & RYSER 1950 for result **2.1.15**. For similar nonexistence theorems for affine designs (related to a more general class of tactical configurations called "divisible partial designs"; cf. Section 7.1), see also SHRIKHANDE 1953b, 1954.

We shall see later (cf. **2.4.34**, **2.4.36** below) that there exist affine designs which are not isomorphic to the system of points and hyperplanes of any finite affine geometry, desarguesian or not. Hence there arises the problem of characterizing these systems among the affine designs. This can be done in a way similar to **2.1.22**; we shall develop this now.

If **D** is a design with a parallelism, then we obtain an *induced parallelism* among the lines of **D**, which are defined as in Section 2.1, as follows. If $(B, \mathfrak{l})$ is an arbitrary block-line pair, define

$$B \parallel \mathfrak{l} \quad \textit{if and only if} \quad \mathfrak{l} \subseteqq (C) \quad \textit{for some block } C \parallel B;$$

here we use the notation (1.1.5) again. Then let

$$(13) \qquad \mathfrak{k} \parallel \mathfrak{l} \quad \textit{if and only if} \quad B \parallel \mathfrak{l} \quad \textit{for every block } B \textit{ with } \mathfrak{k} \subseteqq (B),$$

for any two lines $\mathfrak{k}$ and $\mathfrak{l}$. It follows that

11. (**a**) *The parallelism defined by* (13) *is an equivalence relation in the set of lines of* **D**.

(**b**) *To any point-line pair* $(p, \mathfrak{l})$ *there exists at most one line* $\mathfrak{k}$ *such that* $p \in \mathfrak{k} \parallel \mathfrak{l}$.
(**c**) *If* $\mathfrak{k} \parallel \mathfrak{l}$, *then* $\mathfrak{k} = \mathfrak{l}$ *or* $\mathfrak{k} \cap \mathfrak{l} = \emptyset$.
(**d**) *Any two parallel lines are coplanar; if they are distinct and if* **D** *is smooth, then the plane containing them is unique.*

For the simple proofs of these facts, see DEMBOWSKI 1964a, p. 150. Together with **2.1.21** and **2.1.22**, they yield the following characterizations of the affine designs $\mathbf{A}_{d-1}(d, q)$ among the designs admitting a parallelism:

12. *The following properties of a design* **D** *with parallelism, such that* $m > 2$ [cf. (6)], *are equivalent:*

(**a**) **D** *is isomorphic to the system of points and hyperplanes of a (not necessarily desarguesian) affine geometry.*
(**b**) *Every line meets every non-parallel block.*
(**c**) *Every line consists of m points.*
(**d**) *Every plane is contained in exactly* $\lambda(\lambda - 1)(r - 1)^{-1}$ *blocks.*
(**e**) **D** *is smooth and affine.*

Proof: DEMBOWSKI 1964a, 1967a.[1]) The restriction $m > 2$ is essential: there exist affine designs with $m = 2 \leqq \lambda$ which are not isomorphic to any $\mathbf{A}_{d-1}(d, q)$; for a characterization of these see **2.4.34** below. On the other hand, these designs satisfy condition (**c**): every line

[1]) The statement of the main result in DEMBOWSKI 1964a is not correct: the sesential hypothesis $m > 2$ was omitted there. The correction is in DEMBOWSKI 1967a. For parts (**d**) and (**e**) of **12**, the author is indebted to J. Ryshpan; cf. **2.1.21**.

contains at least two points by definition, and if there is a parallelism, the number of points per line is at most m. However, it is possible to characterize the $\mathbf{A}_{d-1}(d, 2)$ in a fashion similar to **12** as follows:

13. *Let* $\mathbf{D}$ *be a design with parallelism, such that* $m = 2$. *Then* $\mathbf{D}$ *is isomorphic to* $\mathbf{A}_{d-1}(d, 2)$ *for some* $d \geqq 2$ *if, and only if, every plane of* $\mathbf{D}$ *contains four distinct points.*

Proof: DEMBOWSKI 1967a. Note that if $m = 2$, then $|\mathfrak{e}| = 3$ or 4 for any plane $\mathfrak{e}$ of $\mathbf{D}$.

We note another characterization of the projective designs $\mathbf{P}_{d-1}(d, q)$ with $d > 2$.

14. *A projective design* $\mathbf{D}$ *is isomorphic to a projective plane or a* $\mathbf{P}_{d-1}(d, q)$ *with* $d > 2$ *(according as* $\lambda = 1$ *or* $\lambda > 1$*) if, and only if,* $\mathbf{D}^B$ *is an affine design for every block* B *of* $\mathbf{D}$.

The necessity is obvious. Sufficiency: If B and C are two distinct blocks, then any point p incident with B and C is also incident with each of the remaining blocks in the parallel class of $\mathbf{D}^B$ determined by C. As the integer m of (6) for $\mathbf{D}^B$ is

$$\frac{v^{(B)}}{k^{(B)}} = \frac{v-k}{k-\lambda} = \frac{b-\lambda}{r-\lambda} - 1,$$

this means that each of the λ points on B and C is incident with the same $m + 1$ blocks. Hence in the design $\bar{\mathbf{D}}$ dual to $\mathbf{D}$ the line $((\bar{B}, \bar{C}))$ consists of $m + 1 = (b - \lambda)/(r - \lambda)$ points, and as B, C were arbitrary it follows from **2.1.22c** that $\bar{\mathbf{D}}$, and hence also $\mathbf{D}$, consists of points and hyperplanes of a finite projective geometry.[1])

We have now given several characterizations of the designs $\mathbf{P}_{d-1}(d, q)$ and $\mathbf{A}_{d-1}(d, q)$, of points and hyperplanes in finite projective and affine geometries, by their incidence properties. Both these types of designs can also be characterized by a common system of incidence axioms; for this see KANTOR 1969a. In Section 2.3 we shall present more characterizations of the $\mathbf{P}_{d-1}(d, q)$, by properties of their automorphism groups.

We turn to the consideration of internal and external structures $\mathbf{D}_p$ and $\mathbf{D}^p$, respectively, of a design with respect to a point p. These were defined in Section 1.1 as the substructures consisting of the points $\neq p$ and the blocks incident or non-incident, respectively, with p. The next result is analogous to **1**:

15. *The following properties of the design* $\mathbf{D}$ *are equivalent:*

(**a**) $\mathbf{D}$ *is of type* $(1, t)$, *with* $t > 2$.
(**b**) $\mathbf{D}_p$ *is a design for every point* p.
(**c**) $\mathbf{D}^p$ *is a design for every point* p.

---

[1]) This proof is due to W. M. Kantor.

The proof is as simple as that of **1**, cf. DEMBOWSKI 1965b, p. 127. More generally, if **D** is a design of type (s, t), $s \leqq t > 1$, then both $\mathbf{D}_p$ and $\mathbf{D}^p$ are tactical configurations of types $(s', t-1)$ and $(s'', t-1)$ for suitable integers s' and s'', respectively[1]); and the parameters of $\mathbf{D}_p$ and $\mathbf{D}^p$ are given by

$$v_{(p)} = v - 1, \; b_{(p)} = r, \; k_{(p)} = k - 1, \; r_{(p)} = \lambda, \tag{14}$$

$$\textit{and} \quad \lambda_{(p)} = b_3 \quad \textit{if} \quad t > 2.$$

$$v^{(p)} = v - 1, \; b^{(p)} = b - r, \; k^{(p)} = k, \; r^{(p)} = r - \lambda, \tag{15}$$

$$\textit{and} \quad \lambda^{(p)} = \lambda - b_3 \quad \textit{if} \quad t > 2.$$

For $t > 2$, select $t - 2$ distinct points p_i $(i = 1, \ldots, t-2)$ in **D** and consider the sequence of designs $\mathbf{D}_0 = \mathbf{D}$, $\mathbf{D}_{i+1} = (\mathbf{D}_i)_{p_i}$, with $\mathbf{D}_i$ of type $(1, t - i)$, having parameters

$$v^{(i)} = v - i, \; b^{(i)} = b_i, \; k^{(i)} = k - i, \; r^{(i)} = b_{i+1}, \; \lambda^{(i)} = b_{i+2};$$

here the parameters b_j are defined as in Section 1.1. Combination of Fisher's inequality $v^{(i)} \leqq b^{(i)}$ with (1.1.9) now yields a proof of Result **2.1.5**.

We call a tactical configuration **C** *extendable* if there exists a design **D** such that $\mathbf{C} \cong \mathbf{D}_p$ for some point p of **D**. The following is similar to **2** and **3**:

16. *A necessary condition for a design to be extendable is that its parameters satisfy the condition*

$$b(v + 1) \equiv 0 \mod k + 1. \tag{16}$$

This follows at once from (14); see HUGHES 1965a. The condition is not sufficient for extendability, but it is quite useful for proving that certain designs are not extendable:

17. *Suppose that* **D** *is an extendable design.*

(**a**) *If* **D** *is projective, then* $2(\lambda + 1)(\lambda + 2) \equiv 0 \mod k + 1$.
(**b**) *If* **D** *has a parallelism, then* $b(m - 1) \equiv 0 \mod k + 1$.
(**c**) *If* **D** *is affine, then* $\lambda = 1$.

Proofs are in DEMBOWSKI 1965b, pp. 129—131. Parts (**a**) and (**b**) are almost immediate consequences of **16**; see HUGHES 1965a. For part (**c**), one also has to use (10).

We shall now discuss some consequences of **17a** and **17c**. In **17a**, the integer $k + 1$ is bounded by $2(\lambda + 1)(\lambda + 2)$; hence (HUGHES 1965a, Theorem 2.1):

18. *There exist only finitely many extendable projective designs with prescribed value of* λ.

---

[1]) It is natural to ask for the relations between s, s', and s''. If $t > 3$, then $s = s' = s'' = 1$; for $t = 3$ there exist examples with $s = 1$ and $s' = 2$; see Section 2.4.

For the case $\lambda = 1$, further analysis shows that an extendable finite projective plane must be of order 2, 4, or 10. The unique[1]) projective planes $\mathbf{P}(2) = \mathbf{P}_1(2, 2)$ and $\mathbf{P}(4) = \mathbf{P}_1(2, 4)$, of orders 2 and 4, are indeed extendable, as will be seen in Section 2.4. Whether a projective plane of order 10 exists is unknown, so that the question of extendability is in this case rather remote. Another consequence of **17a** is:

19. *None of the projective designs $\mathbf{P}_{d-1}(d, q)$, with $d > 2$ and $q > 2$, is extendable.*

Proof: HUGHES 1965b, Theorem 7; compare the parameters of $\mathbf{P}_{d-1}(d, q)$ in (1.4.9). More generally, one can derive necessary and sufficient conditions for t and d so that the parameters (1.4.9) and (1.4.10) of the $\mathbf{P}_t(d, q)$ and $\mathbf{A}_t(d, q)$ satisfy (16); this will yield necessary conditions for the extendability of these designs. The conditions $d, q > 2$ in **19** are essential, as we have already seen for $d = 2$. The $\mathbf{P}_{d-1}(d, 2)$ are, in fact, all extendable; this is a special case of Result **2.4.34** below.

Result **17c** shows that

20. *The only extendable affine designs are affine planes.*

(DEMBOWSKI 1965e, Lemma 1.) The desarguesian finite affine planes $\mathbf{A}(q) = \mathbf{A}_1(2, q)$ are indeed extendable for all q; the extensions are special "inversive planes", to be defined in Section 2.4 and investigated in detail in Chapter 6. We shall see in 2.4 that for $q = 2^{2k+1}$, $k > 0$, there exist non-isomorphic designs $\mathbf{I}$ and $\mathbf{J}$ such that $\mathbf{I}_x \cong \mathbf{A}(q) \cong \mathbf{J}_y$, for all points x of $\mathbf{I}$ and y of $\mathbf{J}$. Hence extensions of affine planes need not be unique. We shall also see in Chapter 6 that non-desarguesian affine planes (of which many examples will be given in Chapter 5) cannot be extendable if they are of even order. This will prove our earlier assertion that (16) is not sufficient for extendability.

In conclusion and for completeness, we finally mention the analogues of **2**, **3**, and **16** for external structures $\mathbf{D}^p$.

21. *A necessary condition for a design to be isomorphic to $\mathbf{D}^p$, with $\mathbf{D}$ of type $(1, t)$ with $t > 2$, is that its parameters satisfy*

$$\lambda(k - 2) \equiv 0 \mod v + 1 - k.$$

This follows easily from (15); see DEMBOWSKI 1965b, p. 133. As a simple consequence, note that no finite affine or projective plane can be of the form $\mathbf{D}^p$.

---

[1]) See result **3.2.15** below.

2.3 Automorphisms of designs

In this section we consider automorphisms and automorphism groups of designs, as defined in the beginning of Section 1.2. As pointed out earlier (cf. **1.2.2**), the permutation representations on the points and on the blocks, of an arbitrary automorphism group, need not be similar.[1]) Nevertheless, there exist strong interrelations between these representations; these are particularly striking for projective designs. We begin our discussion with some results in this direction.

Let $\mathbf{D} = (\mathfrak{p}, \mathfrak{B}, \mathrm{I})$ be a design with parameters v, b, k, r, λ, and Γ an automorphism group of $\mathbf{D}$. These general hypotheses will not always be repeated.

1. *Let m and m' denote the numbers of point and block orbits of Γ, respectively. Then*

(**a**) $m \leqq m'$,
and
(**b**) *if* $\mathbf{D}$ *is projective, then* $m = m'$.

This is an immediate consequence of **2.1.6**. Result (**b**) is very useful for many investigations in projective designs; independent proofs were given by Parker 1957, Hughes 1957d, and Dembowski 1958. An immediate corollary of **1** is:

2. *If Γ is block transitive, then Γ is also point transitive. The converse also holds if* $\mathbf{D}$ *is projective.*

For the first statement, see also Block 1965.

The *rank* of a transitive permutation group G is the number of orbits of the stabilizer of one of the permuted objects. Thus the rank is always $\geqq 2$, equality holding if and only if G is doubly transitive. In our present situation we may speak of *point rank* and *block rank* when Γ is point (block) transitive. The next result (Kantor 1968b, Theorem 4.4) is an extension of **2**:

3. *Suppose that Γ is point and block transitive, and let s and s' denote point and block rank of Γ, respectively. Then*

(**a**) $s \leqq s'$,
and
(**b**) *if* $\mathbf{D}$ *is projective, then* $s = s'$.

---

[1]) This is clear if the design is non-projective, but there are also examples in the projective case: the elations with fixed axis of $\mathcal{P}(d, q)$, interpreted as automorphisms of $\mathbf{P}_{d-1}(d, q)$, form a group with one fixed block and more than one fixed point. All known examples of this kind are intransitive, so that conceivably, for projective designs, the representations on points and blocks must be similar in the transitive case.

This follows by applying **1** to the stabilizers Γ_p and Γ_B, and using **1.2.11**. As a corollary, we have:

4. *Γ is doubly block transitive if, and only if, $\mathbf{D}$ is projective and Γ is doubly point transitive. If this is the case, then Γ_B is transitive on (B) as well as on $\mathfrak{p} - (B)$, for any $B \in \mathfrak{B}$, and if in addition v and k are relatively prime, the stabilizer Γ_{xy} of any two distinct points x, y is transitive on the blocks X with $x \mathrm{I} X \not\mathrm{I} y$, so that Γ cannot be sharply 2-transitive.*

Here **2.1.1** is also used. For the equivalence of double point and double block transitivity in the projective case, see also DEMBOWSKI 1958, Satz 4. The other claims of **4** follows by applying **1b** to the stabilizer Γ_B: The two point orbits of this group must be (B) and $\mathfrak{p} - (B)$, and if $(v, k) = 1$, then Γ_{Bx}, for $x \mathrm{I} B$, is still transitive on $\mathfrak{p} - (B)$, whence Γ is transitive on the triples (x, X, y) with $x \mathrm{I} X \not\mathrm{I} y$.

An automorphism group which is doubly transitive on points can, of course, exist also in the non-projective case, for example in $\mathbf{P}_t(d, q)$ with $t < d - 1$. In fact, we shall see in Section 2.4 that every doubly transitive permutation group can be interpreted as a doubly point transitive automorphism group for many designs. The following result is a useful supplement to **1** and **4**:

5. *Suppose that the automorphism group Γ of the design $\mathbf{D} = (\mathfrak{p}, \mathfrak{B}, \mathrm{I})$ is doubly transitive on points and transitive on blocks. Let c denote the number of point orbits of Γ_B, where $B \in \mathfrak{B}$. Then:*

(1) $$c \leqq 1 + b v^{-1}.\ ^{1)}$$

For projective designs, **5** gives nothing new, but:

6. *If the affine design $\mathbf{D} = (\mathfrak{p}, \mathfrak{B}, \mathrm{I})$ has an automorphism group Γ which is doubly transitive on points and transitive on blocks, then Γ_B is transitive both on (B) and on $\mathfrak{p} - (B)$, for any $B \in \mathfrak{B}$.*

For in this case, (2.2.10) and (1) give $c = 2$.

If Γ is flag transitive, then Γ is probably primitive on the points. Up to now it has been possible to prove this only under additional hypotheses:

7. *Suppose that Γ is flag transitive. Then each of the following conditions implies that Γ is point primitive:*

(a) $\lambda > (r, \lambda)\,[(r, \lambda) - 1]$, *in particular* $(r, \lambda) = 1$ *and* $\lambda = 1$;
(b) $(n, k) = 1$, *here* $n = r - \lambda$, cf. (2.1.9);
(c) $r > \lambda(k - 3)$;
(d) $\mathbf{D}$ *is of type* $(1, t)$, *with* $t > 2$;
(e) $(v - 1, k - 1) \leqq 2$.

---

[1]) This is an unpublished result of H. Wielandt. A proof of the weaker inequality $c \leqq 1 + (b - 1)(v - 1)^{-1}$ is essentially in ITO 1960. Note that c is also the number of block orbits of Γ_p, where p is an arbitrary point, by **1.2.11**.

In cases (**d**) *and* (**e**), *except possibly when* $(v-1, k-1) = 2$ *and* $v \equiv 1 \bmod 2$ *in case* (**e**), Γ *is even doubly point transitive.*

Proof. Assume Γ imprimitive on points and let $\mathfrak{p}_1 \cup \cdots \cup \mathfrak{p}_s$ be a partition of $\mathfrak{p}$ into classes of imprimitivity, all $\mathfrak{p}_i$ being of the same size w. The flag transitivity implies that the substructures $\mathbf{D}_i$ defined by $\mathfrak{p}_i$ and those blocks X for which $(X) \cap \mathfrak{p}_i \neq \emptyset$ are designs.[1]) Result **2.1.24**, with $\mathbf{D}^{(1)} = \mathbf{D}_i$ and $\mathbf{D}^{(2)} = \mathbf{D}$, now excludes cases (**a**) and (**c**). The incidence structure $\mathbf{D}^* = (\mathscr{P}, \mathfrak{B}, \mathrm{J})$, defined by $\mathscr{P} = \{\mathfrak{p}_1, \ldots, \mathfrak{p}_s\}$ and $\mathfrak{p}_i \, \mathrm{J} \, B \in \mathfrak{B}$ if and only if $(B) \cap \mathfrak{p}_i \neq \emptyset$, is also a design,[1]) with parameters $v^* = s$, k^* a proper divisor of k, and λ^* satisfying

$$\lambda w^2 = \lambda^* (k/k^*)^2;$$

this is verified by simple counting arguments. It follows that k/k^* divides λw, hence also λv. Clearly k/k^* divides k, and now (2.1.5) shows that k/k^* also divides $n = r - \lambda$. Hence $1 < k/k^* \mid (n, k)$; this excludes case (**b**).

For the remaining cases (**d**) and (**e**), consider the internal structure $\mathbf{D}_p$ with respect to an arbitrary point p. By flag transitivity, Γ_p induces a block transitive group in $\mathbf{D}_p$. Using **2** for case (**d**) and **1.2.8** for case (**e**), we conclude that, except when $(v-1, k-1) = 2$ and $v \equiv 1 \bmod 2$ in case (**e**), Γ_p is point transitive on $\mathbf{D}_p$, whence Γ is doubly point transitive. In the excepted case, Γ_p may have two point orbits of equal length $(v-1)/2$, but then Γ is still primitive.[2])

We mention a few similar results.

8. *Suppose that* $(r, \lambda) = 1$.

(**a**) *If* Γ *is doubly point transitive, then* Γ *is flag transitive.*
(**b**) *If* Γ *is sharply flag transitive, then* Γ *acts as a Frobenius group on the points. In this case,* v *is a power of a prime.*

To prove (**a**), consider again $\mathbf{D}_p$ and use the dual of **1.2.8**. For (**b**), it is clear that $\Gamma_p \neq 1$ for every point p, and that Γ is point transitive. The sharp flag transitivity implies that Γ_{pq} permutes the λ blocks through p and q regularly; hence if $\gamma \in \Gamma_{pq}$, then $o(\gamma) \mid \lambda$. On the other hand, $o(\gamma) \mid |\Gamma_p| = r$, whence $(r, \lambda) = 1$ implies $\gamma = 1$. Using **7a**, we see that Γ is a primitive Frobenius group on the points; its sharply transitive Frobenius kernel Φ is, therefore, elementary abelian, and its order $|\Phi| = v$ is a prime power.

---

[1]) These designs may be "degenerate" in the sense that condition (2.1.3) does not hold. However, this possibility does not affect the validity of the proof.

[2]) Parts of the preceding proof, as well as results **8** and **9** below, are due to W. M. Kantor. See Kantor 1968b, Corollary 4.6 and Theorems 4.7, 4.8.

9. *Suppose that* $(b, r, \lambda) = 1$.

(**a**) *If* Γ *is sharply block transitive and* $b > v$, *then* Γ *is a Frobenius group on the points, and* $(b, r) > 1$.
(**b**) *If* Γ *is a Frobenius group on the blocks, then* **D** *is projective.*

The proof of **9a** is essentially the same as that of **8b**; that $(b, r) > 1$ follows from $b > v$ and the dual of **1.2.8**, applied to the Frobenius kernel Φ. To prove (**b**), assume $b > v$ and consider the Frobenius kernel Φ which is now sharply transitive on blocks. Then Φ would be a Frobenius group on the points, by (**a**). But Φ is nilpotent (Thompson 1960a, 1964), and Frobenius groups are not. Hence $b = v$, and **D** is projective.

The fixed elements of an automorphism group are of importance for many investigations. Little can be said in general about the substructure of the Γ-fixed points and blocks, even in the case where **D** is projective. In particular, there is no analogue to **1.4.7** available. Under additional assumptions, however, it is possible to prove various results in this direction. Three such results will be given now; for $\lambda = 1$ see also Section 4.1.

10. *Suppose that* **D** *is projective and that* Γ *is a p-group. Let f and F denote, respectively, the numbers of fixed points and fixed blocks of* Γ.
(**a**) *If* $(v, k) = 1$, *then* $f = 0$ *if and only if* $F = 0$.
(**b**) *If p does not divide* $nk = (k - \lambda)\, k$, *then* $f = F$.
(**c**) *If* $f \leqq p$ *or* $F \leqq p$, *then* $f = F$.
Proof: Dembowski 1958, Satz 10.

Now we restrict our attention to the case where $\Gamma = \langle\gamma\rangle$ is *cyclic*; in other words we study single automorphisms of **D**. Clearly, the permutations induced by γ on $\mathfrak{p}$ and $\mathfrak{B}$ cannot be similar if **D** is not projective. On the other hand:

11. *If* **D** *is projective and* $\Gamma = \langle\gamma\rangle$ *cyclic, then the orbit decomposition of* Γ *is symmetric. In other words,* γ *induces similar permutations on* $\mathfrak{p}$ *and* $\mathfrak{B}$.

This follows from **1.3.12** (incidence matrices of projective designs are nonsingular by **1.3.7**); see Parker 1957. In particular,

12. *Every automorphism of a projective design has equally many fixed points and fixed blocks.*

This was proved by Baer 1947 for $\lambda = 1$, by a matrix-free method.[1]) The common number of fixed points and fixed blocks of γ (for **D** pro-

---

[1]) It is not difficult to generalize Baer's counting argument to the general case, and in fact one may then deduce **11** by induction. This yields a more elementary proof of **11**.

jective) will be denoted by $f(\gamma)$ or simply f. If $\gamma \neq 1$, then there exists a block $B \neq B\gamma$, and a fixed point of γ is either on both or on none of B, $B\gamma$. This shows that

$$f(\gamma) \leqq v - 2(k - \lambda) = v - 2n, \tag{2}$$

for any non-trivial automorphism γ of a projective design. The elations of $\mathbf{P}_{d-1}(d, 2)$ show that $f = v - 2n$ can actually occur.

In Section 2.1, various conditions necessary for the existence of symmetric tactical decompositions in projective designs were given. In view of **11**, these can be interpreted as conditions for the existence of collineations with prescribed orbit lengths. We shall not formulate all these results explicitly, but restrict ourselves to one important application of **2.1.13**:

13. *Suppose that* $\mathbf{D}$ *is projective and that the automorphism* γ *of* $\mathbf{D}$ *has prime order* p. *Then the number of point (and block) orbits of* $\Gamma = \langle\gamma\rangle$ *is*

$$m = f(\gamma) + [v - f(\gamma)]\, p^{-1}, \tag{3}$$

and the diophantine equation

$$n\,x^2 + (-1)^{(m-1)/2}\, p^{f(\gamma)+1}\, \lambda\, y^2 = z^2 \tag{4}$$

has a nontrivial integral solution.

(Hughes 1957d.) If m is even, such a solution must have $y = 0$, and then n must be a square; cf. **2.1.7**. This observation and **2.1.8** yield the following simple corollary:

14. *Let* γ *be an involutorial automorphism of a projective design of non-square order. Then* v *and* $f(\gamma)$ *are both odd, and*

$$f(\gamma) \equiv v \mod 4. \tag{5}$$

We shall now consider special types of automorphisms, namely central, axial, and quasicentral automorphisms, dilatations and translations, homologies and elations.[1]) For the following discussion $\mathbf{D}$ need not be projective, but we shall have to postulate sometimes that two blocks do not have too many common points; see (6) below.

A point c of $\mathbf{D}$ is a *center* of the automorphism γ of $\mathbf{D}$ if every block through c is fixed by γ. Clearly, c is then also fixed. Any automorphism possessing a center is called *central*. *Axes*, and *axial automorphisms*, are defined dually.

15. *Suppose that* $\mathbf{D}$ *satisfies the following condition:*

$$[B, C] \leqq k - 2, \quad \textit{for any two distinct blocks } B, C. \tag{6}$$

[1]) We have already used some of these terms in Section 1.4, but as we shall see, these earlier definitions are merely special cases of the concepts to be introduced presently.

Then a nontrivial automorphism γ of **D** *has at most one center. If* **D** *is projective,* (6) *is automatically satisfied, and then γ also has at most one axis.*

LÜNEBURG (1961a, Satz 2) proved this for **D** projective; his proof carries over to the general case, cf. DEMBOWSKI 1965b, p. 134—135. Note that the $\mathbf{P}_t(d, q)$ with $t < d - 1$ possess nontrivial automorphisms with more than one axis.

16. *If γ is central and $B\gamma \neq B \in \mathfrak{B}$, then all points of $(B, B\gamma) = (B) \cap (B\gamma)$ are fixed by γ. If* **D** *is projective and γ axial, then, dually, for any non-fixed point p, each of the λ blocks through p and $p\gamma$ is fixed by γ.*

(LÜNEBURG 1961a, Hilfssatz 6; DEMBOWSKI 1965b, p. 136.) We mention two more results on central (axial) automorphisms of projective designs; the second of these is a slight improvement of **16.**

17. *Suppose that* **D** *is projective and γ a nontrivial central automorphism of* **D**, *with center c. Then:*

(**a**) *A fixed block $A \not\mathrm{I}\, c$ is an axis of γ.*
(**b**) *There exist two blocks B, B' and a point $p \notin \{c\} \cup (B, B')$ such that γ fixes p and every point of (B, B').*
Dual statements hold if γ is axial.

(LÜNEBURG 1961a, Hilfssatz 5, Satz 3.)

If $\mathbf{D} = \mathbf{P}_{d-1}(d, q)$, then an automorphism is axial if and only if it is central, cf. **1.4.9**. This is a peculiarity of the designs $\mathbf{P}_{d-1}(d, q)$: in general it is not true that central automorphisms of projective designs are axial. For examples, constructed with the help of **2.2.9**, see **2.4.38** below. If a nontrivial automorphism of a projective design has a center c and an axis A, then it is called an *elation* or a *homology* according as $c \,\mathrm{I}\, A$ or $c \not\mathrm{I}\, A$. For convenience, we also include the identical automorphism among both the elations and the homologies.

An automorphism of **D** is *quasicentral* if every point is on some fixed line.[1]) Clearly all central automorphisms are quasicentral. Another class of quasicentral automorphisms is that of the *dilatations* δ, defined by the condition

(7) $p \,\mathrm{I}\, B \,\mathrm{I}\, p\delta$ *implies* $B = B\delta$, *for* $p \in \mathfrak{p}$ *and* $B \in \mathfrak{B}$.

It follows immediately that

[1]) Lines were defined in Section 2.1. Quasicentral automorphisms were first considered for $\lambda = 1$ by BAER 1946b (under the name of "quasiperspectivity"). They do not play an important role in the present section, but they will be more significant in Section 4.1.

18. *Every fixed point of a dilatation is a center,*

and now **15**, **16**, and **18** give:

19. *Suppose that* (6) *is satisfied in* $\mathbf{D}$, *and let* δ *be a nontrivial dilatation of* $\mathbf{D}$. *Then*

(**a**) δ *has at most one fixed point, and*
(**b**) *if* δ *has a fixed point and* B *is a non-fixed block, then* $(B, B\delta) = \emptyset$.

This result, together with **12**, shows that

20. *Projective designs do not admit nontrivial dilatations.*

Despite **20**, there is an important connection between dilatations and axial automorphisms of projective designs: If $\mathbf{D}$ is projective and A a block of $\mathbf{D}$, then the automorphisms with axis A of $\mathbf{D}$ induce dilatations in the external design $\mathbf{E} = \mathbf{D}^A$. Conversely, if $\mathbf{E}$ is embeddable, i.e. isomorphic to $\mathbf{D}^A$ for a block A of a projective design $\mathbf{D}$, then every dilatation of $\mathbf{E}$ is induced by an automorphism of $\mathbf{D}$ with axis A. This follows trivially from **17a** and **18** if the dilatation in question has a fixed point; for the case that it is fixed-point free—such dilatations are called *translations*—see Dembowski 1965b, p. 140. Hence we have:

21. *The set of dilatations of an embeddable design* $\mathbf{E} = \mathbf{D}^A$ *is essentially identical with the group of automorphisms with axis* A *of* $\mathbf{D}$.

In particular, if $\mathbf{E}$ is embeddable, then the dilatations of $\mathbf{E}$ form a group. The same is true for affine (and, trivially, also for projective) designs:

22. *An automorphism of an affine design is a dilatation if and only if it leaves all parallel classes of blocks invariant.*

(Dembowski 1965b, p. 140. The "if" part holds for arbitrary designs with parallelism.)

The set of dilatations of an arbitrary design need not be a group. An example for this will be given in the next section; cf. (2.4.20) and context.

We consider now the following situation: $\mathbf{D}$ is a design satisfying condition (6), and Γ is a group of dilatations of $\mathbf{D}$. For example, $\mathbf{D}$ may be embeddable or affine, and Γ the set of all dilatations of $\mathbf{D}$. Under these circumstances, **19a** implies the following facts:

23. *The translations in* Γ *form a characteristic subgroup* T. *If* $1 \neq \mathsf{T} \neq \Gamma$, *then* T *is nilpotent, and*

(**a**) *the points* p *with* $\Gamma_p \neq 1$ *form a* Γ- *and* T-*orbit* $\mathfrak{c}$ *of length* $|\mathsf{T}|$, *and* Γ *acts as a Frobenius group on* $\mathfrak{c}$;
(**b**) *every other* Γ-*orbit of points has length* $|\Gamma|$.

This is a consequence of well-known facts about Frobenius groups, see HALL 1959, Theorem 16.8.8; THOMPSON 1960a, 1964.

In the two cases that interest us most, more can be said about the structure of the translation group T:

24. *Let* $1 \neq \mathsf{T} \neq \Gamma$ *as in* **23**, *and suppose that one of the following conditions is satisfied:*

(**a**) **D** *is embeddable, every* $\tau \in \mathsf{T}$ *is central as an automorphism of the embedding projective design* (cf. **21**), *and there exist nontrivial translations with distinct centers in* T;

or

(**b**) **D** *is affine, and there exist nontrivial translations with distinct sets of fixed lines in* T.

Then T *is elementary abelian.*

To prove this it is sufficient to show (cf. KONTOROWITSCH 1943) that T has a nontrivial partition into normal subgroups if (**a**) or (**b**) holds. But in case (**a**) such a partition is given by the subgroups $\mathsf{T}(x)$ of the translations with center $x \mathrm{I} A$, and in case (**b**) the subgroups $\mathsf{T}(\mathfrak{x})$, fixing every line of the parallel class $\mathfrak{x}$ of lines, have the desired property. Note that in case (**b**) the fixed lines of any translation $\neq 1$ are precisely those of a parallel class. Also, the partitions of the $\mathsf{T}(x)$ and the $\mathsf{T}(\mathfrak{x})$ coincide if **D** is affine *and* embeddable.

Examples of designs with transitive non-abelian translation groups will be given in the next section; cf. **2.4.29** and context. These are due to SCHULZ 1967 who also proved the following:

25. *Suppose that the design* **D** *admits a point transitive group* T *of translations. Then*:

(**a**) *The set of all dilatations of* **D** *is a group.*

(**b**) *The* T*-orbits of blocks are the classes of a parallelism of* **D** [*in the sense of* (2.2.5)].

(**c**) T *is a* p*-group for some prime* p, *and* v, k *are powers of* p. *Also, the translations of order* $\neq p$ *in* T *generate a proper subgroup of* T.

This subgroup may be the identity. For proofs, see SCHULZ 1967, (4.2), Korollar 3, and Satz 2. Some aspects of these proofs will be discussed in Section 2.4.

26. *Let* **D** *be a projective design and suppose there exists a block* A *of* **D** *and a group* T *of elations with axis* A *and center on* A, *such that, for all points* $x \mathrm{I} A$, *the subgroups* $\mathsf{T}(x)$ *of elations with center* x *have the same order* $h > 1$. *Then*

(**a**) T *is elementary abelian and sharply transitive on the* $v - k$ *points* $\not{\mathrm{I}}\, A$. *In particular,* $v - k$ *is a prime power. Moreover,*

(**b**) *if* $\lambda > 1$, *then* **D** *is isomorphic to some* $\mathbf{P}_{d-1}(d, q)$.

Conversely, for every block A of $\mathbf{P}_{d-1}(d, q)$, the elations with axis A form a group with the properties in question. **26** is again due to LÜNEBURG 1961a, Satz 7 and Satz 6, respectively.[1]) Result **26a** had been proved earlier for $\lambda = 1$ by GLEASON 1956, Lemma 1.6; Lüneburg's proof is in fact an elaboration of Gleason's argument. Result **26b** is proved by showing that condition (**b**) of **2.1.22** is satisfied. The restriction $\lambda > 1$ is essential: there exist many non-desarguesian finite affine planes with transitive translation group, as will be seen in Chapter 5. For an improvement on **26b**, see KANTOR 1969a, Theorem 5.

We list further characterizations of the $\mathbf{P}_{d-1}(d, q)$ which can be deduced from **26**:

27. *Each of the following conditions is necessary and sufficient for the projective design* $\mathbf{D}$ *with* $\lambda > 1$ *to be a* $\mathbf{P}_{d-1}(d, q)$*:*

(**a**) $\mathbf{D}$ *contains a block* A *such that all automorphisms with axis* A *are central, all automorphisms with center on* A *are axial, and every flag* (x, X) *with* $x \mathrm{I} A$ *is the center-axis pair of some non-trivial elation of* $\mathbf{D}$.
(**b**) *Automorphisms of* $\mathbf{D}$ *are central if and only if they are axial; every point is incident with* λ *distinct blocks such that each of these* λ *flags is the center-axis pair of a nontrivial elation; and every block is the axis of a nontrivial elation.*

Proof: LÜNEBURG 1961a, Sätze 10, 11. An infinite class of examples in 5.2 will show that there exist non-desarguesian projective planes satisfying (**a**); hence the hypothesis $\lambda > 1$ is essential in **27a**. In case (**b**), however, $\mathbf{D}$ is a $\mathbf{P}_1(2, q) = \mathbf{P}(q)$ also when $\lambda = 1$. This was proved by WAGNER 1959; for more details concerning the case $\lambda = 1$ see Sections 3.1 and 4.3.

One last result in this direction:

28. *A projective design with* $\lambda > 1$ *is a* $\mathbf{P}_{d-1}(d, q)$ *with* $d > 2 \neq q$ *if, and only if, it contains a block* A *such that*

(**a**) *all automorphisms with axis* A *are central, and*
(**b**) *every point* $x \not\mathrm{I} A$ *is the center of a nontrivial homology with axis* A.

(LÜNEBURG 1961a, Satz 12) Here the restriction $\lambda > 1$ is again essential; **23a** is used for the proof.

---

[1]) Lüneburg's Satz 7 is actually a purely permutation group theoretic result: Let G be a regular [semiregular, in WIELANDT's (1964) terminology] permutation group of degree n, and suppose that G admits a partition into k subgroups of the same order, such that $n + k - 1$ divides $k(k-1)$; then G is transitive. Lüneburg's formulation, incidentally, contains an error: condition (c) of Satz 7 must read "$|\mathfrak{A}| + k - 1$ teilt $k(k-1)$".

For the remainder of this section, we assume that $\mathbf{D}$ is projective and Γ transitive (on points and on blocks; see **2**). As we have seen in Section 1.2, $\mathbf{D}$ can then be represented in the form $\mathbf{C}(\Gamma, \Pi, \Delta)$, where Π is the stabilizer of a point p and Δ is a quotient set

$$\Delta(p, B) = \{\gamma \in \Gamma : p\gamma \mathrel{I} B\},$$

for some block B. The hypothesis that $\mathbf{D} \cong \mathbf{C}(\Gamma, \Pi, \Delta)$ be a projective design may then be expressed by properties of Γ, Π and Δ alone. The only case in which this has been more than superficially investigated is that where Γ is sharply transitive, i.e. $\Pi = 1$.

29. *Let Γ be a finite group and Δ a subset of Γ, with*

$$1 < |\Delta| < |\Gamma| - 1. \tag{8}$$

Then the tactical configuration $\mathbf{C}(\Gamma, 1, \Delta)$ is a design if and only if

(D) *to every $\gamma \neq 1$ in Γ there exist equally many ordered pairs (ξ, η) in $\Delta \times \Delta$ such that $\xi \eta^{-1} = \gamma$.*

If this condition is satisfied, then $\mathbf{C}(\Gamma, 1, \Delta)$ is projective, with parameters

$$v = b = |\Gamma|, \quad k = r = |\Delta|, \quad \lambda = |\Delta|(|\Delta| - 1)/(|\Gamma| - 1). \tag{9}$$

The proof is quite straightforward; see, for example, Bruck 1955, Section 2. A subset Δ of Γ will be called a *difference set* if it satisfies (8) and condition (D) of **29**.[1]) Thus the existence of a difference set Δ in Γ is equivalent to the existence of a projective design with $v = |\Gamma|$ and $k = |\Delta|$ admitting Γ as a sharply transitive automorphism group.

The only non-abelian finite groups Γ known to contain difference sets seem to be those with the relations (1.4.15); here the corresponding projective designs are the $\mathbf{P}_{d-1}(d, q)$ [see **1.4.17**]. In the abelian case, however, there are various methods available to construct difference sets; for these and the corresponding designs see Section 2.4. Here we give some general results which will permit us to prove the nonexistence of difference sets, and hence of projective designs with sharply transitive automorphism groups, in many cases.

A *multiplier* of a quotient set Δ in a group Γ was defined in Section 1.2 as an automorphism μ of Γ, fixing Π and mapping Δ onto a translate $\Delta\gamma$. In our present situation, this reduces to

$$\Delta^{\mu} = \Delta\gamma, \tag{10}$$

for the difference set Δ. The following result gives information about the fixed elements of the design automorphism given by a multiplier.

---

[1]) The reason for this terminology is that in most cases considered in the literature, Γ is abelian and written additively; the quotients $\xi \eta^{-1}$ become differences $\xi - \eta$. Similar reasons hold for the term "multiplier" below. Clearly, difference sets are quotient sets in the sense of Section 1.2.

30. *Let μ be a multiplier of the difference set Δ in Γ, and denote by* **F** *the system of those elements of the design* $\mathbf{D} = \mathbf{C}(\Gamma, 1, \Delta)$ *which are fixed by the design automorphism μ. If every prime divisor of $n = k - \lambda$ is $> \lambda$, then* **F** *satisfies conditions* (R. 1, 2) *and* ($\bar{\text{R}}$. 1, 2) *of Section* 1.1 *and is, therefore, a (possibly degenerate) subdesign, with the same λ as* **D**.

This is Theorem 3.2 of BRUCK 1955.

The subdesign **F** may be degenerate, but it can never be empty, for the point $1 \in \Gamma$ is clearly fixed by all multipliers, without restrictions on the prime divisors of n. Hence **12** implies that every multiplier also has a fixed block (MCFARLAND & MANN 1965, Theorem 1). However, there need not be a common fixed block of all multipliers (examples are among the designs of **2.4.28** below); such a block does exist if Γ is abelian and $(v, k) = 1$ (HALL 1967b, p. 140).

In the abelian case, every permutation of the form $\gamma \to \gamma^t$, with $(t, |\Gamma|) = 1$, is an automorphism of Γ. If this happens to be a multiplier of the difference set Δ in Γ, then we call the integer t a *Hall multiplier* of Δ, after HALL 1947 who first considered multipliers of this form. It is clear that the Hall multipliers form a subgroup of the centre of the group of all multipliers of Δ; hence an arbitrary multiplier permutes the fixed elements of a Hall multiplier among themselves (MCFARLAND & MANN 1965, Theorem 2).

The following result shows the existence of Hall multipliers $\neq 1$ in many cases:

31. *Let Δ be a difference set in the finite abelian group Γ, and let v, k, λ be the parameters of* $\mathbf{C}(\Gamma, 1, \Delta)$, *as given by* (9). *Suppose that $n = k - \lambda$ does not divide v, and let $p_1, \ldots, p_s$ be distinct prime divisors of n, relatively prime to v, such that*

$$\prod_{i=1}^{s} p_i > \lambda. \tag{11}$$

Also, let v_0 be the exponent of Γ, i.e. the least common multiple of the orders of the elements of Γ. Then every integer $t \equiv p_i^{e_i} \bmod v_0$ for suitable $e_i > 0$ $(i = 1, \ldots, s)$ is a Hall multiplier of Δ.

This is the celebrated "multiplier theorem", first proved by HALL 1947 for the special case that Γ is cyclic and $\lambda = 1$. The present version is due to BRUCK 1955, Theorem 4.2.[1]) In fact, Bruck derives **31** from the following more general result for which Γ need not be abelian:

[1]) Bruck assumes a little more, namely $t \equiv p_i^{e_i} \bmod v$ (not v_0), but he proves **31**. For other proofs, of **31** or special cases thereof, see HALL & RYSER 1951; HALL 1956; MENON 1960; NEWMAN 1963; TURYN 1964; MANN 1965b, p. 78—80; DEMBOWSKI 1965b, p. 176—181.

32. *Let μ be an automorphism of the finite group Γ, and suppose that Γ contains a difference set Δ such that the element*

$$(12)\qquad \Big(\sum_{\delta\in\Delta}\delta^{-1}\Big)\Big(\sum_{\delta\in\Delta}\delta^{\mu}\Big)-\lambda\sum_{\gamma\in\Gamma}\gamma,$$

of the group algebra $Q\,\Gamma$ of Γ over the field Q of rational numbers, has only non-negative coefficients. Then μ is a multiplier of Δ.

(Bruck 1955, Theorem 4.2) The use of the group algebra $Q\,\Gamma$, or of the group ring $Z\,\Gamma$ contained in it (Z is the ring of rational integers) has proved highly successful in the investigation of difference sets. Condition (D) of **29** is equivalent to

$$(13)\qquad \Big(\sum_{\delta\in\Delta}\delta\Big)\Big(\sum_{\delta\in\Delta}\delta^{-1}\Big)=n+\lambda\sum_{\gamma\in\Gamma}\gamma,$$

and the automorphism μ of Γ is a multiplier of Δ if and only if

$$(14)\qquad \sum_{\delta\in\Delta}\delta^{\mu}=\gamma\sum_{\delta\in\Delta}\delta,\quad\text{for some}\quad \gamma\in\Gamma.$$

Hence, in the abelian case, t is a Hall multiplier if and only if

$$\sum_{\delta\in\Delta}\delta^{t}=\gamma\sum_{\delta\in\Delta}\delta.$$

Condition (11) of **31**, though necessary for all known proofs, is possibly superfluous: in all known cases, when Γ is abelian, every divisor of n which is prime to v is a Hall multiplier. In some special situations one can prove that (11) can be avoided:

33. *Let Γ, Δ, v, k, λ be as in* **29**, *and suppose that $n=p^e$ or $n=2p^d$, for a prime p not dividing v, and d odd. Then every power of p is a Hall multiplier of Δ.*

(Newman 1963; Turyn 1964.) The case $p=2$ is not excluded.

We list some further results on difference sets in finite abelian groups or, equivalently, on projective designs with sharply transitive abelian automorphism groups.

34. *Let Γ be a finite abelian group, Δ a difference set in Γ, and v, k, λ as given in* (9).

(a) *If t is a Hall multiplier of Δ such that, for some prime divisor p of n, for a divisor $v_1>1$ of v, and for two integers h, e we have*

$$(15)\qquad t^h p^e\equiv -1 \mod v_1,$$

then p cannot divide the squarefree factor of n, and v_1 is not the exponent of Γ.

(b) *If $v=4n$ and $n=p^{2a}$ for some prime p* (cf. **2.1.8**), *then the p-Sylow subgroup of Γ has exponent $\leqq 4p^a$.*

(**c**) *If* Γ *is cyclic and* $(v, n) > 1$, *then for any prime divisor* p *of* (v, n), *for any divisor* v_1 *of* v, *and any integer* e:

$$p^e \equiv -1 \bmod v_1.$$

(**d**) *If* Γ *is elementary abelian of order* p^m, *and if the quadratic residues mod* p *are Hall multipliers of* Δ, *then*

$$x^2 + p\,y^2 = 4n \tag{16}$$

has an integral solution. Furthermore, the number of solutions (x, y) *of* (16) *with* $2k \equiv x \bmod p$, $k + (p-1)\,x/2 \geqq 0$ *and* $k \geqq (x + |y|\,p)/2$ *is* >0, *and if* (x_i, y_i), $i = 1, \ldots, s$, *are these solutions, then one of the systems*

$$k + \frac{p-1}{2} \sum_{i=1}^{s} x_i\, z_i = c$$

$$\sum_{i=1}^{s} z_i = (p^m - 1)\,(p-1)^{-1},$$

with c *either* 0 *or* p^m, *has an integral solution* $z_1, \ldots, z_s$.

These results are due to K. YAMAMOTO 1963; MANN 1964, 1965a; TURYN 1965. All the proofs appear in MANN 1965b, Chapter 7, and depend on the theory of characters of abelian groups. Further character-theoretic results in this context are found in TURYN 1965.

There are various interesting consequences of **34**; only a few of these can be mentioned here. In (**a**), the possibilities $t = 1$ and $e = 0$ are not excluded; hence if $t^h \equiv -1 \bmod v_1$, then either n is a square, or else $n = (n_1\, q)^2\, q$ and $v_1 = q^j$ for some prime q (MANN 1965b, Corollary 7.2.2). Taking $e = 0$ and $h = 1$, we see that if $t = -1$ is a Hall multiplier, then n is a square and Γ is neither cyclic nor elementary abelian; for this see BRUALDI 1964 and MANN 1965a, p. 215. Difference sets with -1 as a Hall multiplier seem to be a rare occurrence. Result (**b**) provides some restrictions on the structure of abelian groups with difference sets: for example, if $v = 4p^{2a}$ ($p = 2$ not excluded), then Γ cannot be cyclic. As a corollary of (**d**), we mention the fact that if $p = 2q + 1$ with a prime q, and if no prime divisor of n is $\equiv 1 \bmod p$, then n is not a prime, and the quadratic residues mod p are Hall multipliers of any difference set in any non-cyclic elementary abelian p-group (MANN 1965a, Theorem 4). For further consequences, the reader is again referred to MANN 1964, 1965a, 1965b, Chapter 7.

We shall see in the next section that there exist many difference sets in finite abelian groups such that the corresponding projective designs are neither projective planes nor of the form $\mathbf{P}_{d-1}(d, q)$. Thus transitivity of the automorphism group is not sufficient to guarantee that a projective design is the system of points and hyperplanes of a finite projective geometry. Even double transitivity is not sufficient:

$\mathbf{P}_{d-1}(d, q)$ and its complementary design $\mathbf{P}_{d-1}(d, q)'$ obviously have the same (doubly transitive) group, but $\mathbf{P}_{d-1}(d, q)'$ is easily seen to be non-isomorphic with any $\mathbf{P}_{d'-1}(d', q')$. A further example is the following: Let $\mathfrak{F} = GF(11)$ and denote by G the multiplicative subgroup of the non-zero squares in $\mathfrak{F}$. Call *points* the elements of $\mathfrak{F}$, *blocks* the point sets $G + a = \{x + a \,\vdots\, x \in G\}$, for each $a \in \mathfrak{F}$, and define incidence by set theoretic inclusion. It is not difficult to verify that this incidence structure, which we call $\mathbf{H}(11)$, is a projective design.[1] Todd 1933 has shown that $\operatorname{Aut}\mathbf{H}(11)$ is $PSL_2(11)$, the simple group of order 660, and that $\operatorname{Aut}\mathbf{H}(11)$ is doubly transitive on $\mathbf{H}(11)$. Since no $\mathbf{P}_{d-1}(d, q)$ has just 11 points, $\mathbf{H}(11)$ is neither $\mathbf{P}_{d-1}(d, q)$ nor $\mathbf{P}_{d-1}(d, q)'$, for any d, q. The complementary $\mathbf{H}(11)'$, which is easily seen to be not isomorphic to $\mathbf{H}(11)$, provides another example.[2]

One further class of projective designs with doubly transitive groups will be presented in Section 2.4. The problem of determining all such designs is unsolved.[3] In conclusion of this section, we collect some results characterizing the doubly transitive designs $\mathbf{P}_{d-1}(d, q)$ and $\mathbf{H}(11)$.

35. *Let* $\mathbf{D}$ *be a projective design with a doubly transitive automorphism group* Γ.

(**a**) Γ *is transitive on ordered triples of noncollinear points if and only if* Γ *is transitive on non-incident point-line pairs. If this is the case, then* $\mathbf{D} \cong \mathbf{P}_{d-1}(d, q)$ *for some* d, q.[4]
(**b**) *If* Γ *contains a nontrivial central or axial automorphism, then* $\mathbf{D} \cong \mathbf{P}_{d-1}(d, q)$ *and* Γ *contains the little projective group* $PSL_{d+1}(d, q)$.
(**c**) *If* Γ *contains no nontrivial central or axial automorphism, and if* v, k *are primes, then* $\mathbf{D} \cong \mathbf{H}(11)$ *and* $\Gamma \cong PSL_2(11)$.
(**d**) *If* k *or* $n = k - \lambda$ *is a prime, then* $\mathbf{D} \cong \mathbf{P}_{d-1}(d, q)$ *or* $\mathbf{D} \cong \mathbf{H}(11)$.
(**e**) *If* $\mathbf{D}$ *has the same parameters as* $\mathbf{P}_{d-1}(d, q)$ *and if either* q *is prime or* Γ_B *doubly transitive on* (B), *for some block* B, *then* $\mathbf{D} \cong \mathbf{P}_{d-1}(d, q)$.

---

[1] We shall see in the next section that $\mathbf{H}(11)$ is a member of an infinite class of projective designs $\mathbf{H}(q)$, $q \equiv 3 \bmod 4$.

[2] Still another example has been found recently by G. Higman 1969: here $v = b = 176$, $k = r = 50$, $\lambda = 14$, Γ is simple of order $44\,352\,000 = 2^9 \cdot 3^2 \cdot 5^3 \cdot 7 \cdot 11$, and probably isomorphic to the group of D. G. Higman & C. W. Sims 1967 (it has not only the same order but also the same character table). In any case, the Higman-Sims group does have a doubly transitive permutation representation of degree 176, as was communicated to the author by C. Hering.

[3] This problem is intimately related to that of determining all doubly transitive permutation groups G of degree v which possess an intransitive subgroup H of index v which is not a stabilizer of a point. Compare here result **2.4.5** and its context below.

[4] This part (**a**) is true without the hypothesis of double transitivity for Γ. But double transitivity is clearly implied by the conditions of (**a**).

(**f**) *If $v \equiv 1 \bmod k$, then every normal subgroup of Γ is also doubly transitive; hence Γ contains a non-abelian simple doubly transitive normal subgroup.*

In case $\lambda = 1$, double transitivity alone, without any of the restrictions above, implies $\mathbf{D} = \mathbf{P}_1(2, q)$ and $\Gamma \geqq PSL_3(q)$; cf. **4.4.20** below. The proofs for $\lambda > 1$ are: DEMBOWSKI & WAGNER 1960 for (**a**), ITO 1967a for (**b**), ITO 1967b for (**c**), and KANTOR 1969b for (**d**)—(**f**).

2.4 Construction of designs

The only designs we have actually encountered so far are the $\mathbf{P}_t(d, q)$ and $\mathbf{A}_t(d, q)$, connected with the finite desarguesian geometries $\mathscr{P}(d, q)$ and $\mathscr{A}(d, q)$, as defined by **1.4.2** and **1.4.3**, and $\mathbf{H}(11)$. The parameters of $\mathbf{P}_t(d, q)$ and $\mathbf{A}_t(d, q)$ were given in (1.4.9) and (1.4.10). The main purpose of this section is to exhibit other classes of designs, and thus to show that the developments of Sections 2.1—2.3 cover an area much broader than that of finite affine and projective geometries. The construction problem for (nondesarguesian) affine and projective planes will be left aside here; for this see Chapter 5. Besides presenting construction techniques, we shall also note some results characterizing the designs so obtained.

We begin with a general method which will yield all possible designs with the following properties:

(1) *There exists an automorphism group transitive on blocks and unordered point pairs.*

(2) *If $(B) = (C)$, then $B = C$, for all blocks B, C.*

We are using here the notation (1.1.5) again. It is clear that a nondegenerate tactical configuration satisfying condition (1) must be a design.

A permutation group Γ on a set $\mathfrak{p}$ is said to be *s-homogeneous* if it permutes the subsets $\mathfrak{x} \subseteq \mathfrak{p}$ with $|\mathfrak{x}| = s$ transitively. Clearly, s-transitivity implies s-homogeneity. The converse is not true in general, but:

1. *If Γ is s-homogeneous on $\mathfrak{p}$, where $4 \leqq 2s \leqq |\mathfrak{p}|$, then Γ is $(s-1)$-transitive. If, moreover, $s > 4$, then Γ is even s-transitive.*

This is the main result of LIVINGSTONE & WAGNER 1965; their proof was substantially simplified by WIELANDT 1967. A weaker result in the same direction was proved by HUGHES 1964a. The following may be deduced from **1**:

2. *If Γ is s-homogeneous for all $s \leqq |\mathfrak{p}|$, then Γ contains the alternating group of $\mathfrak{p}$, except in the following cases:*

(**a**) $|\mathfrak{p}| = 5$ *and* $\Gamma = A_1(5)$

(**b**) $|\mathfrak{p}| = 6$ *and* $\Gamma = PGL_2(5)$

(**c**) $|\mathfrak{p}| = 9$ *and* $\Gamma = PGL_2(8)$ *or* $P\Gamma L_2(8)$.

(BEAUMONT & PETERSON 1955, LIVINGSTONE & WAGNER 1965.) Here $A_1(q)$ is the group of all permutations

$$(3) \qquad x \to xa + c, \qquad a \neq 0$$

of $GF(q)$; $P\Gamma L_2(q)$ consists of the permutations

$$(4) \qquad x \to (x^\alpha a + c)(x^\alpha b + d)^{-1}, \qquad ad \neq bc, \qquad \alpha \in \operatorname{Aut} GF(q)$$

of $GF(q) \cup \{\infty\}$, and $PGL_2(q)$ is the normal subgroup of $P\Gamma L_2(q)$ consisting of those mappings (4) for which $\alpha = 1$.

Suppose now that Γ is 2-homogeneous on a set $\mathfrak{p}$ with $|\mathfrak{p}| = v$. Select a subset B of $\mathfrak{p}$ and define an incidence structure $\mathbf{D}(\mathfrak{p}, \Gamma, B)$ as $(\mathfrak{p}, \mathfrak{B}, \mathrm{I})$, with $\mathfrak{B} = \{B\gamma : \gamma \in \Gamma\}$, and $p \,\mathrm{I}\, B\gamma$ if and only if $p \in B\gamma$, for any $p \in \mathfrak{p}$ and $\gamma \in \Gamma$. It is clear that $\mathbf{D}(\mathfrak{p}, \Gamma, B)$ is a design if, and only if,

$$(5) \qquad 1 < |B| < v - 1;$$

in this case the parameters are

$$(6) \qquad v = |\mathfrak{p}|, \; k = |B|, \; b = |\Gamma : \Gamma_B|,$$

and r, λ determined by (6) and (2.1.5). Here Γ_B denotes the subgroup leaving B invariant as a whole, not pointwise. The following is easily verified:

3. *A design may be represented in the form* $\mathbf{D}(\mathfrak{p}, \Gamma, B)$ *if and only if it satisfies conditions* (1) *and* (2). *In fact,* Γ *is an automorphism group of* $\mathbf{D}(\mathfrak{p}, \Gamma, B)$ *which satisfies* (1).

Note that $\mathbf{D}(\mathfrak{p}, \Gamma, B)$ may be *trivial* in the sense that every k-set of points is a block or, equivalently, that $b = \binom{v}{k}$. Evidently this will be the case if and only if Γ is k-homogeneous on $\mathfrak{p}$. Result **2** shows that every 2-homogeneous permutation group of degree >9 which does not contain the alternating group gives rise to nontrivial designs; in all known cases the design is nontrivial whenever $k > 5$. Another useful observation is:

4. *If* Γ *is s-homogeneous, where* $s > 2$, *then* $\mathbf{D}(\mathfrak{p}, \Gamma, B)$ *is of type* $(1, t)$, *with* $t \geqq s$.

The type of a design was defined in Section 1.1; see in particular **1.1.5**.

The structure of $\mathbf{D}(\mathfrak{p}, \Gamma, B)$ depends to a considerable extent on the choice of the subset B: if B and B' have the same size k and have $k - 1$ points in common, the designs $\mathbf{D}(\mathfrak{p}, \Gamma, B)$ and $\mathbf{D}(\mathfrak{p}, \Gamma, B')$ may be very different. The case in which the number of blocks is as small as possible is often of interest; this amounts to choosing B so that Γ_B

is as large as possible. Fisher's inequality $v \leqq b$, applied to (6), together with **1.1.5** and **2.1.1**, gives the following result:

5. *Suppose that the 2-homogeneous permutation group Γ on $\mathfrak{p}$, where $|\mathfrak{p}| = v$, has an intransitive subgroup Δ of index $|\Gamma:\Delta| \leqq v$. Then $|\Gamma:\Delta| = v$; if Δ has an orbit B satisfying* (5), *then* $\mathbf{D}(\mathfrak{p}, \Gamma, B)$ *is a projective design, and $\Delta = \Gamma_B$. Conversely, every projective design satisfying* (1) *can be interpreted in this fashion.*

Moreover, if Γ is 3-homogeneous, then $\mathbf{D}(\mathfrak{p}, \Gamma, B)$ is not projective, by **4**. It follows that in this case $\Delta = \Gamma_p$ for some $p \in \mathfrak{p}$, and Γ is 2-transitive.[1]

The class of 2-homogeneous permutation groups, and hence that of designs of the form $\mathbf{D}(\mathfrak{p}, \Gamma, B)$, is very large. Note, for example, that $\mathbf{P}_t(d, q)$ and $\mathbf{A}_t(d, q)$ are special cases, cf. **3**. In what follows, we construct various classes of designs by considering special 2-homogeneous groups.

First, we comment briefly on the few known cases of s-homogeneity for $s > 3$. The groups involved are (i) the Mathieu groups M_i of degrees $i = 11, 12, 23, 24$, all 4-transitive, and (ii) five 4-homogeneous but not 4-transitive groups, namely $PSL_2(5)$, $PGL_2(5)$, $PSL_2(8)$, $P\Gamma L_2(8)$, and $P\Gamma L_2(32)$, of degrees 6, 6, 9, 9, 33, respectively (cf. Livingstone & Wagner 1965, p. 403; Kantor 1968a). There are, of course, only finitely many designs connected with these groups; some with small b are found in Witt 1938b; Hughes 1965a. A few other designs of types (1, 4) and (1, 5) are known; these are also in Hughes 1965a. The Mathieu group M_{22}, which is only 3-transitive, gives rise to some interesting designs of type (1, 3); the one with minimal block number among these is in fact an extension of the projective plane $\mathbf{P}(4) = \mathbf{P}_1(2, 4)$; see Witt 1938b, Satz 4. Compare here the remarks following **2.2.18**.

The existence problem for non-trivial designs of type $(1, t)$ with $t > 5$ is unsolved; it is closely connected with the old problem of whether there exists an absolute bound for the degree of transitivity of a non-alternating and nonsymmetric permutation group: If it could be shown that such designs exist only for $t \leqq T$, it would follow that all t-transitive permutation groups with $t > T$ contain the alternating group of the same degree. It is conceivable that $T = 5$; in this context see Wielandt 1960 and Nagao 1966 ($T \leqq 6$ provided Schreier's conjecture that finite simple groups have soluble outer automorphism groups is true). For further results concerning this problem, see Wielandt 1964, p. 21—22.

---

[1]) This is an elementary proof that 3-homogeneity implies 2-transitivity. Some of these results were originally proved by Ito 1960 under the stronger assumption of 2- and 3-transitivity instead of -homogeneity.

Next we present, as an application of **5**, an infinite class of projective designs with doubly transitive automorphism groups.[1]) Let $q = 2^e \geqq 4$ and consider a quadric $\mathfrak{q}$ in the desarguesian projective plane $\mathbf{P} = \mathbf{P}(q) = \mathbf{P}_1(2, q)$. By **1.4.53**, every point of $\mathfrak{q}$ is special, and it is not difficult to verify that the groups of elations with properties (1.4.29)—(1.4.31), for all points of $\mathfrak{q}$, generate a collineation group isomorphic to $PGL_2(q)$ of $\mathbf{P}$ which is, among other things, transitive on the $q(q-1)/2$ lines not meeting $\mathfrak{q}$. Since the common point of the tangents of $\mathfrak{q}$ is clearly fixed, we see, by considering the dual situation, that the affine plane $\mathbf{A} = \mathbf{A}(q) = \mathbf{A}_1(2, q)$ possesses a collineation group $\Delta \cong PGL_2(q)$ with a point orbit B of length $q(q-1)/2$. Let T be the group of all translations of $\mathbf{A}$, and consider $\Gamma = \Delta\mathsf{T}$. Then Γ_x, the stabilizer of a point x in $\mathbf{A}$, is again isomorphic to $PGL_2(q)$, and by a result of LÜNEBURG 1964c (see **4.2.13b** below), Γ is doubly transitive on the points of $\mathbf{A}$. Thus Γ, Δ, and B satisfy the hypotheses of **5**, and it follows that $\mathbf{D}(\mathfrak{p}, \Gamma, B)$, with $\mathfrak{p}$ the set of all points of $\mathbf{A}$, is a projective design. The parameters of this design are

$$v = b = q^2 = 4^e, \quad r = k = q(q-1)/2 = 2^{e-1}(2^e - 1),$$
$$\lambda = q(q-2)/4 = 2^{e-1}(2^{e-1} - 1),$$

and Γ acts as a doubly transitive automorphism group on $\mathbf{D}(\mathfrak{p}, \Gamma, B)$.

Let $\mathfrak{N}$ be a nearfield of order $q \equiv 3 \bmod 4$. Then $\mathfrak{N}$ is regular[2]), i.e. of the form $N(n, q')$, where $q = (q')^n$, and has a unique multiplicative subgroup $S = S(n, q')$ of order $(q-1)/2$. The permutations

$$(7) \qquad \varphi(s, a) : x \to x \cdot s + a$$

of $\mathfrak{N}$, with $s \in S$ and $a \in \mathfrak{N}$, form a Frobenius group $\Phi = \Phi(n, q')$ of odd order $q(q-1)/2$ which is easily seen to be sharply transitive on the unordered pairs of elements in $\mathfrak{N}$, i.e. *sharply 2-homogeneous*. Let G be a nontrivial subgroup of S, so that $1 < |G| \leqq (q-1)/2$. We shall consider designs of the form $\mathbf{D}(\mathfrak{N}, \Phi, G)$, the parameters of which satisfy

$$(8) \quad v = |N| = q, \quad k = |G|, \quad r = (q-1)/2 \equiv 0 \bmod k,$$
$$\lambda = (k-1)/2, \quad b = q(q-1)/2k.$$

---

[1]) These are due to W. M. Kantor. A class of designs with the same parameters appears in the book by MANN 1965b, Theorem 6.3 (cf. **28** below); whether or not Kantor's examples are among these has not yet been determined.

[2]) Regular nearfields were defined in Section 1.4; see in particular (1.4.14). There are only seven finite nearfields which are not regular; these will be exhibited in Section 5.2. Their orders are all odd squares, hence $\equiv 1 \bmod 4$. Some other results on finite nearfields will also be found in Section 5.2.

The sharply 2-homogeneous group Φ is also sharply flag transitive on $\mathbf{D}(N, \Phi, G)$, and this turns out to be a characteristic property of these designs:

6. *Suppose that the parameters of the design* $\mathbf{D}$ *satisfy*

$$r = (v-1)/2, \quad (v, k) = (r, \lambda) = 1, \quad \text{and} \quad v \equiv 3 \bmod 4. \tag{9}$$

Then $\mathbf{D}$ *is of the form* $\mathbf{D}(\mathfrak{N}, \Phi, G)$ *if and only if* $\mathbf{D}$ *has a sharply flag transitive automorphism group.*

Here $(r, \lambda) = 1$ and $v \equiv 3$ mod 4 can be replaced by the condition that the postulated automorphism group be also sharply 2-homogeneous on points. For the proof of **6** see KANTOR 1968b, Theorem 6.3; the main reason for the result is that every sharply 2-homogeneous permutation group of degree $\equiv 3$ mod 4 is of the form $\Phi = \Phi(n, q')$ defined above.[1]

We have seen in Section 1.4 that the regular nearfields are closely related to (and in fact generalizations of) the Galois fields of the same order: $GF(q) = N(1, q)$. In particular, (1.4.14) shows that the groups $\Phi = \Phi(n, q')$ can be considered as subgroups of the group $\Sigma(q)$ consisting of the permutations

$$\sigma(\alpha, s, a) : x \to x^{\alpha} s + a \tag{10}$$

of $GF(q)$, where s is a non-zero square, and $\alpha \in \operatorname{Aut} GF(q)$. Clearly, $\Phi(1, q)$ is the subgroup of the $\sigma(1, s, a)$. It is readily verified that in our case ($q \equiv 3$ mod 4) the group $\Sigma(q)$ is 2-homogeneous but not 2-transitive, and it can be proved[2] that, conversely, every 2-homogeneous but not 2-transitive permutation group may be interpreted as a subgroup of a suitable $\Sigma(q)$ with $q \equiv 3$ mod 4. In the case where $\mathfrak{N} = GF(q)$, $q \equiv 3$ mod 4, and where G is the unique subgroup of order g of $S(1, q)$, we denote $\mathbf{D}(N, \Phi, G)$ more shortly by $\mathbf{D}(q, g)$.

7. *If q is a prime power $p^e \equiv 3$ mod 4, then the group $\Sigma(q)$ of the permutations* (10) *is maximal among the automorphism groups of odd order of* $\mathbf{D}(q, g)$, *for every divisor $g \neq 1$ of $(q-1)/2$.*

Proof: That the $\sigma(\alpha, s, a)$ are automorphisms of $\mathbf{D}(q, g)$ is clear because α leaves every multiplicative subgroup of $GF(q)$ invariant.

---

[1]) Such a group Γ is primitive and of odd order, hence soluble (FEIT & THOMPSON 1963); thus there exists a sharply transitive elementary abelian subgroup Δ, see **1.2.20**. If e is one of the permuted objects, define a new permutation α by $(e^{\delta})^{\alpha} = e^{\delta^{-1}}$ for all $\delta \in \Delta$. The group $\langle \Gamma, \alpha \rangle$ is then sharply 2-transitive, hence similar to the group of all permutations $x \to x \cdot a + c$ of a nearfield $\mathfrak{N}$ (ZASSENHAUS 1935a), with α corresponding to $x \to -x$. Also, $\mathfrak{N}$ is regular because $v \equiv 3$ mod 4 [cf. footnote[2]) of p. 95]. This argument is essentially that of LÜNEBURG 1963, Hilfssatz 4.

[2]) As in footnote[1]). See also LIVINGSTONE & WAGNER 1965, p. 402.

Also e is odd (otherwise $q \not\equiv 3 \bmod 4$), whence $|\Sigma(q)| = e\,q\,(q-1)/2$ is likewise odd. Conversely, if $|\Gamma|$ is odd and $\Sigma(q) \subseteqq \Gamma$, then Γ is 2-homogeneous but not 2-transitive, hence[1] $\Gamma \subseteqq \Sigma(q)$.

In some cases, the designs $\mathbf{D}(q, g)$ are just as general as the $\mathbf{D}(\mathfrak{N}, \Phi, G)$:

8. *Let $q = (q')^n \equiv 3 \bmod 4$, and let G be a multiplicative subgroup of $\mathfrak{N} = N(n, q')$, of order $|G| = g$. If*

(a) $$g = (q-1)/2$$

or

(b) $$g = 3,$$

then $\mathbf{D}(N, \Phi, G) \cong \mathbf{D}(q, g)$.

In case **(a)**, this follows from the fact, mentioned in the context of (7) above, that there is only one subgroup of order $(q-1)/2$ in $N(n, q')^\times$. For case **(b)**, see LÜNEBURG 1963, where the representation $\mathbf{K}(\Gamma, \Pi, B)$ of **1.2.17** is used. There seems to be no known example with $|N| = q$, $|G| = g$ and $\mathbf{D}(\mathfrak{N}, \Phi, G) \not\cong \mathbf{D}(q, g)$; whether or not some restriction on g is necessary is an open question.

Cases **(a)** and **(b)** of **8** are of special interest because then $\mathbf{D}(q, g)$ is an Hadamard design[2] or a Steiner triple system[3], respectively. These two classes of designs, which have received considerable attention, are defined as follows: An *Hadamard design* is one that satisfies

(11) $$v = b = 4\lambda + 3 \quad \text{or, equivalently,} \quad r = k = 2\lambda + 1;$$

in particular, Hadamard designs are projective. It is easily proved from (2.1.5) that a projective design is an Hadamard design if and only if $(v-1, k-1) = 2$.

A *Steiner triple system* is a design with

(12) $$k = 3 \quad \text{and} \quad \lambda = 1.$$

We shall discuss Hadamard designs and Steiner triple systems in general later on; at present we are only interested in the $\mathbf{D}(q, g)$ among them. It follows from (8) and **8a** that the $\mathbf{D}(q, (q-1)/2)$ are precisely the Hadamard (and incidentally also the only projective) designs among the $\mathbf{D}(\mathfrak{N}, \Phi, G)$. These special Hadamard designs were first exhibited by PALEY 1933 and TODD 1933[4]; we shall denote them by $\mathbf{H}(q)$ here, and sometimes refer to them as *Paley designs*. Next, (8) and **8b** show

---

[1]) As in footnote [1]) of p. 96.

[2]) Because of a determinant problem of HADAMARD 1893 which, as will be seen soon, is closely related to the construction of these designs (cf. **31** below).

[3]) STEINER 1853 posed a sequence of problems which, in our present terminology, amount to that of constructing designs of type (s, t) with $k = t + 1$ and $b_t = 1$. The present case is $k = 3$; for $k = 4$ see **21** below.

[4]) However, according to A. BRAUER 1953, they were known to I. Schur at least in 1920.

that the Steiner triple systems among the $\mathbf{D}(\mathfrak{N}, \Phi, G)$ are the $\mathbf{D}(q, 3)$, where $q \equiv 7 \bmod 12$. These seem to be due to NETTO 1893 (see also CARMICHAEL 1937, p. 436); we denote them by $\mathbf{T}(q)$ and call them sometimes *Netto systems*.

Result **6** above yields the following characterizations of the $\mathbf{H}(q)$ and $\mathbf{T}(q)$ among Hadamard designs and Steiner triple systems:

9. *An Hadamard design is of the form* $\mathbf{H}(q)$ *if and only if it admits a sharply flag transitive automorphism group.*

10. *A Steiner triple system is of the form* $\mathbf{T}(q)$ *if and only if it admits a sharply flag transitive automorphism group and satisfies* $v \not\equiv 0 \bmod 3$.

The projective designs $\mathbf{P}_{d-1}(d, 2)$ also satisfy (11) and are, therefore, Hadamard designs. On the other hand, $\mathbf{P}_1(d, 2)$ and $\mathbf{A}_1(d, 3)$ are Steiner triple systems for all $d \geqq 2$. In the following discussion we shall give several intrinsic characterizations of these designs and the $\mathbf{H}(q)$.

11. *The following properties of the Hadamard design* $\mathbf{D}$ *are equivalent:*

(**a**) $\mathbf{D} \cong \mathbf{P}_{d-1}(d, 2)$ *for some* $d \geqq 2$.
(**b**) *Every point is the center of a non-trivial automorphism.*
(**c**) *Every block is the axis of a non-trivial automorphism.*

This follows readily from **2.1.22c** and the observation that $(b - \lambda)(r - \lambda)^{-1} = 3$ for Hadamard designs. The next result shows that the $\mathbf{H}(q)$ are essentially different from the $\mathbf{P}_{d-1}(d, 2)$.

12. $\mathbf{H}(q) \cong \mathbf{P}_{d-1}(d, 2)$ *if and only if* $q = 7$ *and* $d = 2$.

That $\mathbf{H}(7) \cong \mathbf{P}_1(2, 2) = \mathbf{P}(2)$ is clear. Conversely, **9** and **1.4.21** show that this situation is unique. **12** yields non-isomorphic Hadamard designs with the same parameters for all Mersenne primes $q = 2^{d+1} - 1 > 7$.

Result **12** shows that $\operatorname{Aut}\mathbf{H}(7)$ is the simple group $P\Gamma L_3(2)$ of order 168, and we have already mentioned (p. 91) that $\operatorname{Aut}\mathbf{H}(11)$ is the simple group $PSL_2(11)$ of order 660 (TODD 1933). For higher q, however, $\operatorname{Aut}\mathbf{H}(q)$ is less interesting:

13. $\operatorname{Aut}\mathbf{H}(q) = \Sigma(q)$ [see (10) above] for $q \geqq 19$.

For the proof of this surprising result, see KANTOR 1969b. His proof uses Theorem 25.4 of WIELANDT 1964 (a primitive permutation group of composite degree is doubly transitive provided it contains a sharply transitive abelian subgroup with a cyclic Sylow subgroup $\neq 1$) and the obvious fact that the $\varphi(s, 0)$ of (7) form a group which is sharply transitive on the points of the block G; cf. **8**.

14. *The following properties of the projective design* **D** *are equivalent:*

(**a**) $\mathbf{D} \cong \mathbf{H}(q)$, *for some prime power* $q \equiv 3 \mod 4$.
(**b**) **D** *is an Hadamard design and has a flag transitive automorphism group which is not doubly transitive.*
(**c**) **D** *satisfies* $v > 2k$ *and admits a 2-homogeneous automorphism group of odd order.*

The basic idea underlying the proof is the same as that of **6**; see KANTOR 1968b. Theorem 1.1. If the condition $v > 2k$ is removed from (**c**), the only alternative to (**a**) is that **D** is isomorphic to the complementary design $\mathbf{H}(q)'$; this is Corollary 5.1 of KANTOR 1968b.

We turn now to characterizations of the three classes of Steiner triple systems $\mathbf{T}(q)$, $\mathbf{P}_1(d, 2)$, and $\mathbf{A}_1(d, 3)$. We have the following analogue to **12**:

15. *The only instance in which* $\mathbf{T}(q)$ *is isomorphic to* $\mathbf{P}_1(d, 2)$ *or* $\mathbf{A}_1(d, 3)$ *is* $\mathbf{T}(7) = \mathbf{H}(7) = \mathbf{P}(2)$.

This follows from **10** and **1.4.21**. Again the case where q is a Mersenne prime $2^{d+1} - 1 > 7$ yields nonisomorphic Steiner triple systems with the same parameters.

16. *Let* **D** *be a design with* k *a prime dividing* v, *and with* $r = (v-1)/2$ *odd and relatively prime to* λ. *Then* **D** *admits a sharply flag transitive automorphism group if and only if* $\mathbf{D} = \mathbf{A}_1(d, 3)$ *for some odd* $d > 1$.

This is an improvement of a result of LÜNEBURG 1963 (Satz 2, 1. Fall); see KANTOR 1968b, Theorem 6.4. Results **10** and **16** together give a complete classification of Steiner triple systems with sharply flag transitive groups: such a system **S** exists if and only if v is a prime power such that $v(v-1) \equiv 6 \mod 12$, and then $\mathbf{S} \cong \mathbf{T}(v)$ if $v \equiv 1 \mod 3$ and $\mathbf{S} \cong \mathbf{A}_1(d, 3)$ with odd d if $v \equiv 0 \mod 3$ (LÜNEBURG 1963).

17. *Suppose that the Steiner triple system* **S** *admits an automorphism group* Γ *of even order, with a subgroup of order* v, *such that the identity is the only automorphism in* Γ *with more than one fixed point. Then* $\mathbf{S} \cong \mathbf{A}_1(d, 3)$, *for some* $d \geqq 2$.

Proof: LÜNEBURG 1965b, Satz 1. In particular, $\mathbf{S} \cong \mathbf{A}_1(d, 3)$ if Γ is sharply 2-transitive on points. Furthermore, Lüneburg shows that the subgroup of order v mentioned in **17** is the group of all translations.

The only known Steiner triple systems with a 2-transitive group are the $\mathbf{P}_1(d, 2)$ and the $\mathbf{A}_1(d, 3)$, and there is reason to conjecture that there are no others. We present some results supporting this conjecture. A *subsystem* of a Steiner triple system **S** is a set of points and blocks in **S** which form a Steiner triple system themselves. **S** is called *minimal* if there are no proper subsystems in **S**. Clearly, **P**(2) and **A**(3) are minimal Steiner triple systems.

18. *Let* **S** *be a Steiner triple system.*

(**a**) *Every point of* **S** *is the only fixed point of some involutorial automorphism of* **S** *if, and only if, every minimal subsystem of* **S** *is isomorphic to* **A**(3).
(**b**) *If every block B of* **S** *is the axis of an involutorial automorphism fixing no point not on B, then every minimal subsystem is isomorphic to* **P**(2) *or* **A**(3).

These are Theorems 3.1 and 3.2 of HALL 1960a. Hall also points out that the converse of (**b**) is false: The minimal subsystems of $\mathbf{P}_1(d, 2)$ are all **P**(2), but for $d > 3$ every involutorial automorphism of $\mathbf{P}_1(d, 2)$ has at least seven fixed points, for the fixed points of an involution of $\mathscr{P}(d, 2)$ form a subspace of dimension $\geqq [(d+1)\, 2^{-1}] - 1$.

19. *Suppose that the Steiner triple system* **S** *admits a doubly point transitive automorphism group* Γ. *Then:*

(**a**) **S** *contains a minimal subsystem isomorphic to* **P**(2) *or* **A**(3).
(**b**) *If* Γ *is transitive on ordered triangles (triples of points not on one block), then* $\mathbf{S} \cong \mathbf{P}_1(d, 2)$ *or* $\mathbf{A}_1(d, 3)$, *for some* $d \geqq 2$.

These results are also due, essentially, to HALL 1960a, Section 4. In the case that $\mathbf{A}(3) \subseteqq \mathbf{S}$, for (**b**), Hall assumes a condition stronger than triangle transitivity. It can be shown, however, by an argument due to R. H. Bruck, that the present hypothesis is sufficient; cf. HALL 1967a, pp. 134—136. This approach involves an algebraization of Steiner triple systems by certain loops, which is of interest apart from **19**. We give a brief outline.[1]

In an arbitrary Steiner triple system **S**, choose a point e and define the *interior loop* $I = I_e(\mathbf{S})$ as follows: The elements of I are the points of **S**, and multiplication in **S** is defined by the rules $ex = xe = x$ and

$$(13)\quad xy = \begin{cases} \textit{third point on the block through } e \textit{ and } x, & \textit{if } x = y,\\ \textit{third point on the block through } xx \textit{ and } yy, & \textit{if } x \neq y, \end{cases}$$

for $x, y \neq e$. It is easily seen that I is a commutative, and therefore power associative, loop of exponent 3. Moreover, $\mathbf{S} \cong \mathbf{A}_1(d, 3)$ if and only

[1]) Further results and ideas along these lines can be found in BRUCK 1963c, BELOUSOV 1960, FISCHER 1964.

if $I_e(\mathbf{S})$ is the elementary abelian group of order 3^d, for every point e.[1]) There is a similar construction characterizing the $\mathbf{P}_1(d, 2)$: The *exterior loop* $E = E(\mathbf{S})$ of the Steiner triple system $\mathbf{S}$ consists of the points of $\mathbf{S}$ and a new symbol 0, and addition is defined by the rules $x + 0 = 0 + x = x$ and

$$(14) \quad x + y = \begin{cases} 0 & \textit{if } x = y \\ \textit{third point on the block through } x \textit{ and } y, & \textit{if } x \neq y. \end{cases}$$

The condition of triangle transitivity in **19b** implies that the group of all automorphisms of S with a given block B as axis is transitive on the points not on B. This is the special case $k = 3$ of a transitivity property considered by JORDAN 1871 who proved that if a permutation group Γ is primitive on a finite set $\mathfrak{p}$ and contains a subgroup Δ fixing $k \geqq 1$ points and transitive on the remaining points, then Γ is doubly transitive (cf. WIELANDT 1964, Theorem 13.1). HALL 1960a calls such a permutation group Γ a *Jordan group* if it is not triply transitive. Let B denote the set of points fixed by Δ; then the design $\mathbf{D}(\mathfrak{p}, \Gamma, B)$ admits, for each block X, a group of automorphisms with axis X which is contained in Γ and transitive on the points not on X. Clearly, the Steiner triple systems $\mathbf{P}_1(d, 2)$ and $\mathbf{A}_1(d, 3)$ are of this form. For all known Jordan groups, the design $\mathbf{D}(\mathfrak{p}, \Gamma, B)$ is of the form $\mathbf{P}_t(d, q)$ or $\mathbf{A}_t(d, q)$; result **19b** shows that there are no others if $k = 3$. It seems plausible to presume that analogous results hold for other k as well, and it is even conceivable that no restrictions on k are necessary at all.[2])

We shall return later to more results concerning Hadamard designs, Steiner triple systems, and other special designs. Now we discuss a method of constructing designs from $\mathbf{D}(\mathfrak{p}, \Gamma, B)$ whenever Γ has a *transitive extension*. This means that $\Gamma = \Delta_\infty$, where Δ is a transitive permutation group of $\mathfrak{p} \cup \{\infty\}$, with ∞ a new symbol.[3]) Suppose that Γ has such a transitive extension $\Delta = \Gamma^*$. As we have assumed Γ to be 2-homogeneous, it follows that Γ^* must be 3-homogeneous

---

[1]) The proof of the affine case in **19b** can now be finished as follows: First it is shown, using the fact that every triangle of $\mathbf{S}$ generates a subsystem isomorphic to $\mathbf{A}(3)$, that each $I_e(\mathbf{S})$ is a commutative Moufang loop, i.e. $(x\,y)\,(z\,x) = [x\,(y\,z)]\,x$ for all x, y, z. Since I is finitely generated, it follows from BRUCK 1958, Theorem 10.1 on p. 157, that I is centrally nilpotent, and hence not simple. On the other hand, the 2-transitivity on points of Γ implies that each $I_e(\mathbf{S})$ has an automorphism group transitive on the elements $\neq e$. A theorem of BRUCK 1951b then guarantees that I is an elementary abelian 3-group. (Note that triangle transitivity is more than is needed here.) See also HALL 1967a, pp. 133—136.

[2]) Result **2.3.35b** of ITO 1967 shows that if $\mathbf{D} = \mathbf{D}(\mathfrak{p}, \Gamma, B)$ is projective, then $\mathbf{D} \cong \mathbf{P}_{d-1}(d, q)$. For other results on Jordan groups, also for higher degrees of transitivity, see KANTOR 1969a.

[3]) For the case that Γ is doubly transitive, a theory of transitive extensions appears in WITT 1938a.

on $\mathfrak{p}^* = \mathfrak{p} \cup \{\infty\}$. Put $B^* = B \cup \{\infty\}$ and consider the design $\mathbf{D}^* = \mathbf{D}(\mathfrak{p}^*, \Gamma^*, B^*)$ of type $(1, t)$, with $t \geqq 3$. The parameters are

$$(15) \qquad \begin{aligned} v^* &= v + 1, \quad k^* = k + 1, \quad b^* = b(v+1)\,j^{-1}, \\ r^* &= b(k+1)\,j^{-1}, \qquad \lambda^* = r(k+1)\,j^{-1}, \end{aligned}$$

and in fact $b_i = b_{i-1}(k+1)\,j^{-1}$ for $i \leqq t$; here we have written

$$|\Gamma^*_{B^*} : \Gamma_B| = j.$$

For the proof of (15) see HUGHES 1962.

If p is any point of $\mathbf{D}^*$, then the internal structure $\mathbf{D}^*_p$ has the same parameters v and k as the original design $\mathbf{D} = \mathbf{D}(\mathfrak{p}, \Gamma, B)$ [cf. (2.2.14)], but in general $b \leqq b^*_{(p)}$ because $j \leqq k + 1$. In fact, we have:

20. *The following conditions on* $\mathbf{D} = \mathbf{D}(\mathfrak{p}, \Gamma, B)$ *and* $\mathbf{D}^* = \mathbf{D}(\mathfrak{p}^*, \Gamma^*, B^*)$, *where* Γ^* *is a transitive extension of* Γ, *are equivalent:*

(**a**) $\mathbf{D}^*_p \cong \mathbf{D}$, *for any point* p *of* $\mathbf{D}^*$.
(**b**) $j = k^* = k + 1$.
(**c**) Γ^* *is flag transitive on* $\mathbf{D}^*$.

This is contained in HUGHES 1965a; for the proof see also DEMBOWSKI 1965e, Lemma 2. Hence if Γ^* is flag transitive, then $\mathbf{D}$ is extendable in the sense of Section 2.2, whence $b(v+1) \equiv 0 \bmod k+1$, by **2.2.16**. On the other hand, flag transitivity of Γ^* can often be deduced from that of Γ: if Γ is flag transitive, then $j = 1$ or $j = k + 1$, and the first alternative may be numerically impossible. For further details see HUGHES 1965a; this paper lays the combinatorial foundations for a proof that most of the finite classical groups, considered as permutation groups of suitable point sets in $\mathcal{P}(d, q)$, do not possess transitive extensions.[1])

The groups $PGL_2(q)$, of the permutations (4) with $\alpha = 1$, are transitive extensions of the $A_1(q)$ described by (3). For $q \equiv 3 \bmod 4$, the subgroup $PSL_2(q)$ of $PGL_2(q)$, consisting of all permutations (4) with $\alpha = 1$ for which $ad - bc$ is a square in $GF(q)$, is a transitive extension of the group $\Phi(1, q)$ given by (7). Thus we get new designs $\mathbf{D}^*$, in the fashion described above, from the designs $\mathbf{D}(\mathfrak{p}, \Gamma, B)$ with $\mathfrak{p} = GF(q)$ and Γ either $A_1(q)$ or $\Phi(1, q)$. We discuss some special cases.

---

[1]) That $PSL_n(q)$ has no transitive extension except when $q = 2$ or $(d, q) = (2, 4)$ was first proved by ZASSENHAUS 1935c; the result is, of course, closely related to **2.2.19**. Theorems on the non-extendability of finite affine, unimodular, symplectic and unitary groups appear in HUGHES 1965b and DEMBOWSKI 1965e; similar ideas were used earlier by WITT 1938a, b to show that the 5-transitive Mathieu groups M_{12} and M_{24} do not have transitive extensions.

Let $\Gamma = PGL_2(q^d)$, $\mathfrak{p} = GF(q^d) \cup \{\infty\}$, and $B = GF(q) \cup \{\infty\}$. Then $\mathbf{D}(\mathfrak{p}, \Gamma, B)$ is a design of type $(1, 3)$, which we will denote by $\mathbf{E}(d, q)$. The parameters are

$$(16) \qquad \begin{aligned} &v = q^d + 1, \qquad k = q + 1, \qquad b = q^{d-1} \sum_{i=0}^{d-1} q^{2i}, \\ &r = q^{d-1} \sum_{i=0}^{d-1} q^i, \qquad \lambda = \sum_{i=0}^{d-1} q^i, \qquad b_3 = 1. \end{aligned}$$

Also, Γ is triply transitive on points and flag transitive on $\mathbf{E}(d, q)$, and since Γ is a transitive extension of $A_1(q^d)$, it follows that

$$(17) \qquad \mathbf{E}(d, q)_p \cong \mathbf{A}_1(d, q),$$

for any point p of $\mathbf{E}(d, q)$.

The special case $q = 3$ yields a class of *Steiner quadruple systems*,[1] which are defined to be the designs of type $(1, 3)$ satisfying the condition

$$(18) \qquad k = 4 \quad \text{and} \quad b_3 = 1,$$

i.e. every block contains four points, and there is a unique block through any three points. For any point p of a Steiner quadruple system $\mathbf{Q}$, the internal design $\mathbf{Q}_p$ is a Steiner triple system, and in our present situation it follows from (17) that $\mathbf{E}(d, 3)_p \cong \mathbf{A}_1(d, 3)$.

Any Netto system $\mathbf{T}(q)$, for $q \equiv 7 \bmod 12$, can also be extended to a Steiner quadruple system $\mathbf{Q}$; this follows from the fact, mentioned above, that the groups $\Phi(1, q)$ of the permutations (7) of $GF(q) = N(1, q)$ have $PSL_2(q)$ as transitive extension. The following is similar to the earlier characterization of Steiner triple systems with sharply flag transitive groups (see **10** and **17** above):

21. *The Steiner quadruple systems $\mathbf{Q}$ which admit a flag transitive automorphism group Γ such that every $\gamma \neq 1$ in Γ has at most two fixed points are, with precisely one exception, those for which $\mathbf{Q}_p \cong \mathbf{A}_1(d, 3)$ or $\mathbf{Q}_p \cong \mathbf{T}(q)$, for an arbitrary point p of $\mathbf{Q}$. The exception is $\mathbf{Q} \cong \mathbf{A}_2(5, 2)$; here $\mathbf{Q}_p \cong \mathbf{P}_1(4, 2)$.*

Proof: Lüneburg 1965b, Satz 2. For the exceptional case, compare also **1.4.22**.

Another important subclass of the $\mathbf{E}(d, q)$ occurs when $d = 2$. A design of type $(1, 3)$ with parameters of the form

$$(19) \qquad \begin{aligned} &v = n^2 + 1, \qquad k = n + 1, \qquad b = n(n^2 + 1), \\ &r = n(n + 1), \qquad \lambda = n + 1, \qquad b_3 = 1 \end{aligned}$$

[1] These are solutions of the Steiner 1853 problem [see footnote [3] of p. 97] for $k = 4$.

is called a (finite) *inversive plane*. These will be discussed in detail in Chapter 6; in particular we shall give a more geometrical definition in Section 6.1. At this point, we note that $\mathbf{E}(2, q)$ has parameters (19), with $n = q$ a prime power, and is therefore an inversive plane. These inversive planes $\mathbf{E}(2, q)$ will be denoted by $\mathbf{M}(q)$ later on.[1])

We use this occasion to define the only other known class of finite inversive planes. Let $q = 2^{2m+1}$, $m > 0$, and $\Gamma = Sz(q)$, the group of SUZUKI (1960; 1962, Section 13) of order $q^2(q^2+1)(q-1)$. As was pointed out in Section 1.4, this group Γ has a doubly transitive representation on the q^2+1 points of the ovoid $\mathfrak{p} = \mathfrak{t}(\psi)$ of $\mathcal{P}(3, q)$ defined by (1.4.34). Consider a plane $\mathfrak{x}$ in $\mathcal{P}(3, q)$ which has more than one point in common with $\mathfrak{p}$, and let $B = \mathfrak{p} \cap \mathfrak{x}$. Then $\mathbf{D}(\mathfrak{p}, \Gamma, B)$ is a design $\mathbf{S}(q)$, and it turns out that $\mathbf{S}(q)$ is an inversive plane,[2]) again with $n = q$. Clearly, $\mathbf{M}(q)$ and $\mathbf{S}(q)$ have the same parameters; also

$$\mathbf{M}(q)_x = \mathbf{S}(q)_y = \mathbf{A}_1(2, q)$$

for arbitrary points $x \in \mathbf{M}(q)$ and $y \in \mathbf{S}(q)$, but $\mathbf{M}(q)$ and $\mathbf{S}(q)$ are not isomorphic. Compare here the comments following **2.2.20**.

Next, we mention two series of designs with parameters of the form

$$(20) \qquad v = s^3 + 1, \quad k = s + 1, \quad b = s^2(s^2 - s + 1), \quad r = s^2, \quad \lambda = 1.$$

As was mentioned in Section 1.4, the absolute points and non-absolute lines of a unitary polarity π in $\mathcal{P}(2, s^2)$ form such a design, which we call $\mathbf{U}(s)$; for this reason designs with parameters (20) are called *unitals*. In our present context, $\mathbf{U}(s)$ can be written as $\mathbf{D}(\mathfrak{p}, \Gamma, B)$, with $\mathfrak{p}$ the set of all π-absolute points, B the set of π-absolute points on some non-absolute line, and $\Gamma = PSU_3(s^2)$. Of course we need here (cf. **1.4.60**) that Γ is doubly transitive on $\mathfrak{p}$. Note also that Γ is generated by elations of $\mathcal{P}(2, s^2)$ whose center-axis flags are corresponding pairs of π-absolute points and lines. These elations induce central dilatations in $\mathbf{U}(s)$; it is evident that they do not form a group. (Compare here the remarks after **2.3.22**.)

There is one other known class of unitals which is intimately connected with the groups $R_1(s)$ of REE 1961b of order $s^3(s^3+1)(s-1)$; here $s = 3^{2m+1}$, $m > 0$. Let P denote the normalizer of a Sylow 3-subgroup of $R_1(s)$; then P is maximal in $R_1(s)$, of index $s^3 + 1$. Hence

[1]) The $\mathbf{M}(q)$ are the only finite inversive planes in which the "Theorem of Miquel" holds, cf. Section 6.1. The $\mathbf{M}(q)$ will therefore be called *miquelian* in Chapter 6. We remark here also that an arbitrary design with parameters (19) need not be of type (1, 3) and hence not an inversive plane. Examples of such designs, with $n = 2^e$, were given by LÜNEBURG 1965c, Section 1.

[2]) This definition of $\mathbf{S}(q)$ can be restated more geometrically as follows: $\mathbf{S}(q)$ is the system of points and plane sections of $\mathfrak{t}(\psi)$. A similar definition can be given for the $\mathbf{M}(q)$ also; see Section 6.1. A more algebraic definition of $\mathbf{S}(q)$ appears in HUGHES 1962.

the permutation representation Γ of $R_1(s)$ on the set $\mathfrak{p}$ of right cosets of P is primitive of degree s^3+1. The representation of $R_1(s)$ by TITS (1960, p. 12) shows that Γ is even doubly transitive, that the stabilizer Γ_{xy}, of order $s-1$, is cyclic and hence contains exactly one involution, and that the total number of fixed points of this involution is $s+1$. Let B be the set of these $s+1$ fixed points; then $\mathbf{R}(s)=\mathbf{D}(\mathfrak{p},\Gamma,B)$ is the desired unital. These unitals $\mathbf{R}(s)$ were discovered by LÜNEBURG 1966b, who also proved that the $\mathbf{R}(s)$, unlike the $\mathbf{U}(s)$, cannot be embedded into a projective plane $\mathbf{P}$ of order s^2 (desarguesian or not) such that the automorphism group $R_1(s)$ is induced by a collineation group of $\mathbf{P}$.

As a final remark on unitals, we mention the fact that they constitute a particularly homogeneous class of finite *hyperbolic planes*, which are defined as incidence structures of points and lines such that (i) any two distinct points p,q determine a unique line $p+q$, (ii) to a non-incident point-line pair (p,L), there exists more than one line through p and not intersecting L, and (iii) if a point set $\mathfrak{S}$ contains a triangle and has the property that $p,q\in\mathfrak{S}$ implies that all points of $p+q$ are in $\mathfrak{S}$, then $\mathfrak{S}$ consists of all points. Other finite hyperbolic planes [not all designs, since (i)—(iii) do not imply that all lines carry equally many points] may be found in GRAVES 1962; OSTROM 1962b; SZAMKOŁOWICZ 1962, 1963; PUHAREV 1963; CROWE 1965, 1966.

This concludes our survey of designs satisfying conditions (1) and (2). We turn now to projective designs with a sharply transitive abelian automorphism group; these can be described by difference sets in abelian groups, see **2.3.25**. In what follows we present several classes of such difference sets.

Let q be a prime power and d an integer >1. Then $\mathfrak{F}=GF(q^{d+1})$ can be considered as a vector space V of rank $d+1$ over $\mathfrak{K}=GF(q)$, and the multiplicative group $\mathfrak{F}^\times$ of $\mathfrak{F}$ is essentially identical with the group of permutations

$$\mu(t): x\to xt \qquad 0\neq t\in\mathfrak{F}$$

of $\mathfrak{F}$. Every such $\mu(t)$ is a linear transformation of V; hence $\mathfrak{F}^\times$ induces a transitive collineation group Γ of $\mathscr{P}(d,q)$ [cf. **1.4.2**], and $\Gamma=\mathfrak{F}^\times/\mathfrak{K}^\times$ is cyclic of order $v=(q^{d+1}-1)/(q-1)$. This shows that every finite projective geometry $\mathscr{P}(d,q)$ has a point transitive cyclic collineation group (SINGER 1938, cf. **1.4.18**). Hence the projective design $\mathbf{P}_{d-1}(d,q)$ can be represented by a difference set in Γ; in fact such a difference set Δ consists of all $\gamma\in\Gamma$ mapping a given point of $\mathbf{P}_{d-1}(d,q)$ onto one of the points of a given hyperplane. More algebraically, Δ can be described as follows: Let L be a non-zero linear functional on V, i.e. a linear transformation with values in $\mathfrak{K}$ and hence with kernel a subspace of rank d in V. Also, let c be a primitive element (generator of the multiplicative group) of $\mathfrak{F}$. Then Γ may be identified with the additive group of residue classes mod v, and Δ consists of those residue classes R whose members r satisfy

$$(21) \qquad L(c^r) = 0.$$

Two difference sets D and D' in the additively written abelian group G are called *equivalent*[1]) if there exists an integer t with $(t, |G|) = 1$ and an element $g \in G$ such that $D' = \{t\,x + g, x \in D\}$. By a variation of the method leading to the description (21) of a difference set, the following can be shown:

22. *If* $d + 1 = nm$, *where* $n > 2$ *and* $m = \prod_{i=1}^{s} p_i$, *with* $p_1, \ldots, p_s$ *not necessarily distinct primes, then the cyclic group of order* $(q^{d+1} - 1)(q - 1)^{-1}$, *where* q *is a prime power* p^e, *contains at least* 2^s *mutually inequivalent difference sets, each consisting of* $(q^d - 1)(q - 1)^{-1}$ *elements, and each having precisely the powers of* p *as Hall multipliers.*

For the proof see GORDON, MILLS & WELCH 1962. The designs corresponding to these difference sets all have the same parameters as $\mathbf{P}_{d-1}(d, q)$, but some of them, for example in the minimal case $q = 2$, $n = 3$, $m = 2$, are not isomorphic to $\mathbf{P}_{d-1}(d, 2)$. In particular there exist Hadamard designs $\not\cong \mathbf{P}_{d-1}(d, 2)$ among them. A similar result will be **36** below.

Considering the definition of the Paley designs $\mathbf{H}(q)$ above, we see:

23. *For every Galois field* $\mathfrak{K} = GF(q)$ *with* $q \equiv 3 \bmod 4$, *the non-zero squares of* $\mathfrak{K}$ *form a difference set in the (elementary abelian) additive group of* $\mathfrak{K}$. *The corresponding projective design is* $\mathbf{H}(q)$.

In view of this it is natural to ask, more generally, under what circumstances the elements of a given multiplicative subgroup of $GF(q)$ form a difference set in the additive group. In general this problem is unsolved; it leads into difficult cyclotomic arithmetic. But some partial answers can be given. For an arbitrary Galois field $\mathfrak{K}$, let $\mathfrak{K}^n$ denote the multiplicative subgroup of the non-zero n^{th} powers in $\mathfrak{K}$. The additive group is denoted by $\mathfrak{K}^+$. Three results in the indicated direction are:

24. *Let* $\mathfrak{K} = GF(q)$. *In each of the following cases, the set* D *is a difference set in* $\mathfrak{K}^+$:

(a) $q = 4x^2 + 1 \equiv 5 \bmod 8$, $D = \mathfrak{K}^4$

(b) $q = 4x^2 + 9 \equiv 5 \bmod 8$, $D = \mathfrak{K}^4 \cup \{0\}$.

(c) $q = x^2 + 27 \equiv 7 \bmod 12$, $D = \mathfrak{K}^3 \cup 3\,\mathfrak{K}^6$.

[1]) This is a somewhat misleading terminology: equivalent difference sets define isomorphic designs, but the same design may be representable by inequivalent difference sets.

Here $3\mathfrak{K}^6$ denotes the multiplicative coset of $\mathfrak{K}^6$ which contains $3 \in \mathfrak{K}$. These results are due to LEHMER 1953[1]) and HALL 1956. The parameters of the corresponding designs are uniquely determined by q and $|D|$; see (2.1.5'). In cases (**a**), (**b**), (**c**), we have $|D| = (q-1)/4$, $(q+3)/4$, and $(q-1)/2$, respectively. In particular, (**c**) yields Hadamard designs with the same parameters as $\mathbf{H}(q)$, but in general not isomorphic to $\mathbf{H}(q)$.[2])

We turn now to difference sets with v not a prime power. The following construction is due to STANTON & SPROTT 1958.[3]) Let q and q' be two prime powers, satisfying

$$(22) \qquad q' = q + 2 \quad \text{and } q > 2.$$

Let $\mathfrak{K} = GF(q)$ and $\mathfrak{K}' = GF(q')$, and consider the ring $R = \mathfrak{K} \times \mathfrak{K}'$, with addition and multiplication defined as follows:

$$(x, x') + (y, y') = (x + y, x' + y'), \; (x, x')(y, y') = (xy, x'y').$$

Let D_0 be the subring of all $(x, 0)$ and D_1 the cyclic multiplicative group in R which is generated by (c, c'), with c, c' primitive elements in $\mathfrak{K}$, $\mathfrak{K}'$, respectively. Then $D = D_0 \cup D_1$ can be shown to be a difference set in the additive group of R, which is of odd order qq' and abelian, but not necessarily cyclic. Hence:

25. *If Γ is the direct sum of two elementary abelian groups of orders q, q' satisfying* (22), *then there exists a difference set in Γ such that the corresponding projective design is an Hadamard design with qq' points.*

For the proof see STANTON & SPROTT 1958. For q and q' primes, result **25** is a special case of a result of WHITEMAN 1962, Theorem 1. Suppose that p and p' are distinct primes with

$$(23) \quad (p-1, p'-1) = d \quad \text{and} \quad p' = (d-1)p + 2, \quad p > 2.$$

Whiteman's theorem gives a necessary and sufficient condition for a certain set D, constructed similarly to $D_0 \cup D_1$ above, to be a difference set in the cyclic additive group of order pp'. For $d = 2$, this gives **25** with primes q, q'; and for $d = 4$ it follows that

1) In this paper, conditions for $\mathfrak{K}^n$ to be a difference set in $\mathfrak{K}^+$ are given, for the case that q is a prime. Besides (**a**) and (**b**) for $q = p$, these conditions yield also that $\mathfrak{K}^8$ is a difference set in $GF(p)$ if $p = 8x^2 + 1 = 64y^2 + 9$, with x, y odd, and that $\mathfrak{K}^8 \cup \{0\}$ is a difference set in $GF(p)$ if $p = 8x^2 + 49 = 64y^2 + 441$. For $n = 16$, see LEHMER 1954 and WHITEMAN 1957.

2) This has been checked only for small values of q, by HALL 1956, who determined all difference sets Δ with $3 \leq |\Delta| \leq 50$ and the corresponding designs. (12 cases left undecided by Hall were shown to be impossible by K. YAMAMOTO 1963, see also MANN 1965b, p. 86.) The difference sets in **24c** are always nonequivalent to those of **23**; this follows from a multiplier argument of HALL 1956.

3) The basic idea is also in A. BRAUER 1953.

26. *If p, p' are primes satisfying* (23) *such that $(p p' - 1)/4$ is an odd square, then the cyclic group of order pp' contains a difference set D with $|D| = (pp' - 1)/4$.*

For more details, the reader is referred to WHITEMAN 1962; in particular, **26** is Whiteman's Theorem 4.

If D is a difference set in the finite group G, then the complement $G - D$ is likewise a difference set: the design corresponding to $D' = G - D$ is the complementary design of that corresponding to D. This simple observation allows the construction of difference sets in certain abelian groups of even order. Let D_i be a difference set in G_i $(i = 1, 2)$, and consider the subset

$$D = (D_1 \times D_2') \cup (D_1' \times D_2) \tag{24}$$

of the direct product $G = G_1 \times G_2$. Then

27. *If G_i is abelian and of order $4n_i^2$, and if $|D_i| = n_i(2n_i - 1)$, $i = 1, 2$, then the set D given by* (24) *is a difference set in $G = G_1 \times G_2$, with $|D| = 2n_1 n_2 (4n_1 n_2 - 1)$.*

Proof: MANN 1965b, p. 68—69. As a corollary, we note:

28. *The direct product of m groups of order* 4 *contains a difference set such that the corresponding design has parameters*

$$v = b = 4^m, \; k = r = 2^{m-1}(2^m - 1), \; \lambda = 2^{m-1}(2^{m-1} - 1). \tag{25}$$

The preceding results **22**—**28** were concerned with sharply transitive groups of projective designs only. Such groups can, however, exist also in the non-projective case. Designs of this sort can be described within the group by means of several quotient sets (cf. Section 1.2); some constructions of this kind appear in BOSE 1939. Here we give a brief survey of recent results of SCHULZ 1967, on designs with a transitive group T of *translations*, as defined in Section 2.3. Such a group is automatically sharply point transitive, whence $v = |\mathsf{T}|$. Moreover, as was mentioned earlier (cf. **2.3.25c**), v and k must be powers of the same prime in this case.

Let G be a finite group. By a *quasi-partition*[1]) of G we mean a set $\mathfrak{Q}$ of subgroups of G such that

$$\text{all } Q \in \mathfrak{Q} \text{ have the same order } k < |G|, \tag{26}$$

[1]) This concept is closely related to that of a partition of a group, as defined in Section 1.2. But while quasi-partitions with $\lambda = 1$ are clearly partitions, not every partition is a quasi-partition. Hence the term, due to SCHULZ 1967, is somewhat misleading. — The idea of using partitions for the construction of incidence structures with transitive translation groups is due to ANDRÉ 1954; see also Section 3.1.

and

(27) *each element* $\neq 1$ *in* G *is contained in equally many, say* λ, *subgroups in* $\mathcal{Q}$.

Consider now the incidence structure $\mathbf{J}(G, \mathcal{Q})$ of **1.2.21**, defined by $\mathcal{Q}$. It is clear that $\mathbf{J}(G, \mathcal{Q})$ is a design with $v = |G|$, $r = |\mathcal{Q}|$, and k, λ as in (26), (27). In fact:

29. *The designs isomorphic to* $\mathbf{J}(G, \mathcal{Q})$, *for suitable* G *and* $\mathcal{Q}$, *are precisely those that admit a transitive group of translations. The translation group of* $\mathbf{J}(G, \mathcal{Q})$ *is isomorphic to* G.

(Schulz 1967, Satz 1.) After this it is easy to construct designs with non-abelian transitive translation groups. One only has to take G as a non-abelian p-group of exponent p (this implies $p > 2$); the subgroups of order p then form a quasi-partition $\mathcal{Q}$ with $\lambda = 1$. If $|G| = p^e$, the parameters of $\mathbf{J}(G, \mathcal{Q})$ are

$$\begin{aligned} &v = p^e, \quad b = p^{e-1}(p^e - 1)(p - 1)^{-1}, \quad k = p, \\ &r = (p^e - 1)(p - 1)^{-1}, \quad \lambda = 1; \quad \text{with} \quad p, e > 2. \end{aligned} \tag{28}$$

These designs also include examples for which there exists no nontrivial central dilatation; cf. Schulz 1967, (4.7), (4.8).

If the transitive translation group T is abelian, then it is actually elementary abelian (Schulz 1967, p. 76). The designs of this class turn out to be just the following (Schulz 1967, Satz 3): Points are those of an affine geometry $\mathcal{A}(d, q)$, blocks are precisely those t-dimensional subspaces of $\mathcal{A}(d, q)$ which intersect the ideal hyperplane $\mathcal{P}(d-1, q)$ in a member of a given set $\mathcal{S}$ of $(t-1)$-dimensional subspaces which, together with all points of $\mathcal{P}(d-1, q)$, form a tactical configuration. Note that the configuration $\mathbf{W}(q)$ of **1.4.55** is of this kind; other examples for $\mathcal{S}$ can easily be constructed with the help of symplectic polarities in higher dimensional finite projective geometries.

In the remainder of this section, we discuss general construction techniques for designs, in which group theoretical methods play only a minor role. These methods are of three basic kinds: First, we have *direct methods*, where a design is constructed from some given combinatorial or algebraic structure, secondly *composition methods* which combine two or more given designs to construct a larger one out of them, and thirdly *alteration methods*, where the incidences in a given design are changed to give another design, with the same parameters as the given one, but not isomorphic to it. Many of the references to be quoted below contain combinations of direct and composition techniques. Alteration methods have been studied less extensively because they yield nothing for the existence problem of designs with prescribed parameters.

A fundamental paper for direct methods is BOSE 1939. His basic method (of "symmetrically repeated differences" and "initial blocks") is a generalization of the representation by difference sets; it yields designs with various sets of parameter values, possessing an abelian automorphism group Γ whose point orbits are either fixed points or of length $|\Gamma|$; the difference set case is that in which Γ is transitive. Details cannot be given here; we mention only that Bose's paper generalizes and unifies several earlier methods for the construction of Steiner triple and quadruple systems. On the other hand, BOSE 1939 has been the basis for various other results in this direction. The references are too numerous to be given here; we mention only the survey article of GUÉRIN 1965, where many construction techniques are summarized. For further references, see the Bibliography.

The STEINER 1853 problem[1]) for $k = 3$ had been solved by KIRKMAN 1847; Steiner was obviously not aware of this. Another solution was given by REISS 1859; see also NETTO 1893 and MOORE 1893. These papers contain several composition methods, which were generalized and extended by HANANI 1960a, 1961. The first paper of Hanani contains the first complete solution of the Steiner problem for $k = 4$. A more general result, without Steiner's stipulation that $b_3 = 1$, is the following:

30. *The conditions*

$$(29)\qquad \lambda(v-1) \equiv 0 \bmod k-1 \quad \textit{and} \quad \lambda v(v-1) \equiv 0 \bmod k(k-1),$$

necessary for the existence of a design with parameters v, k, b, r, λ, are also sufficient if $k \leqq 4$.

Proof: HANANI 1960a, 1961. Note that the restriction $k \leqq 4$ is best possible: the parameters $v = 15$, $b = 21$, $r = 7$, $\lambda = 2$ satisfy (29), but a design with these parameters cannot exist, for in view of **2.2.5** such a design would have to be embeddable into a projective design with 22 points, of order $n = r - \lambda = 5$; this is impossible because of **2.1.8**. However, HANANI 1961, 1965 has shown that (29) is sufficient if $k = 5$ and $\lambda = 1,4$, or 20.

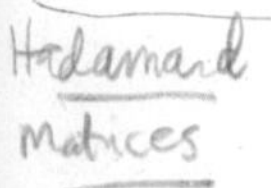

After these brief comments on general methods, we give a more detailed survey of the present status of the existence problem for Hadamard designs. We shall see (cf. **31** below) that this is equivalent to a determinant problem of HADAMARD 1893; we formulate this now.

Let $\mathscr{A}(n)$ denote the set of all real n by n matrices $A = (a_{ij})$ with $|a_{ij}| \leqq 1$ for all i, j, and define

$$(30)\qquad f(n) = \sup\{\det A : A \in \mathscr{A}(n)\}.$$

[1]) See footnote [3]) on p. 97.

It is readily verified that there exists $A \in \mathscr{A}(n)$ with $a_{ij} = \pm 1$ such that $f(n) = \det A$. For this A, let $B = AA^T$. By Sylvester's law of inertia, the eigenvalues $e_1, \ldots, e_n$ of B are all positive, and the arithmetic-geometric mean inequality gives $f(n)^2 = \prod_{i=1}^{n} e_i \leqq (n^{-1} \operatorname{tr} B)^n = n^n$, whence

$$f(n) \leqq n^{n/2}. \tag{31}$$

Equality holds if and only if $e_1 = \cdots = e_n = n$; in this case $Q^{-1} BQ = n\, I$ for some nonsingular Q, and this means

$$AA^T = n\, I. \tag{32}$$

An n by n matrix $A = (a_{ij})$ with $a_{ij} = \pm 1$ satisfying (32) is called an *Hadamard matrix* of order n, and it is *normalized* if $a_{i1} = a_{1i} = 1$ for $i = 1, \ldots, n$. Clearly every Hadamard matrix can be transformed into a normalized one by multiplying suitable rows and columns by -1.

The Hadamard problem is to determine all values of n for which Hadamard matrices exist.[1] The connection with Hadamard designs is the following:

31. *Let A be a normalized n by n Hadamard matrix. Delete the first row and the first column of A, and in the remaining $n-1$ by $n-1$ matrix replace every -1 by 0. Then the resulting matrix is the incidence matrix of an Hadamard design. Conversely, every incidence matrix of an Hadamard design can be obtained in this way.*

Hence there is an Hadamard design with v points if and only if there is an Hadamard matrix of order $v+1$. The simple but fundamental observation **31** seems to be due to Todd 1933; for the proof see, for example, Ryser 1963, p. 107.

The Hadamard problem is not yet completely solved. The equation $f(n) = n^{n/2}$ can, of course, hold only if n is even, and a little further analysis shows:

32. *An n by n Hadamard matrix exists only if $n = 2$ or $n \equiv 0 \bmod 4$.*

Proof: Paley 1933; Ryser 1963, pp. 105—106. It has been conjectured that the converse of **32** is also true, i.e. that there exist Hadamard matrices of every order $n = 4m$ or, equivalently, Hadamard designs for every $v = 4m - 1$. The various construction techniques so far developed suggest strongly that this conjecture is true. The present state of our knowledge can be summarized as follows. Let the integer

[1] Equivalent problems are: (i) For what n is there an n-dimensional regular simplex whose vertices are contained among those of an n-dimensional cube? (Coxeter 1933) and (ii) For what n does the n-dimensional cube have n mutually orthogonal diagonals? (Gruner 1939).

n be called an *H-number* if $n > 1$ and there exists an n by n Hadamard matrix. Then

33. *Each of the following conditions implies that n is an H-number:*

(**a**) *n is a product of H-numbers;*
(**b**) $n = 2^e, e > 0$;
(**c**) $n = p^e + 1 \equiv 0 \bmod 4$, *with p a prime;*
(**d**) $n = h(p^e + 1)$, *with h an H-number and p a prime;*
(**e**) $n = m(m - 1)$, *with m a product of numbers of the forms* (**b**) *and* (**c**)*;*
(**f**) $n = m(m + 3)$, *with m and $m + 4$ products of numbers of the forms* (**b**) *and* (**c**)*;*
(**g**) $n = h_1 h_2 p^e(p^e + 1)$, *with h_1, h_2 both H-numbers and p a prime;*
(**h**) $n = h_1 h_2 s(s + 3)$, *with h_1, h_2 both H-numbers, and $s, s + 4$ both of the form $p^e + 1$, where p is a prime;*
(**i**) $n = (q + 1)^2$, *with both q and $q + 2$ odd prime powers;*
(**j**) $n = (m - 1)^3 + 1$, *with m as in* (**e**)*;*
(**k**) $n = (m - 1)^2$, *with $m - 2$ a prime power* $\equiv 1 \bmod 4$, *and $m + 1$ a product of numbers of the form* (**b**) *and* (**c**)*;*
(**l**) $n = 92, 116, 156, 172$, or 232.

Some of these are contained in earlier results. For example, the $\mathbf{P}_{d-1}(d, 2)$ yield (**b**); this was proved already by SYLVESTER 1867. The Paley designs $\mathbf{H}(q)$ yield (**c**), and the designs of **25** yield (**i**). That (**a**) holds is a consequence of the simple but important fact that the Kronecker product of two Hadamard matrices is again an Hadamard matrix (PALEY 1933; this may be considered a composition technique for the construction of Hadamard designs). Results (**d**)—(**h**) are due to WILLIAMSON 1944, 1947; they are partly generalizations of earlier methods of SCARPIS 1898. Result (**i**) for q and $q + 2$ primes was found by A. BRAUER 1953, (**j**) is due to GOLDBERG 1966, and (**k**) to EHLICH 1965. For a long time, $n = 92$ was the least order for which no Hadamard matrix was known; such a matrix was then found by BAUMERT, GOLOMB & HALL 1962. BAUMERT 1966 gave examples for $n = 116, 232$, BAUMERT & HALL 1965a found an example for $n = 156$, and in WILLIAMSON 1944 there is one with $n = 172$. It is not clear whether these belong to larger classes of Hadamard matrices. The only order $\leqq 200$ for which the existence of an Hadamard matrix is in doubt today (1967) is $n = 188$.

In conclusion of this section, we present some special results which will provide the examples promised in Sections 2.2 and 2.3. This will also include some alteration techniques.[1]) We begin with the following:

---

[1]) Some other constructions by alteration are due to BARLOTTI 1962a, in connection with a generalization of affine spaces by SPERNER 1960.

34. *Every Hadamard design is extendable in exactly one way, and the extension is an affine design of type* (1, 3) *with exactly two blocks in every parallel class. Conversely, if* **A** *is such an affine design, then* $\mathbf{A}_p$ *is an Hadamard design for every point* p *of* **A**.

Compare here the remarks after **2.2.10** and **2.2.19**; the extension of $\mathbf{P}_{d-1}(d, 2)$ is $\mathbf{A}_d(d+1, 2)$. For the proof of **34** (a special case of which is contained in BOSE 1947b), adjoin a new point ∞ to the Hadamard design **H**, let ∞ be incident with every block in **H**, and define new blocks as the complements of the blocks in this extended system. This yields an affine design with the desired properties; all other claims are straightforward.

If **H**(11) is extended in the manner indicated in **34**, the resulting affine design **A** of type (1, 3) has parameters $v = 12$, $k = 6$, $\lambda = 2$, and Aut **A** is the Mathieu group M_{11} in its triply transitive representation of degree 12. The stabilizer of a point of **A** is Aut **H**(11) $= PSL_2(11)$, in its doubly transitive representation of degree 11. (Compare here result **20** above.) This is an essentially unique situation:

35. *Let* **A** *be an affine design with* $m = 2$, *i.e. with exactly two blocks in every parallel class. Suppose that* λ *is even. If* **A** *has an automorphism group* Γ *wich is triply transitive on points, then* $\lambda = 2$, *hence* $\mathbf{A}_p \cong \mathbf{H}(11)$ *for any point* p, *and* $\Gamma = M_{11}$.

This is Theorem 2 of NORMAN 1968. Norman also conjectured that if the hypothesis "λ even" of **35** is replaced by "λ odd", the only possibilities for **A** might be the $\mathbf{A}_d(d+1, 2)$, whose automorphism groups are triply transitive extensions of $PGL_{d+1}(2) = \mathrm{Aut}\,\mathbf{P}_{d-1}(d, 2)$.

36. *For every* $d > 2$ *and every prime power* q, *there exist projective and affine designs with the same parameters as, but not isomorphic to,* $\mathbf{P}_{d-1}(d, q)$ *and* $\mathbf{A}_{d-1}(d, q)$, *respectively.*[1])

This is proved with the help of **2.2.9**: Every one-one correspondence ω between the parallel classes of the affine design $\mathbf{A}_{d-1}(d, q)$ and the blocks of the projective design $\mathbf{P}_{d-2}(d-1, q)$ yields an embedding of $\mathbf{A}_{d-1}(d, q)$ into a projective design **P** with the same parameters as $\mathbf{P}_{d-1}(d, q)$; we then have $\mathbf{P}^W = \mathbf{A}_{d-1}(d, q)$ for a well-determined block W of **P**. Now ω can be so chosen that **P** contains a line $\mathfrak{l}$ not meeting W; it then follows that $\mathbf{P} \ncong \mathbf{P}_{d-1}(d, q)$, see **2.1.22**. Also, there exists a block $B \neq W$ in **P** such that $\mathbf{P}^B \ncong \mathbf{A}_{d-1}(d, q)$ if ω is chosen suitably; here **2.2.12** is used. This proves **36**, but the argument shows more:

[1]) For this observation and the following proofs, the author is indebted to W. M. Kantor. For special cases, see TODD 1933 and LÜNEBURG 1961a. Other examples for **36**, though only for odd d, were given by HIGMAN 1964, p. 155. Higman's designs have flag transitive automorphism groups, and precisely two points on every line.

37. *With the assumptions of* **36** *for d and q, there exist designs with the same parameters as $\mathbf{A}_{d-1}(d, q)$ which do not admit a parallelism.*

For if all $\mathbf{P}^B$ had a parallelism, they would all be affine, because of **2.2.8**. But then $\mathbf{P} \cong \mathbf{P}_{d-1}(d, q)$ by **2.2.14**, whereas we have seen that we can have $\mathbf{P} \ncong \mathbf{P}_{d-1}(d, q)$. Next:

38. *With the assumptions of* **36** *and* **37** *for d and q, there exist designs with the same parameters as $\mathbf{P}_{d-1}(d, q)$, admitting axial automorphisms which are not central.*

For all translations of $\mathbf{A}_{d-1}(d, q)$ extend to axial automorphisms of $\mathbf{P}$, by **2.3.18**, but as $\mathbf{P} \ncong \mathbf{P}_{d-1}(d, q)$, it follows from **2.3.26b** that they cannot all be central.

Finally, the reader will verify without difficulty that the same idea of construction leads to designs with the same parameters as $\mathbf{P}_{d-1}(d, q)$ in which not all lines have the same size.

3. Projective and affine planes

This chapter and the following two are in a relationship similar to the three parts of Chapter 2: the combinatorial part will be covered here, automorphisms (collineations) follow in Chapter 4, and constructions in Chapter 5.

Section 3.1, however, does not quite fit into this scheme. In that section we collect results from the general theory of projective planes, not depending on finiteness assumptions but relevant for later applications in the finite case. Section 3.1 also contains a few selected results on collineations and constructions of (not necessarily finite) projective and affine planes.

Section 3.2 is concerned with different systems of axioms for finite planes, with the connections between finite planes, certain sets of permutations, nets, and Latin squares, and with special substructures of finite planes, primarily subplanes, arcs, and ovals.

In Section 3.3, the rather few known results on dualities and polarities of finite planes are collected.

Section 3.4, finally, presents results on projectivities in projective planes. We include again some theorems for infinite planes, but the main part of 3.4 is concerned with results characterizing the finite desarguesian plane $\mathbf{P}(q) = \mathbf{P}_1(2, q)$ by properties of its group of projectivities.

3.1 General results

Projective and affine planes were defined in Section 1.4; we give equivalent definitions now. A *projective plane* is an incidence structure of points and lines satisfying the following conditions:

(1) *To any two distinct points, there exists a unique line incident with both of them.*
If $p \neq q$, then the unique line joining p and q is denoted by pq or qp.[1])
(2) *To any two distinct lines, there exists a unique point incident with both of them.*

[1]) This is a more convenient notation than $p + q$, which was used in Section 1.4 because of the connection with vector space addition.

If $L \neq M$, then the unique common point of L and M is denoted by LM or ML.

(3) *There exist four points of which no three are incident with the same line.*

Such a set of four points is called a *quadrangle*. A set of points is *collinear* if there exists a line incident with every point of the set, and dually a set of lines is *concurrent* if there is a common point of all lines in the set. Hence a quadrangle may be defined as a set of four points no three of which are collinear, and dually a *quadrilateral* is a set of four lines no three of which are concurrent.

Conditions (1) and (2) are dual to each other, and together with (3) they imply the dual of (3): there exist quadrilaterals. Thus the theory of projective planes is self-dual in the sense that the dual structure $\overline{\mathbf{P}}$ (defined in Section 1.1) of an arbitrary projective plane $\mathbf{P}$ is again a projective plane. We shall see later that $\mathbf{P}$ and $\overline{\mathbf{P}}$ need not be isomorphic.

An *affine plane* is an incidence structure of points and lines satisfying (1) and the following conditions:

(4) *To any non-incident point-line pair p, L, there exists a unique line through p which has no point in common with L.*

(5) *There exist three non-collinear points.*

Such a set of three points is a *triangle*, and dually a set of three non-concurrent lines is a *trilateral*.

The dual of (5) is easily proved in any affine plane, but the dual of (4) is clearly false. Hence the dual of an affine plane is never an affine plane.[1])

In an arbitrary affine plane $\mathbf{A}$, define a relation $\parallel$ of *parallelism* among the lines as follows:

(6) $$L \parallel M \quad \textit{if and only if} \quad L = M \quad \textit{or} \quad [L, M] = 0.$$

The notation is that of (1.1.6) again. It is easily seen that $\parallel$ is an equivalence relation; the classes of this relation are called *ideal points*. If incidence in $\mathbf{A}$ is retained, if the ideal points are defined to be incident with exactly those lines of $\mathbf{A}$ which are contained in them, and if an *ideal line* W is introduced which is incident with all ideal points and no point of $\mathbf{A}$, then the extended incidence structure so defined is a projective plane $\mathbf{P}$, and $\mathbf{A}$ is the external structure $\mathbf{P}^W$ of $\mathbf{P}$ with respect to the ideal line W [see Section 1.1 for definitions]. Conversely, if L is

---

[1]) Some authors, however, speak of the "dual" $\tilde{\mathbf{A}}$ of an affine plane $\mathbf{A} = \mathbf{P}^W$ in the following sense: Consider the dual $\overline{\mathbf{P}}$ of $\mathbf{P}$, which is a projective plane. Then $\tilde{\mathbf{A}} = (\overline{\mathbf{P}})^L$, where L is a line of $\overline{\mathbf{P}}$ through the point W of $\overline{\mathbf{P}}$. The line L of $\overline{\mathbf{P}}$ is usually unambiguously defined.

any line of a projective plane **P**, then $\mathbf{P}^L$ is an affine plane **A**, and **P** can be interpreted as constructed from **A** by means of ideal elements, as described above. On the other hand, if L, M are two distinct lines of **P**, then the affine planes $\mathbf{P}^L$ and $\mathbf{P}^M$ need not be isomorphic.

These considerations show that the theories of projective and affine planes are closely interrelated: an affine plane is essentially the same as a projective plane with a distinguished line. In some investigations, the projective point of view is more appropriate, and in others the affine point of view is. We shall encounter both situations later on.

A *subplane* of a projective plane **P** is a subset **S** of points and lines which is itself a projective plane, relative to the incidence relation given in **P**. This means that

(7) $p \in \mathbf{S}$ *and* $q \in \mathbf{S}$ *implies* $pq \in \mathbf{S}$,
(8) $L \in \mathbf{S}$ *and* $M \in \mathbf{S}$ *implies* $LM \in \mathbf{S}$,
(9) **S** *contains a quadrangle.*

Subsets **S** which satisfy (7) and (8) but not necessarily (9) will prove to be of interest later on. We call such subsets *closed*;[1]) hence subplanes are those closed subsets which satisfy (9).

Intersections of closed subsets are again closed subsets. The closed subset *generated* by the arbitrary subset **S** of the projective plane **P** is defined as the intersection of all closed subsets containing **S**, and denoted by $\langle \mathbf{S} \rangle$, i.e.

$$\langle \mathbf{S} \rangle = \bigcap_{\substack{\mathbf{S} \subseteqq \mathbf{C} \\ \mathbf{C} \text{ closed}}} \mathbf{C}. \qquad (10)$$

It is easily verified that $\mathbf{S} \to \langle \mathbf{S} \rangle$ is a closure operation, i.e.

$$\langle\langle \mathbf{S} \rangle\rangle = \langle \mathbf{S} \rangle, \quad \mathbf{S} \subseteqq \langle \mathbf{S} \rangle, \quad \text{and} \quad \mathbf{S} \subseteqq \mathbf{T} \to \langle \mathbf{S} \rangle \subseteqq \langle \mathbf{T} \rangle.$$

Hence[2]) the closed subsets of an arbitrary projective plane **P** form a complete lattice $\mathscr{C}(\mathbf{P})$, and the subplanes, together with the empty set, form a complete lattice $\mathscr{S}(\mathbf{P})$. Little seems to be known about these lattices in general.[3])

A *prime plane* is a projective plane **P** which does not possess proper subplanes, i.e. one for which $|\mathscr{S}(\mathbf{P})| = 2$. Clearly,

1. *A projective plane is a prime plane if and only if it is generated by each one of its quadrangles.*

---

[1]) Closed subsets which are not subplanes are sometimes called "degenerate subplanes". For a complete classification of all possible types of these, see HALL 1943, or PICKERT 1955, p. 13.

[2]) cf. BIRKHOFF 1948, p. 49—50.

[3]) An example in HALL 1943 shows that $\mathscr{S}(\mathbf{P})$ need not be modular. Note that $\mathscr{S}(\mathbf{P})$ need not be a sublattice of $\mathscr{C}(\mathbf{P})$ in the sense of BIRKHOFF 1948, p. 19.

A subplane of a projective plane $\mathbf{P}$ is called *minimal* if it is a prime plane, i.e. an atomic element[1] of $\mathscr{S}(\mathbf{P})$. A *maximal closed subset*, or a *maximal subplane*, is an anti-atomic element $\neq \emptyset$ of $\mathscr{C}(\mathbf{P})$ or $\mathscr{S}(\mathbf{P})$, respectively. A closed subset $\mathbf{C}$ of $\mathbf{P}$ is called a *Baer subset*[2], or a *Baer subplane* in case it is a subplane, if it satisfies the following conditions:

(11) *Every point of* $\mathbf{P}$ *is incident with a line of* $\mathbf{C}$.
(12) *Every line of* $\mathbf{P}$ *is incident with a point of* $\mathbf{C}$.

Clearly every Baer subset is a maximal closed subset, but there are examples of maximal subplanes which are not Baer subplanes. On the other hand:

2. *A closed subset which is not a subplane is maximal if and only if it is a Baer subset; these Baer subsets are precisely the sets*

$$\mathbf{B}(p, L) = \{p, L\} \cup (p) \cup (L)$$

for some point-line pair p, L [*notation as in* (1.1.5)].

This is easily verified; the cases $p \mathrel{\mathrm{I}} L$ and $p \not\mathrel{\mathrm{I}} L$ are both admissible. Baer subsets and subplanes will be encountered frequently in the finite case.

A *subplane of an affine plane* $\mathbf{A} = \mathbf{P}^W$ is a subset $\mathbf{S}$ of points and lines which is itself an affine plane, relative to the incidence given in $\mathbf{A}$, such that

(13) *If two lines are parallel in* $\mathbf{S}$, *then they are also parallel in* $\mathbf{A}$.

One could, of course, define affine subplanes without this condition,[3] but then there would be no connection between the subplanes of $\mathbf{A} = \mathbf{P}^W$ and those of $\mathbf{P}$. Condition (13) guarantees such a connection:

3. *The subplanes of the affine plane* $\mathbf{A} = \mathbf{P}^W$ *are precisely the affine planes* $\mathbf{S}^W$, *where* $\mathbf{S}$ *ranges over all projective subplanes of* $\mathbf{P}$ *which contain the line* W.

It will be clear what we mean by minimal, maximal, and Baer subplanes of affine planes: we require that these subplanes, when interpreted as projective planes, contain the ideal line W.

We turn now to *collineations* of projective planes. These are automorphisms as defined in Section 1.2, i.e. incidence preserving per-

[1]) Atomic and anti-atomic elements need not exist in an arbitrary lattice $\mathscr{S}(\mathbf{P})$, but clearly such elements exist if $\mathbf{P}$ is finite.

[2]) After BAER 1946b, who first realized the importance of these subsets as systems of fixed elements with respect to certain collineations. This will be discussed in detail further below.

[3]) Cf. OSTROM & SHERK 1964; RIGBY 1965.

mutations which map points onto points and lines onto lines.[1]) The group of all collineations of a projective plane $\mathbf{P}$ will be denoted by $\operatorname{Aut}\mathbf{P}$, as in Section 1.2; a *collineation group* of $\mathbf{P}$ is any subgroup of $\operatorname{Aut}\mathbf{P}$. The full collineation group of an affine plane $\mathbf{A} = \mathbf{P}^W$ is essentially identical with the stabilizer of W in $\operatorname{Aut}\mathbf{P}$.

Our main interest here lies in central collineations of projective planes, as defined in Section 1.4. A *center* of a collineation $\alpha \in \operatorname{Aut}\mathbf{P}$ is a point c such that $X\alpha = X$ for all lines $X \mathrm{I} c$. Dually, an *axis* of α is a line A with $x\alpha = x$ for all $x \mathrm{I} A$.

4. *A collineation α of a projective plane has a center if and only if it has an axis. If $\alpha \neq 1$, then the center and the axis of α are unique.*

Proof: PICKERT 1955, p. 62—65; see also **1.4.8** and **1.4.9** above. As in 1.4, we define *elations* and *homclogies* as central collineations with incident and non-incident center-axis pairs, respectively. A simple but useful fact about central collineations is the following:

5. *Let α be a collineation with center c and axis A, and suppose that $c \neq p \not\mathrm{I} A$ for some point p. Then every subplane containing $c, A, p, p\alpha$ is left invariant by α.*

For the proof, see LÜNEBURG 1964c, p. 446—447.

The set $\mathbf{F} = \mathbf{F}(\alpha)$ of the elements fixed by the central collineation $\alpha \neq 1$, with center c and axis A, is the Baer subset $\mathbf{B}(c, A)$ described in **2**. A collineation whose fixed elements form a Baer subset will be called *quasicentral*. Hence a collineation which is quasicentral but not central has a Baer subplane of fixed elements. Such collineations will be of interest in the finite case; here we note only that

6. *Involutorial collineations are quasicentral.*

For the easy proof, see BAER 1946b, p. 275.

Let Γ be an arbitrary collineation group of the projective plane $\mathbf{P}$, and let c, p be points and A, L lines of $\mathbf{P}$. Then each of the following sets is a subgroup of Γ:

$$(14)\quad \begin{cases} \Gamma(c, A) = \{\gamma \in \Gamma : \gamma \textit{ has center } c \textit{ and axis } A\}, \\ \Gamma(L, A) = \bigcup\limits_{x \mathrm{I} L} \Gamma(x, A), \quad \Gamma(c, p) = \bigcup\limits_{X \mathrm{I} p} \Gamma(c, X) \\ \Gamma(A) = \bigcup\limits_{x} \Gamma(x, A), \quad \Gamma(c) = \bigcup\limits_{X} \Gamma(c, X). \end{cases}$$

Here the last two unions are to be taken over all points x and all lines X of $\mathbf{P}$, respectively. It follows from **4** that the unions in (14) are in

[1]) The requirement of section 1.2, that inverses be incidence-preserving [cf. footnote [1]) of p. 8] is easily proved for projective and affine planes.

fact partitions. The groups listed in (14) consist of central collineations, either with common axis or with common center. The product of two central collineations is, in fact, usually not central unless they have the same center or axis:

7. *Let* $1 \neq \alpha \in \Gamma(a, A)$ *and* $1 \neq \beta \in \Gamma(b, B)$, *and suppose that* $a \neq b$ *and* $A \neq B$. *Then:*

(**a**) *A fixed point of* $\alpha\beta$ *is either* AB *or incident with* ab.
(**b**) $\alpha\beta$ *is a central collineation if and only if* α *and* β *are homologies such that*

(15) $$a \mathrm{I} B,\ b \mathrm{I} A, \text{ and } x\alpha = x\beta^{-1} \text{ for every } x \mathrm{I} ab.$$

If this is the case, then $\alpha\beta$ *is a homology with center* AB *and axis* ab.

Proof[1]: That (**a**) holds is easily verified; in fact if $AB \neq x \,\bar{\mathrm{I}}\, ab$, then x, $x\alpha$ and $x\alpha\beta$ are non-collinear points. Hence if $\alpha\beta$ is central, its axis must be $C = ab$, whence $x\alpha = x\beta^{-1}$ for all $x \mathrm{I} C$. Putting $x = a$ and using the fact that $a \neq b$, we get $a \mathrm{I} B$, and $b \mathrm{I} A$ follows in the same fashion. As $A \neq B$, at least one of α, β, say α, is a homology. If β were an elation, then $B = ab = C$, and $\alpha = (\alpha\beta)\beta^{-1}$ would also have axis $A = C$, against the hypothesis. The remainder is clear.

It follows from **7** that if a collineation group Γ consists of central collineations with neither the same center nor the same axis, then Γ is the non-cyclic group of order 4, and its non-trivial elements are three involutorial homologies whose centers and axes are the vertices, and opposite sides, of a triangle. The first part of the next result shows that this situation can actually occur:

8. *Let* α *and* β *be two involutorial homologies, in* $\Gamma(a, A)$ *and* $\Gamma(b, B)$, *respectively, and put* $\gamma = \alpha\beta$.

(**a**) *If* $a \mathrm{I} B \neq A \mathrm{I} b \neq a$, *then* γ *is an involutorial homology in* $\Gamma(AB, ab)$.
(**b**) *If* $a \neq b$ *and* $A = B$, *then* γ *is an elation in* $\Gamma(A(ab), A)$.

Proof: Ostrom 1956, Lemmas 4 and 6.[2] Note that in case (**a**), both $\Gamma(a, A)$ and $\Gamma(b, B)$ contain no involution $\neq\alpha$ or β, respectively. In case (**b**), we can draw a similar conclusion:

9. *If the group* $\Gamma(A)$ *contains nontrivial homologies with different centers, then, for any point* $c \,\bar{\mathrm{I}}\, A$, *the group* $\Gamma(c, A)$ *contains at most one involution.*

---

[1]) This result is well known, but the author has been unable to locate a convenient reference.

[2]) Both results are, however much older. For example, a proof of (**b**) was given by Baer 1944, p. 103.

For the proof, the following simple but important fact is needed:

10. *If α is in the normalizer of Γ in* Aut **P**, *then*

$$\alpha^{-1}\,\Gamma(c, A)\,\alpha = \Gamma(c\,\alpha, A\,\alpha) \tag{16}$$

for any point-line pair (c, A) of **P**.

The proof of **10** is straightforward. We prove **9**: Let ϱ and σ be involutorial homologies in $\Gamma(A)$ and assume that they have the same center $p \,\not{\mathrm{I}}\, A$. Then also $\varrho\,\sigma \in \Gamma(p, A)$. On the other hand, there exists $\alpha \neq 1$ in $\Gamma(A)$ with center $\neq p$, and $\tau = \alpha^{-1}\varrho\,\alpha$ is, by **10**, an involutorial homology with axis A and center $p\,\alpha \neq p$. But then **8b** shows that $\varrho\,\tau$ and $\tau\,\sigma$ are both elations, whence $\varrho\,\sigma = \varrho\,\tau \cdot \tau\,\sigma$ is likewise an elation. This is compatible with $\varrho\,\sigma \in \Gamma(p, A)$ only if $\varrho\,\sigma = 1$ or $\varrho = \sigma$.

If c and A are fixed by Γ, then $\Gamma(c, A)$ is a normal subgroup of Γ, because of **10**. It follows that

$$\Gamma(A, A) \lhd \Gamma(A) \quad \textit{and} \quad \Gamma(c, c) \lhd \Gamma(c) \tag{17}$$

for any A, c. Also, $\Gamma(c, A) \lhd \Gamma(A)$ and $\Gamma(c, A) \lhd \Gamma(c)$ whenever $c \,\mathrm{I}\, A$. This can also be concluded from (17) and the following result:

11. *If $\Gamma(A, A)$ contains nontrivial elations with different centers (on A), then $\Gamma(A, A)$ is abelian.*

The proof is again easy, see Pickert 1955, p. 199. By a very similar argument, one may in fact prove more about commuting central collineations:

12. *Let (a, A) and (b, B) be two distinct point-line pairs in* **P** *and $1 \neq \alpha \in \Gamma(a, A)$, $1 \neq \beta \in \Gamma(b, B)$. Then $\alpha\,\beta = \beta\,\alpha$ if and only if $a \,\mathrm{I}\, B$ and $b \,\mathrm{I}\, A$.*

Note the connection with **7**.

We shall now be interested in the case where $\alpha\,\beta \neq \beta\,\alpha$. Hence let (a, A) and (b, B) be distinct point-line pairs of **P** such that

$$a \,\not{\mathrm{I}}\, B \quad \text{or} \quad b \,\not{\mathrm{I}}\, A. \tag{18}$$

For any $\alpha \neq 1$ in $\Gamma(a, A)$ and $\beta \neq 1$ in $\Gamma(b, B)$ we consider the mappings

$$\alpha^* : \xi \to \xi^{-1}\,\alpha^{-1}\,\xi\,\alpha \quad \text{and} \quad \beta^* : \eta \to \beta^{-1}\,\eta^{-1}\,\beta\,\eta \tag{19}$$

from $\Gamma(b, B)$ or $\Gamma(a, A)$, respectively, into Γ. It follows from **12** and (18) that both α^* and β^* are one-one. Now **10** shows that

$$\beta^{-1}\,\alpha^{-1}\,\beta\,\alpha \in \Gamma(a\,\beta, A\,\beta)\,\Gamma(a, A) \cap \Gamma(b, B)\,\Gamma(b\,\alpha, B\,\alpha).$$

Hence if $a \,\mathrm{I}\, B$, which implies $B\,\alpha = B$ and $a\,\beta = a$, then

$$\beta^{-1}\,\alpha^{-1}\,\beta\,\alpha = \beta^{\alpha^*} = \alpha^{\beta^*} \in \Gamma(a) \cap \Gamma(B) = \Gamma(a, B).$$

As α^* and β^* are one-one, we can conclude:

13. *Let* $a \mathrm{I} B$ *and* $b \not\mathrm{I} A$.

(a) *If* $\Gamma(a, A) \neq 1$, *then* $|\Gamma(a, B)| \geqq |\Gamma(b, B)|$.
(b) *If* $\Gamma(b, B) \neq 1$, *then* $|\Gamma(a, B)| \geqq |\Gamma(a, A)|$.

Thus if $\Gamma(a, A)$ or $\Gamma(b, B)$ is nontrivial, then so is $\Gamma(a, B)$. In a special case, we can say more:

14. *If* $B \neq A \mathrm{I} a \mathrm{I} B \mathrm{I} b \neq a$ *and* $\Gamma(a, A) \neq 1 \neq \Gamma(b, B)$, *then* $\Gamma(a, B)$ *contains subgroups (which may coincide) isomorphic to* $\Gamma(a, A)$ *and* $\Gamma(b, B)$, *respectively.*

Proof. One verifies first that the mapping α^* from $\Gamma(b, B)$ into Γ satisfies

$$(\xi \eta)^{\alpha^*} = \xi^{\alpha^*} \eta^{\alpha^*}$$

if and only if $\Gamma(b, B)^{\alpha^*}$ is in the centralizer of $\Gamma(b, B)$ in Γ. But under the hypothesis of **14**, we have $\Gamma(b, B)^{\alpha^*} \subseteqq \Gamma(a, B) \neq 1$, and $\Gamma(B, B)$ is abelian by **11**, so that this centralizer condition is satisfied. Hence α^* is an isomorphism into $\Gamma(a, B)$, and $\Gamma(a, B)$ contains an isomorphic copy of $\Gamma(b, B)$. The remainder of **14** follows from a dual argument.[1]

A central collineation is uniquely determined by the image of any one of its non-fixed points. More precisely:

15. *Let* (c, A) *be a point-line pair in* **P** *and* x, y *two points such that* $x \neq c \neq y \not\mathrm{I} A \not\mathrm{I} x$ *and* $cx = cy$. *Then there is at most one* $\gamma \in \Gamma(c, A)$ *with* $x\gamma = y$.

Proof: Pickert 1955, p. 66.

The group Γ will be called (c, A)-*transitive* if the "at most" in **15** can be replaced by "exactly", in other words if $\Gamma(c, A)$ is transitive on the non-fixed points of any line $\neq A$ through c. (If this is so for one such line, then it can be proved for all others as well; see Pickert 1955, p. 66.) Also, (c, A)-transitivity may be defined dually by transitivity of $\Gamma(c, A)$ on the non-fixed lines through any point $\neq c$ on A.

Next, we say that a projective plane **P** is (c, A)-*transitive* if its full collineation group Aut**P** is (c, A)-transitive. This concept, due to Baer 1942, has proved to be a very useful classifying principle for projective planes; we shall discuss this now at some length. First, there is a close connection with a special case of Desargues' theorem. We say that a projective plane **P** is (p, L)-*desarguesian* if every central couple[2] $(p_1 p_2 p_3)$, $(p_1', p_2', p_3',)$ of triangles with $p_i p_i' \mathrm{I} p, (i = 1, 2, 3)$ and $(p_1 p_2)(p_1' p_2') \mathrm{I} L \mathrm{I} (p_2 p_3)(p_2' p_3')$ is axial. It is not difficult to prove that

[1] Compare here Hering 1963, p. 156. We remark that the mapping $\alpha \to \alpha^*$ is never a homomorphism; in fact $(\alpha \beta)^* \neq \alpha^* \beta^*$ for all $\alpha, \beta \in \Gamma(b, B)$.

[2] For the definition of central and axial couples of triangles, see Section 1.4.

16. **P** *is* (p, L)*-transitive if and only if* **P** *is* (p, L)*-desarguesian.*

A proof appears in PICKERT 1955, pp. 76—78; three other equivalent properties are also given there.

We proceed now to a complete classification of the various possibilities for distinct (p, L)-transitivities in an arbitrary collineation group. The following preliminary results are basic for this classification.

17. *If* Γ *is* (c, A)*-transitive and if* γ *is in the normalizer of* Γ *in* Aut **P**, *then* Γ *is also* $(c\gamma, A\gamma)$*-transitive.*

This follows immediately from **10**. We say that Γ is (L, A)*-transitive* if Γ is (c, A)-transitive for every $c \mathrel{I} L$; dually, Γ is (c, p)*-transitive* if Γ is (c, A)-transitive for every $A \mathrel{I} p$. Also, (L, A)- and (c, p)-transitivity for **P** is defined as (L, A)- resp. (c, p)-transitivity for Aut **P**.

18. *Suppose that* **P** *has more than three points per line. If* Γ *is* (c, A)*- and* (c', A)*-transitive for two distinct points* c, c', *then* Γ *is also* (cc', A)*-transitive. Dually,* (c, A)*- and* (c, A')*-transitivity for* $A \neq A'$ *implies* (c, AA')*-transitivity.*

Proof: BAER 1942; PICKERT 1955, pp. 67—68. Note that if $\Gamma = \text{Aut}\,\mathbf{P}$, then **18** holds also for the plane $\mathbf{P}(2)$ with only three points per line. This is a first instance of the situation that stronger conclusions can be drawn from certain (p, L)-transitivities of **P** than from those of Γ. A more interesting case for this situation is:

19. *A projective plane* **P** *is* (p, q)*-transitive if and only if it is* (q, p)*-transitive. Dually,* (L, A)*- and* (A, L)*-transitivity for* **P** *are equivalent.*

(GINGERICH 1945; PICKERT 1955, p. 103.) Clearly, **19** ceases to be true if "**P**" is replaced by "Γ".[1]

Now we present the classification mentioned above.

20. *For any collineation group* Γ *of a projective plane* **P** *with more than five*[2] *points per line, define*

$$\mathbf{T} = \mathbf{T}(\Gamma) = \{(x, X) : \Gamma \text{ is } (x, X)\text{-transitive}\}. \tag{20}$$

[1]) It should be mentioned that while **16**—**18** are quite elementary, the simplest proof of **19** seems to require the use of coordinates. For this, see p. 127 and result **22e** below.

[2]) This hypothesis, more restrictive than that of **18**, is essential. For example, if π is a unitary polarity of $\mathbf{P}(4)$, then $\Gamma_0(\pi) \cong PGU_3(4)$ [cf. p. 47] is (p, L)-transitive if and only if p is non-absolute and $L = p\pi$. This corresponds to none of the types I.1—VII.2 below.

Then $\mathbf{T}$ *is of one of the following types:*

I.1. $\mathbf{T} = \emptyset$.

I.2. $\mathbf{T} = \{(c, A)\}, \quad c \not\mathrm{I} A$.

I.3. $\mathbf{T} = \{(c, A), (c', A')\}, \quad c \not\mathrm{I} A \mathrm{I} c' \not\mathrm{I} A' \mathrm{I} c$.

I.4. $\mathbf{T} = \{(a, A), (b, B), (c, C)\}$, *(vertices and opposite sides of a triangle).*

I.5. $\mathbf{T} = \{(x, x^\sigma p) : x \mathrm{I} L\}$, $p \not\mathrm{I} L$, σ *a fixed point free involution of* (L).

I.6. $\mathbf{T} = \{(x, x^\sigma) : p \neq x \mathrm{I} L\}$, $p \mathrm{I} L$, σ *one-one from* $(L) - \{p\}$ *onto* $(p) - \{L\}$.

I.7. $\mathbf{T} = \{(p, L)\} \cup \{(x, x^\sigma p) : x \mathrm{I} L\}$, p, L, σ *as in* I.5.

I.8. $\mathbf{T} = \{(x, x^\pi)\}$, π *a polarity of* $\mathbf{P}$, *without absolute points.*

II.1. $\mathbf{T} = \{(c, A)\}, \quad c \mathrm{I} A$.

II.2. $\mathbf{T} = \{(c, A), (c', A')\}, \quad A' \neq A \mathrm{I} c' \neq c = A A'$.

II.3. $\mathbf{T} = \{(p, L)\} \cup \{(x, x^\sigma) : p \neq x \mathrm{I} L\}, \quad p, L, \sigma$ *as in* I.6.

II.4a. $\mathbf{T} = \{(x, A) : x \mathrm{I} L\}, \quad L \neq A$.

II.5a. $\mathbf{T} = \{(p, L)\} \cup \{(x, A) : x \mathrm{I} L\}, \quad L \neq A \mathrm{I} p \neq AL$.

II.4b *and* II.5b *are dual to* II.4a *and* II.5a, *respectively.*

III.1. $\mathbf{T} = \{(x, p x) : x \mathrm{I} L\}$,

III.2. $\mathbf{T} = \{(p, L)\} \cup \{(x, p x) : x \mathrm{I} L\}$,

III.3. $\mathbf{T} = \{(x, p y) : x, y \mathrm{I} L\}$,

III.4. $\mathbf{T} = \{(p, L)\} \cup \{(x, p y) : x, y \mathrm{I} L\}$,

(III.1–III.4: $p \not\mathrm{I} L$.)

IVa.1. $\mathbf{T} = \{(x, A) : x \mathrm{I} A\}$,

IVa.1′. $\mathbf{T} = \{(x, A) : x \mathrm{I} A\} \cup \{(p, Y) : Y \mathrm{I} q\}, \quad p \neq q;\ p, q \mathrm{I} A$.

IVa.2. $\mathbf{T} = \{(x, A) : x \mathrm{I} A\} \cup \{(p, Y) : Y \mathrm{I} q\} \cup \{(q, Z) : Z \mathrm{I} p\}$, $p \neq q;\ p, q \mathrm{I} A$.

IVa.3. $\mathbf{T} = \{(x, Y) : x \mathrm{I} A;\ Y \mathrm{I} x^\sigma\}$, σ *a fixed point free involution of* (A).

IVa.4. $\mathbf{T} = \{(x, A)$, *all* $x\}$.

IVa.5. $\mathbf{T} = \{(x, A)$, *all* $x\} \cup \{(p, Y) : Y \mathrm{I} q\}, \quad p \neq q;\ p, q \mathrm{I} A$.

IVa.6. $\mathbf{T} = \{(x, A)$, *all* $x\} \cup \{(p, Y) : Y \mathrm{I} q\} \cup \{(q, Z) : Z \mathrm{I} p\}$, $p \neq q;\ p, q \mathrm{I} A$.

IVa.7. $\mathbf{T} = \{(x, A)$, *all* $x\} \cup \{(y, Y) : y \mathrm{I} A;\ Y \mathrm{I} y^\sigma\}$, σ *as in* IVa.3.

IVb.1.—IVb.7. *are dual to* IVa.1.—IVa.7., *respectively.*

V.1. $\mathbf{T} = \{(x, A) : x \mathrm{I} A\} \cup \{(c, Y) : Y \mathrm{I} c\}$,
V.2. $\mathbf{T} = \{(z, Z) : z \mathrm{I} A; Z \mathrm{I} c\}$,
V.3a. $\mathbf{T} = \{(x, A),\ \mathit{all}\ x\} \cup \{(c, Y) : Y \mathrm{I} c\}$,
V.4. $\mathbf{T} = \{(x, A),\ \mathit{all}\ x\} \cup \{(c, Y),\ \mathit{all}\ Y\}$,
V.5a. $\mathbf{T} = \{(x, A),\ \mathit{all}\ x\} \cup \{(y, Y) : y \mathrm{I} A; Y \mathrm{I} c\}$,
V.6. $\mathbf{T} = \{(x, A),\ \mathit{all}\ x\} \cup \{(c, Y),\ \mathit{all}\ Y\} \cup (z, Z) : z \mathrm{I} A; Z \mathrm{I} c\}$,
} $c \mathrm{I} A$.

V.3b *and* V.5b *are dual to* V.3a *and* V.5a, *respectively.*

VIa.1. $\mathbf{T} = \{(x, Y) : A \mathrm{I} x \mathrm{I} Y\}$,
VIa.2. $\mathbf{T} = \{(x, A),\ \mathit{all}\ x\} \cup \{(y, Y) : A \mathrm{I} y \mathrm{I} Y\}$,
VIa.3. $\mathbf{T} = \{(x, Y),\ \mathit{all}\ Y; x \mathrm{I} A\}$,
VIa.4. $\mathbf{T} = \{(x, A),\ \mathit{all}\ x\} \cup \{(y, Y),\ \mathit{all}\ Y; y \mathrm{I} A\}$,
} *for some line* A.

VIb.1—VIb.4 *are dual to* VIa.1—VIa.4, *respectively.*

VII.1. $\mathbf{T} = \{(x, X) : x \mathrm{I} X\}$,
VII.2. $\mathbf{T} = \{(x, X),\ \mathit{all}\ x, X\}$.

This theorem is proved by multiple application of **17** and **18** above, in the spirit of LENZ 1954 and BARLOTTI 1957b.[1]) Consequently, we refer to the 53 possibilities for $\mathbf{T}(\Gamma)$, listed in **20**, as the *Lenz-Barlotti types for collineation groups* of projective planes.

With the exception of I.8, there actually exist collineation groups Γ for each of the types in **20**.[2]) In many cases it can be proved, however, that Γ cannot be the full collineation group Aut**P**. This leads to the problem of determining the possibilities for **T**(Aut**P**); these are called the *Lenz-Barlotti types for projective planes* **P**. The possibilities are much more limited here; this follows from **19** and other considerations, of which many will be at least outlined further below. The following

[1]) These authors were only interested in the case Γ = Aut**P**; here **19** is also applicable and leads to the exclusion of many of the types listed in **20**. Furthermore, Lenz only determined the subsets of *incident* point-line pairs in **T**(Aut**P**). There are only seven such types, referred to by the Roman numerals in **20**. Barlotti's work consisted in refining Lenz's classification so as to include also non-incident center-axis pairs. (Incidentally, the reason that we have called one of the types IVa. 1′ is only that we wanted to retain Barlotti's numbering; there is no mathematical reason for it.) Another refinement of the Lenz-Barlotti classification, taking into account also certain correlations of the plane (cf. Section 3.3), was given by JÓNSSON 1963.

[2]) The reason for including I.8 here is that the nonexistence proof for such collineation groups requires rather more than **17** and **18**; cf. **4.3.32** below. For the finite case, see also **3.3.2**.

table represents the present state of knowledge on the existence problem for Lenz-Barlotti types (of groups and of planes):

Table 1: *Lenz-Barlotti types*

Type	Finite case: Group exists	Finite case: Plane exists	Infinite case: Group exists	Infinite case: Plane exists
I.1	yes	yes; cf. 5.4	yes	yes (HILBERT 1899)
I.2	yes	?	yes	yes (SPENCER 1960)
I.3	yes	?	yes	yes (YAQUB 1961 b)
I.4	yes	?	yes	yes (NAUMANN 1954)
I.5	yes; $n \leqq 9$	no	no	no
I.6	yes; $n \leqq 4$ cf. 4.3	no	?	?
I.7	yes; $n \leqq 5$	no	no	no
I.8	no	no	no	no
II.1	yes	yes; cf. 5.4	yes	yes (SPENCER 1960)
II.2	yes	?	yes	yes[1]
II.3	yes; $n \leqq 4$ (cf. 4.3)	no	no	no (SPENCER 1960)
II.4 a	yes	no	yes	no
II.5 a	yes	no by **19**	yes	no by **19**
III.1	yes	?	yes	yes (YAQUB 1961 a)
III.2	yes	no; cf. 4.3	yes	yes (MOULTON 1902)
III.3	yes	no	yes	no
III.4	yes	no by **19**	yes	no by **19**
IV a.1	yes	yes; cf. 5.2	yes	yes
IV a.1′	yes	no, by **19**	yes	no, by **19**
IV a.2	yes	yes; cf. 5.2	yes	yes
IV a.3	yes; $n \leqq 9$ (cf. 4.3)	yes; $n = 9$	no	no
IV a.4	yes	no	yes	no
IV a.5	yes	no by **19**	yes	no by **19**
IV a.6	yes	no	yes	no
IV a.7	yes	no	yes	no
V.1	yes	yes; cf. 5.3	yes	yes
V.2	yes	no	yes	no
V.3 a	yes	no	yes	no
V.4	yes	no by **19**	yes	no by **19**
V.5 a	yes	no	yes	no
V.6	yes	no	yes	no
VI a.1	yes	no; cf. **22**	yes	no; cf. **22**
VI a.2	yes	no	yes	no
VI a.3	yes	no by **19**	yes	no by **19**
VI a.4	yes	no	yes	no
VII.1	yes	no; cf. **22**	yes	yes (MOUFANG 1933)
VII.2	yes	yes	yes	yes

[1]) No example of this seems to have been published. For the following construction, the author is indebted to H. Salzmann. Call points the pairs (x, y) of real numbers, lines the point sets $L(a, b) = \{(x, (x^\alpha a^\alpha)^{\alpha^{-1}} + b)\}$ and $L(c) = \{(c, y)\}$, where α is a non-identity order preserving permutation of the reals, and define incidence by set theoretical inclusion. This yields an affine plane, and the corresponding projective plane is of Lenz-Barlotti type II.2.

The main tool for the proofs of many of the results in Table 1 is the introduction of *coordinates* in a projective plane. We give a brief account of this now, following essentially HALL 1943. (See also HALL 1959, Section 20.3.)

Let o, e, u, v be an (ordered) quadrangle in the projective plane $\mathbf{P}$, put $W = u\,v$, and let $\mathbf{A}$ denote the affine plane $\mathbf{P}^W$. We consider a set $\mathfrak{T}$ with the same cardinality as a line of $\mathbf{A}$; for example, $\mathfrak{T}$ may consist of the lines $\neq W$ through u. Two distinct elements in $\mathfrak{T}$ are called 0 and 1. We set up a one-one correspondence between the points of $\mathbf{A}$ and the ordered pairs (x, y) of elements in $\mathfrak{T}$ as follows: The points

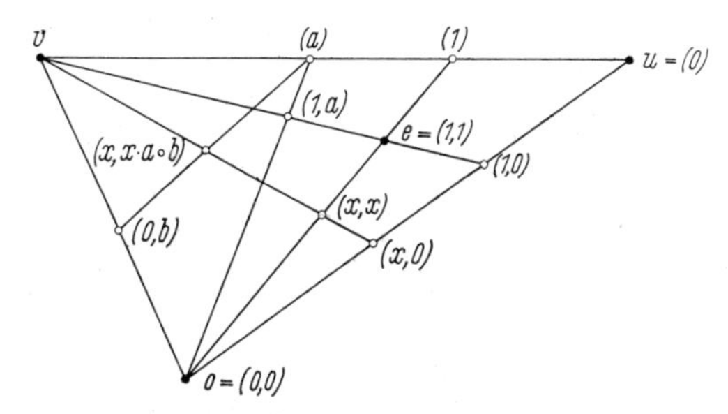

Fig. 2. Coordinates

$\neq u$ of ou are assigned the pairs $(x, 0)$, $x \in \mathfrak{T}$, such that $o = (0, 0)$ and $(ou)\,(ev) = (1, 0)$. The points $\neq v$ of ov are then assigned the pairs $(0, y)$, $y \in \mathfrak{T}$, by the following rule:

$$(0, y) = (o\,v)\,\big(u[(o\,e)\,\{(y, 0)\,v\}]\big).$$

In particular, $(0, 1) = (o\,v)\,(u\,e)$. Finally, we put

$$(x, y) = [(x, 0)\,v]\,[(0, y)\,u]; \tag{21}$$

note that this assigns the pairs (x, x) to the points of the line oe, and in particular $e = (1, 1)$. Next, we label the points of the line W which are different from v:

$$(t) = [(0, 0)\,(1, t)]\,W. \tag{22}$$

Now we define a ternary operation in $\mathfrak{T}$ as follows (cf. Fig. 2):

$$(x, x \cdot a \circ b) = [(x, 0)\,v]\,[(a)\,(0, b)]. \tag{23}$$

In other words, the line through (a) and $(0, b)$ is represented by the equation $y = x \cdot a \circ b$. The axioms (1)—(3) for $\mathbf{P}$ imply the following properties of this operation:

$$(24)\quad \begin{cases} \text{(a)} & x \cdot 0 \circ b = 0 \cdot x \circ b = b, \quad \textit{for all } x, b \in \mathfrak{T}. \\ \text{(b)} & x \cdot 1 \circ 0 = 1 \cdot x \circ 0 = x, \quad \textit{for all } x \in \mathfrak{T}. \\ \text{(c)} & \textit{Given } x, y, a \in \mathfrak{T}, \textit{ there is a unique } b \in \mathfrak{T} \\ & \quad \textit{such that } y = x \cdot a \circ b. \\ \text{(d)} & \textit{Given } x_i, y_i \in \mathfrak{T}\ (i = 1, 2) \textit{ with } x_1 \neq x_2, \\ & \quad \textit{there is a unique ordered pair } (a, b) \textit{ such that} \\ & \qquad y_i = x_i \cdot a \circ b\ (i = 1, 2). \\ \text{(e)} & \textit{Given } a_i, b_i \in \mathfrak{T}\ (i = 1, 2) \textit{ with } a_1 \neq a_2, \textit{ there is a} \\ & \quad \textit{unique } x \textit{ such that } x \cdot a_1 \circ b_1 = x \cdot a_2 \circ b_2. \end{cases}$$

Any set $\{0, 1, \ldots\}$ with a ternary operation satisfying (24) will be called here a *ternary field.*[1]) Thus every ordered quadrangle o, e, u, v of a projective plane **P** gives rise to a ternary field $\mathfrak{T} = \mathfrak{T}(o, e, u, v)$, and:

21. *$\mathfrak{T}(o, e, u, v)$ and $\mathfrak{T}(o', e', u', v')$ are isomorphic if and only if there exists a collineation mapping o onto o', e onto e', u onto u', and v onto v'.*

Proof: Pickert 1955, p. 37—38.

Given any ternary field $\mathfrak{T}$, the incidence structure $\mathbf{A} = \mathbf{A}(\mathfrak{T})$ defined as follows is an affine plane: points of **A** are the ordered pairs (x, y) with $x, y \in \mathfrak{T}$, lines are the point sets

$$(25)\quad \begin{aligned} L(a, b) &= \{(x, x \cdot a \circ b) \,\vdots\, x \in \mathfrak{T}\}, \\ L(c) &= \{(c, y) \,\vdots\, y \in \mathfrak{T}\}, \end{aligned}$$

and incidence is set theoretic inclusion. In the projective plane **P** corresponding to **A**, let $o = (0, 0)$, $e = (1, 1)$, $u = (0)$, and $v =$ ideal point of $L(0)$. Then the ternary field $\mathfrak{T}(o, e, u, v)$ is essentially identical with the given $\mathfrak{T}$. Thus there is a canonical correspondence between ternary fields and projective planes with distinguished ordered quadrangles.

In any ternary field $\mathfrak{T}$, *addition* and *multiplication* are defined as follows:

$$(26)\quad \begin{aligned} a + b &= a \cdot 1 \circ b \\ a\, b &= a \cdot b \circ 0. \end{aligned}$$

[1]) The more customary term is "planar ternary ring" (Hall 1943, 1959). The present terminology follows Pickert 1955 ("Ternärkörper"), the reason being that there are no proper homomorphisms between ternary fields, and hence no ideals. This follows from (24).

With respect to addition, $\mathfrak{T}$ is a loop with neutral element 0; this loop will be denoted by $\mathfrak{T}^+$. The set $\mathfrak{T} - \{0\}$ is a loop with respect to multiplication, with neutral element 1, which will be denoted by $\mathfrak{T}^\times$. A ternary field is *linear* if it satisfies

(27) $$x \cdot a \circ b = x\,a + b \qquad \textit{for all } x, a, b \in \mathfrak{T}.$$

It can be verified that linearity of $\mathfrak{T}(o, e, u, v)$ is equivalent to a common special case of $(v, u\,v)$- and $(u, o\,v)$-Desargues, as defined in the context of **16** (cf. PICKERT 1955, p. 98). Thus if a projective plane $\mathbf{P}$ is (v, uv)- or (u, L)-transitive, for two distinct points u, v and a line $L \neq uv$ through v, then $\mathfrak{T}(o, e, u, v)$ is linear for any choice of $o \mathrm{I} L$ and e.

More than this can be proved. In fact, each of the possible Lenz-Barlotti types for projective planes corresponds to a system of algebraic laws which must be satisfied by certain well-defined ternary fields of any plane of that type. We treat only the more important cases here; other situations will occur, for the finite case, in Section 4.3 below.

A linear ternary field is called a *cartesian group*[1]) if its additive loop is associative and thus a group. Note that in a cartesian group the mappings

(28) $$x \to -x\,a + x\,b \quad \textit{and} \quad x \to a\,x - b\,x$$

must be permutations whenever $a \neq b$. A *quasifield*[1]) is a cartesian group $\mathfrak{T}$ satisfying

(29) $$(x + y)\,z = x\,z + y\,z \qquad \textit{for all } x, y, z \in \mathfrak{T};$$

it is not difficult to prove (see PICKERT 1955, p. 91) that addition in any quasifield is commutative. A *semifield*[1]) is a quasifield satisfying also

(30) $$x\,(y + z) = x\,y + x\,z \qquad \textit{for all } x, y, z \in \mathfrak{T},$$

and a *planar nearfield*[1]) is a quasifield whose multiplicative loop is associative and hence a group. An *alternative field* is a semifield satisfying

(31) $$x^2\,y = x\,(x\,y) \quad \textit{and} \quad x\,y^2 = (x\,y)\,y \qquad \textit{for all } x, y \in \mathfrak{T},$$

[1]) The terminologies vary widely here. Cartesian groups (PICKERT 1952, 1955) are called "cartesian number systems" by BAER 1942. Instead of "quasifield" (PICKERT 1955), the terms "left Veblen-Wedderburn system" and "right Veblen-Wedderburn system" are customary, after VEBLEN & WEDDERBURN 1907, the choice of "left" or "right" depending on whether (29) is called the left or the right distributive law. Similarly, (32) is sometimes called the "right inverse property". Instead of "semifield" (KNUTH 1965), PICKERT 1955 uses "distributive quasifield", and many other authors say "division algebra" or "division ring". Planar nearfields, as defined here, are of course nearfields in the sense of Section 1.4, but not every nearfield is planar (cf. ZEMMER 1964), because (28) may not be satisfied. However, it can be shown that every nearfield of finite rank over its kernel (cf. p. 132 below) is planar.

where x^2 is defined to be xx. Finally, we say that $\mathfrak{T}$ (more precisely, the multiplicative loop of $\mathfrak{T}$) has the *left inversive property*[1]) if

(32) $xx' = 1$ *implies* $x(x'y) = y$, *for all* $x, y \in \mathfrak{T}$.

Note that (32) implies that the right inverse x' of x is also a left inverse. Thus right and left inverses coincide if $\mathfrak{T}$ has the left inverse property; without (32) this need not be so

22. *Let* o, e, u, v *be a quadrangle in a projective plane* **P**, *and let* $\mathfrak{T}$ *be the ternary field* $\mathfrak{T}(o, e, u, v)$. *Also, let* Γ *denote the full collineation group* Aut**P** *of* **P**. *Then:*

(a) **P** *is* (v, uv)*-transitive if and only if* $\mathfrak{T}$ *is a cartesian group. In this case,* $\Gamma(v, uv)$ *is isomorphic to* $\mathfrak{T}^+$.
(b) **P** *is* (u, ov)*-transitive if and only if* $\mathfrak{T}$ *is linear with associative multiplication. In this case,* $\Gamma(u, ov)$ *is isomorphic to* $\mathfrak{T}^\times$.
(c) **P** *is* (uv, uv)*-transitive if and only if* $\mathfrak{T}$ *is a quasifield.*
(d) **P** *is* (v, v)*-transitive if and only if* $\mathfrak{T}$ *is a cartesian group satisfying* (30).
(e) **P** *is* (u, v)*-transitive if and only if* $\mathfrak{T}$ *is a planar nearfield.*
(f) **P** *is* (uv, uv)*- and* (v, v)*-transitive if and only if* $\mathfrak{T}$ *is a semifield.*
(g) **P** *is* (uv, uv)*- and* (ov, ov)*-transitive if and only if* $\mathfrak{T}$ *is a semifield with the left inversive property.*
(h) **P** *is* (L, L)*-transitive for every line* L *if and only if* $\mathfrak{T}$ *is an alternative field.*
(i) **P** *is desarguesian if and only if* $\mathfrak{T}$ *is a (not necessarily commutative) field.*

For the proofs of these results, see Pickert 1955, Sections 3.5 and 6.1. Some of them are easy consequences of others; for example, **(e)** and **(f)** follow immediately from **(b)**, **(c)**, **(d)**. Result **(e)** provides a proof of **19** above. The left inverse property (32), for a semifield, implies the identity

(33) $[x(yx)]z = x[y(xz)]$ *for all* $x, y, z \in \mathfrak{T}$

(Moufang 1935; Pickert 1955, p. 160; Hall 1959, p. 370); and it can be shown[2]) that (33) implies (31). Thus it follows that there exists no plane of any one of the Lenz-Barlotti types VI of **20**, as claimed in Table 1. Also, a theorem of E. Artin (Zorn 1931; Pickert 1955,

[1]) See footnote [1]) on p. 129.

[2]) That the first equation (31) follows from (33) is obvious: put $y = 1$. But the second equation is difficult to derive. If the semifield in question has characteristic $\neq 2$, then the second equation (31) follows from the first (Skorynakov 1951, Kleinfeld 1953). If the characteristic is 2, this is no longer true; an example (due to R. H. Bruck) is given by San Soucie 1955. But in the same paper it is shown that (33) does imply the second equation (31), also in case of characteristic 2.

p. 161—162; HALL 1959, p. 376—382) shows that a finite alternative field is associative and hence a field; this proves the claim in Table 1 that there exist no finite planes of Lenz-Barlotti type VII.1.

Let **A** be an affine plane and **P** the corresponding projective plane: $\mathbf{A} = \mathbf{P}^W$ for some line W of **P**. A collineation of **A** is called a *dilatation* if it has axis W when regarded as a collineation of **P**. A *translation* of **A** is a fixed point free dilatation or the identity (compare here the terminology of Section 2.3), i.e. an elation of **P** with axis W. We call **A** a *translation plane* if the group of all translations of **A** is transitive on the points of **A**, in other words if **P** is (W, W)-transitive. By **22c**, the translation planes are precisely those affine planes which can be coordinatized by a quasifield. Translation planes have been well investigated, and almost all known finite planes are either translation planes or closely related to them, as will be seen in Chapter 5. For this reason we devote the remainder of this section to some general results on translation planes and related concepts.

Let **A** be a translation plane and T its (full) translation group. Again we put $\mathbf{A} = \mathbf{P}^W$ and consider T also as the group of all elations with axis W of the projective plane **P**. By (14), we have $\mathsf{T} = \mathsf{T}(W, W) = \bigcup_{x \,\mathrm{I}\, W} \mathsf{T}(x, W)$, and as the $x \,\mathrm{I}\, W$ are just the parallel classes of **A**, we can write

$$\mathsf{T} = \bigcup_{\mathfrak{X}} \mathsf{T}(\mathfrak{X}), \tag{34}$$

where $\mathfrak{X}$ ranges over the parallel classes of **A** and $\mathsf{T}(\mathfrak{X})$ denotes the subgroup of T which fixes every line of $\mathfrak{X}$. It follows from **4** that

$$\mathsf{T}(\mathfrak{X}) \frown \mathsf{T}(\mathfrak{Y}) = 1 \qquad \textit{if } \mathfrak{X} \neq \mathfrak{Y}, \tag{35}$$

so that the $\mathsf{T}(\mathfrak{X})$ form a *partition* of T, as defined in Section 1.2. Furthermore, this is even a *congruence partition* of T, in the sense that

$$\textit{if } \mathfrak{X} \neq \mathfrak{Y}, \textit{ then } \mathsf{T}(\mathfrak{X})\,\mathsf{T}(\mathfrak{Y}) = \mathsf{T}.^{1)} \tag{36}$$

Conversely, if T is an abstract group possessing a nontrivial congruence partition, i.e. a set $\mathscr{C}$ of proper subgroups $\mathsf{T}(\mathfrak{X})$ satisfying (34)—(36) [here $\mathfrak{X}$ ranges over some index set of cardinality $\geqq 2$], then T may be regarded as the full translation group of a translation plane. In fact, the incidence structure $\mathbf{J} = \mathbf{J}(\mathsf{T}, \mathscr{C})$ of Section 1.2, whose points are the elements of T and whose blocks are the cosets[2]) of the $\mathsf{T}(\mathfrak{X})$, is a translation plane and T its full translation group. Hence:

---

[1]) This implies, of course, that $\mathsf{T}(\mathfrak{X})$ and $\mathsf{T}(\mathfrak{Y})$ generate T. There are examples showing that (36) cannot be replaced by this weaker property if the following converse is to hold.

[2]) Left and right cosets are identical here, for (34)—(36) imply that T is abelian (ANDRÉ 1954a, Satz 7).

23. *There is a canonical correspondence*[1]) *between translation planes and congruence partitions.*

Proof: ANDRÉ 1954a, Satz 9.

The *kernel of the translation plane* **A** is the set $\mathfrak{K}(\mathbf{A})$ of all endomorphisms α of T with

$$(37) \qquad \mathsf{T}(\mathfrak{X})^{\alpha} \subseteqq \mathsf{T}(\mathfrak{X}), \qquad \textit{for all parallel classes } \mathfrak{X}.$$

With the usual addition and multiplication of endomorphisms, $\mathfrak{K}(\mathbf{A})$ is a ring, and it can be shown (ANDRÉ 1954a, Satz 10) that $\mathfrak{K}(\mathbf{A})$ is even a field[2]). The *kernel of a quasifield* $\mathfrak{Q}$ is the set $\mathfrak{K}(\mathfrak{Q})$ of all $k \in \mathfrak{Q}$ with

$$(38) \quad k(xy) = (kx)\,y \quad \textit{and} \quad k(x+y) = k\,x + k\,y, \qquad \textit{for all } x, y \in \mathfrak{Q}.$$

Clearly, $\mathfrak{K}(\mathfrak{Q})$ is also a field, and in fact

24. *If* $\mathfrak{Q}$ *is any*[3]) *coordinatizing quasifield of the translation plane* **A**, *then* $\mathfrak{K}(\mathfrak{Q}) \cong \mathfrak{K}(\mathbf{A})$.

Proof: ANDRÉ 1954a, p. 174—176. This result justifies the choice of the same term "kernel" for two seemingly unrelated concepts.

By (38), a quasifield $\mathfrak{Q}$ may be regarded as a left vector space over any subfield $\mathfrak{F}$ of its kernel $\mathfrak{K}(\mathfrak{Q})$. Also, the group T may be regarded as the additive group of a vector space over $\mathfrak{F}$, and if $[\mathfrak{Q}:\mathfrak{F}]$ and $[\mathsf{T}:\mathfrak{F}]$ denote the respective ranks of these vector spaces, then

$$(39) \qquad [\mathsf{T}:\mathfrak{F}] = 2[\mathfrak{Q}:\mathfrak{F}]$$

(ANDRÉ 1954a, p. 181). Thus if one of these ranks is finite, then they both are, and $[\mathsf{T}:\mathfrak{F}]$ is even. Clearly, $\mathfrak{K}(\mathfrak{Q}) = \mathfrak{Q}$ if and only if $\mathfrak{Q}$ is a field; hence **22i** shows that

25. **A** *is desarguesian if and only if* $[\mathsf{T}:\mathfrak{K}(\mathbf{A})] = 2$.

The subgroups $\mathsf{T}(\mathfrak{X})$ of the congruence partition associated with the translation plane **A** are, by (37), subspaces of the vector space T over $\mathfrak{F} \subseteqq \mathfrak{K}(\mathbf{A})$. Hence in the corresponding desarguesian projective

[1]) In **23** and **26**, the term "canonical correspondence" is to be understood in the following sense: The isomorphism classes of translation planes [in **26** with condition (**a**)] can be put into a one-one correspondence with the isomorphism classes of groups with congruence partitions [of projective $(2t+1)$-spaces over $\mathfrak{K}$ with t-spreads].

[2]) As always in this book, "fields" need not be commutative. The field $\mathfrak{K}(\mathbf{A})$ was first considered in a more special situation by ARTIN 1940.

[3]) Two such quasifields are usually not isomorphic; cf. **21**. In general, it is a difficult question to decide whether or not two quasifields with isomorphic kernels coordinatize the same translation plane. Complicated necessary and sufficient conditions for this were given by SKORNYAKOV 1949. Some of the questions involved will be discussed in a more special situation further below; cf. **32** and **34**.

geometry $\mathscr{P} = \mathscr{P}(\mathsf{T})$ of the vector space T [cf. **1.4.2**], these subgroups define a *spread,* i.e. a collection $\mathscr{S}$ of mutually disjoint subspaces covering all of $\mathscr{P}$ [cf. **1.4.6**]; and $\mathscr{S}$ has the further property that any two distinct subspaces of $\mathscr{S}$ span $\mathscr{P}$. Conversely, every spread of $\mathscr{P}$ with this property defines a congruence partition of the additive group of the vector space underlying $\mathscr{P}$, and hence a translation plane, by **23**. In the finite-dimensional case, the situation is as follows (terminology as in Section 1.4):

26. *Let $\mathfrak{F}$ be a field and t a positive integer. Then there is a canonical correspondence*[1]) *between*

(**a**) *the translation planes* $\mathbf{A}$ *with* $\mathfrak{F} \subseteqq \mathfrak{K}(\mathbf{A})$ *and* $[\mathfrak{Q} : \mathfrak{F}] = t + 1$, *for an arbitrary coordinatizing quasifield* $\mathfrak{Q}$ *of* $\mathbf{A}$, *and*

(**b**) *the t-spreads of the* $(2t+1)$*-dimensional projective geometry over* $\mathfrak{F}$.

Proof: André 1954a, p. 182, where it is also pointed out that different choices of $\mathfrak{F} \subseteqq \mathfrak{K}(\mathbf{A}) \cong \mathfrak{K}(\mathfrak{Q})$ yield different spread representations of $\mathbf{A}$. Result **26** was rediscovered independently by Bruck & Bose 1964 and Segre 1964.

The following results are concerned with representations of certain types of collineations in a translation plane $\mathbf{A}$ with coordinatizing quasifield $\mathfrak{Q}$.

27. *The translations of* $\mathbf{A}$ *are the mappings*

$$\tau(s,t)\colon (x,y) \to (x+s, y+t), \qquad s,t \in \mathfrak{Q}; \tag{40}$$

and the dilatations with center $(0,0)$ *are the mappings*

$$\delta(k)\colon (x,y) \to (kx, ky), \qquad k \in \mathfrak{K}(\mathfrak{Q}). \tag{41}$$

The proof [André 1954a, (7) and (9)] is straightforward. As a consequence of (41), we note:

28. *For any point p of a translation plane* $\mathbf{A}$, *the group of dilatations with center p is isomorphic to the multiplicative group of the kernel* $\mathfrak{K}(\mathbf{A})$.

We consider now axial collineations in $\mathbf{A}$. Such a collineation is central in the corresponding projective plane $\mathbf{P}$ with $\mathbf{A} = \mathbf{P}^W$, and since W must stay fixed, the center is on W. Thus:

29. *A nontrivial collineation φ of an affine plane* $\mathbf{A}$ *which has an axis in* $\mathbf{A}$ *has no center in* $\mathbf{A}$. *Instead, φ fixes every line of a unique parallel class.*

If the axis belongs to this parallel class, we call φ a *shear,* otherwise a *strain.* We give now representations of shears with axis ov (hence

[1]) See footnote [1]) on p. 132.

with center v) and of strains with axes ov and ou (and corresponding centers u, v), in the translation plane **A** over the quasifield $\mathfrak{Q}$. The *distributor* of $\mathfrak{Q}$ is the set $\mathfrak{D} = \mathfrak{D}(\mathfrak{Q})$ of all $d \in \mathfrak{Q}$ with

(42) $$x(d + y) = x\,d + x\,y \qquad \textit{for all } x, y \in \mathfrak{Q}.$$

The *middle* and *right nucleus*[1]) of $\mathfrak{Q}$ are the sets $\mathfrak{M}(\mathfrak{Q})$ and $\mathfrak{R}(\mathfrak{Q})$ of those $m \in \mathfrak{Q}$ and $r \in \mathfrak{Q}$ for which

(43) $$(x\,m)\,y = x\,(m\,y)$$

or, respectively,

(44) $$(x\,y)\,r = x\,(y\,r)$$

holds, for all $x, y \in \mathfrak{Q}$. We then have the following results.

30. *The shears with axis ov, of the translation plane* **A** *over the quasifield $\mathfrak{Q}$, are the mappings*

(45) $$\sigma(d): \quad (x, y) \to (x, xd + y), \qquad d \in \mathfrak{D}(\mathfrak{Q}).$$

The strains with center u and axis ov are the mappings

(46) $$\mu(m): \quad (x, y) \to (xm, y), \qquad m \in \mathfrak{M}(\mathfrak{Q}),$$

and the strains with center v and axis ou are the mappings

(47) $$\varrho(r): \quad (x, y) \to (x, yr), \qquad r \in \mathfrak{R}(\mathfrak{Q}).$$

The proofs are again straightforward; see André 1955, (4), (7), (7′). These results generalize parts of **22**: the projective plane **P** with $\mathbf{A} = \mathbf{P}^W$ is (uv, uv)-transitive by hypothesis, and it is (v, v)-transitive it and only if $\mathfrak{D}(\mathfrak{Q}) = \mathfrak{Q}$, i.e. if $\mathfrak{Q}$ is a semifield (cf. **22c**, **d**). Also, **P** is (u, ov)-transitive if and only if $\mathfrak{M}(\mathfrak{Q}) = \mathfrak{R}(\mathfrak{Q}) = \mathfrak{Q}$, in which case it must then also be (v, ou)-transitive (cf. **19** and **22b**); $\mathfrak{Q}$ is then clearly a planar nearfield.

We give a few more details for those special cases where $\mathfrak{Q}$ is a semifield or a planar nearfield. First, it is straightforward to prove that

31. *The nuclei $\mathfrak{M}(\mathfrak{Q})$ and $\mathfrak{R}(\mathfrak{Q})$ of a semifield $\mathfrak{Q}$ are fields.*

Thus **30** shows that the groups of (u, ov)- and (v, ou)-strains are isomorphic to the multiplicative groups of these fields. If $\mathfrak{F}$ is the intersection of the fields $\mathfrak{K}(\mathfrak{Q})$, $\mathfrak{M}(\mathfrak{Q})$ and $\mathfrak{R}(\mathfrak{Q})$, then $\mathfrak{Q}$ may be regarded as a left vector space over $\mathfrak{F}$. Suppose that $\mathfrak{Q}$ and $\mathfrak{S}$ are two semifields with the same $\mathfrak{F}$. Then by an *isotopism* from $\mathfrak{Q}$ to $\mathfrak{S}$ is meant

[1]) The *left nucleus* $\mathfrak{L}(\mathfrak{Q}) = \{l \in \mathfrak{Q} : (l\,x)\,y = l\,(x\,y) \text{ for all } x, y \in \mathfrak{Q}\}$ bears less significance in the present context than $\mathfrak{M}(\mathfrak{Q})$ and $\mathfrak{R}(\mathfrak{Q})$. Note that $\mathfrak{L}(\mathfrak{Q})$ contains $\mathfrak{K}(\mathfrak{Q})$, and in fact $\mathfrak{L}(\mathfrak{Q}) = \mathfrak{K}(\mathfrak{Q})$ if and only if $\mathfrak{Q}$ is a semifield.

a triple (α, β, γ) of nonsingular linear transformations from $\mathfrak{Q}$ onto $\mathfrak{S}$ (both considered as vector spaces over $\mathfrak{F}$) such that

$$x^\alpha \cdot y^\beta = (x\,y)^\gamma \qquad \textit{for all } x, y \in \mathfrak{Q}.$$

Here multiplication is written $x\,y$ in $\mathfrak{Q}$ and $x \cdot y$ in $\mathfrak{S}$. If there exists an isotopism from $\mathfrak{Q}$ onto $\mathfrak{S}$, then $\mathfrak{Q}$ and $\mathfrak{S}$ are said to be *isotopic.*

32. *Two semifields coordinatize isomorphic translation planes if and only if they are isotopic.*[1])

Proof: ALBERT 1960, Section 9. An isotopism of $\mathfrak{Q}$ onto itself is called an *autotopism.* If φ is a collineation which fixes the points $o = (0, 0)$, $u = (0)$ and v of the translation plane **A** over the semifield $\mathfrak{Q}$, then $(x, 0)\varphi = (x^\alpha, 0)$, $(a)\varphi = (a^\beta)$ [cf. (22)], and $(0, y)\varphi = (0, y^\gamma)$, for three well-defined permutations α, β, γ of $\mathfrak{Q}$. It follows easily that (α, β, γ) is an autotopism of $\mathfrak{Q}$, and in fact one may prove now:

33. *The autotopisms of a semifield $\mathfrak{Q}$ form a group isomorphic to the stabilizer of the points o, u, v, in the full collineation group of the translation plane over $\mathfrak{Q}$.*

This stabilizer has a complement in the group Γ of all collineations fixing v and uv. (Unless $\mathfrak{Q}$ is a field, Γ is the full collineation group, because the plane must then be of Lenz-Barlotti type V.1.) This complement is the subgroup Σ generated by all translations (40) and shears (45). For an elegant representation of the plane within this metabelian group Σ, see CRONHEIM 1965.

Now let **A** be a translation plane over a planar nearfield $\mathfrak{Q}$ which is not a field. Then **A** may be coordinatized also by quasifields not isomorphic to $\mathfrak{Q}$. No such quasifield can be also a nearfield:

34. *If $\mathfrak{Q}$ and $\mathfrak{Q}'$ are planar nearfields coordinatizing the same translation plane, then $\mathfrak{Q}$ and $\mathfrak{Q}'$ are isomorphic.*

Proof: ANDRÉ 1955, Satz 7. A nondesarguesian translation plane can be coordinatized by a planar nearfield if and only if it is of Lenz-Barlotti type IVa.2 or IVa.3; this shows that any collineation fixing v must also fix u, so that in particular there are no shears with axis ov in such a plane. In view of **30**, we can conclude:

35. *The distributor $\mathfrak{D}(\mathfrak{Q})$ of any planar nearfield $\mathfrak{Q}$ which is not a field consists of* 0 *only.*

Further results on planes over planar nearfields, particularly their collineation groups, are found in ANDRÉ 1955. Some of these results (for the finite case) will be given in Section 5.2.

---

[1]) Compare here footnote [3]) on p. 132. There is a generalization of **34** to arbitrary ternary fields; see KNUTH 1965, Theorem 3.3.2; cf. also SANDLER 1964.

We have remarked above that a translation plane is essentially the same as a (W, W)-transitive projective plane. We shall now make a few remarks about the dual concept. Let $\mathbf{P}$ be a projective plane which is (v, v)-transitive for some point v. Let W be a line through v; then the affine plane $\mathbf{A} = \mathbf{P}^W$ is a *shears plane* in the sense that

(48) *There is a parallel class $\mathfrak{B}$ of lines (with ideal point v) such that, for every $X \in \mathfrak{B}$, the group of shears with axis X is transitive on the ideal points $\neq v$.*

It follows from **22d** that every shears plane can be coordinatized by a cartesian group satisfying (30). If $\mathfrak{Q}$ is an arbitrary quasifield, then the following system $\mathfrak{Q}^*$ is clearly such a cartesian group: Addition is the same in $\mathfrak{Q}$ and $\mathfrak{Q}^*$, and the multiplication $*$ in $\mathfrak{Q}^*$ is related to that in $\mathfrak{Q}$ by

(49) $$x * y = y x \qquad \text{for all } x, y \in \mathfrak{Q}.$$

Also, it is clear that every cartesian group satisfying (30) must be of this form $\mathfrak{Q}$. Moreover:

36. *For any quasifield $\mathfrak{Q}$, the projective planes coordinatized by $\mathfrak{Q}$ and $\mathfrak{Q}^*$ are dual to each other.*

For the proof of this, see PICKERT 1955, pp. 41, 91.

Finally, we discuss a class of affine planes closely related to translation planes. We say that the affine plane $\mathbf{A}$ is a *semi-translation plane*[1]) if it contains a Baer subplane $\mathbf{B}$ [in the affine sense, as defined by (13)], such that

(50) $\mathbf{B}$ *is a translation plane,*

and

(51) *Every translation of* $\mathbf{B}$ *is induced by a translation of* $\mathbf{A}$.

Given a semi-translation plane $\mathbf{A} = \mathbf{P}^W$ with Baer subplane $\mathbf{B} = \mathbf{Q}^W$, select o, e, u, v in $\mathbf{Q}$, with $u, v \,\mathrm{I}\, W$, and consider the ternary field $\mathfrak{T} = \mathfrak{T}(o, e, u, v)$ of $\mathbf{P}$. It is clear that $\mathfrak{T}$ will contain a ternary subfield $\mathfrak{Q}$ coordinatizing $\mathbf{Q}$, and $\mathfrak{Q}$ is a quasifield, because of (50). Also, (51) implies that the translations of $\mathbf{B}$ extend to translations

(52) $$(x, y) \to (x + s, y + t) \qquad x, y \in \mathfrak{T};\ s, t \in \mathfrak{Q}$$

[1]) OSTROM 1964a uses this term in a slightly different sense; his definition applies only to finite planes, but even then ours is more special. PICKERT 1965a calls the planes considered here "normal" semi-translation planes. Note that the term is somewhat unfortunate insofar as a translation plane is not necessarily a semi-translation plane. For example, the desarguesian affine planes $\mathbf{A}(q)$, for $q = p^e$ with odd e, do not contain Baer subplanes and are therefore not semi-translation planes.

of **A**; here the addition is that of $\mathfrak{T}$, as defined by (26). The fact that (52) describes a collineation yields the following properties of $\mathfrak{T}$:

$$(53) \qquad \left.\begin{array}{r} (x+s)+y = x+(s+y) \\ (x+y)+s = x+(y+s) \\ (x+y)\,t = x\,t+y\,t \end{array}\right\} \textit{for } x, y \in \mathfrak{T};\ s, t \in \mathfrak{Q};$$

here multiplication is that defined by (26). Also, it follows that

(54) *if* $z \in \mathfrak{T} - \mathfrak{Q}$, *then* $(s, t) \to s\,z + t$ *is one-one from* $\mathfrak{Q} \times \mathfrak{Q}$ *onto* $\mathfrak{T}$.

Hence much of the structure of $\mathfrak{T}$ is determined by addition and multiplication. $\mathfrak{T}$ will be termed $\mathfrak{Q}$-*linear*[1]) provided that

$$(55) \qquad x \cdot s \circ y = x\,s + y \qquad \textit{for } x, y \in \mathfrak{T} \textit{ and } s \in \mathfrak{Q}.$$

Now the following converse holds:

37. *Let $\mathfrak{T}$ be an algebraic system with two binary operations, addition and multiplication, and suppose that $\mathfrak{T}$ contains a subsystem $\mathfrak{Q}$ which is a quasifield with respect to these operations, such that* (53) *and* (54) *are satisfied. Define*

$$(56) \qquad x \cdot y \circ z = \begin{cases} x\,y + z & \textit{if } y \in \mathfrak{Q} \\ (x+s)\,y + t & \textit{if } y \notin \mathfrak{Q} \textit{ and } z = s\,y + t; \quad s, t \in \mathfrak{Q}. \end{cases}$$

If this turns $\mathfrak{T}$ into a ternary field, then $\mathfrak{T}$ is $\mathfrak{Q}$-linear and coordinatizes a semi-translation plane **A**, *with Baer subplane* **B** *coordinatized by $\mathfrak{Q}$.*

For the proof see Ostrom 1964a; Pickert 1965a, Satz 2. Further results in this direction are found in Morgan & Ostrom 1964. Result **37** will be used in Section 5.4 for the construction of finite semi-translation planes which are not translation planes.

3.2 Combinatorics of finite planes

The definitions in Sections 2.1, 2.2, and 3.1 imply immediately that the finite projective [affine] planes are precisely the projective [affine[2])] designs with $\lambda = 1$. We shall present some combinatorial properties of these designs in this section. The blocks will always be referred to as lines; this is consistent with the definition of lines in arbitrary designs, as given in Section 2.1.

By definition (2.1.9), the *order* of a finite projective or affine plane is the integer n determined by the condition that the number of lines

---

[1]) Whether or not $\mathfrak{T}$ must be automatically $\mathfrak{Q}$-linear for every semi-translation plane seems to be an open question. (It appears rather likely that this need not be the case.)

[2]) In the affine case it suffices to demand that the design have a parallelism; cf. (2.2.5). That we must then have an affine design, and hence an affine plane, follows from result **2.2.6**.

through any point is $n+1$. Note that if **P** is a finite projective plane and $\mathbf{A} = \mathbf{P}^W$ for some line W of **P**, then **P** and **A** have the same order. The following facts are easily verified:

1. *If* $\mathbf{P} = (\mathfrak{p}, \mathfrak{L}, \mathrm{I})$ *is a projective plane of order* n, *then*

$$|\mathfrak{p}| = n^2 + n + 1, \tag{1}$$

$$|\mathfrak{L}| = n^2 + n + 1, \tag{2}$$

$$[p] = n + 1, \quad \text{for every } p \in \mathfrak{p}, \tag{3}$$

$$[L] = n + 1, \quad \text{for every } L \in \mathfrak{L}, \tag{4}$$

$$[p, q] = 1, \quad \text{for } p, q \in \mathfrak{p} \text{ and } p \neq q; \tag{5}$$

$$[L, M] = 1, \quad \text{for } L, M \in \mathfrak{L} \text{ and } L \neq M. \tag{6}$$

2. *If* $\mathbf{A} = (\mathfrak{p}, \mathfrak{L}, \mathrm{I})$ *is an affine plane of order* n, *then* (3) *and* (5) *hold, and*

$$|\mathfrak{p}| = n^2, \tag{7}$$

$$|\mathfrak{L}| = n(n+1), \tag{8}$$

$$[L] = n, \quad \text{for every } L \in \mathfrak{L}. \tag{9}$$

The converses of **1** and **2** are likewise true; in fact (1)—(6) and (3), (5), (7)—(9) are then redundant sets of conditions, as will be seen in **3** and **4** below. Before stating these results, we make the following convention: For $1 \leqq i \leqq 9$, let (i') and (i'') stand for condition (i), with "$=$" replaced by "$\leqq$" or "$\geqq$", respectively.

3. *Each of the following conditions on the integer* $n > 1$ *is necessary and sufficient for the nondegenerate incidence structure* $(\mathfrak{p}, \mathfrak{L}, \mathrm{I})$ *to be a projective plane of order* n:

(a) (3′), (4), (5), *or dually* (3), (4′), (6);
(b) (4), (5), (6), *or dually* (3), (5), (6);
(c) (1), (4), (5), *or dually* (2), (3), (6);
(d) (2), (4″), (5), *or dually* (1), (3″), (6);
(e) (1), (4), (5″), (6′), *or dually* (2), (3), (5′), (6″);
(f) (1″), (4′), (5), (6″), *or dually* (2″), (3′), (5″), (6);
(g) (1″), (4′), (5″), (6), *or dually* (2″), (3′), (5), (6″);
(h) (2), (4″), (5″), (6′), *or dually* (1), (3″), (5′), (6″);
(i) (2″), (4), (5), (6″), *or dually* (1″), (3), (5″), (6);
(j) (2″), (4), (5″), (6), *or dually* (1″), (3), (5), (6″);
(k) (1′), (2″), (4″), (5′), *or dually* (1″), (2′), (3″), (6′);
(l) (1″), (2′), (4′), (5″), *or dually* (1′), (2″), (3′), (6″);
(m) (1″), (2′), (3″), (5′), *or dually* (1′), (2″), (4″), (6′);
(n) (2″), (3′), (4), (5″), *or dually* (1″), (3), (4′), (6″);
(o) (2), (3″), (4″), (5′), *or dually* (1), (3″), (4″), (6′);

(p) (2''), (3'), (4''), (5), *or dually* (1''), (3''), (4'), (6);
(q) (2''), (3), (4'), (6''), *or dually* (1''), (3'), (4), (5'');
(r) (2), (3''), (4''), (6'), *or dually* (1), (3''), (4''), (5');
(s) (2''), (3''), (4), (6), *or dually* (1''), (3), (4''), (5);
(t) (1''), (3''), (4'), (5'), (6''), *or dually* (2''), (3'), (4''), (5''), (6');
(u) (1''), (3'), (4'), (5''), (6''), *or dually* (2''), (3'), (4'), (5''), (6'');
(v) (1''), (3'), (4''), (5''), (6'), *or dually* (2''), (3''), (4'), (5'), (6'');
(w) (1'), (2''), (3'), (4''), (5''), *or dually* (1''), (2'), (3''), (4'), (6'').

For the proof, see Corsi 1963; parts of **3** are also in Hall 1959, p. 392 and Barlotti 1962. Conditions (**c**)—(**w**) constitute a complete list of those systems taken from (1)—(6'') which (i) axiomatize finite projective planes and (ii) cannot be weakened by the omission of further conditions (1)—(6'') without losing property (i).[1]

A similarly thorough investigation for finite affine planes does not seem to exist. We note here the following analogue:

4. *Each of the following conditions on the integer $n > 1$ is necessary and sufficient for the nondegenerate incidence structure $(\mathfrak{p}, \mathfrak{L}, \mathrm{I})$ to be an affine plane of order n:*

(a) (3), (5), (9);
(b) (5), (7), (9);
(c) (5'), (8), (9);
(d) (3'), (5''), (7''), (9');
(e) (3''), (5'), (7'), (9'');
(f) (3'), (5''), (8''), (9);
(g) (3''), (5'), (8'), (9);
(h) (5'), (7''), (8'), (9);
(i) (5''), (7'), (8''), (9).

The proofs consist in simple generalizations of arguments in Ostrom 1964, Lemma 8, and Dembowski & Ostrom 1968, Lemma 11.

Next, we give two characterizations of finite projective planes by certain sets of permutations. The first of these is the special case $\lambda = 1$ of **2.1.18**:

5. *A projective plane of order n exists if and only if there exists a set Σ of permutations of a set $\mathfrak{p}$ with $|\mathfrak{p}| = n^2 + n + 1$, satisfying*

(a) $x^\varrho = x^\sigma$ *for some* $x \in \mathfrak{p}$ *and* $\varrho, \sigma \in \Sigma$ *implies* $\varrho = \sigma$, *and*

[1] In Corsi's paper, conditions (**a**) and (**b**) are not considered because they are also satisfied in the degenerate case $\mathfrak{p} = \mathfrak{L} = \emptyset$ and hence cannot serve as axioms for finite projective planes. This is the reason for including the word "nondegenerate" in the hypothesis of **3**.

(**b**) *If $x, y \in \mathfrak{p}$ and $x \neq y$, then there is a unique pair ϱ, σ of permutations in Σ such that $x^{\varrho\sigma^{-1}} = y$.*

Note that (**a**) and (**b**) imply that Σ cannot be a group, and $|\Sigma| = n + 1$.

The second characterization involves permutations of degree n rather than $n^2 + n + 1$:

6. *A projective plane of order n exists if and only if there exists a sharply 2-transitive set of permutations of degree n.*

Note that this set need not be a group; it must consist of $n(n-1)$ permutations. Result **6** can be found in Section IV of WITT 1938b; see also HALL 1943, Theorem 5.2, and Appendix I. For the proof, label the lines $\neq u\,v$ through u and the lines $\neq u\,v$ through v (where u, v are distinct points of a plane **P** of order n) by the integers $1, \ldots, n$, and call (i, j) the intersection point of line i through v and line j through u. Then any line $L\,\mathrm{I}\!\!\!/\, u, v$ of the affine plane $\mathbf{P}^{uv}$ consists of the points (i, i^π), $i = 1, \ldots, n$, where $\pi = \pi(L)$ is a permutation of $\{1, \ldots, n\}$, well defined by L. The set Π of all these $\pi(L)$ is sharply 2-transitive on $\{1, \ldots, n\}$, and conversely, every such sharply 2-transitive set can be interpreted in this fashion. If coordinates are introduced in **P** with u, v as in Section 3.1, then the permutations $\pi(L)$ are essentially the same as the mappings

$$x \to x \cdot a \circ b, \quad \text{with} \quad a \neq 0, \tag{10}$$

of the ternary field $\mathfrak{T}(o, e, u, v)$ onto itself, for any admissible choice of o and e. In fact, if L has equation $y = x \cdot a \circ b$, then (10) is $\pi(L)$.

Contrary to the situation in **5**, a sharply 2-transitive set Π of permutations of degree n may well be a group. In fact, if this is the case, Π may be identified with the set of permutations

$$x \to x\,a + b, \quad \text{with} \quad a \neq 0, \tag{11}$$

of a finite nearfield.[1]) It can be shown (HALL 1943, Theorem 5.7) that this nearfield is essentially identical with $\mathfrak{T}(o, e, u, v)$. Thus:

7. *A sharply 2-transitive set Π of permutations of degree n is a group if and only if the ternary field given by* (10) *is a nearfield; in this case the plane determined by Π is of Lenz-Barlotti type* IVa.2, IVa.3, *or* VII.2.

We shall now discuss several types of substructures of finite projective and affine planes; these will be important either for the problem of constructing such planes or for dualities or collineations of them.

---

[1]) This seems to have been first proved by CARMICHAEL 1931b; see also CARMICHAEL 1937, Chapter 13; ZASSENHAUS 1935a, b; and HALL 1943.

A *net*[1]) is a nondegenerate partial plane satisfying the parallel axiom (3.1.4); in other words, a net is an incidence structure of points and lines such that

(12) *There exist points and lines, and to every point (line) there exist two lines (points) not incident with it.*
(13) $[p, q] \leqq 1$ *for any two distinct points* p, q.
(14) *If* $p \not\mathrel{\mathrm{I}} L$, *then there exists one and only one line* $M \mathrel{\mathrm{I}} p$ *such that* $[M, L] = 0$.

Condition (14) permits the introduction of a *parallelism*[2]) in any net:

$$L \parallel M \quad \text{if and only if} \quad L = M \quad \text{or} \quad [L, M] = 0.$$

This is an equivalence relation among the lines of the net, with the following property:

(15) *Every point is on exactly one line of each parallel class.*

If $\mathbf{A}$ is an affine plane and $\mathfrak{U}$ a union of complete parallel classes of lines in $\mathbf{A}$, then the points of $\mathbf{A}$ and the lines of $\mathfrak{U}$ obviously form a net. In particular, $\mathbf{A}$ is itself a net. Conversely, the question arises under what circumstances a given net may be interpreted in this fashion as a union $\mathfrak{U}$ of parallel classes in an affine plane. A net which can be described in this way will be called *imbeddable*; we shall see that there exist many non-imbeddable finite nets.

8. *Every finite net is a tactical configuration whose parameters satisfy*

$$v = k^2, \quad b = r\,k, \quad r \leqq k + 1. \tag{16}$$

Furthermore, the number of parallel classes is r, *and every parallel class consists of* k *lines. A finite net is an affine plane if and only if* $r = k + 1$.

When speaking of finite nets, we shall henceforth often denote the parameter k by n and call it the *order* of the net. In case of imbeddability, $k = n$ is clearly the order of any imbedding affine plane. The parameter r, i.e. the number of parallel classes, is of obvious importance for the problem of imbeddability; for this reason we shall often refer to a net with parameters (16) as an *r-net.*

For the following results concerning the imbeddability problem, it is convenient to define

$$f(x) = \frac{x}{2}\,[x^3 + 3 + 2x(x + 1)]. \tag{17}$$

---

[1]) There is an extensive literature on (not necessarily finite) nets: BLASCHKE & BOL 1938; BAER 1939, 1940; PICKERT 1955, Kap. 2. For the topics discussed here, see LEVI 1942; BRUCK 1951, 1963a; OSTROM 1964c, 1965b, 1966b, 1968.

[2]) Cf. (2.2.5).

We then have:

9. *Let* **N** *be an r-net of order n.*

(**a**) *If* $n > f(n - r)$, *then* **N** *is imbeddable.*

(**b**) *If* **N** *is imbeddable and* $n > (n - r)^2$, *then the imbedding affine plane is unique.*

Also, since $f(x) > x^2$ for all positive x, the imbedding is unique in case (**a**). These results are due to BRUCK 1963a, Theorems 4.3 and 3.1. (See also BOSE 1963b.) That (**b**) is a best possible result was established by OSTROM 1964c: he showed that if $n = (n - r)^2$, then there are at most two nonisomorphic imbeddings, and he gave examples where there are actually two.

We shall now set up a fundamental relationship between r-nets and sets of $r - 2$ mutually orthogonal Latin squares. A *Latin square* of order n is an (n, n)-matrix $L = (l_{ij})$ whose entries are the integers $1, \ldots, n$,[1]) such that

(18) *Each of the integers* $1, \ldots, n$ *occurs exactly once in every row and every column of* L.

Two Latin squares $L = (l_{ij})$ and $L' = (l'_{ij})$ are called *orthogonal* to each other provided that

(19) *Each of the* n^2 *ordered pairs* (s, t), *where* $1 \leqq s, t \leqq n$, *occurs exactly once among the ordered pairs* (l_{ij}, l'_{ij}), $1 \leqq i, j \leqq n$.

The correspondence between nets and Latin squares is the following. Given an r-net **N** of order n, select two parallel classes $\mathfrak{R}$ and $\mathfrak{C}$ of **N**, and number their lines in an arbitrary but fixed fashion:

$$\mathfrak{R} = \{R_1, \ldots, R_n\}, \qquad \mathfrak{C} = \{C_1, \ldots, C_n\}.$$

An arbitrary point p of **N** is then on exactly one line R_i and exactly one line C_j; hence p may be denoted by (i, j). If $\mathfrak{X}$ is any one of the remaining $r - 2$ parallel classes of **N**, number its lines

$$\mathfrak{X} = \{X_1, \ldots, X_n\}$$

and define $L = L(\mathfrak{X}) = (l_{ij})$ as follows: $l_{ij} = m$ if the unique line of $\mathfrak{X}$ through (i, j) is X_m $(1 \leqq m \leqq n)$. It is straightforward to check that each $L(\mathfrak{X})$ is a Latin square, and that these $r - 2$ Latin squares are mutually orthogonal. Conversely, any set of $r - 2$ mutually orthogonal Latin squares of order n gives rise to an r-net of order n (BRUCK 1951; for the case $r = n + 1$ see BOSE 1938). Thus:

10. *An r-net of order n exists if and only if a set of $r - 2$ mutually orthogonal Latin squares of order n exists. In particular, there exist (pro-*

---

[1]) That $l_{ij} \in \{1, \ldots, n\}$ is not essential but convenient. It is sufficient to have any n distinct symbols as entries of L.

jective and affine) planes of order n if and only if there exist $n-1$ mutually orthogonal Latin squares of order n.

In view of this result, the maximal number $N(n)$ of mutually orthogonal Latin squares of order n is of obvious importance for the existence problem of finite nets and planes. This function is one of the oldest objects of study in combinatorial mathematics.[1] We list some of the more important properties of this function.

11. *The maximal number $N(n)$ of mutually orthogonal Latin squares of order n satisfies:*

(**a**) $N(n) \leqq n-1$;

(**b**) $N(q) = q-1$ *if q is a prime power;*

(**c**) $N(nm) \geqq \min(N(n), N(m))$.

For proofs see, for example, Ryser 1963, p. 80—84. Note that, in view of **10**, result **11b** follows immediately from the existence of projective planes for every prime power order q, viz., the desarguesian planes $\mathbf{P}(q)$. Combination of **11b** and **11c** gives

$$N\left(\prod_{i=1}^{m} p_i^{e_i}\right) \geqq \min\{p_i^{e_i} - 1 : i = 1, \ldots, m\} \tag{20}$$

(MacNeish 1922). MacNeish also conjectured that equality holds in (20); that this is false was first shown by Parker 1959a by the following theorem, whose proof depends on **6**:

12. *If there exists a projective plane of order n, and if there exists a design with parameters $v, b, r, k = n, \lambda = 1$, then $N(v) \geqq n-2$.*

For example, if n and $n+1$ are both prime powers (i.e. $n = 8$ or n a Mersenne prime or $n+1$ a Fermat prime), then $N(n^2+n+1) \geqq n-1$, and in this particular situation Parker (1959a, Theorem 2) has shown that even $N(n^2+n+1) \geqq n$, while (20) would yield only $N(n^2+n+1) \geqq 2$. Result **12** was the starting point for a more thorough investigation of Bose & Shrikhande 1959, 1960; Parker 1959b; and Bose, Shrikhande & Parker 1960. In the last paper it is shown that

$$N(n) > 1 \quad \textit{for all } n \neq 1, 2, 6. \tag{21}$$

[1] Euler 1782 investigated the "problem of the 36 officers": *Is it possible that 36 officers, of 6 different ranks and from 6 different regiments, stand in a square of 6 rows and 6 columns in such a way that every rank and every regiment is represented exactly once in every row and every column?* This is clearly equivalent to finding two orthogonal Latin squares of order 6. Euler conjectured correctly that the answer to this problem is negative (Tarry 1900), and incorrectly [see (21) below] that the corresponding problem for $n \equiv 2 \bmod 4$, instead of $n = 6$, is also unsolvable [Euler 1782, p. 183, § 144]. His conjecture was generalized by MacNeish 1922; see the context of (20) below. Euler was the first to prove that $N(n) > 1$ if $n \not\equiv 2 \bmod 4$.

Note that $N(1) = N(2) = 1$ is trivial; $N(6) = 1$ was proved by TARRY 1900, 1901. In the papers of Bose, Shrikhande and Parker just mentioned, much more than (21) is shown, but for these and other results on Latin squares the reader must be referred to the literature.[1])

The last results gave lower bounds for the function $N(n)$. There are also some upper bounds, distinct from the trivial **11a**: Result **2.1.15**, when restricted to the case $\lambda = 1$, gives the following nonexistence theorem for finite planes:

13. *If $n \equiv 1$ or 2 mod 4 and if n is not the sum of two squares [i.e. if the square-free factor of n has a prime divisor $p \equiv 3$ mod 4],[2]) then there exists no (projective or affine) plane of order n.*

This is the celebrated Theorem of BRUCK & RYSER 1949. It follows immediately that no finite plane has an order $\equiv 6$ mod 8. In view of **9, 10** and **13**, we can now conclude:

14. *If n satisfies the conditions of* **13**, *then*

$$n \leqq f(n - 2 - N(n)) < \tfrac{1}{2}(n - 1 - N(n))^4, \tag{22}$$

where f is the polynomial defined by (17).

(BRUCK 1963a.) Results **10, 11b** and **13** constitute our complete knowledge on the existence problem of finite projective planes with prescribed order. All known planes have prime power order. The smallest orders for which the existence problem is undecided are $n = 10, 12, 15, 18, 20, 24, 26, 28, 34, 35, 36$.

We shall see in Chapter 5 that for any $n = p^e > 8$, with p a prime and $e > 1$, there exist at least two nonisomorphic projective planes of order n. Whether this is true for $e = 1$ also is an open problem. But for small orders there is at most one projective plane:

15. *Every projective plane of order $n \leqq 8$ is desarguesian, i.e. isomorphic to a* $\mathbf{P}(q)$ *as defined in Section* 1.4.

For $n \leqq 5$, this was proved by MACINNES 1907; see also PICKERT 1955, p. 302. By **13**, there is no plane of order 6. For the uniqueness of $\mathbf{P}(7)$, cf. BOSE & NAIR 1941; HALL 1953, 1954b; PIERCE 1953; PICKERT 1955, pp. 319—325. The uniqueness of $\mathbf{P}(8)$ was determined by an

[1]) We mention here the following references: LEVI 1942; MANN 1942, 1943, 1950; CHOWLA, ERDÖS & STRAUS 1960. Latin squares can be interpreted as multiplication tables of finite quasigroups; for results in this context see JOHNSON, DULMAGE & MENDELSOHN 1961, also BRUCK 1958. Complete tables of Latin squares are given by TARRY 1900, 1901 and PETERSEN 1902 for $n = 6$, and by NORTON 1939 (with omissions) and SADE 1951 for $n = 7$.

[2]) The equivalence of these conditions is well known; see for example HARDY & WRIGHT 1962, p. 299.

electronic computer; see Hall, Swift, Walker 1956. That there exist nondesarguesian planes of order 9 (three such planes are known) will be seen in Chapter 5.

Despite the existence of non-desarguesian planes, it can be shown that every finite projective plane contains many Desargues configurations, i.e. subsystems of points and lines isomorphic to the incidence structure shown in Fig. 1 (p. 26). More precisely:

16. *Let* (c, A) *be a point-line pair in a finite projective plane* $\mathbf{P}$, *let* M_i $(i = 1, 2, 3)$ *be three lines* $\neq A$ *through* c, *and let* r_1, r_2 *be two points* $\neq AM_i$ $(i = 1, 2, 3)$ *on* A. *Then* $\mathbf{P}$ *contains a Desargues configuration in which* c, A, M_i, r_j *have the same significance as in Fig.* 1.

This was proved by a simple but ingenious counting argument, by Ostrom 1957.

Few general results are known about the number of points necessary to generate a finite projective plane [cf. (3.1.10)]; this is known only in the desarguesian case:

17. *The desarguesian projective plane* $\mathbf{P}(q)$ *is generated by any one of its quadrangles (i.e. it is a prime plane) if, and only if,* q *is a prime. If* q *is not a prime, then* $\mathbf{P}(q)$ *can be generated by any quadrangle* $\mathfrak{q}$ *and a suitable point on one of the sides of* $\mathfrak{q}$.

This is a simple consequence of the fact that the multiplicative group of a finite field is cyclic.

It is conceivable that every nondesarguesian finite projective plane (i) can be generated by some quadrangle and (ii) contains the seven-point plane $\mathbf{P}(2)$ as a subplane. This has been found true in all planes which have been investigated in this respect; see Wagner 1956, Cofman 1964, Killgrove 1964, for (i), and Lenz 1953, H. Neumann 1955, for (ii). On the other hand, the existence of too many subplanes $\mathbf{P}(2)$ implies Desargues' theorem (Gleason 1956; cf. result **3.4.23** below).

We turn to a brief discussion of the possible orders of subplanes of a finite projective plane. The following is a useful lemma:

18. *Let* $\mathbf{P}$ *be a projective plane of order* n, *and* $\mathbf{Q}$ *a proper subplane of order* m. *Then*

(**a**) $m^2 = n$ *if and only if* $\mathbf{Q}$ *is a Baer subplane,*
and
(**b**) $m^2 + m \leqq n$ *if* $\mathbf{Q}$ *is not a Baer subplane.*

The proof is again by simple counting; see Bruck 1955, Lemma 3.1, and Hall 1959, p. 398. It is an unsolved problem whether (**b**) can hold with equality; clearly this would imply the existence of planes of orders

$m(m+1)$ which are not prime powers.[1]) We shall see in Section 4.1 that $m^2 + m \neq n$ at least in the case where $\mathbf{Q}$ is the system of fixed elements of some collineation group.

A little more information on subplanes of finite planes can be obtained by considering the tactical decomposition defined by a subplane (DEMBOWSKI 1958, Section 2.4). We shall not present this here, but for the convenience of the reader, and for later reference, we restate the basic equations for tactical decompositions, viz. (1.1.14—16) and (2.1.16) and its dual, for the case the incidence structure under consideration is a projective plane of order n. If $\mathfrak{x}$ is a point class and $\mathfrak{X}$ a line class, then $(\mathfrak{x}\,\mathfrak{X})$ denotes the number of points of $\mathfrak{x}$ on any line of $\mathfrak{X}$, and $(\mathfrak{X}\,\mathfrak{x})$ has the dual meaning. Then:

$$(\mathfrak{x}\,\mathfrak{X})\,|\mathfrak{X}| = (\mathfrak{X}\,\mathfrak{x})\,|\mathfrak{x}| \qquad \textit{for all } \mathfrak{x},\ \mathfrak{X}; \tag{23}$$

$$\sum_{\mathfrak{x}} |\mathfrak{x}| = \sum_{\mathfrak{X}} |\mathfrak{X}| = n^2 + n + 1; \tag{24}$$

$$\sum_{\mathfrak{x}} (\mathfrak{x}\,\mathfrak{C}) = \sum_{\mathfrak{X}} (\mathfrak{X}\,\mathfrak{c}) = n + 1 \qquad \textit{for all } \mathfrak{c},\ \mathfrak{C}; \tag{25}$$

$$\sum_{\mathfrak{X}} (\mathfrak{c}\,\mathfrak{X})\,(\mathfrak{X}\,\mathfrak{c}') = |\mathfrak{c}| + n\,\delta(\mathfrak{c}, \mathfrak{c}') \qquad \textit{for all } \mathfrak{c},\ \mathfrak{c}'; \tag{26}$$

$$\sum_{\mathfrak{x}} (\mathfrak{C}\,\mathfrak{x})\,(\mathfrak{x}\,\mathfrak{C}') = |\mathfrak{C}| + n\,\delta(\mathfrak{C}, \mathfrak{C}') \qquad \textit{for all } \mathfrak{C},\ \mathfrak{C}'. \tag{26'}$$

Here $\delta(x, y) = 1$ or 0 according as $x = y$ or $x \neq y$.

We mention a rather isolated application, showing that the concept of homomorphism is of little value for finite planes.[2]) In fact:

19. *Every epimorphism* φ *of a projective design* $\mathbf{D}$ *onto a projective plane* $\mathbf{P}$ *is an isomorphism.*[3])

Hence $\mathbf{P}$ and $\mathbf{D}$ are then isomorphic projective planes, and in particular there exist no proper epimorphisms of one finite projective plane onto another.[4]) The proof of **19** appears in DEMBOWSKI 1959; it is shown there that the *cosets* of φ, i.e. the pre-images of the elements in $\mathbf{P}$ under φ, must form a tactical decomposition Δ of $\mathbf{D}$ such that the quotient structure $\mathbf{D}/\Delta$, as defined in the end of Section 1.1, is isomorphic to $\mathbf{P}$. But:

---

[1]) An interesting but so far unsuccessful idea for the construction of such planes, with the help of certain systems of 1-spreads of $\mathcal{P}(3, m)$ (for definitions see Section 1.4), has been pointed out by BRUCK 1963b, Section 9.

[2]) It is true that every projective plane is a homomorphic image of a *free plane* (cf. HALL 1943, or PICKERT 1955, Section 1.3) but this fact has, so far, not yielded much for the theory of finite planes.

[3]) The definitions for epi- and isomorphisms of incidence structures are in Section 1.2. Result **19** is in contrast to some theorems in Section 7.2 below: if a certain generalization of the concept of "design" is admitted, then there do exist proper homomorphisms onto finite projective planes.

[4]) This more special result was also proved by HUGHES 1960b and CORBAS 1964.

20. *If a projective design* $\mathbf{D}$ *has a tactical decomposition* Δ *such that* $\mathbf{D}/\Delta$ *is a projective plane* $\mathbf{P}$, *then* Δ *is the trivial decomposition each of whose classes has only one element, and hence* $\mathbf{D} \cong \mathbf{P}$.

This is Satz 6 of DEMBOWSKI 1958, and proves **19**.

For the remainder of this section, we shall be concerned with point sets in projective planes which contain no three collinear points. Such a set will be called an *arc*, and an arc which consists of k points will be called a *k-arc*. For example, the 3- and 4-arcs are just the triangles and quadrangles. An arbitrary line L meets an arc $\mathfrak{c}$ in at most two points; L will be called *secant, tangent,* or *exterior* according as $|(L) \cap \mathfrak{c}| = 2, 1,$ or 0. Note that

21. *In a projective plane of order* n, *every k-arc* $\mathfrak{c}$ *has* $\binom{k}{2}$ *secants,* $k(n+2-k)$ *tangents, and* $\binom{n}{2} + \binom{n+2-k}{2}$ *exterior lines. Through every point of* $\mathfrak{c}$ *there are* $k-1$ *secants and* $n+2-k$ *tangents.*

An *oval*[1]) is an arc $\mathfrak{o}$ of a (not necessarily finite) projective plane such that

(27) *Through every point of* $\mathfrak{o}$, *there is exactly one tangent.*

Hence **21** implies:

22. *The ovals in finite planes of order* n *are precisely the* $(n+1)$*-arcs.*

Every arc in a finite projective plane gives rise to a system of diophantine equations which will now be displayed.[2]) Let $\mathfrak{c}$ be a fixed k-arc in a projective plane of order n. For any point $x \notin \mathfrak{c}$, define $t(x)$ as the number of tangents of $\mathfrak{c}$ passing through x. The number of secants through x is then clearly $[k - t(x)]/2$, whence

$$t(x) \equiv k \bmod 2, \qquad \text{for every point } x \notin \mathfrak{c}. \tag{28}$$

Now define e_i as the number of those points $x \notin \mathfrak{c}$ for which $t(x) = i$; simple counting then yields the following equations:

$$\left\{ \begin{aligned} \sum_{i=0}^{k} e_i &= n^2 + n + 1 - k, \\ \sum_{i=0}^{k} i\, e_i &= k\, n(n+2-k), \\ \sum_{i=0}^{k} i(i-1)\, e_i &= k(k-1)\,(n+2-k)^2. \end{aligned} \right. \tag{29}$$

---

[1]) This is what we have called an "ovoid" in Section 1.4. The term "oval" is more customary for projective planes. The reader must be warned that our terminology differs somewhat from that of SEGRE 1961.

[2]) These may be interpreted as consequences of (23)—(26′), for a certain tactical decomposition associated with a k-arc by LOMBARDO-RADICE 1962. The special case where $k = n+1$ or $n+2$ is in DEMBOWSKI 1958, Section 2.5.

Furthermore, for any line L, let $f_i(L)$ denote the number of points $x \notin \mathfrak{c}$ on L for which $t(x) = i$. Then

$$(30) \qquad \sum_{i=0}^{k} f_i(L) = \begin{cases} n-1 & \textit{if } L \textit{ is secant,} \\ n & \textit{if } L \textit{ is tangent,} \\ n+1 & \textit{if } L \textit{ is exterior,} \end{cases}$$

and

$$(31) \qquad \begin{cases} \sum_{i=0}^{k} i\, f_i(L) = \begin{cases} (k-2)(n+2-k) & \textit{if } L \textit{ is secant,} \\ k(n+2-k) & \textit{if } L \textit{ is exterior,} \end{cases} \\ \sum_{i=0}^{k} (i-1)\, f_i(L) = (k-1)(n+2-k) \quad \textit{if } L \textit{ is tangent.} \end{cases}$$

Furthermore,

$$(32) \quad \sum_{L} f_i(L) = \begin{cases} (k-i)e_i/2 \\ i\, e_i \\ [n+1-(k+i)/2]\, e_i \end{cases} \textit{if } L \textit{ ranges over all} \begin{cases} \textit{secants} \\ \textit{tangents} \\ \textit{exterior lines,} \end{cases}$$

for $i = 0, \ldots, k$. The proofs for (29)—(32) are all given by MARTIN 1967a; some of these equations are also in SEGRE 1959a, b; 1961, Chapter 17. See also BARLOTTI 1965.

An arc $\mathfrak{c}$ of the projective plane $\mathbf{P}$ is called *complete* if it is not properly contained in another arc of $\mathbf{P}$ or, equivalently, if every point of $\mathbf{P}$ is on a secant of $\mathfrak{c}$. In the finite case, we can say that a k-arc is complete if and only if $e_k = 0$. The following completeness results may all be proved from (28)—(32), the main tool being (29).

23. *Let $\mathfrak{o}$ be an oval in a projective plane of order n. Then there are $n+1$ tangents, one through each point of $\mathfrak{o}$, $\binom{n+1}{2}$ secants, and $\binom{n}{2}$ exterior lines. Moreover,*

(a) *If n is odd, then $\mathfrak{o}$ is complete, with $e_2 = \binom{n+1}{2}$, $e_0 = \binom{n}{2}$, and $e_i = 0$ for $i \neq 0, 2$.*

(b) *If n is even, then $\mathfrak{o}$ is not complete; in fact $e_{n+1} = 1$, $e_1 = n^2 - 1$, and $e_i = 0$ for $i \neq 1$, $n+1$.*

This result is due to QVIST 1952. It shows that the concept of an oval is self-dual if n is odd: the tangents then form an oval in the dual plane. This is not so, however, if n is even: in that case there exists a unique point $k = k(\mathfrak{o})$, the knot of the oval $\mathfrak{o}$, which is on all the $n+1$ tangents of $\mathfrak{o}$, and conversely every line through $k(\mathfrak{o})$ is a tangent. The set $\mathfrak{o} \cup k(\mathfrak{o})$ is then a (necessarily complete) $(n+2)$-arc. As a consequence, we note:

24. *If a plane of order n contains a k-arc, then*

$$k \leqq \begin{cases} n+1 & \text{if } n \text{ is odd,} \\ n+2 & \text{if } n \text{ is even.} \end{cases}$$

This was first proved by BOSE 1947a.

The following results show that an oval $\mathfrak{o}$ and a k-arc $\mathfrak{c} \nsubseteq \mathfrak{o}$ cannot have too many common points:

25. *Let $\mathfrak{o}$ be an oval and $\mathfrak{c}$ a k-arc in a projective plane of order n.*

(a) *If n is odd, suppose that $|\mathfrak{c} \cap \mathfrak{o}| > (n+3)/2$ or that $|\mathfrak{c} \cap \mathfrak{o}| > (n+1)/2$ and $k > 3(n+3)/4$. Then $\mathfrak{c} \subseteqq \mathfrak{o}$.*

(b) *If n is even, suppose that $|\mathfrak{c} \cap (\mathfrak{o} \cup \{k(\mathfrak{o})\})| > (n+2)/2$. Then $\mathfrak{c} \subseteqq \mathfrak{o} \cup \{k(\mathfrak{o})\}$.*

For the proof, see MARTIN 1967a; a slightly weaker version is in BARLOTTI 1965, Section 2.5, and the case where $\mathfrak{c}$ is also an oval is contained in QVIST 1952. LOMBARDO-RADICE 1956 has exhibited a class of complete $[(q+3)/2]$-arcs in $\mathbf{P}(q)$, where q is a prime power $\equiv 3 \bmod 4$; this shows that the first result of **25a** is best possible. The second result of **25a** is also best possible, but there cannot exist examples showing this in a desarguesian finite projective plane: SEGRE's [1954, 1955a] Theorem **1.4.50** states that every oval in a desarguesian finite projective plane of odd order is a conic, and it is well known that, in any projective geometry over a commutative field, a conic is determined by five points. However, there exist two ovals in a nondesarguesian plane[1]) of order 9 with five common points (MARTIN 1967a). Other examples of ovals in nondesarguesian planes appear in WAGNER 1959; RODRIGUES 1959; BARLOTTI 1965, Section 2.9.

Several papers, mostly by Italian authors, have been devoted to completeness criteria for k-arcs with $k < n+1$; the principal reference for these is Chapter 17 of SEGRE 1961. We survey some of these results here. It is not difficult to see that

26. *k-arcs with $\binom{k-1}{2} < n$ are not complete;*

and for small k this may be further improved: 5-arcs are never complete, 4-arcs are incomplete if $n > 3$, 6-arcs are incomplete if $n > 10$, and 7-arcs are incomplete if $n > 13$. As completeness is equivalent to $e_k = 0$, it is desirable to have inequalities involving e_k. In this context

[1]) This plane, the "smallest Hughes plane", will be defined and discussed in Section 5.4.

we mention SCE's [1958] inequalities:

$$(33)\qquad \frac{(k-1)(k-2)^2}{2} \leqq e_k - n^2 + \frac{n(k-1)(k-2)}{2} \leqq \frac{(k-1)(k-2)}{2}\left[\frac{k(k-3)}{4}+1\right]$$

for any k-arc in a projective plane of order n.

An arc is called *uniform* if there is at most one $i > 2$ for which $e_i \neq 0$. Ovals and $(n+2)$-arcs in planes of order n are uniform.

27. *Uniform k-arcs in planes of order n are complete, except when $k = 4$ or n if k is even, or when $k = 5$ or $n - 1$ if k is odd.*

For the proof, see MARTIN 1967a; there exist counterexamples in the excepted cases.

Finally, we mention briefly n-arcs in planes of order n. SEGRE 1955b has shown that

28. *In the desarguesian plane* $\mathbf{P}(q)$ *of order q, there exists no complete q-arc.*

For the proof, see also SEGRE 1961, no. 175. The result cannot be extended to nondesarguesian planes: there exist complete 9-arcs in a nondesarguesian projective plane[1]) of order 9 which are uniform with $e_5 = 6$, and hence complete by **27**; see BARLOTTI 1965, Section 2.9, and MARTIN 1967a who also gives various conditions necessary and sufficient for an n-arc to be complete.

The concept of k-arc has been generalized to that of (k, m)-*arc*; these are point sets $\mathfrak{c}$ with $|\mathfrak{c}| = k$ containing m, but not $m+1$, collinear

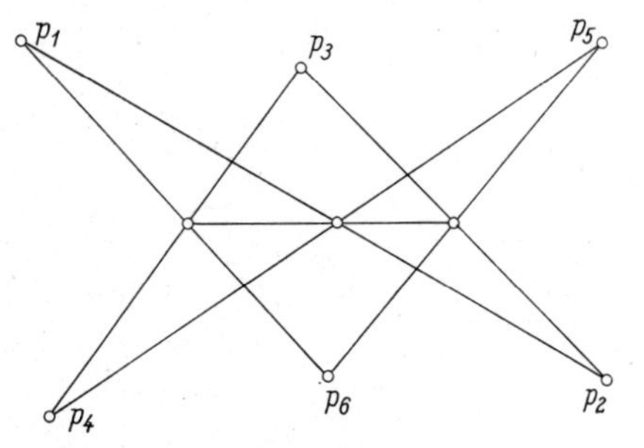

Fig. 3 Pascalian hexagon

points [hence k-arcs are $(k, 2)$-arcs]. We do not discuss these here, but mention only three references: BARLOTTI 1956; COSSU 1961; and BARLOTTI 1965, Chapter 3, where there are also more references.

In conclusion, we mention a beautiful theorem of BUEKENHOUT 1966b on pascalian ovals which holds also in the infinite case. By a

[1]) See footnote [1]) on p. 149.

hexagon, in an arbitrary projective plane $\mathbf{P}$, we mean a sextuple $(p_1, \ldots, p_6)$ of distinct points such that the lines $p_i p_{i+1}$ (subscripts mod 6) are all distinct. A hexagon is *pascalian* if the three points $(p_i p_{i+1})(p_{i+3} p_{i+4}), i = 1, 2, 3$ reduced mod 6, are collinear (see Fig. 3). The classical Theorem of Pascal says that in a projective plane over a commutative field, a hexagon is pascalian if it is contained in a conic, i.e. a nondegenerate quadric in the language of Section 1.4. Buekenhout's theorem is the following extension of this result:

29. *Suppose that the projective plane* $\mathbf{P}$ *contains an oval* $\mathfrak{o}$ *such that every hexagon of points in* $\mathfrak{o}$ *is pascalian. Then* $\mathbf{P}$ *is the desarguesian plane over some commutative field* $\mathfrak{K}$, *and* $\mathfrak{o}$ *is a conic in* $\mathbf{P}$.

The main ideas of the proof are the following. First it is shown that the permutations $\sigma(p)$ of $\mathfrak{o}$, defined for every point $p \notin \mathfrak{o}$ by

$$(34) \qquad x^{\sigma(p)} = \begin{cases} \text{the point} \neq x \text{ of } \mathfrak{o} \text{ on } px, & \text{if } px \text{ is secant to } \mathfrak{o}, \\ x & \text{if } px \text{ is tangent to } \mathfrak{o}, \end{cases}$$

generate a sharply 3-transitive group Σ of permutations of $\mathfrak{o}$. The condition that the hexagons in $\mathfrak{o}$ are pascalian also implies that Σ is faithfully induced by a collineation group of $\mathbf{P}$ which may also be called Σ. This allows the reconstruction of $\mathbf{P}$ within Σ; for example, the points $\notin \mathfrak{o}$ of $\mathbf{P}$ may be identified with the involutions. On the other hand, a result of Tits 1952 shows that $\Sigma \cong PGL_2(\mathfrak{K})$ for some commutative field $\mathfrak{K}$. If $\mathbf{Q}$ is the desarguesian plane over $\mathfrak{K}$ and $\mathfrak{c}$ any conic in $\mathbf{Q}$, one may consider the permutations (34) with $\mathfrak{c}$ and $\mathbf{Q}$ instead of $\mathfrak{o}$ and $\mathbf{P}$; the resulting group is well known to be $PGL_2(\mathfrak{K})$ also. Thus the isomorphy of Σ and $PGL_2(\mathfrak{K})$ permits us to set up an isomorphism from $\mathbf{P}$ onto $\mathbf{Q}$ which maps $\mathfrak{o}$ onto $\mathfrak{c}$. For more details, the reader is referred to Buekenhout 1966b.

3.3 Correlations and polarities

By a *correlation* is meant here an anti-automorphism of a projective plane $\mathbf{P} = (\mathfrak{p}, \mathfrak{L}, \mathrm{I})$, i.e. a permutation ϱ of $\mathfrak{p} \cup \mathfrak{L}$ such that $\mathfrak{p}\varrho = \mathfrak{L}$, $\mathfrak{L}\varrho = \mathfrak{p}$, and $p \,\mathrm{I}\, L$ if and only if $p\varrho \,\mathrm{I}\, L\varrho$ for all $p \in \mathfrak{p}$, $L \in \mathfrak{L}$ (cf. Section 1.2). A *polarity* is a correlation of order 2. Clearly, a correlation ϱ which is not a polarity has the property that ϱ^2 is a nontrivial collineation.

In this section, we consider correlations, and mostly polarities, of finite projective planes. Let $\mathbf{P}$ be such a finite plane, and write its order as

$$(1) \qquad n = n^* s^2,$$

where n^* is square-free. We denote by $\mathfrak{a} = \mathfrak{a}(\varrho)$ and $\mathfrak{A} = \mathfrak{A}(\varrho)$ the sets of absolute points and lines, respectively, of the correlation ϱ, and by $\mathbf{A} = \mathbf{A}(\varrho)$ the substructure of $\mathbf{P}$ defined by $\mathfrak{a}$ and $\mathfrak{A}$; see (1.1.4) for definitions. Obviously $|\mathfrak{a}| = |\mathfrak{A}|$; we put

(2) $$a(\varrho) = |\mathfrak{a}(\varrho)| = |\mathfrak{A}(\varrho)|.$$

The starting point of our discussion is the following corollary (for $\lambda = 1$) of result **2.1.17**:

1. *Every correlation ϱ of a finite projective plane has absolute points. In fact,*

(3) $$a(\varrho) \equiv 1 \mod n^* s.$$

Moreover, if ϱ is a polarity, then there exists a nonnegative integer r such that

(4) $$a(\varrho) = n + 1 + 2r\sqrt{n}.$$

Equations (3) and (4) are due to Ball 1948 and Hoffman, Newman, Straus, Taussky 1956. That $r \geqq 0$ is not immediate from **2.1.17**, but it is not difficult to prove (for example, with the help of **5** below). We note the following consequences of (4):

2. *Let π be a polarity of a projective plane of order n. Then*

(5) $$a(\pi) \geqq n + 1;$$

equality holds whenever n is not a square.

This was first proved by Baer 1946a, Section 1, Theorems 5 and 6.

Note that **2** implies the nonexistence of collineation groups of Lenz-Barlotti type I.8 in finite projective planes; cf. **3.1.20.**[1])

A power ϱ^i of a correlation ϱ is, of course, a collineation or a correlation according as i is even or odd. The case where i is even will be discussed at the end of Section 4.1; for i odd we have the following result:

3. *Let ϱ be a correlation of a plane of order n. For any odd prime p not dividing n, for any odd integer j prime to p, and for any integer $i \geqq 0$,*

$$\text{if } \left\{\begin{matrix} \left(\frac{n}{p}\right) = 1 \\ \left(\frac{n}{p}\right) = -1 \end{matrix}\right\}, \text{ then } a(\varrho^{jp^i}) \equiv \left\{\begin{matrix} a(\varrho^{jp^{i+1}}) \\ 2(n+1) - a(\varrho^{jp^{i+1}}) \end{matrix}\right\} \bmod p^{i+1}.$$

Here $\left(\frac{n}{p}\right)$ is the Legendre symbol. For the proof of **3**, see Ball 1948, Proposition 2.1. From this, and from the Dirichlet theorem that there are infinitely many primes of the form $ex + d$ whenever e and d are relatively prime [see, for example, LeVeque 1956, vol. II, Chapter 6], Ball has derived the following result:

---

1) On the other hand, it will be shown in Section 4.3 that in the infinite case such groups cannot exist either.

4. *If n is a square and i relatively prime to the order of ϱ (which implies that i is odd), then $a(\varrho) = a(\varrho^i)$.*

(BALL 1948, Theorem 3.1) Ball also gives some results for the case where n is not a square (Theorem 3.2 of the quoted paper). Some of Ball's results have been proved in another way by HOFFMAN, NEWMAN, STRAUS, TAUSSKY 1956; these authors have also given more conditions for $a(\varrho) = n + 1$. We remark further that (5) is not true for arbitrary correlations; an example with $a(\varrho) = 1$ appears in BALL 1948, p. 931.

For the remainder of this section, let π be a polarity of the projective plane **P** of order n. Result **1.2.2** shows that every absolute point (line) of π is incident with exactly n non-absolute lines (points) of π. If L is a non-absolute line, then the mapping $x \to (x\pi)L$ is a well-defined involutorial permutation of the point set (L), and the fixed points of this permutation are just the absolute points of π on L. Hence:

5. *The number of non-absolute points (lines) incident with a non-absolute line (point) is even.*

From this and $a(\pi) > 0$ it is not difficult to derive (5); cf. the remark after **1** above. Furthermore, if (5) holds with equality, **5** yields satisfactory information about the substructure $\mathbf{A}(\pi)$ of the π-absolute elements of **P**:

6. *Let π be a polarity with $a(\pi) = n + 1$, of a projective plane of order n.*

(**a**) *If n is odd, then $\mathbf{A}(\pi)$ consists of the points and tangents of some oval.*

(**b**) *If n is even, then $\mathbf{A}(\pi)$ consists of the points of a distinguished non-absolute line L and the lines through $L\pi$.*

(BAER 1946a, p. 82, Corollary 1.)

Next, we restate some results of Section 1.4, on polarities of finite desarguesian projective planes:

7. *Let π be a polarity of the desarguesian projective plane $\mathbf{P}(q)$ of order $q = p^e$. Then either $a(\pi) = q + 1$ (if π is orthogonal) or $a(\pi) = q^{3/2} + 1$ (if π is unitary). In the second case, which cannot occur unless $q = s^2$ is a square, the substructure of the absolute points and non-absolute lines of π is a design with parameters*

$$(6) \qquad v = s^3 + 1, \quad b = s^2(s^2 - s + 1), \quad k = s + 1, \quad r = s^2 = q, \quad \lambda = 1.$$

This substructure is clearly what we have called a unital[1]) in Section 2.4.

[1]) Compare here (2.4.20) and context, where this unital was denoted by $\mathbf{U}(s)$, and the remarks preceding **1.4.60**. For a generalization, see **9** below.

We discuss now a more general situation where $\mathbf{P} = (\mathfrak{p}, \mathfrak{L}, \mathrm{I})$ is finite of order n but not necessarily desarguesian. We call the polarity π of $\mathbf{P}$ *regular* (with BAER 1946a) if, for some integer $s = s(\pi)$, the number of absolute points on a non-absolute line (absolute lines through a non-absolute point) is either 0 or $s+1$. We can then restate **7** as follows: every polarity of a desarguesian finite projective plane is regular, and s is either 1 or $\sqrt{n}$. For an arbitrary regular polarity, the non-absolute lines fall into two disjoint classes: the set $\mathfrak{O} = \mathfrak{O}(\pi)$ of the *outer lines* which carry no absolute point, and the set $\mathfrak{J} = \mathfrak{J}(\pi)$ of the *inner lines*, each carrying $s+1$ absolute points. Dually, $\mathfrak{i} = \mathfrak{i}(\pi)$ is the set of *inner points*, carrying no absolute line, and $\mathfrak{o} = \mathfrak{o}(\pi)$ is the set of *outer points*, each carrying $s+1$ absolute lines. Note that $\mathfrak{O}\pi = \mathfrak{i}$ and $\mathfrak{J}\pi = \mathfrak{o}$. It is almost immediate that the classes $\mathfrak{O}$, $\mathfrak{J}$, $\mathfrak{o}$, $\mathfrak{i}$, together with the classes $\mathfrak{A} = \mathfrak{A}(\pi)$ and $\mathfrak{a} = \mathfrak{a}(\pi)$ of the absolute elements, form a tactical decomposition of $\mathbf{P}$, in the sense defined in Section 1.1 Using the basic equations (3.2.23)–(3.2.26′) for tactical decompositions of finite projective planes, and putting

$$\mathfrak{p}_1 = \mathfrak{a}, \quad \mathfrak{p}_2 = \mathfrak{o}, \quad \mathfrak{p}_3 = \mathfrak{i}, \quad \mathfrak{B}_1 = \mathfrak{A}, \quad \mathfrak{B}_2 = \mathfrak{J}, \quad \mathfrak{B}_3 = \mathfrak{O},$$

one derives the matrix $C = ((\mathfrak{p}_i\, \mathfrak{B}_j))$, defined by (1.3.1):

$$C = \begin{pmatrix} 1 & s+1 & 0 \\ n & \dfrac{s(n-1)}{s+1} & \dfrac{sn+1}{s+1} \\ 0 & \dfrac{n-s^2}{s+1} & \dfrac{n+s}{s+1} \end{pmatrix}; \tag{7}$$

furthermore:

$$a(\pi) = sn+1, \quad |\mathfrak{o}| = |\mathfrak{J}| = \frac{n(sn+1)}{s+1}, \quad \textit{and} \quad |\mathfrak{i}| = |\mathfrak{O}| = \frac{n(n-s^2)}{s+1}. \tag{8}$$

As all these numbers must be integers, we can conclude the following:

8. *Let π be a regular polarity of a projective plane of order n, and put $s(\pi) = s$. Then* (8) *holds, and*

$$s \equiv n \bmod 2, \tag{9}$$

$$n - 1 \equiv 0 \bmod s+1, \tag{10}$$

$$1 \leqq s^2 \leqq n. \tag{11}$$

Also, $s = 1$ if n is not a square, $s^2 = n$ if n is even, and if $s^2 < n$, then $s^2 + s + 1 \leqq n$.

These results, which bear a certain resemblance to **3.2.18**, are due to BAER 1946a, Section 2. In analogy with the desarguesian case, we may call a polarity *unitary* if it is regular with $s^2 = n$. We can then infer from (8):

9. *The absolute points and nonabsolute lines of a unitary polarity of a finite projective plane form a unital.*[1])

It seems to be unknown whether there exist, necessarily in nondesarguesian finite planes, regular polarities with $1 < s^2 < n$. There do exist non-regular polarities; examples will be given in Section 5.3.

Two lines are called *perpendicular*,[2]) with respect to the polarity π, if each of them is incident with the pole of the other. Perpendicularity is clearly a symmetric relation, and any line is perpendicular to itself if and only if it is absolute. Also:

10. *If L is any line and $p \neq L\pi$, then $p(L\pi)$ is the only line through p which is perpendicular to L.*

In the regular case, we call a point *elliptic*[2]) if it is the intersection of two inner lines which are perpendicular to each other. Consider the following conditions:

(12) *Every inner point is elliptic.*

(13) *No outer point is elliptic.*

(14) *If $p \mathrel{\mathrm{I}} L$, with p elliptic and L inner, then the line through p perpendicular to L is also inner.*

The following result gives the relationships between these conditions:

11. *For any regular polarity π of a projective plane of order n,*

(a) $s(\pi) > 1$ *implies* (12).
(b) (13) *implies* (12), (14), $s(\pi) = 1$, *and* $n \equiv 3 \bmod 4$.
(c) (12), (14), *and* $s(\pi) = 1$ *together imply* (13).

This is essentially Theorem 6 of Baer 1946a, p. 88. Note that for unitary polarities (12) holds, but not (13).

We conclude this discussion with another result of Baer, showing that a projective plane with a certain familiar type of polarity is necessarily infinite. For every polarity π with the property that some line carries no absolute point, it is possible to divide the nonabsolute points into two disjoint sets $\mathfrak{i}$ and $\mathfrak{e}$ ("interior" and "exterior" points) such that

(15) $$\text{if } p \in \mathfrak{i} \text{ and } L \mathrel{\mathrm{I}} p, \text{ then } L\pi \in \mathfrak{e}.$$

For example, $\mathfrak{i}$ may consist of a single point p carrying no absolute line, and $\mathfrak{e}$ of all other nonabsolute points. Now π is called *hyperbolic* if

[1]) Cf. footnote [1]) on p. 153.

[2]) We follow here (and further below) the terminology of Baer 1946a. See also Liebmann 1934.

such a division can be made in such a way that the following condition is also satisfied:

(16) *If L and M are perpendicular lines in $\mathfrak{e}\pi$, then $LM \in \mathfrak{i}$.*

The result then reads:

12. *No finite projective plane possesses a hyperbolic polarity.*

We outline a proof. (See BAER 1948 and, for a weaker preliminary version of the theorem, BAER 1946a, p. 91.) Assume that **P** is a projective plane with a hyperbolic polarity. Firstly, (15) and (16) show that both $\mathfrak{i}$ and $\mathfrak{e}$ are non-empty. Secondly, it follows easily from (15) and **1.2.2** that not all points on a line $L \in \mathfrak{e}\pi$ can be absolute. Hence every such L carries a non-absolute point p, and either p or $L(p\,\pi)$ is in $\mathfrak{i}$, by (16). Thus every line of $\mathfrak{e}\pi$ carries points of $\mathfrak{i}$. Next:

(17) *All lines in $\mathfrak{e}\pi$ carry equally many points of $\mathfrak{i}$.*

For let $L, M \in \mathfrak{e}\pi$. If L and M are non-perpendicular, then the mapping $x \to x' = L[x(M\pi)]$ is one-one and sends the points $x \in \mathfrak{i}$ on M onto the points $x' \in \mathfrak{i}$ on L [both (15) and (16) are used here]. If L and M are perpendicular, then $LM \in \mathfrak{i}$ by (16), and any line $X \mathrm{I} LM$ is likewise in $\mathfrak{e}\pi$, by (15). If $\mathfrak{X} \neq L, M$, then X is perpendicular to neither L nor M, and both L and M carry the same number of points of $\mathfrak{i}$ as does X. This proves (17).

(18) *There are equally many points of $\mathfrak{e}$ and $\mathfrak{i}$ on any line of $\mathfrak{e}\pi$.*

To see this, define $x^0 = (x\,\pi)\,L$, for any $x \mathrm{I} L \in \mathfrak{e}\pi$. Then $x^{00} = x$, and $x^0 = x$ if and only if x is absolute. Furthermore, (16) shows that if $x \neq x^0$, exactly one of x, x^0 is in $\mathfrak{e}$, the other in $\mathfrak{i}$. This proves (18).

So far, we have not used finiteness, so that (17) and (18) are true for any hyperbolic polarity. Now assume that **P** is finite of order n. From (17) and (18) we conclude that, for some integer m, every line of $\mathfrak{e}\pi$ carries exactly m points of $\mathfrak{e}$ and exactly m points of $\mathfrak{i}$. Then (15) shows that

$$|\mathfrak{i}| = 1 + (n+1)(m-1).$$

On the other hand, the dual of (15) shows that the n nonabsolute lines through an absolute point (see **5**) are in $\mathfrak{e}\,\pi$, and all points of $\mathfrak{i}$ are on these lines. Hence

$$|\mathfrak{i}| = n\,m,$$

and we conclude that $m = n$. But $2m \leqq n+1$ by (18); this gives $n \leqq 1$, a contradiction proving **12**.

The polarity associated with the classical hyperbolic plane (Klein's model) is hyperbolic in our sense. Result **12** shows, therefore, that there is no finite analogue of this classical situation. In particular, a finite "hyperbolic plane" in the sense of GRAVES 1962 [cf. the remarks on p. 105], if it can be interpreted as the system of "interior points" and polars of "exterior points" with respect to a polarity of a finite projective plane, cannot satisfy condition (16).

The preceding results provide no information as to when a given finite projective plane actually admits a polarity. The following is a result in this direction which, incidentally, provides another proof that the desarguesian finite projective planes do possess polarities.

13. *A finite projective plane* $\mathbf{P}$ *which is* (v, W)*-transitive for some flag* (v, W) *admits a polarity* π *with* $W = v\pi$ *if, and only if, it can be coordinatized by a cartesian group* $\mathfrak{C}$ *which possesses an involutorial permutation* α *such that*

$$(x+y)^\alpha = y^\alpha + x^\alpha \quad \text{and} \quad (x\,y)^\alpha = y^\alpha\, x^\alpha. \tag{19}$$

In fact, any cartesian group $\mathfrak{C} = \mathfrak{T}(o, e, u, v)$ with

$$u \,\mathrm{I}\, W, \quad v\pi = W, \quad o\pi = o\,u, \quad \text{and} \quad (e\,v)\pi = (o\,v)\,W \tag{20}$$

will have this property, and if $\mathfrak{C}$ is a cartesian group satisfying (19), then the mapping π defined by

$$(u, v)\,\pi = (y = x\,u^\alpha - v^\alpha) \tag{21}$$

is a polarity satisfying (20). For the proof of these results, see DEMBOWSKI & OSTROM 1968, Lemma 1. The following result is also proved in this paper (DEMBOWSKI & OSTROM 1968, Lemma 2):

14. *Let* $\mathfrak{C}$ *be a cartesian group of finite order* n *with commutative addition, and suppose that the plane over* $\mathfrak{C}$ *admits a polarity* π *satisfying* (20), *such that* $\mathfrak{A}(\pi)$ *is an oval. Then* n *is odd, and multiplication in* $\mathfrak{C}$ *is also commutative.*

In fact, it can be shown that the permutation α of (17) is the identity in this case.

3.4 Projectivities

Let (p, L) be a nonincident point-line pair in a projective plane $\mathbf{P}$. We define a one-one mapping $\pi(L, p)$ of the set (L) of all points on L onto the set (p) of all lines through p by

$$x^{\pi(L,p)} = x\,p \qquad \text{for every } x \,\mathrm{I}\, L. \tag{1}$$

The inverse mapping will be denoted by $\pi(p, L)$:

$$\pi(L, p)^{-1} = \pi(p, L). \tag{2}$$

A *projectivity*[1]) of $\mathbf{P}$ is a mapping of some set (L) or (p) onto some other set (M) or (q) which can be written as a product of mappings (1)

---

[1]) This term is often reserved only for mappings of lines onto lines, and later in this section we shall be only concerned with these.

and (2). Note that there are four distinct types of projectivities; here we shall consider mostly the type $(L) \to (M)$, which can be written in the form

$$\pi = \prod_{i=1}^{l} \pi(L_{i-1}, p_i)\, \pi(p_i, L_i), \tag{3}$$

where $L_0 = L$ and $L_l = M$. In order to avoid trivialities, we may assume also that

$$p_{i-1} \neq p_i \quad \text{and} \quad L_{i-1} \neq L_i \quad (i = 1, \ldots, l-1), \tag{4}$$

because of (2). The integer l is called the *length* of the representation (3) of π, and if π cannot be represented by a product of type (3) with less than $2l$ factors, then l is also called the length of π. A projectivity of length 1 is called a *perspectivity*.

We state now some well known results on projectivities. For proofs, the reader is referred to VEBLEN & YOUNG 1910, Sections 23—26, and HESSENBERG 1930.

1. *If $p_i \mathrm{I} L$ and $q_i \mathrm{I} M$, where the p_i as well as the q_i are distinct $(i = 1, 2, 3)$, then there exists a projectivity π of length $\leqq 3$ such that $p_i^\pi = q_i$ $(i = 1, 2, 3)$. If $L \neq M$, such a projectivity exists with length $\leqq 2$.*

2. *If p_1, p_2, q_1, q_2 are four distinct points on L, then there exists a projectivity π of length $\leqq 3$ such that $p_1^\pi = p_2$, $p_2^\pi = p_1$ and $q_1^\pi = q_2$, $q_2^\pi = q_1$.*

In an arbitrary projective plane, the projectivities of **1** and **2** need not be unique. We shall now formulate a condition which guarantees this uniqueness.[1] A projective plane is called *pappian* if every hexagon $(p_1, \ldots, p_6)$ with $p_5 \mathrm{I}\, p_1 p_3$ and $p_6 \mathrm{I}\, p_2 p_4$ is pascalian,[2] in other words if the following condition is satisfied:

(5) *If $L \neq M$, if p_1, p_3, p_5 are distinct points on L and p_2, p_4, p_6 distinct points on M, and if $p_i \neq LM$ for $i = 1, \ldots, 6$, then*

$$q_1 = (p_1 p_2)(p_4 p_5), \quad q_2 = (p_2 p_3)(p_5 p_6) \quad \text{and} \quad q_3 = (p_3 p_4)(p_6 p_1)$$

are collinear points (Fig. 4).

This condition is known as the "Theorem of Pappus"; it is the first in a sequence of *incidence propositions* to be considered in this section. The "Theorem of Desargues" [Condition (**D**) of Section 1.4] is another example.[3]

[1]) The word "uniqueness" refers here to the actual mapping defined by a projectivity, not the manner of its representation: Many different products of the form (3) may define the same mapping.

[2]) For the definition of pascalian hexagons, see the context of **3.2.29**.

[3]) Instead of "incidence proposition", the terms "configuration theorem" and "Schliessungssatz" are customary. We avoid the use of the word "theorem" in this context.

The basic relationship between condition (5) and projectivities is expressed in the following lemma.

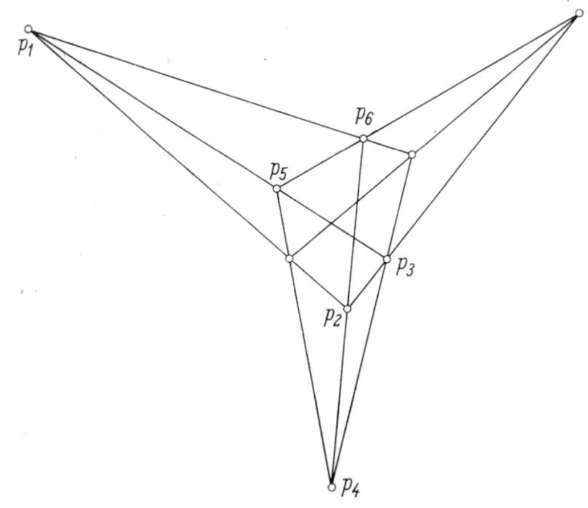

Fig. 4 Theorem of Pappus

3. *A projective plane is pappian if and only if it satisfies the following condition:*

(6) *Let L_0, L_1, L_2 be three nonconcurrent lines and suppose that the projectivity $\pi = \pi(L_0, p_1)\,\pi(p_1, L_1)\,\pi(L_1, p_2)\,\pi(p_2, L_2)$ satisfies $(L_0 L_2)^\pi = L_0 L_2$. Then π is equal to a perspectivity $\pi(L_0, p_0)\,\pi(p_0, L_2)$.*

The proof is straightforward; see, for example, HESSENBERG 1930, p. 68. Repeated application of **3** yields the following theorem:

4. *Every pappian projective plane is desarguesian.*

(HESSENBERG 1905) For the proof,[1] see CRONHEIM 1953 or PICKERT 1955, Sections 5.1, 5.2. Next, we translate (5) into conditions on central collineations:

5. *The following conditions for a projective plane* **P** *are equivalent:*

(**a**) **P** *is pappian.*
(**b**) **P** *is (p, L)-transitive for every point-line pair p, L, and for $p \not\mathrel{I} L$ the group of all (p, L)-homologies is abelian.*[2]
(**c**) **P** *is (p, L)-transitive for all p, L, and if σ, τ are homologies of* **P** *with the same axis but distinct centers, then $\sigma^{-1}\tau^{-1}\sigma\tau$ is an elation.*

The proof of this is straightforward. In view of **3.1.22i**, it follows from (**b**) that

[1]) Hessenberg's original proof is not complete; he disregarded the possibility that certain additional incidences may occur in a Desargues configuration.

[2]) That the group of (p, L)-elations $(p \mathrel{I} L)$ is abelian follows from **3.1.11**.

6. *A desarguesian projective plane is pappian if and only if its coordinatizing field is commutative.*

But every finite field is commutative (WEDDERBURN 1904, WITT 1931, ZASSENHAUS 1952, BRANDIS 1964), hence

7. *A finite projective plane is desarguesian if and only if it is pappian.*

We return to projectivities. If in (6) the word "nonconcurrent" is replaced by "concurrent", then (6) remains true even in every desarguesian plane. In view of **4**, we therefore have the following improvement of **3**:

8. *A projective plane is pappian if and only if, for any two distinct lines L and M, any projectivity of length $\leqq 2$ from (L) onto (M) which fixes LM is a perspectivity.*

With this result as the main tool, one can now prove the so-called "Fundamental Theorem of Projective Geometry", saying that in a pappian plane the projectivity π of **1** is unique, in other words that every projectivity in a pappian plane is determined by its action on three distinct points (or lines). We shall give an equivalent formulation here. The set of all projectivities of a line L onto itself is clearly a group, in any projective plane. If this group is called $\Pi(L)$ and if $M \neq L$, then

$$\Pi(M) = \pi^{-1}\,\Pi(L)\,\pi,$$

for any projectivity π from L on to M. Hence $\Pi(L)$ and $\Pi(M)$ are similar as permutation groups and in particular isomorphic. We can therefore write $\Pi(\mathbf{P})$ or simply Π instead of $\Pi(L)$; this group will be called the *group of projectivities* of $\mathbf{P}$. We can now summarize the results collected so far:

9. $\Pi(\mathbf{P})$ *is a triply transitive group, and it is sharply 3-transitive if and only if* $\mathbf{P}$ *is pappian.*

The second statement of **9** is equivalent to the "Fundamental Theorem" mentioned above. Little is known about Π if $\mathbf{P}$ is not desarguesian. In the finite case, triply transitive permutation groups are comparatively rare, and it seems plausible to conjecture that, if $\mathbf{P}$ is nondesarguesian of order n, then $\Pi(\mathbf{P})$ contains the alternating group of degree $n+1$. In a few special cases (with $n = 9, 16$) this was proved by BARLOTTI 1959, 1964. In particular, an example with $n = 16$ shows that Π need not be the full symmetric group of degree $n+1$.

We mention two recent improvements of **9**.

Let W be a fixed line of the projective plane $\mathbf{P}$, and let L be any line distinct from W. Denote by $\Pi^W(L)$ the group of all those permutations of $L - \{WL\}$ which are induced by projectivities (3) with $L_0 = L_l = L$ and $p_i \mathrm{I} W$ $(i = 1, \ldots, l)$. As in the case of $\Pi(L)$, this permutation group does not depend on the choice of the line $L \neq W$; it will therefore be denoted be $\Pi^W(\mathbf{P})$ or simply by Π^W.

10. *Let* **P** *be a projective plane which need not be finite.*

(**a**) *If* $\Pi_{uvxyz} = 1$ *for any five distinct points* $u, \ldots, z$, *then* $\Pi_{xyz} = 1$, *and* **P** *is pappian.*[1])
(**b**) *There exist three distinct points* x, y, z *such that* $\Pi^W_{xy} = \Pi^W_{xyz}$ *if, and only if,* $\mathbf{P}^W$ *is a translation plane with kernel* $\neq GF(2)$ [*cf.* **3.1.24**].
(**c**) *If there exist four distinct points* x, y, z, u *such that* $\Pi_{xyz} = \Pi_{xyzu}$, *then* **P** *is of Lenz-Barlotti type* VII.1 *or* VII.2, *and therefore pappian if finite.*

Result (**a**) is the main theorem of SCHLEIERMACHER 1967, and (**b**) is a relatively simple consequence of Theorem 1 of LÜNEBURG 1967b. As $\Pi^W(L)$ is in the stabilizer of WL in $\Pi(L)$, result (**c**) is a consequence of (**b**). In the finite case, we have:

11. *Let* **P** *be finite of order* n, *and suppose that* Π_x, *for an arbitrary point* x, *has a normal subgroup of order* n. *Then* **P** *is pappian.*

This was proved by LÜNEBURG 1967b, Theorem 3, for odd n and by YAQUB 1968 for even n. Yaqub's argument uses result **23** below.

We discuss now some connections between projectivities and collineations. Central collineations clearly induce perspectivities on nonfixed lines; hence:

12. *In a desarguesian projective plane, any projectivity of one line onto another is induced by a product of at most three central collineations.*

For a more detailed discussion, see PICKERT 1955, p. 114; also LENZ 1965, p. 30.

The following considerations, due to GLEASON 1956, give connections between projectivities and central collineations in a more general situation. Let L, X be two distinct lines, and $LX = c$. Also, let u, v be two points, not incident with L or X. Define

$$\lambda_X(u, v) = \pi(L, u)\, \pi(u, X)\, \pi(X, v)\, \pi(v, L) \tag{7}$$

and put

$$\Lambda_L(u, v) = \{\lambda_X(u, v) \mid c \,\mathrm{I}\, X \,\not\mathrm{I}\, u, v\}. \tag{8}$$

Then the following is easily verified:

13. $\Lambda_L(u, v)$ *is a set of permutations of* (L), *fixing* c *and* $(uv)L$, *and sharply transitive on the remaining points of* L. *Moreover,* $\Lambda_L(u, v)$ *is a group if and only if the following condition holds:*

---

[1]) The analogue of this for six points is not true; cf. BARLOTTI 1964c.

(9) *Let p_1, p_2 be two distinct points $\neq c, (uv)L$ on L, and X, Y two distinct lines $\neq L$ and $\not{\mathrm{I}}\; u, v$, through c. Define $x_i = X(p_i u)$, $y_i = Y(p_i v)$, and $z_i = (x_i v)(y_i u)$, for $i = 1, 2$. Then c, z_1 and z_2 are collinear* (cf. Fig. 5).

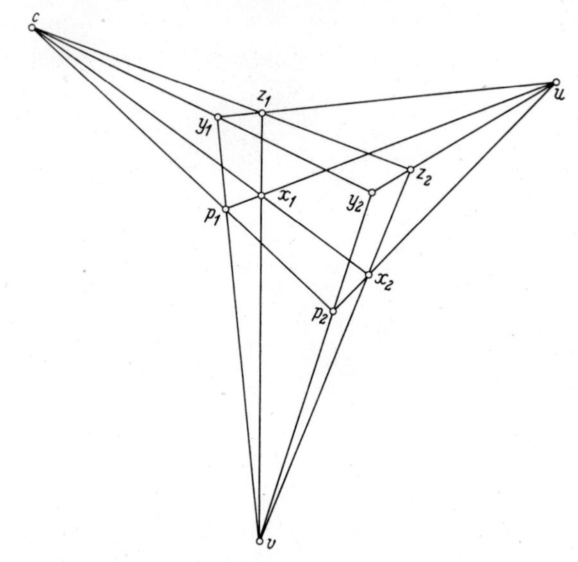

Fig. 5 Reidemeister condition

Condition (9) is called the *Reidemeister condition*, after REIDEMEISTER 1929. It is easily proved in any desarguesian plane[1]), and conversely KLINGENBERG 1955 has shown that

14. *If the Reidemeister condition is satisfied in a projective plane* **P**, *then* **P** *is desarguesian.*

Combining results **13** and **14**, we can say that a projective plane is desarguesian if and only if each of the permutation sets $\Lambda_L(u, v)$ as defined in (8) is a group. The object of the following considerations is to give an essential improvement of this result in the finite case. We begin with some preparatory lemmas which are valid also in infinite planes.

15. *Let (c, A) be an arbitrary point-line pair in the projective plane* **P**, *and let L be an arbitrary line $\neq A$ through c.*

[1]) As a matter of fact, the Reidemeister condition with $c \not{\mathrm{I}}\; uv$ is easily seen to be equivalent to associativity of multiplication, and that with $c \mathrel{\mathrm{I}} uv$ to associativity of addition, in any ternary field of the plane under consideration.

(a) *A permutation of the point set* (L) *commutes with every permutation in the set*

$$\Lambda = \bigcup_{L \not\mathrm{I}\, u,\, v\, \mathrm{I}\, A} \Lambda_L(u, v)$$

if, and only if, it is induced by a collineation with center c *and axis* A.
(b) *If the sets* $\Lambda_L(u, v)$ *are groups for all* $u, v \neq c$ *on* A, *then*

$$\Lambda_L(u, v) = \Lambda_L(v, u), \quad \textit{and} \tag{10}$$

$$\Lambda_L(v, w) \subseteqq \Lambda_L(u, v)\, \Lambda_L(u, w) = \Lambda_L(u, w)\, \Lambda_L(u, v). \tag{11}$$

For the proof, see GLEASON 1956, Lemma 2.1 and p. 805; the additional assumption $c \mathrm{I} A$ made there is unnecessary. These proofs are quite straightforward [for example, (**b**) follows from $\lambda_X(u, v) = \lambda_X(v, u)^{-1}$ and $\lambda_X(u, v)^{-1} \lambda_X(u, w) = \lambda_X(v, w)$]. We emphasize, however, that **15a** is of fundamental importance for the sequel.

16. P *is* (c, A)*-transitive if and only if* $\Lambda_L(u, v) = \Lambda_L(u, w)$, *for all lines* $L \neq A$ *through* c *and all points* $u, v, w \neq c$ *on* A.

This was proved by KEGEL & LÜNEBURG 1963, p. 10, again under the superfluous condition $c \mathrm{I} A$. The equivalence of (c, A)-transitivity and (c, A)-Desargues is used here (cf. **3.1.16**).

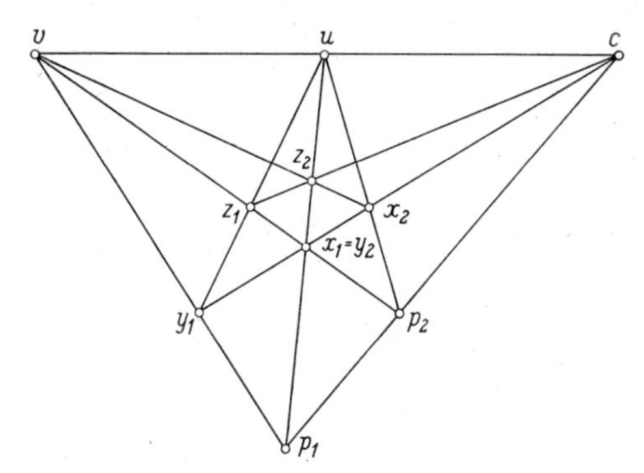

Fig. 6 Little hexagonality condition

We shall now be concerned with special cases of (9). The *little Reidemeister condition* is (9) with the additional assumption that u, v, c are collinear. By **13**, the little Reidemeister condition holds if and only if $\Lambda_L(u, v)$ is a group whenever $c \mathrm{I}\, uv$. The *hexagonality condition* is (9) with $x_1 = y_2$, and the *little hexagonality condition* (Fig. 6) is the hexagonality condition with $c \mathrm{I}\, uv$. It follows from theorems of MOUFANG 1931 that

17. *The little hexagonality condition is satisfied in a projective plane* **P** *if and only if every quadrangle in* **P** *generates a pappian prime subplane.*

For the proof, see PICKERT 1955, Chapter 11.[1]) Hence we can conclude that, in a plane satisfying the little Reidemeister condition, every quadrangle generates a subplane which is isomorphic either to a finite desarguesian plane $\mathbf{P}(p)$ for some prime p, or to the plane over the rational numbers.

Now we restrict ourselves to the finite case. Let $\mathbf{P}$ be a finite projective plane of order n, and suppose that the little Reidemeister condition is satisfied for $\mathbf{P}$. Then **17** implies that for every prime divisor p of n there exists a subplane isomorphic to $\mathbf{P}(p)$ in $\mathbf{P}$, and that every permutation in the group $\Gamma = \Lambda_L(u, v)$ [from now on always with $c \mathrm{I} uv$] has prime order. It follows that the cyclic subgroups of Γ form a *normal partition* of Γ, i.e. a set $\mathscr{C}$ consisting of full conjugate classes of subgroups, such that $\Gamma = \bigcup_{\Delta \in \mathscr{C}} \Delta$ and $\Delta \cap \Delta' = 1$ for any two distinct subgroups Δ, Δ' in $\mathscr{C}$ (cf. Section 1.2). For the following it is convenient to collect some results on normal partitions of finite groups:

18. *Let $\mathscr{C}$ be a normal partition of the finite group G.*

(**a**) *At most one conjugate class in $\mathscr{C}$ consists of subgroups which are their own normalizers in G.* [If there is such a conjugate class, $\mathscr{C}$ is called a *Frobenius partition.*[2])]

(**b**) *If $\mathscr{C}$ is a Frobenius partition and $F = F(G)$ the Fitting subgroup of G [i.e. the product of all nilpotent normal subgroups], then $X \in \mathscr{C}$ implies either $X \subseteqq F$ or $X \cap F = 1$ and $XF = G$. Also, $F \in \mathscr{C}$ unless F is a p-group.*

(**c**) *If G is not a p-group and $\mathscr{C}$ not a Frobenius partition, then every normal subgroup K satisfying $X \subseteqq K$ or $X \cap K = 1$ for all $X \in \mathscr{C}$ is itself in $\mathscr{C}$ and has index $[G:K]$ a prime divisor of $|K|$.*

(**d**) *If $X \in \mathscr{C}$ is not its own normalizer in G, then X is nilpotent.*

(**e**) *If G is non-soluble and $\mathscr{C}$ not a Frobenius partition, then G is isomorphic to $PGL_2(p^e)$ with odd p or $PSL_2(p^e)$ or $Sz(2^{2m+1})$, for suitable p, e, m.*

The proofs of (**a**)—(**c**) are in BAER 1961, Lemma 1.6, p. 343—345, and Satz 5.1. (**d**) is due to KEGEL 1961, Satz 2.[3]) Result (**e**) is the main theorem of SUZUKI 1961a; the symbol $Sz(q)$ stands, as on p. 52, for the Suzuki group over $GF(q)$, see SUZUKI 1960, 1962a, Section 13.

We return to the situation discussed above, and show:

---

[1]) In fact it is shown there that both conditions of **17** are equivalent to a special case of the theorem of Desargues (sometimes called condition D_8). In this context compare also DEMARIA 1959.

[2]) The reason for this is that the Frobenius groups (having a faithful transitive permutation representation which is not regular but in which only 1 fixes more than one symbol) are precisely the groups admitting such a partition.

[3]) This result will not be used here directly. We have listed it mainly because it is an important tool in the proof of (**e**).

19. *If a finite projective plane satisfies the little Reidemeister condition, then its order is a power of a prime.*

Proof (LÜNEBURG 1961 b; LÜNEBURG & KEGEL 1963): Let $\Gamma = \Lambda_L(u, v)$ for some line L and $uv \mathrm{I}\, c \,\mathrm{I}\, L$; cf. (8). Also, let $\mathscr{C}$ be the normal partition of Γ into its cyclic subgroups of prime order. Assume first that $\mathscr{C}$ is a Frobenius partition. If the Fitting subgroup $F(\Gamma)$ of Γ were in $\mathscr{C}$, then $|F(\Gamma)| = p$ and $n = |\Gamma| = pq$, for two distinct primes p and q, by **18b**. Hence there would exist two prime subplanes of orders p, q in $\mathbf{P}$, and as n cannot be a square, **3.2.18** would imply that $p^2 + p \leqq n = pq \geqq q^2 + q$, which leads to a contradiction. Hence $F(\Gamma) \notin \mathscr{C}$, and **18b** gives $|F\Gamma| = p^m$, so that $n = |\Gamma| = p^m q$, again with p and q distinct primes. The groups $\Lambda_L(u, x)$, with $c \neq x \,\mathrm{I}\, uv = A$, commute in pairs, because of (11); by a result of WIELANDT [1951, Satz 8], the same is true for their Fitting subgroups $\Phi_L(u, x) = F(\Lambda_L(u, x))$. Hence the product of all the $\Phi_L(u, x)$ is a p-group Σ of permutations of $(L) - \{c\}$, and all Σ-orbits have length $\geqq p^m$. As $p^{m+1} \nmid n$, some Σ-orbit $\mathfrak{c} \subseteqq (L) - \{c\}$ must have exact length p^m, and $\mathfrak{c}$ is then a $\Phi_L(u, x)$-orbit for every $x \neq c$ on $A = uv$. Let s, t be two points in $\mathfrak{c}$ and $x \neq c, u$ on A. Then the subplane $\langle s, t, u, x\rangle$ must be of order p, and there exists $\lambda \in \Lambda_A(s, t)$, of order $o(\lambda) = p$, such that $u^\lambda = x$. This means that a Sylow p-group of $\Lambda_A(s, t)$ must be transitive on the points $\neq c$ of A, and now the regularity of $\Lambda_A(s, t)$ implies $n = p^m$, a contradiction. This shows that $\mathscr{C}$ cannot be a Frobenius partition.

Now assume that **19** is false, i.e. that n has two distinct prime divisors. Then **18c** shows that Γ must be simple, for the existence of a proper normal subgroup would imply $|\Gamma| = n = p^2$. Hence we can use **18e** and conclude from known properties of $PGL_2(q)$, $PSL_2(q)$ [DICKSON 1901, p. 285) and $Sz(q)$ [SUZUKI 1962a, Section 13] that the only remaining possibility is $n = 60$ and $\Gamma \cong PSL_2(4) \cong A_5$, the alternating group of degree 5. In order to exclude this also, consider besides $\Gamma = \Lambda_L(u, v)$ the groups $\Delta = \Lambda_L(u, w)$ and $\Phi = \Lambda_L(v, w)$. All these groups are isomorphic to A_5, and if Σ denotes the group generated by Γ and Δ, then

$$\Sigma = \Gamma\Delta = \Delta\Gamma \supset \Phi \tag{12}$$

because of **15b** and **16**, so that in particular $|\Sigma| > 60$. Now we use the following fact:

20. *A group of order* > 60 *which is the product* $AB = BA$ *of two copies* A, B *of* A_5 *is isomorphic to either* $A \times B$ *or* A_6.

Proof: KEGEL & LÜNEBURG 1963, Satz B. In view of this, it will now be sufficient to prove that Σ cannot be either of A_6 or $A_5 \times A_5$.

Assume first $\Sigma \cong A_6$. Then there are precisely two conjugate classes of subgroups isomorphic to A_5, and two subgroups in the same class have intersection of order $\geqq 12$ [DICKSON 1901, Chapter 12]. But from (12) and $|\Sigma| = 360 = |A_5|^2/10$ it follows that $|\Gamma \cap \Delta| = |\Delta \cap \Phi| = |\Phi \cap \Gamma| = 10$, so that we have the desired contradiction.

Finally assume $\Sigma \cong A_5 \times A_5$. Then we consider the *diagonals* of Σ, i.e. the subgroups isomorphic to $\{(x, x^\alpha) : x \in A_5\}$, with $\alpha \in \operatorname{Aut} A_5$. Any subgroup of $A_5 \times A_5$ which is not a direct factor but isomorphic to A_5 must be such a diagonal. As every automorphism of A_5 fixes some element $\neq 1$, two distinct diagonals must have intersection $\neq 1$. But under our present assumption we have

$$|\Gamma \cap \Delta| = |\Delta \cap \Phi| = |\Phi \cap \Gamma| = 1,$$

so that we can assume without loss in generality that $\Sigma = \Gamma \times \Phi$ and $\Delta = \Delta_\alpha = \{\gamma \gamma^\alpha : \gamma \in \Gamma\}$, with α an isomorphism from Γ onto Φ. But then a stabilizer Σ_x of any point $x \neq c$ on L has order 60 and trivial intersection with Γ and Φ; hence it must be a diagonal $\Delta_\beta = \{\gamma \gamma^\beta : \gamma \in \Gamma\}$. But the above remark implies that then $\Delta \cap \Sigma_x \neq 1$, and this contradicts the regularity of Δ on $(L) - \{c\}$. This completes the proof of **19**.

We can now give the improvement of **14** mentioned above.

21. *If the little Reidemeister condition is satisfied in a finite projective plane* $\mathbf{P}$, *then* $\mathbf{P}$ *is desarguesian.*

In view of **19**, it suffices to prove this under the additional assumption that the order n of $\mathbf{P}$ is a prime power p^e. The following argument is due to GLEASON 1956, Theorem 2.5. Let Π denote the permutation group generated by all the groups $\Lambda_L(u, x)$ of order n, with u a fixed point $\neq c$, and x ranging over all points $\neq u, c$ of $A = uc$. By (11), the p-groups $\Lambda_L(u, x)$ commute pairwise; thus Π as the product of these groups, is also a p-group. Consequently, Π has a nontrivial centre. As Π contains the set Λ of **15a**, again by (11), it follows from **15a** that there exist nontrivial (c, A)-elations in $\mathbf{P}$. This is true for every flag (c, A) of $\mathbf{P}$, and an appeal to **2.3.27b** finally shows that $\mathbf{P}$ is desarguesian.[1])

---

[1]) The case $\lambda = 1$ was actually excluded in **2.3.27**; it was only mentioned there that **2.3.27b** holds also if $\lambda = 1$ (see **4.3.22a** below). The proof of **21** may be finished without **2.3.27b** as follows (GLEASON 1956, Lemma 1.6): Let Σ denote the group generated by all elations of $\mathbf{P}$. As $\Sigma(c, A) \neq 1$ for every flag (c, A), result **3.1.14** shows that the subgroups $\Sigma(x, A)$ of $\Sigma(A, A)$ all have the same order $h > 1$, for any line A; hence $\Sigma(A, A)$ has order

$$1 + (h - 1)(n + 1).$$

Also, as $\Sigma(A, A)$ is regular on the n^2 points not on A, the order of $\Sigma(A, A)$ divides n^2. Hence $n^2 = m[1 + (h - 1)(n + 1)]$ for some integer $m > 0$. But this equation shows (i) that $m \equiv 1 \bmod n + 1$ and (ii) that $m < n$, because $h > 1$. Hence $m = 1$; this means $|\Sigma(A, A)| = n^2$ or (A, A)-transitivity of $\mathbf{P}$. As A was an arbitrary line, result **3.122h** (and its context) show that $\mathbf{P}$ is desarguesian.

We note a fairly immediate consequence of **21**. If condition (5), with the additional restriction that $q_2 \mathrm{I} p_1 p_4$,[1]) holds in a projective plane **P**, then every ternary field of **P** has a commutative additive loop. But by a result of BOL 1938 and BAER 1939, these loops are then also associative [see also PICKERT 1955, p. 49], and this is equivalent to the little Reidemeister condition in **P**. Thus (5) with $q_2 \mathrm{I} p_1 p_4$ implies this condition. Combining this and the dual argument with **7**, **14**, **21**, and **3.1.22i**, we can summarize our results as follows:

22. *The following properties of a finite projective plane* **P** *are equivalent:*

(**a**) **P** *is desarguesian.*
(**b**) **P** *is* (L, L)*-transitive for all lines* L.
(**c**) **P** *is* (L, L)*- and* (M, M)*-transitive for two distinct lines* L *and* M.
(**d**) **P** *satisfies the Reidemeister condition* (9).
(**e**) **P** *satisfies the little Reidemeister condition, i.e.* (9) *whenever* $c \mathrm{I} u v$.
(**f**) **P** *satisfies condition* (5) *whenever* $q_2 \mathrm{I} p_1 p_4$.
(**g**) **P** *satisfies condition* (5) *whenever* $(p_1 p_3)(p_2 p_4) \mathrm{I} q_1 q_2$.
(**h**) **P** *is pappian, i.e. satisfies* (5) *without restriction.*

A more direct proof that (**f**) implies (**a**) is in LÜNEBURG 1960.

We conclude this section with some remarks on *planes with characteristic*. Let **P** be a projective plane and $\mathfrak{q} = (o, e, u, v)$ an ordered quadrangle in **P**. Consider the permutation $\lambda = \lambda_{ev}((o\, e)(u\, v), u)$ of $(o\, v) - \{v\}$, as defined by (7), and let $\mathfrak{c}$ be the λ-cycle containing o. If $\mathfrak{c}$ is infinite, we say that $\mathfrak{q}$ has *characteristic* 0; clearly this cannot happen in a finite plane. Otherwise, the *characteristic* of $\mathfrak{q}$ is the integer $|\mathfrak{c}|$. Note that this definition depends on the ordering of the four points in $\mathfrak{q}$, and that in the desarguesian case the characteristic of any quadrangle coincides with the characteristic of the coordinatizing field of the plane, hence it is either 0 or a prime number.

A projective plane **P** is said to be of *characteristic* m if every ordered quadrangle in **P** has characteristic m. It is easy to see that if a finite plane of order n has characteristic m, then m is a divisor of n. Some other simple relations between m and n were proved by LOMBARDO-RADICE 1955a.

The only known finite planes with characteristic are the desarguesian ones; here m is a prime number. It has been conjectured that there are no others, but so far this can be proved only in the minimal case:

23. *A finite projective plane of characteristic* 2 *is desarguesian.*

---

[1]) This is also known as the *axial*, and its dual (cf. **22g** below) as the *central, little theorem of Pappus;* see PICKERT 1955, p. 153.

Note that a projective plane is of characteristic 2 if and only if each of its quadrangles has collinear diagonal points;[1]) cf. Fig. 7. For the proof of **23** (GLEASON 1956, Theorem 3.5), consider the set $\Lambda_L(u, v)$ for arbitrary line L, and with $c \mathrm{I} u v$. The requirement of characteristic 2 implies that all permutations $\neq 1$ in $\Lambda_L(u, v)$ are of order 2, and the product of any two distinct permutations in $\Lambda_L(u, v)$ is likewise an involution. It follows that the elements of $\Lambda_L(u, v)$ commute pairwise, so that $\Lambda_L(u, v)$ is an abelian group. But then **13** shows that the little Reidemeister condition is satisfied, and **21** yields **23**.

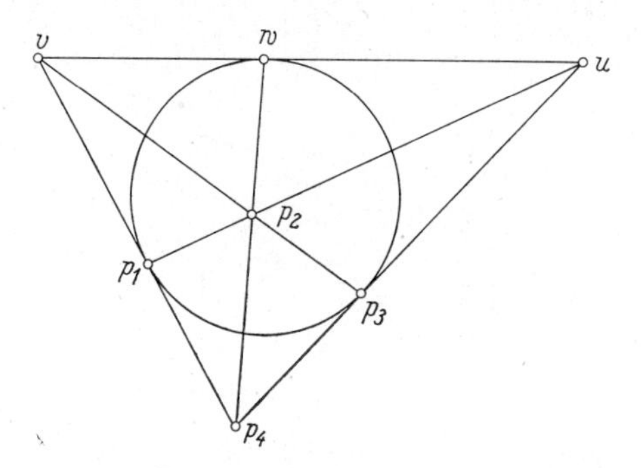

Fig. 7 The plane **P**(2)

Little beyond **23** is known about finite planes with characteristic. Note that **17** implies:

24. *If the little hexagonality condition holds in a finite plane of characteristic m, then m is a prime number, and every quadrangle generates a subplane isomorphic to* $\mathbf{P}(m)$.

This was rediscovered by DEMARIA 1959. LOMBARDO-RADICE 1955a has shown that **24** can be improved in the case of characteristic 3:

25. *A projective plane is of characteristic 3 if and only if every quadrangle generates a subplane isomorphic to* $\mathbf{P}(3)$.

Thus the hexagonality condition is superfluous here. Further results on planes with characteristic may be found in KEEDWELL 1963, 1964.

---

[1]) GLEASON 1956 calls such a plane a "Fano plane", despite the fact that the condition known as the Axiom of FANO [1896] forbids the occurence of quadrangles with collinear diagonal points. For this reason, ZADDACH 1956, 1957, calls the same planes "Anti-Fano planes". He shows in these papers that certain infinite planes of characteristic 2 (for example those generated by a quadrangle and a point on one of its sides) are also desarguesian.

4. Collineations of finite planes

In this chapter we give a fairly complete account of the known results on automorphisms of finite projective and affine planes. As before, we shall use the term "collineation" rather than "automorphism".

Section 4.1 is mostly of a rather elementary nature; it is concerned with relations between the order of the plane on the one hand, and the order, the fixed point (line) number, and orbit number of a given collineation group on the other. The case where this group is cyclic will receive particular attention.

The main theme of Sections 4.2—4.4 will be to determine the structure of a finite plane about whose collineation group some information is given. In 4.2, we make assumptions on the abstract structure of the collineation group, in 4.3 we postulate the existence of sufficiently many central collineations (for example, that there exist collineation groups of Lenz-Barlotti types I.5, I.6, I.7, III.1, or III.2), and in 4.4 we assume certain transitivity properties. In each of these three situations we shall give, among other things, characterizations of the finite desarguesian planes again.

Some of the results in these three sections, particularly in 4.3, require rather deep theorems on finite groups. In most cases, we will quote these in full, without indication of proof, and outline how they come into play for the proof of the result on finite planes under consideration.

4.1 Fixed elements and orders

Throughout this section, $\mathbf{P}$ will denote a projective plane of order n. For an arbitrary set Φ of collineations of $\mathbf{P}$, we define $\mathbf{F} = \mathbf{F}(\Phi)$ as the substructure of those points and lines in $\mathbf{P}$ which are fixed by every $\varphi \in \Phi$. In all later applications, Φ will either be a group or consist of a single collineation φ; in the latter case we write $\mathbf{F}(\varphi)$ instead of $\mathbf{F}(\{\varphi\})$. Incidentally, we have $\mathbf{F}(\varphi) = \mathbf{F}(\langle\varphi\rangle)$, where $\langle\varphi\rangle$ denotes the cyclic group generated by φ.

The uniqueness of intersection points and joining lines in a projective plane shows that $p, q \in \mathbf{F}(\Phi)$ implies $pq \in \mathbf{F}(\Phi)$, and the dual property.[1]) In other words:

1. $\mathbf{F}(\Phi)$ *is a closed subset of* $\mathbf{P}$.

We use here the terminology of Section 3.1; cf. (3.1.7) and (3.1.8). The numbers of points [resp. lines] in $\mathbf{F}(\Phi)$ will be denoted by $f(\Phi)$ [resp. $F(\Phi)$]. If Φ is a group of elations with axis A, containing nontrivial elations with distinct centers, then $f(\Phi) = n + 1$ and $F(\Phi) = 1$; this is the simplest example showing that $f(\Phi)$ and $F(\Phi)$ may be distinct. Hence it is of interest to have conditions which will guarantee $F = f$. We discuss some such conditions now; the most important of these is:

2. *If* Φ *is a cyclic group, then* $F(\Phi) = f(\Phi)$. *Equivalently,* $F(\varphi) = f(\varphi)$ *for every collineation* φ *of* $\mathbf{P}$.

This is the special case $\lambda = 1$ of **2.3.12**; the result is due to Baer 1947.

3. *If* Φ *is a* p*-group with* $p \nmid n$, *then* $F(\Phi) = f(\Phi)$.

Proof: Dembowski 1958, Satz 10. More about p-groups of collineations will be said in **16** below; compare here also Result **2.3.10**.

4. *Suppose that every subplane of the form* $\mathbf{F}(\Gamma)$, *where* Γ *is any collineation group of* $\mathbf{P}$, *has an order dividing* n.[2]) *Then* $F(\Phi) = f(\Phi)$ *for every soluble collineation group* Φ *whose order is prime to* n.

This is Theorem 2 of Roth 1964a.

We use now result **2** to classify the collineations φ of $\mathbf{P}$ by the nature of $\mathbf{F}(\varphi)$. We call φ *planar* if $\mathbf{F}(\varphi)$ is a subplane. A *generalized elation*[3]) is a collineation φ with $f(\varphi) > 0$ which is either the identity or has all its fixed points on a (necessarily unique) fixed line A and therefore, dually, all its fixed lines through a unique fixed point $c \mathrel{I} A$. A *generalized homology*[3]) is either the identity or a nonplanar collineation φ such that $\mathbf{F}(\varphi)$ contains a nonincident point-line pair (c, A) with every fixed point $\neq c$ on A and, dually, every fixed line $\neq A$ through c. Note that

---

[1]) This is a characteristic property of partial planes. It is easy to find automorphisms of designs with $\lambda > 1$ (for example, of $\mathbf{P}_t(d, q)$ with $t > 1$) which fix two points but none of the λ blocks through them.

[2]) This condition seems to be satisfied by all known finite planes; see Chapter 5. Roth 1964a, Theorem 5, has shown that it holds in any finite plane which is (c, A)-transitive for some flag (c, A).

[3]) This terminology (Baer 1947) is self-explanatory in view of the definitions in Section 3.1. Baer uses slightly different definitions, allowing a collineation φ with $\mathbf{F}(\varphi)$ a single flag to be either a generalized elation or a generalized homology.

here c and A are uniquely determined if and only if $f(\varphi) \neq 3$. Also, φ is simultaneously a generalized elation and a generalized homology if and only if either $\varphi = 1$ or $f(\varphi) = 2$. This ambiguity will not lead to confusion, mainly because of **17a** below. It follows from these definitions that the only collineations which are neither planar nor generalized elations or homologies are those with $\mathbf{F}(\varphi) = \emptyset$.

Consider an arbitrary power $\varphi^k \neq 1$ of the collineation φ. It is clear that $\mathbf{F}(\varphi) \subseteqq \mathbf{F}(\varphi^k)$, so that

$$f(\varphi) \leqq f(\varphi^k). \tag{1}$$

Also, φ leaves $\mathbf{F}(\varphi^k)$ invariant, hence induces an automorphism of the incidence structure $\mathbf{F}(\varphi^k)$. It follows that if φ^k is either a generalized elation or a generalized homology with $f(\varphi^k) \neq 3$, then the special fixed elements c, A of φ^k are also fixed by φ. Some corollaries of this observation are:

5. *Suppose that* $\varphi^k \neq 1$.

(a) *If φ^k is a generalized elation, then so is φ.*
(b) *Let φ^k be a generalized homology. If $f(\varphi^k) \neq 3$, or if $f(\varphi^k) = 3$ and $(o(\varphi)\, o(\varphi^k)^{-1}, 3) = 1$, then φ is also a generalized homology.*
(c) *If $f(\varphi^k) = 3$, then $f(\varphi) \neq 2$, and $f(\varphi) = \begin{Bmatrix} 1 \\ 0 \end{Bmatrix}$ implies $\begin{Bmatrix} f(\varphi^2) \\ f(\varphi^3) \end{Bmatrix} = 3$ and $k \equiv 0 \bmod \begin{Bmatrix} 2 \\ 3 \end{Bmatrix}$.*

For the proofs of these statements, see Baer 1947, pp. 658—659. The symbol $o(\varphi)$ in **(b)** denotes, as usual, the order of φ.

If φ is planar, then the order m of the subplane $\mathbf{F}(\varphi)$ is of interest. We know from **3.2.18** that either $m^2 = n$ or $m(m+1) \leqq n$. This can be improved here:

6. *If Φ is a collineation group such that $\mathbf{F}(\Phi)$ is a subplane of order m, then either $m^2 = n$ or $m(m+1) \leqq n - 2$.*

For the proof see Roth 1964a, Section 3. It suffices to prove this for Φ cyclic of order a prime p. Roth proved that then, if $m^2 \neq n$, we have $m(m+1) \leqq n - d$ with $d \geqq 2$; moreover $d = 2$ implies $p = 3$ or 7.

In the case $m^2 = n$ of **6**, the subplane $\mathbf{F}(\Phi)$ is a Baer subplane in the sense of (3.1.11) and (3.1.12); consequently Φ consists of quasicentral collineations in this case (these were defined in the context of **3.1.6**). The next result is a characterization of the quasicentral collineations by the numbers of their fixed elements.

7. *The following conditions for the collineation $\varphi \neq 1$ of* **P** *are equivalent:*

(**a**) *φ is quasicentral.*
(**b**) *$f(\varphi) = n + 1$, $n + 2$, or $n + \sqrt{n} + 1$.*
(**c**) *$f(\varphi) \geqq n$.*

That (**a**) implies (**b**) follows from **3.1.2** and **3.2.18a**, and that (**c**) follows from (**b**) is trivial. To show that (**c**) implies (**a**), assume [with BAER 1946b] that there exists a point w which is on no fixed line. By **1**, any line with two fixed points is itself fixed; hence none of the lines joining the $f(\varphi) \geqq n$ fixed points can pass through w. This implies that w is not fixed and that each of the $n + 1$ lines through w carries at most one fixed point. Let $L = w(w\varphi)$. If there were a fixed point on L, then $L\varphi = L$, i.e. L would be a fixed line through w. Thus L must be free of fixed points. It follows that $f(\varphi) = n$ and L is the only fixed point free line through w. Similarly, $L\varphi$ is the only fixed point free line through $w\varphi$. But both L and $L\varphi$ pass through $w\varphi$, whence $L = L\varphi$, a contradiction proving **7**.

Result **7** shows that $f(\varphi)$ is either $<n$ or one of $n + 1$, $n + 2$, $n + \sqrt{n} + 1$, for any collineation φ of **P**. In particular,

$$f(\varphi) \neq n. \tag{2}$$

Result **7**, together with **3.1.2** and **3.2.18a**, allows us to classify the quasicentral collineations as follows:[1])

8. *Let φ be a nontrivial collineation of* **P**. *Then*

(**a**) *φ is an elation if and only if $f(\varphi) = n + 1$. In this case $o(\varphi) \mid n$.*
(**b**) *φ is a homology if and only if $f(\varphi) = n + 2$. In this case $o(\varphi) \mid n - 1$.*
(**c**) *φ is a planar quasicentral collineation if and only if $f(\varphi) = n + \sqrt{n} + 1$. In this case n is a square, and $o(\varphi) \mid \sqrt{n}(\sqrt{n} - 1)$.*

The statements concerning the order $o(\varphi)$ are immediate consequences of the easily proved fact that a quasicentral collineation permutes its non-fixed points in orbits of length $o(\varphi)$.

We have remarked earlier (see **3.1.6**) that involutions (collineations of order 2) are always quasicentral. Hence **8** gives:

9. *An involutorial collineation φ of* **P** *is either*

(**a**) *planar, in which case n is a square and* **F**(φ) *a Baer subplane of* **P**, *or*
(**b**) *an elation, in which case n is even, or*
(**c**) *a homology, in which case n is odd.*
In any case, $f(\varphi)$ must be odd.

[1]) Compare this with result **1.4.12** of NORMAN 1965.

We shall now present some results which are supplementary to **9a** and **9b**.

10. *If* **P** *admits a* 2-*group* Φ *of collineations such that* $\mathbf{F}(\Phi)$ *is a subplane of order* m, *then* n *is a power of* m; *in fact*

$$n = m^{2^k} \quad \textit{for some integer } k \geqq 0. \tag{3}$$

This is Theorem 6 of Ostrom & Wagner 1959. For the proof, let ζ be an involution in the centre of Φ. If $\mathbf{F}(\Phi) = \mathbf{F}(\zeta)$, then (3) holds with $k = 1$. The alternative is $\mathbf{F}(\Phi) \subset \mathbf{F}(\zeta)$; in this case Φ induces a collineation group $\Phi_0 \neq 1$ in $\mathbf{F}(\zeta)$, and Φ_0 is essentially the quotient group Φ/Σ, with Σ the subgroup of those $\varphi \in \Phi$ which fix every point of $\mathbf{F}(\zeta)$. An obvious inductive argument then completes the proof.

The next result will be useful in Section 4.4.

11. *Let* α *and* β *be two planar involutions of* **P**, *and suppose* $\alpha\beta = \beta\alpha$ *and* $\mathbf{F}(\alpha) \neq \mathbf{F}(\beta)$. *Then* α *and* β *induce involutorial collineations* α_0 *and* β_0 *in* $\mathbf{F}(\beta)$ *and* $\mathbf{F}(\alpha)$, *respectively.*

(**a**) *If one of* α_0, β_0 *is central, then so is the other.*

(**b**) *If* **P** *has no subplane of order* 2 *and if one of* α_0, β_0 *is a homology, then* $\alpha\beta$ *is an involutorial homology.*

For the proof see Wagner 1962a, Lemmas 2 and 3.

The following is an improvement of **9b**:

12. *If* **P** *admits an involutorial elation, then* $n = 2$ *or* $n \equiv 0 \bmod 4$.

This is due to Hughes 1957e; the following proof is essentially that of Hall 1959, pp. 414—416. The n^2 points not on the axis A of the elation φ are precisely the non-fixed points of φ; they fall into $m = n^2/2$ point orbits of length 2. Similarly, the lines not through the center $c \mathrel{\mathrm{I}} A$ fall into m orbits of length 2. For each of these orbits $\mathfrak{x}$, $\mathfrak{X}$, choose representative elements $x \in \mathfrak{x}$, $X \in \mathfrak{X}$, and define

$$h(\mathfrak{x}, \mathfrak{X}) = \begin{cases} 1 & \text{if } x \mathrel{\mathrm{I}} X, \\ -1 & \text{if } x \not\mathrel{\mathrm{I}} X \quad \text{but} \quad x \mathrel{\mathrm{I}} X\varphi, \\ 0 & \text{otherwise.} \end{cases}$$

It is then verified without difficulty that

$$\sum_{\mathfrak{X}} h(\mathfrak{p}, \mathfrak{X})\, h(\mathfrak{q}, \mathfrak{X}) = n\, \delta(\mathfrak{p}, \mathfrak{q}) \tag{4}$$

for arbitrary point orbits $\mathfrak{p}$, $\mathfrak{q}$ of length 2; here $\delta = 0$ or 1 according as the arguments differ or coincide. Let L be an arbitrary line $\neq A$ through c. Then $L\varphi = L$, and we write $\mathfrak{x} \mathrel{\mathrm{I}} L$ if the two points of the orbit $\mathfrak{x}$ are both on L. Define

$$k(L, \mathfrak{X}) = \sum_{\mathfrak{x} \mathrel{\mathrm{I}} L} h(\mathfrak{x}, \mathfrak{X}),$$

for any line orbit $\mathfrak{X}$ of length 2; then $k(L, \mathfrak{X}) = \pm 1$. We number the L and $\mathfrak{X}$ in an arbitrary fashion and consider the (n, m)-matrix $M = (k(L_i, \mathfrak{X}_j))$. It then follows from (4) that

$$(5) \qquad MM^T = m I_n,$$

i.e. the scalar product of any row with itself is m, while two distinct rows are orthogonal.

Now suppose $n > 2$. Then M has at least 3 rows, and by a suitable choice of representatives, as well as by possibly renumbering the columns of M, we may assume that the first three rows of M are as follows:

$$\begin{array}{cccc} +1 \cdots\cdots\cdots\cdots\cdots\cdots\cdots\cdots\cdots\cdots\cdots +1 \\ +1 \cdots\cdots\cdots\cdots\cdots +1 \quad -1 \cdots\cdots\cdots\cdots\cdots -1 \\ \underbrace{+1 \cdots +1}_{r} \quad \underbrace{-1 \cdots -1}_{s} \quad \underbrace{+1 \cdots +1}_{t} \quad \underbrace{-1 \cdots -1}_{u}. \end{array}$$

Here r, s, t, u denote the numbers of columns with the respective distributions of $+1$ and -1. But (5) implies $r = s = t = u$, so that $m = n^2/2 \equiv 0 \bmod 4$. This implies $n \equiv 0 \bmod 4$, proving **12**. Corollary:

13. *If there exists a projective plane* **P** *of order* $n > 2$ *with* $n \equiv 2 \bmod 4$, *then its full collineation group is of odd order.*

(HUGHES 1957e.) For $n \equiv 2 \bmod 4$ cannot be a square, so that by **9** every involution would be an elation, in conflict with **12**.

We mention another property of involutorial central collineations φ: if p and q are non-fixed points such that pq is not incident with the center of φ, then the points $p, q, p\varphi, q\varphi$ form a quadrangle with two diagonal points on the axis. Since the third diagonal point is obviously the center, it follows in particular that

14. *If* **P** *admits involutorial elations, then* **P** *contains subplanes of order* 2.

The following discussion will be concerned with arithmetical relations between the order n of **P** and the integers $o(\varphi)$ and $f(\varphi)$, for an arbitrary collineation φ. We mention first that

$$(6) \qquad o(\varphi) \leqq n^2 + n + 1;$$

this was mentioned without proof by BAER 1947; a proof will be indicated in the next section (cf. **4.2.9**).

If π is any permutation of a finite set, and if $f(\pi)$ denotes the number of elements fixed by π, then, for any prime p,

$$(7) \qquad f(\pi^{p^i}) \equiv f(\pi^{p^j}) \bmod p^{i+1} \qquad \textit{whenever } 0 \leqq i < j.$$

For the proof see BAER 1947, pp. 659—660. A straightforward application of (7) is:

15. *If φ is a collineation of prime power order $o(\varphi) = p^e$, then, for $0 \leqq i < e$,*

(a) $f(\varphi^{p^i}) \equiv n^2 + n + 1 \mod p^{i+1}$,
and
(b) *for any line L of $\mathbf{F}(\varphi)$, the number of points in $\mathbf{F}(\varphi^{p^i})$ incident with L is congruent to $n+1 \mod p^{i+1}$.*

These results are the basis for a detailed discussion of collineations of prime power order, by BAER 1947, Section 3. Several of Baer's results have later been extended to p-groups of collineations, by DEMBOWSKI 1958 and ROTH 1964a. We list some of these now, and in view of **6** and **10** we leave aside the planar case.

16. *Suppose that Φ is a p-group of collineations of $\mathbf{P}$, with $\mathbf{F}(\Phi)$ not a subplane.*

(a) *If $f(\varphi) = 0$ for some $\varphi \in \Phi$, then*

$$n^2 + n + 1 \equiv 0 \mod p. \tag{8}$$

In this case either $p = 3$ or $p \equiv 1 \mod 3$. Conversely, if (8) holds and $p \neq 3$, then $\mathbf{F}(\Phi) = \emptyset$.
(b) *p divides n if and only if $\mathbf{F}(\Phi)$ is nonempty and all points of $\mathbf{F}(\Phi)$ are collinear. Only in this case can $f(\Phi)$ be distinct from $F(\Phi)$.*
(c) *If $\mathbf{F}(\Phi)$ is a nonincident point-line pair, then $p \mid n+1$. The converse holds if $p \neq 2$.*
(d) *If $\mathbf{F}(\Phi)$ contains three noncollinear points, then $p \mid n-1$. The converse holds if $p \neq 2, 3$.*

For proofs, see DEMBOWSKI 1958, Satz 10 (c'), and ROTH 1964a, Theorem 4; for the case where Φ is cyclic, the results are mostly contained in BAER 1947, Section 3. Note that **(a)** implies $f(\varphi) > 0$ for all $\varphi \in \Phi$ whenever $p \equiv 2 \mod 3$; this is simply the fact that (8) implies $p = 3$ or $p \equiv 1 \mod 3$ [see, for example, NAGELL 1951, Theorem 94]. For $p = 2$, we see again (cf. **9**) that involutions must have fixed points.

Of the many refinements and special cases of **16** discussed in BAER 1947, we mention here only the following:

17. *Let φ be a collineation of prime power order p^e. Then:*

(a) $f(\varphi) \neq 2$.
(b) *If $f(\varphi) = 3$ and φ is a generalized elation, then $p = 2$.*
(c) *If $f(\varphi) = 4$, then $p = 3$ and φ is a generalized elation.*
(d) *If $f(\varphi) = 5$, then $p = 2$.*
(e) *If $f(\varphi) = 6$, then either $p = 5$ and φ is a generalized elation, or $p = 3$ and φ is a generalized homology.*

This is part of Theorem 7 in BAER 1947, p. 673.

18. *Suppose that Φ is a collineation group of* **P** *such that no $\varphi \neq 1$ in Φ is planar. Then*

(**a**) $|\Phi|$ *divides* $n^3(n^3-1)(n^2-1)$.
(**b**) *If* $\mathbf{F}(\Phi) \neq \emptyset$, *then* $|\Phi|$ *divides* $n^3(n+1)(n-1)^2$.
(**c**) *If* $\mathbf{F}(\Phi)$ *contains two distinct points (or lines), then* $|\Phi|$ *divides* $n^2(n-1)^2$.
(**d**) *If* $\mathbf{F}(\Phi)$ *contains a triangle, then* $|\Phi|$ *divides* $(n-1)^2$.

This is Theorem 1 of BAER 1947, p. 676.[1])

For the remainder of this discussion, we suppose that $o(\varphi)$ is not only a prime power, but actually a prime.

19. *If* $o(\varphi) = p > n+1$, *then* (8) *holds, whence* $p \equiv 1 \bmod 3$.

For $o(\varphi) = p > n+1$ implies $f(\varphi) = 0$, so that **16a** applies.

If $o(\varphi) = p$, we can utilize HUGHES's 1957d result **2.3.13**, in addition to **15**—**18**.[2]) We denote the number of $\langle\varphi\rangle$-orbits of points by $m = m(\varphi)$, so that equation (2.3.3) becomes

$$f(\varphi) + p[m(\varphi) - f(\varphi)] = n^2 + n + 1. \tag{9}$$

This implies again (cf. **9**) that $f(\varphi)$ is odd if $p = 2$; furthermore $m(\varphi)$ is odd if $p > 2$. We give now two less trivial relations between n, $f(\varphi)$, $o(\varphi)$, and $m(\varphi)$. Let n^* denote the square-free factor of n, i.e. the product of all those primes which divide n an odd number of times.

20. *Suppose that φ is a collineation of prime order $p > 2$, such that $p \nmid n^*$. Then:*

(**a**) $\left(\frac{p}{q}\right)^{f+1} = (-1)^{(m-1)(q-1)/4}$, *for every odd prime divisor q of n^*. In particular, if f is odd and not a square, then $(m-1)(q-1) \equiv 0 \bmod 8$ for any such q.*
(**b**) *If f is even, then* $\left(\frac{n^*}{p}\right) = 1$.

Here we have written f for $f(\varphi)$ and m for $m(\varphi)$, and $\left(\frac{p}{q}\right)$ is, of course, the Legendre quadratic residue symbol. Result **20** is due to HUGHES 1957d; it is in fact a simple consequence of the equation

$$n x^2 + (-1)^{(m-1)/2} p^{f+1} y^2 = z^2 \tag{10}$$

which is the case $\lambda = 1$ of (2.3.4): We can assume x, y, z relatively prime in (10); this implies that if $q \mid n^*$ then $q \nmid y$, so that there exists

[1]) If **P** is desarguesian and Φ the full projective group of **P**, as defined in Section 1.4, then the word "divides" in (**a**) may be replaced by "equals". Conversely, if $|\Phi| = n^3(n^3-1)(n^2-1)$ [this is the number of all quadrangles in **P**], then **P** is desarguesian under the hypotheses of **18**. This will be proved in Section 4.4.

[2]) It may be not too difficult to extend this from p to p^e; this would result in further improvements of **17**, **20**, and **21**.

an integer j with $jy \equiv 1 \mod q$. After multiplication with j^2, we get from (10) that $\left(\frac{(-1)^{(m-1)/2} p^{f+1}}{q}\right) = 1$, which is equivalent to (**a**). The proof of (**b**) is similar.

21. *If n^* and $f(\varphi)$ are both even and φ of prime order $p \nmid n$, then $p \equiv \pm 1 \mod 8$.*

Proof. Put $n^* = 2q_1 \dots q_t \cdot q_{t+1} \dots q_k$, where the q_i are primes $\equiv -1$ or $+1 \mod 4$ according as $i \leqq t$ or $i > t$. Result **20b** and the quadratic reciprocity law then give

$$1 = \left(\frac{n^*}{p}\right) = \left(\frac{2}{p}\right) \prod_{i=1}^{k} \left(\frac{q_i}{p}\right) = (-1)^{(p^2-1)/8} \prod_{i=1}^{k} \left(\frac{p}{q_i}\right) (-1)^{(p-1)(q_i-1)/4}.$$

Furthermore, as f is even, **20a** gives

$$\left(\frac{p}{q_i}\right) = (-1)^{(m-1)(q_i-1)/4}.$$

As $p + m$ is even, it follows that

$$(-1)^{(p^2-1)/8} = (-1)^{t(p+m-2)/2}$$

because the q_i with $i > t$ give contribution 1. Thus we have

$$(11) \qquad (p^2 - 1)/8 \equiv t(p + m - 2)/2 \mod 2.$$

If t is even, the assertion of **21** follows now at once. Hence suppose t odd, so that t can be cancelled from (11). If we put $p = 2s + 1$, a short calculation transforms (11) into

$$(12) \qquad s(s-1) \equiv m - 1 \mod 4.$$

On the other hand, since t is odd, there actually exists a prime divisor $\equiv -1 \mod 4$ of n^*. Hence the Bruck-Ryser Theorem **3.2.13**, together with the hypothesis $2 \mid n^*$, implies $n \equiv 0 \mod 4$. As $f(p-1)$ is also divisible by 4, we can infer from (9) that

$$(13) \qquad pm \equiv 1 \mod 4.$$

Finally, (12) and (13) together show that $p \equiv 1 \mod 8$ if $m \equiv 1 \mod 4$, and $p \equiv -1 \mod 8$ if $m \equiv -1 \mod 4$. This proves our contention.[1]

We conclude this section with a brief discussion of collineations which are powers of correlations. The order of a correlation δ of **P**, as a permutation of the points and lines, must be even:

$$o(\delta) = 2s;$$

[1] The preceding results are quite useful in determining the possible orders of collineations in a hypothetical plane of order not a prime power. For example, if there exists a plane of order 10, then the only possible prime orders of a collineation in such a plane are 3 and 5, and further analysis shows that none of these can be central collineations. (This is an unpublished result of E. T. Parker.)

and $\delta^2 = \varphi$ is a collineation of order s. Clearly $\varphi = 1$ if and only if δ is a polarity. The following result gives, among other things, a relation between the number $f(\varphi)$ of fixed points of φ and the number $a(\delta)$ of absolute points of δ.

22. *Let δ be a correlation of* **P**, *and $\varphi = \delta^2$.*

(**a**) *For any odd prime p not dividing n, let j be a positive integer, not a multiple of p, and i an arbitrary integer. Then*

$$f(\varphi^{jp^{i+1}}) \equiv f(\varphi^{jp^i}) \bmod p^{i+1}. \tag{14}$$

(**b**) *If n is odd and $i \geqq 2$, then*

$$f(\varphi^{2^i}) \equiv f(\varphi^{2^{i-1}}) \bmod 2^{i+1}, \tag{15}$$

and

$$f(\varphi) \equiv a(\delta) + 1 \bmod 2.$$

(**c**) $$\textit{If } n \equiv \begin{Bmatrix} 1 \\ 2 \\ 3 \end{Bmatrix} \bmod 4, \textit{ then } \begin{cases} f(\varphi^2) \equiv f(\varphi) \bmod 4, \\ f(\varphi) \equiv 1 \bmod 2, \\ f(\varphi) + f(\varphi^2) \equiv 2 \bmod 4. \end{cases}$$

This is Proposition 2.2 of Ball 1948; the congruence (7) is used here again.

4.2 Collineation groups

In this section we will be concerned with the following general situation. Suppose there is given a projective plane **P** of order n, an abstract finite group G, and the information that **P** possesses a collineation group $\Gamma \cong G$. We shall be concerned with questions of the following kind: How does n influence the structure of G? [For example, it can be shown that

$$|G| \leq n^{3+\log_2 n}(n^3 - 1)(n^2 - 1); \tag{1}$$

cf. Hall 1959, p. 403.] Conversely, how does G influence **P**, and in particular, under what conditions on G can we conclude that **P** is desarguesian? In dealing with this problem, it will often be necessary to determine first how $\Gamma \cong G$ must operate on **P**. Not too many results have been obtained so far in this area, and further research seems rather promising.

We begin our discussion with a few results on 2-groups of collineations. Some of these are closely related to parts of Section 4.1, the difference being that here we make no assumptions on how Γ operates on **P**. Result **4.1.13** can be rephrased to say that if $2 < n \equiv 2 \bmod 4$, then a 2-group of collineations of **P** is trivial. The next results will show that if $n \equiv -1 \bmod 4$, the structure of a 2-group on **P** is also quite restricted. First we determine the subsets of fixed elements, in the case where the group is abelian.

1. *Let* **P** *be of order* $n \equiv -1 \bmod 4$ *and* Γ *a non-trivial abelian 2-group of collineations of* **P**.

(**a**) *If* $|\Gamma| = 2$, *then the unique non-trivial collineation in* Γ *is a homology.*
(**b**) *If* $|\Gamma| > 2$ *and* Γ *is elementary abelian, then* $|\Gamma| = 4$, *and* $\mathbf{F}(\Gamma)$ *consists of the vertices and sides of a triangle.*
(**c**) *If* Γ *contains collineations of order* > 2, *then* $\mathbf{F}(\Gamma)$ *is a nonincident point-line pair* (c, A). *In this case, if* 2^e *is the exponent of* Γ, *then* $n \equiv -1 \bmod 2^{e-1}$.

Proof. As n cannot be a square, every involution of Γ is a homology (cf. **4.1.9**). This proves (**a**). Also, for every non-incident point-line pair (p, L), the subgroup $\Gamma(p, L)$ is a 2-group and has order dividing $n - 1 \equiv 2 \bmod 4$; hence $|\Gamma(p, L)| \leq 2$, and it follows that distinct involutions have distinct center-axis pairs. Now (**b**) is a simple consequence of **3.1.7** and **3.1.8a**. For the proof of (**c**), the reader is referred to Hering 1967a, Hilfssatz 2. Hering treats the cyclic case only, but his proof is easily modified to the present situation. As a matter of fact, Γ must actually be cyclic in case (**c**); this will follow from result **4** below. Before we state this, however, it will be convenient to remind the reader of the definitions of the *dihedral, semi-dihedral*, and *generalized quaternion groups*. Each of these is generated by two elements β and γ, and γ always generates a cyclic normal subgroup of index 2:

$$\gamma^r = 1, \quad r \geqq 2. \tag{2}$$

In the dihedral and semi-dihedral case, the other generator is an involution:

$$\beta^2 = 1. \tag{3}$$

The dihedral groups are defined by (2), (3), and

$$\beta^{-1} \gamma \beta = \gamma^{-1}. \tag{4}$$

Here the exponent r need not be a power of 2; the dihedral group defined by (2)—(4) is of order $2r$. For the semi-dihedral and generalized quaternion groups, however, we require

$$r = 2^e, \quad e > 1. \tag{5}$$

Then the semi-dihedral groups are defined by (2), (3), (5), and

$$\beta^{-1} \gamma \beta = \gamma^{2^{e-1}-1}, \tag{6}$$

and the generalized quaternion groups by (2), (4), (5), and

$$\beta^2 = \gamma^{2^{e-1}}. \tag{7}$$

Both the generalized quaternion and the semi-dihedral groups are of order $2r = 2^{e+1}$. The significance of these groups in our context lies mainly in the following facts:

2. *A finite 2-group which contains only one involution is either cyclic or a generalized quaternion group.*

(BURNSIDE 1911, p. 132; the converse is obviously true also.)

3. *A finite 2-group which contains an involution whose centralizer has order 4 is dihedral, semi-dihedral, or cyclic of order 4.*

(SUZUKI 1951, Lemma 4. Note that the non-cyclic group of order 4 is dihedral.)

We state now the main result on collineation 2-groups of planes of order $\equiv -1 \bmod 4$:

4. *If* **P** *is of order* $n \equiv -1 \bmod 4$, *then every 2-group* Γ *of collineations of* **P** *is cyclic, dihedral, semi-dihedral, or a generalized quaternion group. Also,* $|\Gamma|$ *divides* $4(n+1)$.

For the proof (HERING 1967a, Satz 1), let ζ be an involution in the centre of Γ, and let c and A be center and axis of ζ. In view of **2**, we can assume that Γ contains another involution $\eta \neq \zeta$. Then the centralizer $C(\eta)$ has order at least 4. On the other hand, if (c', A') is the center-axis pair of η, then $A \mathrel{\mathrm{I}} c'$ and $c \mathrel{\mathrm{I}} A'$; this follows essentially from the argument given above for the proof of **1b**. Consequently, $\mathbf{F}(C(\eta))$ contains the triangle c, c', AA'. But then **1c** shows that $C(\eta)$ cannot contain collineations of order >2, so that $C(\eta)$ is elementary abelian and thus of order 4, by **1b**. The result now follows from **3**. The following result on affine planes is proved similarly:

5. *If* **A** *is an affine plane of order* $n \equiv -1 \bmod 4$, *then every collineation group of* **A** *has a normal subgroup of index* $\leqq 2$ *whose Sylow 2-groups are cyclic or generalized quaternion groups.*

(HERING 1967a, Satz 2.)

The dihedral groups, not necessarily of order a power of 2, are of interest also from another point of view. Suppose that α and β are two involutions generating the finite group G. If $\gamma = \alpha\beta$ and $o(\gamma) = r$, then (2)—(4) are satisfied, so that G is the dihedral group of order $2r$. Furthermore, if r is odd, any two involutions in G are conjugate while if r is even, say $r = 2s$, there are three conjugate classes of involutions. One of these consists of γ^s alone, and the other two contain s involutions each, α being contained in one of these classes, and β in the other. We have remarked above that G is the elementary abelian group of order 4 if $s = 1$; for $s > 1$, the centre of G consists of 1 and γ^s alone. These facts (for the proofs, see e.g. COXETER & MOSER 1957, p. 6) are the basis for several results on central collineations, for example:

6. *Let α and β be involutorial elations of a finite projective plane, with axes A, B and centers a, b, respectively, such that $A \neq B$ and $a \neq b$. Then the cyclic group $\langle\alpha\beta\rangle$ operates regularly on the points $\neq AB$ which are not on ab.*

This is Lemma 5 of PIPER 1966. (Compare here result **3.1.7a** which has a much weaker conclusion.) Moreover, it may be proved without difficulty that, under the hypotheses of **6**, the order r of $\alpha\beta$ is odd if $a \not\mathrm{I}\, B$ and $b \not\mathrm{I}\, A$, and that $(\alpha\beta)^2$ is an involutorial (a, B)-elation if $a \,\mathrm{I}\, B$ (this also implies $b \not\mathrm{I}\, A$). Similarly, one can prove analogous results on groups generated by two involutorial homologies.

We turn now to results of a different nature; these will include some group theoretical characterizations of the desarguesian projective planes $\mathbf{P}(q)$. We begin this discussion by restating result **2.3.1b** for $\lambda = 1$, which will be fundamental for the rest of this chapter:

7. *A collineation group of a finite projective plane has equally many point and line orbits.*

A useful supplement to this is the following lemma, valid also in infinite planes:

8. *Every finite collineation group of a projective plane operates faithfully on at least one (point or line) orbit.*[1]

For the simple proof, see DEMBOWSKI 1965c, Satz 2. As a matter of fact, Γ is faithful on any orbit $\mathfrak{X}$ whose length at least equals that of each orbit of elements incident with elements of $\mathfrak{X}$.

Suppose now that Γ is *quasiregular* in the sense that the constituents of Γ on all point and line orbits are regular. This means that, for any $\gamma \in \Gamma$,

(8) *If $p\gamma = p$, then $x\gamma = x$ for all $x \in p\Gamma$, and if $L\gamma = L$, then $X\gamma = X$ for all $X \in L\Gamma$.*

This condition is satisfied in particular when Γ is abelian or hamiltonian (cf. HALL 1959, p. 190), for it is easy to prove that all permutation representations of a group G are quasiregular if and only if every subgroup of G is normal. We infer from **8**:

[1]) That one cannot expect a group Γ to be faithful on a point *and* a line orbit is seen by considering any group of elations with fixed axis which contains elations $\neq 1$ with different centers. Conversely, PIPER 1968a has shown that if Γ is not faithful on any line orbit, then Γ has exactly one fixed line L, and all fixed points of Γ are on L. Piper has also constructed other examples for such groups.

9. *Every quasiregular (and in particular every abelian or hamiltonian) finite collineation group Γ of a projective plane* $\mathbf{P}$ *has a point or line orbit of length* $|\Gamma|$. *If* $\mathbf{P}$ *is finite of order* n, *it follows that*

$$|\Gamma| \leqq n^2 + n + 1. \tag{9}$$

Note that this provides, for the case that Γ is cyclic, a proof of the inequality (4.1.6).

If $\mathbf{P}$ is of order n and

$$|\Gamma| > \tfrac{1}{2}(n^2 + n + 1), \tag{10}$$

then there is clearly at most one point orbit of length $|\Gamma|$. Hence, in the quasiregular case, we may assume (by reasons of duality and because of **9**) that there is exactly one such point orbit. This uniqueness is one of the main tools in the following determination of all quasiregular groups Γ satisfying (10).

10. *Let* Γ *be a quasiregular collineation group of the projective plane* $\mathbf{P}$ *of order* n. *(In particular,* Γ *may be abelian or hamiltonian.) Denote by* $m = m(\Gamma)$ *the number of point (or line; cf.* **7**) *orbits of* Γ, *and by* $\mathbf{F} = \mathbf{F}(\Gamma)$ *the substructure of the elements fixed by* Γ. *If* (10) *holds, then there are only the following possibilities:*

(a) $|\Gamma| = n^2 + n + 1$, $m = 1$, *and* $\mathbf{F} = \emptyset$. *Here* Γ *is transitive.*
(b) $|\Gamma| = n^2$, $m = 3$, *and* $\mathbf{F}$ *is a flag.*
(c) $|\Gamma| = n^2$, $m = n + 2$, *and* $\mathbf{F}$ *is either a unique line* A *and all* $x \mathrm{I} A$, *or dually a unique point* c *and all* $X \mathrm{I} c$. *In this case,* Γ *is elementary abelian and consists entirely of elations.*
(d) $|\Gamma| = n^2 - 1$, $m = 3$, *and* $\mathbf{F}$ *is a unique non-incident point-line pair.*
(e) $|\Gamma| = n^2 - \sqrt{n}$, $m = 2$, $\mathbf{F} = \emptyset$. *In this case, one point and one line orbit together form a Baer subplane, of order* $\sqrt{n}$.
(f) $|\Gamma| = n(n-1)$, $m = 5$, *and* $\mathbf{F}$ *consists of two points* u, v, *the line* uv, *and one line* $\neq uv$ *through one of* u, v.[1])
(g) $|\Gamma| = (n-1)^2$, $m = 7$, *and* $\mathbf{F}$ *consists of the vertices and sides of a triangle.*
(h) $|\Gamma| = (n - \sqrt{n} + 1)^2$, $m = 2\sqrt{n} + 1$, $\mathbf{F} = \emptyset$. *In this case, there are* $2\sqrt{n}$ *disjoint subplanes of order* $\sqrt{n} - 1$ *whose point sets constitute* $2\sqrt{n}$ *orbits, each of length* $n - \sqrt{n} + 1$. *Furthermore,* $n - \sqrt{n} + 1$ *is a prime* p, *and* Γ *is elementary abelian.*

---

[1]) Results **(a)**—**(f)** are actually consequences of more general theorems on tactical decompositions; cf. DEMBOWSKI 1965c, Satz 1, and DEMBOWSKI & PIPER 1967, Theorem 4.

For the proof in cases (**b**), (**c**), see DEMBOWSKI 1965c, Sätze 1, 3; for the remaining cases, except for the last claim of (**h**),[1]) see DEMBOWSKI & PIPER 1967, Corollary 4.9 and Theorem 5. Some of these results are also contained in two papers of HUGHES 1956, 1957c, with more specialized hypotheses.[2])

Examples of quasiregular groups of types (**a**)—(**g**) are known, for example in finite desarguesian planes. However, the only known case for type (**e**) occurs when $n = 4$; there may be no other. Whether case (**h**) can occur at all is an open problem; many values of n are excluded by the fact that $n - \sqrt{n} + 1$ is a prime and that both n and $\sqrt{n} - 1$ must be orders of projective planes (cf. **3.2.13**).

In case (**c**), the group Γ is either (A, A) or (c, c)-transitive, and in this case the structure of **P** is determined in a satisfactory fashion (cf. **3.1.22c, d**). In case (**b**), on the other hand, Γ is transitive on the points not on A as well as on the points $\neq c$ on A, and dually on the lines not through c as well as on the lines $\neq A$ through c. All known examples for this second alternative belong likewise to (c, c)- or (A, A)-transitive planes, but it is not known whether this must always be so. In an effort to solve this problem, DEMBOWSKI & OSTROM 1968 have proved the following partial results:

11. *Suppose that* **P**, *of order* n, *possesses an abelian collineation group* Γ *of order* n^2 *which permutes the points as well as the lines of* **P** *in* 3 *orbits. Let* (c, A) *be the fixed flag of* Γ (cf. **10b**). *Then*

(**a**) **P** *admits* n^2 *polarities* π *with* $\gamma^\pi = \gamma^{-1}$ *for every* $\gamma \in \Gamma$. *In particular,* **P** *is self-dual.*

---

[1] For the proof of this, pointed out to the author by C. Hering, consider a prime divisor p of $n - \sqrt{n} + 1$ and a Sylow p-group Σ of Γ. Then $|\Sigma| = p^{2e}$ for some $e > 0$, and $|\Sigma_{\mathfrak{x}}| = p^e$ for each Γ-orbit $\mathfrak{x}$ with $|\mathfrak{x}| = n - \sqrt{n} + 1$. If $\mathfrak{x}$ and $\mathfrak{y}$ are distinct such orbits, then $\Sigma_{\mathfrak{x}} \cap \Sigma_{\mathfrak{y}} = 1$. This implies, since the number of these orbits is $2\sqrt{n}$, that $p^{2e} = |\Sigma| \geqq 1 + 2\sqrt{n}(p^e - 1)$, whence $p^e \geqq 2\sqrt{n} - 1$. Now assume that $n - \sqrt{n} + 1$ has a prime divisor $q \neq p$; then the same argument shows that $q^f \geqq 2\sqrt{n} - 1$, where q^f is the highest power of q dividing $n - \sqrt{n} + 1$. But this would imply

$$(n - \sqrt{n} + 1)^2 = |\Gamma| \geqq p^{2e} q^{2f} > (2\sqrt{n} - 1)^4,$$

which is impossible. Hence $n - \sqrt{n} + 1 = p^e$, and results of LJUNGGREN 1942 and BRUCK 1955 [see (4.4.3) below] imply that $e = 1$. Thus $|\Gamma| = p^2$, and since Γ has more than one subgroup of order p, Γ is elementary abelian.

[2]) In a subsequent paper by PIPER 1968a, Section 4, it is shown that if the hypothesis (10) of **10** is relaxed by postulating only that there be only one point (line) orbit on which Γ is faithful, then the only possibilities besides (**a**)—(**h**) occur either for $n \leqq 4$ or for $\mathbf{F}(\Gamma)$ consisting of $k\,(\geqq 2)$ collinear points and the line carrying them (or the dual).

(b) *If the (normal) subgroup* Δ *of the* (c, A)*-elations in* Γ *is a direct factor of* Γ, *then* n *is odd.*[1])

(c) *If* $o, e \mathrel{\rlap{/}I} A \, I \, u \neq c$ *and* $c = v$, *then the ternary field* $\mathfrak{T}(o, e, u, v)$ *is a cartesian group with commutative addition. If* n *is odd, multiplication is also commutative.*

For proofs, see DEMBOWSKI & OSTROM 1968, Theorem 2 and Lemmas 2, 9. The paper also contains characterizations of the cartesian groups involved, but the question as to whether or not these must be quasifields remains open.

Results **10b, c** lead to the following characterization of the finite desarguesian planes.

12. *Let* **P** *be a projective plane of order* n *and* Γ *a collineation group of* **P**. *Then the following conditions are equivalent:*

(a) **P** *is desarguesian and* Γ, *of order* n^3, *is contained in the stabilizer of a flag in the little projective group*[2]) *of* **P**.

(b) n *is a prime power* p^e, *and* Γ *is isomorphic to the Sylow* p*-group* $S(n)$ *of* $PGL_3(n)$.[3])

(c) Γ *fixes a flag of* **P**, *has order* $> n^2$, *and is the set theoretical union of its abelian subgroups of order* n^2.

For the proof, see DEMBOWSKI 1965c, Satz 6 and Korollar 5.1. The main burden of the proof lies in showing that **(c)** implies **(a)**; only here **10** is used.

Result **12** shows that the desarguesian finite projective plane $\mathbf{P}(q)$ is characterized by the abstract structure of the stabilizer of a flag within its little projective group; this stabilizer contains $S(q)$. The analogous statement for the stabilizer of a non-incident point-line pair [viz., the group $SL_2(q)$ of (2, 2)-matrices with determinant 1 over $GF(q)$, of order $q(q^2-1)$] is likewise true:

13. *A projective plane* **P** *of prime power order* $q = p^e$ *is desarguesian if and only if it admits a collineation group* $\Gamma \cong SL_2(q)$. *If this is the case, then every involution in* Γ *is a central collineation, and* Γ *has* 3 *orbits. If* $p > 2$, *then*

---

[1]) This remains true in the non-abelian case, provided Γ is quasiregular and Δ is a semi-direct factor of Γ.

[2]) This is by definition (see Section 1.4) the group generated by all elations of **P**. It is isomorphic to $PSL_3(n)$.

[3]) It is well known that the group $S(n)$ is isomorphic to the group of matrices $\begin{pmatrix} 1 & u & v \\ 0 & 1 & w \\ 0 & 0 & 1 \end{pmatrix}$, with $u, v, w \in GF(n)$.

(a) *Γ fixes a non-incident point-line pair (c, L) and contains all (x, cx)-elations, for any $x \mathrel{I} L$.*
If $p = 2$, then either **(a)** *holds, or*
(b) *Γ fixes a point k, one of the two other point orbits is a conic with knot k, and the third point orbit consists of the centers of the elations in Γ, or Γ operates in the fashion dual to* **(b)**. *Both cases* **(a)** *and* **(b)** *can actually occur.*

For the proof, see LÜNEBURG 1964c, Satz 2 and Korollar 1.[1]) This proof is considerably more complicated, and it uses far deeper tools, than that of the similar result **12**. Some of the main ideas of Lüneburg's proof are of importance also for other investigations; we shall therefore give an outline of these, for the case $p > 2$.[2])

Let **P** be of order $q = p^e \equiv 1 \bmod 2$, and let $\Gamma \cong SL_2(q)$ be a collineation group of **P**. The first step consists in proving that Γ must fix a non-incident point-line pair c, L. The group $SL_2(q)$ has centre of order 2; let σ be the unique involution in this centre. If σ were planar, Γ would induce a collineation group Γ^* in the Baer subplane $\mathbf{F}(\sigma)$, and either $\Gamma^* = 1$ or $\Gamma^* \cong PSL_2(q)$. Both alternatives prove to be impossible; so σ must be a homology, and the center c and axis L of σ form the desired fixed point-line pair.

In the second step it is shown that the Sylow p-groups of Γ consist of elations with center on L and axis through c; hence Γ is of Lenz-Barlotti type III.1 or III.2. The proof of this, and in fact that of most remaining results in this section, requires a rather intimate knowledge of the structure of $SL_2(q)$; this may be found, essentially, in DICKSON 1901, Chapter 12.

The third step is the following lemma which will also be used in Section 4.3:

14. *Let Γ be a collineation group fixing a non-incident point-line pair c, L of a projective plane* **P**. *If Γ is of Lenz-Barlotti type* III.1 *or* III.2, *then Γ is transitive on the flags (x, Y) with $c \neq x \mathrel{I} L$ and $L \neq Y \mathrel{I} c$.*

The proof of this is easy; see LÜNEBURG 1964c, p. 434. Hence Γ can be considered as a flag transitive automorphism group of the incidence structure $\mathbf{P}^{Lc}$ consisting of those points x and lines Y of **P** which satisfy the conditions stated in **14**. Note that $\mathbf{P}^{Lc} = \mathbf{P}^{cL}$ determines **P** uniquely.[3])

---

[1]) Lüneburg makes the additional assumption that, in case $p = 2$, every involution of $SL_2(q)$ operates as an elation. That this can actually be proved was shown by YAQUB 1966a.

[2]) The case $p = 2$ is more complicated, mainly because of the different ways in which Γ can operate. The main ideas, however, are the same as in the case $p > 2$.

[3]) A proof of this simple fact will be given in a more general context in Section 7.4.

The last step is an application of Result **1.2.17**. This allows us to describe $\mathbf{P}^{Lc}$ completely within the group $\Gamma = SL_2(q)$. But the stabilizer of a non-incident point-line pair (c', L') in the desarguesian plane $\mathbf{P}(q)$ also contains $SL_2(q)$, and $SL_2(q)$ is also flag transitive on $\mathbf{P}(q)^{L'c'}$. It turns out that the representations of $\mathbf{P}^{Lc}$ and $\mathbf{P}(q)^{L'c'}$ in the form $\mathbf{K}(\Gamma, \Pi, \mathrm{B})$ of **1.2.17** are precisely the same, so that $\mathbf{P}^{Lc} \cong \mathbf{P}(q)^{L'c'}$. As remarked above, this implies $\mathbf{P} = \mathbf{P}(q)$, so that $\mathbf{P}$ is desarguesian.

Along the same line of ideas, the following results can be proved:

15. *A projective plane* $\mathbf{P}$ *of prime power order* $q = p^e$ *admits a collineation group* $\Gamma \cong PSL_2(q)$ *if, and only if,* $\mathbf{P}$ *is desarguesian.*

That $PSL_2(q) \subset \mathrm{Aut}\,\mathbf{P}(q) \cong P\Gamma L_3(q)$ follows, for example, from the fact that $PGL_2(q)$ is the group leaving a conic of $\mathbf{P}(q)$ invariant (cf. **1.4.51e** and footnote). The converse was proved by Lüneburg 1964c, Satz 3.[1]) Lüneburg also determines (Korollar 2) how $PSL_2(q)$ must operate on $\mathbf{P}$ in case $p > 2$.

In another paper of Lüneburg 1965c, the question is investigated how the Suzuki group $Sz(q)$, with $q = 2^{2m+1} \geqq 8$, can operate on a projective plane of order q^2. At this point, it is not even clear whether $Sz(q)$ can operate faithfully on a plane of order q^2 at all; note that such a plane cannot be desarguesian because $\mathrm{Aut}\,\mathbf{P}(q^2) = P\Gamma L_3(q^2)$ does not contain $Sz(q)$. We shall see, however, in Section 5.2, that case (**a**) of the following result can really occur (Lüneburg 1965c, Section 6).

16. *If* $\mathbf{P}$ *is of order* $q^2 = 2^{2(2m+1)} \geqq 64$ *and admits a collineation group* $\Gamma \cong Sz(q)$, *then every involution in* Γ *is an elation, and* Γ *has* 4 *orbits.* Γ *must operate in one of the following ways:*

(**a**) Γ *fixes a non-incident point-line pair* c, L *and is doubly transitive on the points of* L *as well as on the lines through* c. *The two point orbits* $\neq \{c\}$, (L) [*line orbits* $\neq \{L\}$, (c)] *have lengths* $(q^2+1)(q-1)$ *and* $(q^2+1)\,q\,(q-1)$.

(**b**) Γ *fixes a point* k, *and one of the other point orbits is an oval* $\mathfrak{o}$, *with knot* k, *on which* Γ *operates doubly transitively. The remaining two point orbits have lengths* $(q^2+1)(q-1)$ *and* $(q^2+1)\,q\,(q-1)$. *The tangents and secants of* $\mathfrak{o}$ *each form a complete line orbit, while the exterior lines fall into two orbits of lengths* $(q \pm \sqrt{2q} + 1)\,q^2(q-1)/4$.

($\bar{\mathbf{b}}$) *Dual to* (**b**).

The proof is essentially[2]) that in Lüneburg 1965c (Satz 9). The result is of the same nature as **13**, but is not as satisfactory since it is not clear whether or not case (**b**) can actually occur.

[1]) Lüneburg makes the additional assumption that Γ has an orbit of length $q+1$. That this can actually be proved was shown by Yaqub 1966a.

[2]) The conclusion that the involutions act as elations is a hypothesis of Lüneburg's proof. That this can actually be shown was proved by Dembowski 1966.

4.3 Central collineations

Central collineations, as defined in Sections 2.3 and 3.1 above, are of fundamental importance in the theory of projective planes, mainly because of their close connection with the theorem of Desargues; cf. **3.1.16**. We have already encountered various results involving central collineations in earlier sections; here we collect these, gather further information on groups of central collineations, and determine the structure of finite planes which, in one sense or another, admit "many" central collineations. In particular, we shall collect the known facts on the existence problem for finite planes of prescribed Lenz-Barlotti types.

If a collineation group Γ of a projective plane **P** consists entirely of central collineations, then the collineations in Γ have, in general[1]), either the same center or the same axis. By reasons of duality, we may assume that they have the same axis W. Hence we are dealing essentially with a group of dilatations (as defined in Section 2.3 and repeated in Section 3.1) of the affine plane $\mathbf{P}^W$. Thus, suppose that **A** is an affine plane of order n, and that Γ is a group of dilatations of **A**. We recall the following facts from Section 2.3:

1. *If* $1 \neq \gamma \in \Gamma$ *and* $L\gamma = L$ *for some line* L, *then either*

(**a**) γ *fixes every line parallel to* L*; in this case* γ *is fixed point free, i.e. a translation; or*

(**b**) γ *fixes exactly one point* c*; this point is then on* L, *and all other lines through* c *are also fixed, so that* c *is the center of* γ.

This follows immediately from the definitions of dilatations and translations in 2.3.

2. *The translations in* Γ *form a characteristic subgroup* T. *If* $1 \neq \mathrm{T} \neq \Gamma$, *then* T *is nilpotent, and*

(**a**) *the points* c *with* $\Gamma_c \neq 1$ *form a* T*-orbit* $\mathfrak{c}$ *of length* $|\mathrm{T}|$, *and* Γ *acts as a Frobenius group on* $\mathfrak{c}$, *with Frobenius kernel* T.

(**b**) *Every other point orbit of* Γ *has length* $|\Gamma|$.

This is a literal repetition of result **2.3.23**. Note that (**a**) implies:

3. *If* $\Gamma_c \neq 1$, *then* Γ_c *is its own normalizer in* Γ.

This is easily proved directly; cf. André 1954b, Satz 1.

Let $\mathfrak{X}$ be a parallel class of lines in **A**, and denote by $\Gamma(\mathfrak{X}) = \mathrm{T}(\mathfrak{X})$ the elementwise stabilizer of $\mathfrak{X}$, as in Section 3.1. Then **1** implies $\Gamma(\mathfrak{X}) = \mathrm{T}(\mathfrak{X}) = \mathrm{T}_L$, for any line $L \in \mathfrak{X}$. Also:

---

[1]) The only exception is that where Γ is the non-cyclic group of order 4, consisting of the identity and three involutorial homologies whose centers form a triangle; cf. **3.1.7** and **3.1.8a**.

4. (**a**) *Every* $\mathsf{\Gamma}(\mathfrak{X}) = \mathsf{T}(\mathfrak{X})$ *is a normal subgroup of* $\mathsf{\Gamma}$.
(**b**) *If* $\mathfrak{X} \neq \mathfrak{Y}$ *and* $\mathsf{T}(\mathfrak{X}) \neq 1 \neq \mathsf{T}(\mathfrak{Y})$, *then* T *is elementary abelian.*
(**c**) *If* $\mathsf{T}(\mathfrak{X})$ *has the same order* $h > 1$ *for all parallel classes* $\mathfrak{X}$, *then* $h = n$ *and* $|\mathsf{T}| = n^2$. *In this case,* **A** *is a translation plane and* T *its full translation group. Also,* n *is a power of a prime.*

The proof of (**a**) is straightforward, and (**b**) and (**c**) follow immediately from **2.3.24b** and **2.3.26a**, respectively.[1])

We discuss now the case of an arbitrary translation group T of **A**, which need not be transitive on the points. Let t denote the number of point orbits of T; then $n + t$ is that of the line orbits, by **4.2.7**.

5. *If* $|\mathsf{T}(\mathfrak{X})| = h$, *then the parallel class* $\mathfrak{X}$ *falls into* $th\,n^{-1}$ *line orbits, each consisting of* $n^2 (th)^{-1}$ *lines.*

This result (Dembowski 1965d, Lemma 1) is a simple consequence of

(1) $$t|\mathsf{T}| = n^2,$$

which follows from the regularity of T on the n^2 points of **A**. Let $h_1, \ldots, h_m$ denote the distinct orders of the subgroups $\mathsf{T}(\mathfrak{X})$, and s_i the number of parallel classes $\mathfrak{X}$ for which $\mathsf{T}(\mathfrak{X})$ has order h_i $(i = 1, \ldots, m)$. Then

(2) $$\sum_{i=1}^{m} s_i = n + 1 \quad \textit{and} \quad \sum_{i=1}^{m} s_i h_i = n + |\mathsf{T}|.$$

Also, **5** implies the following congruences:

(3) $$th_i \equiv 0 \bmod n \quad \textit{and} \quad n^2 \equiv 0 \bmod th_i \qquad (i = 1, \ldots, m).$$

The following is a consequence of (1)—(3):

6. $|\mathsf{T}| \geqq n$ *if and only if* T *is transitive on the lines of some parallel class. If this is the case and if* $\mathsf{T} \neq \mathsf{T}(\mathfrak{X})$ *for any* $\mathfrak{X}$, *then* n *and* t *are powers of the same prime, and* $n \geqq t$.

For the proof, see Dembowski 1965d, Lemma 2 and Teorema 1. Note that

7. $|\mathsf{T}| > n$ *if and only if* $\mathsf{T}(\mathfrak{X}) \neq 1$ *for all* $\mathfrak{X}$.

(Ostrom 1964a, Theorem 1.) It is clear that if $\mathsf{T}(\mathfrak{X}) \neq 1$ for all $\mathfrak{X}$, then $\mathsf{T} \neq \mathsf{T}(\mathfrak{X})$ for any $\mathfrak{X}$. Hence, by **6** and **7**, $|\mathsf{T}| > n$ implies that n is a prime power. The following result is a slight improvement of **7**:

8. *If the* $\mathsf{T}(\mathfrak{X})$ *are nontrivial with one possible exception, then they are all nontrivial.*

[1]) A proof of (**c**), due to Gleason 1956, Lemma 1.6, was given in footnote [1]) of p. 166.

For the proof, see PIPER 1963, p. 249.[1]) The next result, a more complicated consequence of (1)—(3) and **6**, improves considerably on **8** in a special situation:

9. *If* T *is intransitive and distinct from all* $\mathsf{T}(\mathfrak{X})$, *and if all* $\mathsf{T}(\mathfrak{X}) \neq 1$ *have the same order* $h > 1$, *then the number* s *of those parallel classes* $\mathfrak{Y}$ *for which* $\mathsf{T}(\mathfrak{Y}) = 1$ *is divisible by* h, *and either* $s > \sqrt{n}$ *or* $h = s = 2$ *and* $n = |\mathsf{T}| = 2^e$.

Proof: DEMBOWSKI 1965d, Teorema 2. Note that the case $h = s = 2$ can actually occur: the translations $(x, y) \to (x + u, y + u^2)$ of the desarguesian plane of order 2e provide an example for this situation.

The following investigation, due to ANDRÉ 1958a and based on an idea of ARTIN 1940, shows that if $\mathsf{T} \neq \mathsf{T}(\mathfrak{X})$ for any $\mathfrak{X}$ (in other words if T contains translations in at least two different directions), then **A** usually also admits central dilatations, so that T cannot be the full dilatation group of **A**. We know from **4b** that T is elementary abelian in this case; consider the ring $\mathfrak{K}(\mathsf{T})$ of all endomorphisms $\varkappa$ of T which satisfy $\mathsf{T}(\mathfrak{X})^\varkappa \subseteqq \mathsf{T}(\mathfrak{X})$ for all parallel classes $\mathfrak{X}$. We call $\mathfrak{K}(\mathsf{T})$ the *kernel*[2]) of T. This kernel can be shown to be a Galois field $GF(q)$ [ANDRÉ 1958a, Satz 2.1], and for any point p of **A**, the group of dilatations with center p is isomorphic to a subgroup of the multiplicative group of $\mathfrak{K}(\mathsf{T})$. This implies:

10. *Let* Γ *be a group of dilatations of a finite affine plane* **A**, *containing nontrivial translations in at least two different directions. Then, for any point* p *of* **A**, *the stabilizer* Γ_p *is cyclic.*

(ANDRÉ 1958a, Satz 2.4). Along the same line of ideas, the following result can be proved:

11. *Let* Δ *be the group of all dilatations of a finite affine plane* **A**. *Then either* Δ *is transitive, in which case* T *is also transitive and* **A** *a translation plane, or else the number of point orbits of* Δ *is at least* 3.

Proof: ANDRÉ 1958a, Satz 5.1. The desarguesian plane of order 2 certainly admits a translation group with 2 point orbits; this shows the necessity of assuming in **11** at least that all translations of **A** are in Δ. Examples where Δ has more than one orbit will appear in Section 5.3.

We turn now to projective planes. We consider a finite plane **P** of order n and a collineation group Γ of **P** containing central collineations;

---

[1]) Piper ascribes this result to D. R. Hughes.

[2]) By definition in Section 3.1, this kernel coincides with the kernel of **A** if T happens to be transitive.

for most of the following discussion we only need that Γ contains non-trivial elations.

12. *Let A, B be two distinct lines, $a = AB$, and $a \neq b \mathrm{I} B$. Suppose that the subgroups $\Gamma(a, A)$ and $\Gamma(b, B)$ of Γ are nontrivial. Then $\Gamma(a, B)$ is also nontrivial; in fact $\Gamma(a, B)$ is an elementary abelian group containing subgroups (which may coincide) isomorphic to $\Gamma(a, A)$ and $\Gamma(b, B)$, respectively. Consequently, all elations $\neq 1$ in $\Gamma(a, A)$ and $\Gamma(b, B)$ have the same prime order.*

This follows without difficulty from **3.1.14** and **4b** above (cf. HERING 1963, p. 156). Now define, for any collineation group Γ of $\mathbf{P}$, the substructure $\mathbf{E} = \mathbf{E}(\Gamma)$ of $\mathbf{P}$ as the set of all those points and lines in $\mathbf{P}$ which are centers or axes, respectively, of nontrivial elations in Γ. Then **12** implies that $\Gamma(c, A) \neq 1$ for every flag (c, A) in $\mathbf{E}$. The incidence structure $\mathbf{E}$ need not be *connected* in the sense that any two of its elements can be joined by an alternating sequence of points and lines of $\mathbf{E}$ such that consecutive elements of this sequence are incident. But certainly $\mathbf{E}$ is the disjoint union of maximal connected substructures of $\mathbf{E}$; these will be called the *components* of $\mathbf{E}$. A component is *trivial* if it consists of one flag only, and *proper* if it contains more than one point and more than one line. It follows by a simple inductive argument from **12** that

13. *If two nontrivial elations of Γ have their centers in the same nontrivial component of $\mathbf{E}$, then they have the same prime order.*

(PIPER 1965, Lemma 4.) To know that all elations in Γ have the same order is often very useful. This condition is satisfied for all known finite planes, but the present state of our knowledge does not allow a proof in general.

For the next results, and in various places later on, the following simple but useful fact is needed:

14. *For any permutation group Γ on a finite set M, and for any integer $t > 1$, the subset M_t of those $x \in M$ which are fixed by a permutation in Γ all of whose other cycles have length t is either empty or a Γ-orbit.*

Proof: If M_t is non-empty, then it clearly falls into full orbits, so that it suffices to show that M_t is contained in an orbit. If this were not so, let X be an orbit in M_t and $y \in M_t - X$. Considering $\gamma \in \Gamma_y$ with all cycles $\neq \{y\}$ of length t, we see that $|X| \equiv 0 \bmod t$. On the other hand, a similar argument with $x \in X$ shows $|X| \equiv 1 \bmod t$, a contradiction proving **14**. This is an obvious modification of an argument often used in the proof that all Sylow p-groups of a finite group are conjugate (GLEASON 1956, Lemma 1.7):

15. *Let M be a finite set and Γ a permutation group of M such that, for some prime p, every element of M is fixed by a permutation in Γ which has order p and fixes no other element. Then Γ is transitive on M.*

We prove a Corollary of **4c**, **12**, and **15** (GLEASON 1956, Theorem 1.8):

16. *Suppose that the finite projective plane* **P** *contains a line A such that, for some collineation group Γ,*

(**a**) $\Gamma(A, A) \neq 1$
and
(**b**) *through every $c \mathrel{I} A$ there exists a line $L \neq A$ such that $\Gamma(c, L) \neq 1$.*

Then Γ is (A, A)-transitive, and Γ_A is transitive on the lines $\neq A$.

For (**a**) and (**b**) imply, by **12**, that all $\Gamma(c, c)$, $c \mathrel{I} A$, are elementary abelian p-groups, for the same prime p. Then **15** shows that Γ_A is transitive on the points of A, so that all $\Gamma(c, A)$ have the same order > 1. But this implies, by **4c**, that Γ is (A, A)-transitive. Finally, the transitivity of Γ_A on the lines $\neq A$ follows from the fact that Γ_A has only 2 point orbits; cf. **4.2.7**.

Next, we use **14** in another situation. Let Γ be a collineation group of **P** and assume that Γ contains distinct elations of the same prime order. Consider two points $x, y \in \mathbf{E} = \mathbf{E}(\Gamma)$. If $xy \notin \mathbf{E}$, then **14** implies that there exists $\gamma \in \Gamma$ with $x\gamma = y$. If $xy \in \mathbf{E}$ and there are lines X, $Y \neq xy$ through x, y in **E**, respectively, then the same argument applies, and it follows now without difficulty that if Γ has no fixed elements in **E**, then Γ must be transitive on the points of **E**. A little further analysis shows more than this:

17. *Let* **P** *be a finite projective plane and Γ a collineation group of* **P** *all of whose elations have the same prime order p. Suppose also that $\mathbf{F}(\Gamma)$ and $\mathbf{E}(\Gamma)$ are disjoint. Then:*

(**a**) *Γ is flag transitive on* **E**, *so that* **E** *is a tactical configuration.*
(**b**) *If* **E** *has $k \geqq 2$ points per line and $r \geqq 2$ lines per point, then $k = r$.*

These results are due to PIPER 1965, Lemmas 1—3. The following can be concluded without the hypothesis of all elations having the same order:

18. *Let* **P** *be a finite projective plane and Γ a collineation group containing nontrivial elations. Suppose that $\mathbf{E} = \mathbf{E}(\Gamma)$ contains a line A carrying more than one point of* **E**. *Then:*

(**a**) *If $\mathbf{F}(\Gamma) = \emptyset$ and if x, y are distinct points of* **E** *on A, then $\Gamma(x, A)$ is transitive on the lines $\neq A$ of* **E** *through y, and $\Gamma(A, A)$ is transitive on those points $\not\mathrel{I} A$ of* **E** *which are on a line $L \neq A$ of* **E** *through x.*

(**b**) *If* $\mathbf{F}(\Gamma) = \emptyset$ *and if* A *is contained in a proper component, then any* $\Gamma(A, A)$*-orbit of points not on* A, *together with the points of* $\mathbf{E}$ *on* A, *is the point set of a projective subplane of* $\mathbf{P}$.
(**c**) *If* $\mathbf{E}$ *contains a point* $p \mathrel{\mathrm{I}} A$ *through which there exists a line* $\neq A$ *of* $\mathbf{E}$, *and if* $p \notin \mathbf{F}(\Gamma)$, *then the component containing* A *is a desarguesian projective subplane of* $\mathbf{P}$.

For the proofs, see Theorems 1—3 of PIPER 1965. We mention here only that the subplanes of (**b**) are (A, A)-transitive, i.e. their affine subplanes with respect to A are translation planes. In case (**c**), the component of A is (X, X)-transitive for each of its lines X, and hence desarguesian. The following is an easy corollary of (**c**):

19. *If a component of* $\mathbf{E}$ *contains a triangle and its sides, then this component is a desarguesian subplane.*

(PIPER 1965, Lemma 5). Further analysis leads to the following main result of PIPER 1965 (Theorems 4, 5):

20. *Let* $\mathbf{P}$ *be a finite projective plane and* Γ *a collineation group of* $\mathbf{P}$ *which contains nontrivial elations. Suppose that* $\mathbf{F}(\Gamma) = \emptyset$ *and that* $\mathbf{E}(\Gamma) = \mathbf{E}$ *contains a proper component* $\mathbf{C}$ *such that* $|\Gamma(x, X)| > 2$ *for every flag* (x, X) *in* $\mathbf{C}$. *Then:*
(**a**) $\mathbf{E} = \mathbf{C}$, *i.e.* $\mathbf{E}$ *is connected.*
(**b**) $\mathbf{E}$ *is a desarguesian projective subplane of* $\mathbf{P}$.
(**c**) *All elations of* $\mathbf{P}$ *have the same prime order.*

The case $\mathbf{F}(\Gamma) \neq \emptyset$, though in a rather more special situation, is included in an investigation of OSTROM 1964a, Section II, where the following is proved:

21. *Let* $\mathbf{P}$ *be a projective plane of square order* $n = q^2 > 4$, *and let* Γ *be a collineation group of* $\mathbf{P}$. *Suppose that the set* $\mathfrak{A}$ *of all those lines* A *in* $\mathbf{P}$ *for which*

$$(4) \qquad |\Gamma(A, A)| = q^2 \quad \textit{and} \quad \Gamma(x, A) = 1 \textit{ or } q, \quad \textit{for} \quad x \mathrel{\mathrm{I}} A$$

satisfies $|\mathfrak{A}| > 1$. *For any* $A \in \mathfrak{A}$, *let* $\mathfrak{c}(A)$ *denote the set of all those points* $x \mathrel{\mathrm{I}} A$ *for which* $|\Gamma(x, A)| = q$.[1]) *Then* $\mathbf{P}$ *contains a desarguesian Baer subplane* $\mathbf{B}$ *such that*

(**a**) *The points of* $\mathbf{B}$ *are precisely those of* $\bigcup_{A \in \mathfrak{A}} \mathfrak{c}(A)$, *and*
(**b**) $\mathfrak{A}$ *consists either of all lines of* $\mathbf{B}$ *or precisely of those lines of* $\mathbf{B}$ *that pass through a distinguished point* c *of* $\mathbf{B}$.

[1]) There must be just $q + 1$ such points in any $\mathfrak{c}(A)$; hence for any $A \in \mathfrak{A}$ the affine plane $\mathbf{P}^A$ is a semi-translation plane in the sense of Section 3.1.

Note that the first alternative of (**b**) agrees with **20b**, whereas the second (which can actually occur; cf. **5.4.9** below) shows that the hypothesis $\mathbf{F}(\Gamma) = \emptyset$ in **20** is essential.

Theorem 4 of Piper 1965 asserts a little more than **20a** and **20b**: in addition one can say that Γ, when restricted to the subplane $\mathbf{E}$, contains the little projective group of $\mathbf{E}$. This follows from an earlier result of Piper 1963, Theorem 1:

22. *Suppose that every point of* $\mathbf{P}$ *is the center, or that every line of* $\mathbf{P}$ *is the axis, of a nontrivial elation in* Γ.

(**a**) *If* $\mathbf{F}(\Gamma) = \emptyset$, *then* $\mathbf{P}$ *is desarguesian, and* Γ *contains all elations of* $\mathbf{P}$.
(**b**) *If* $\mathbf{F}(\Gamma) \neq \emptyset$, *then* $\mathbf{F}(\Gamma)$ *is either a point* c *or a line* A. *In this case,* Γ *has only two orbits and is* (c, c)- *or* (A, A)-*transitive, respectively.*

Variations of this result (with the stronger hypothesis that either every flag of $\mathbf{P}$ is a center-axis pair, or that every point is the center *and* every line is the axis of an elation in Γ) were given by Gleason 1956, Theorem 1.8, and Wagner 1959, Theorem 1; here case (**b**) does not arise. Another variant, with weaker hypothesis than in **22**, is the following:

23. *Let* $\mathbf{P}$ *be a projective plane of order* n *and* Γ *a collineation group of* $\mathbf{P}$ *such that, for any flag* (x, X) *in* $\mathbf{P}$, *at least one of the groups* $\Gamma(x, x)$, $\Gamma(X, X)$ *is nontrivial (i.e. every flag is fixed by an elation* $\neq 1$).

(**a**) *If* $\mathbf{F}(\Gamma) = \emptyset$, *then* $\mathbf{P}$ *is desarguesian. Moreover, the only case where* Γ *does not contain all elations of* $\mathbf{P}$ *is* $n = 4$; *the points of* $\mathbf{P} - \mathbf{E}$ *then form a 6-arc, and* $\Gamma \cong S_6$ *or* A_6.
(**b**) *If* $\mathbf{F}(\Gamma) \neq \emptyset$, *then* $\mathbf{F}(\Gamma)$ *is either a point* c *or a line* A. *In this case,* Γ *has only two orbits and is* (c, c)- *or* (A, A)-*transitive, respectively.*

This is Theorem 2 of Piper 1963.[1]) In particular, $\mathbf{E}(\Gamma)$ is connected in all cases, so that all elations in Γ again have the same prime order.

Result **22** can be considered as the special case $\mathbf{E} = \mathbf{P}$ of **20**. However, the hypothesis $|\Gamma(x, X)| > 2$ of **20** does not enter into **22**, so that the question arises whether this is essential. This question was investigated in another paper of Piper 1966a; it turns out that the case $|\Gamma(x, X)| = 2$ is truly exceptional if $\mathbf{E} \neq \mathbf{P}$:

[1]) Piper's proof for (**a**) contains an error (on p. 254), but this can be corrected by using either **3.1.14** or **9**. Note also that **16** is used in the proof of (**b**). — In view of **22** and **23**, one might ask whether the hypothesis of **23** may be weakened further. Certainly it is not enough to postulate that every point be fixed by some elation (or homology) $\neq 1$; for minimal examples of such groups see Dembowski 1967b.

24. *Let* **P** *be a finite projective plane and* Γ *a collineation group of* **P** *which contains involutorial elations. Suppose that* $\mathbf{F}(\Gamma) = \emptyset$, *and that* $\mathbf{E} = \mathbf{E}(\Gamma)$ *contains a proper component* **C** *such that* $|\Gamma(x, X)| = 2$ *for every flag* (x, X) *in* **C**. *Then either*

(**a**) **C** *is a projective subplane of order 2; in this case* Γ *induces the little projective group of* **C**;

or else

(**b**) **C** *is a subplane of order 4 from which a 6-arc and its exterior lines are removed. In this case* $\mathbf{C} = \mathbf{E}$, *every elation of* **P** *is involutorial, every elation in* Γ *has center and axis in* **C**, *and* $\Gamma \cong S_6$ *or* A_6.

(PIPER 1966a, Theorems 1—3.)

Another question arises in connection with results **22** and **23**, namely whether the conclusion that **P** is desarguesian holds also in case (**b**) of either **22** or **23**. The answer to this is *no*, examples being a class of planes exhibited by LÜNEBURG 1965c, Section 6. These were already mentioned at the end of Section 4.2, and they will be properly defined in Section 5.2. The following theorem is noteworthy in this context:

25. *If every flag of a projective plane* **P** *of order* n *is fixed by a non-trivial elation, if* Γ *is the group generated by all elations fixing the line* A, *and if every* $\gamma \in \Gamma$ *fixes either all or at most two points on* A, *then either*

(**a**) **P** *is desarguesian, and* Γ_p, *for any* $p \,\mathrm{I}\!\!\!\!/\, A$, *is either dihedral of order* $2(n+1)$ [*here* $n = 2^e$] *or isomorphic to* $SL_2(q)$, *with* $q = n$ *or* $q = 5$ *and* $n = 9$, *or else*

(**b**) **P** *is non-desarguesian,* $n = 2^{2(2r+1)} \geqq 64$, *and* $\Gamma_p \cong Sz(\sqrt{n})$.

The planes of LÜNEBURG 1965c, Section 6, are examples for case (**b**). The proof of **25** (LÜNEBURG 1965c, Satz 7) is long and complicated. It uses some highly nontrivial results in finite group theory which will be of value also for later investigations. We outline the main steps of the proof.

By **23b**, the group Γ is (A, A)-transitive and transitive on the points of A; in particular, n is a prime power p^e. Let Γ^* denote the permutation group induced by Γ on the points of A. By hypothesis, only the identity of Γ^* has more than two fixed points.

Case 1: $p = 2$. Then Γ^* satisfies the hypotheses for G in the following theorem:

26. *Let* G *be a finite group of even order in which the centralizer of every involution is a 2-group. Suppose further that any two distinct Sylow 2-groups of* G *have trivial intersection. Then either*

(**a**) G *admits a faithful representation as a doubly transitive permutation group such that only the identity fixes more than two letters,*

or else

(**b**) *For any Sylow 2-groups S of G, either $S \lhd G$, or G is a Frobenius group with Frobenius complement S.*

Proof: SUZUKI 1961b, Theorem 5; LÜNEBURG 1965e, (3.4). Note that in the second alternative of (**b**), there is only one involution in S, so that S is cyclic or a generalized quaternion group (cf. **4.2.2**).— Suppose now that case (**a**) holds for Γ^*. Then the structure of Γ^* is determined by the following result:

27. *Let Π be a doubly transitive permutation group of degree $m+1$, such that only $1 \in \Pi$ fixes more than 2 letters. Then one of the following is true:*

(**a**) *Π is sharply 2-transitive, i.e. isomorphic to the group of all permutations $x \to xa + b$, where $a \neq 0$, of a finite nearfield.*
(**b**) *Π is isomorphic to the group of all permutations $x \to x^\alpha a + b$ of $GF(2^q)$, where q is a prime, $a \neq 0$, and $\alpha \in \mathrm{Aut}\, GF(2^q)$.*
(**c**) *Π contains $PSL_2(m)$ as a normal subgroup of index $\leqq 2$.*
(**d**) *$\Pi = Sz(\sqrt{m})$; here $m = 2^{2(2r+1)} \geqq 64$.*

This is the essence of three papers by FEIT 1960, ITO 1962, and SUZUKI 1962a; for the sharply 3-transitive case, which is part of (**c**), see ZASSENHAUS 1935b. Note that $m+1$ is a prime power in cases (**a**), (**b**) (these are the only cases where Π contains a regular normal subgroup), and that m is a prime power in cases (**c**), (**d**).

In the situation of the proof of **25**, we have $\Pi = \Gamma^*$, and it can be shown that (**a**) and (**b**) cannot occur. In case (**c**) it follows that $\Gamma_x \cong SL_2(n)$ if $x \,\mathrm{I}\, A$, whence $\mathbf{P}$ is desarguesian by **4.2.13**. Finally, $\mathbf{P}$ cannot be desarguesian in case (**d**) because $Sz(\sqrt{n})$ is not contained in $\mathrm{Aut}\,\mathbf{P}(n) = P\Gamma L_3(n)$.

There remains the possibility **26b** for Γ^*. In this case it can be shown that $\mathbf{P}$ must be desarguesian; for details see LÜNEBURG 1965c, pp. 51—52.

Case 2: $p > 2$. In this case, if Γ^* is soluble, then $n = p = 3$. (Here a result of FOULSER 1964b is used; this will be discussed in more detail in Section 4.4.) Hence $\mathbf{P}$ is again desarguesian, and we can assume now that Γ^* is non-soluble. It can then be shown that Γ^* has no normal subgroup of odd order >1 (Γ^* is primitive of even degree) and that the Sylow 2-groups of Γ^* are dihedral. This is the most difficult part of the proof; see LÜNEBURG 1965c, pp. 54—57. Then the following theorem can be applied:

28. *Let G be a finite group whose Sylow 2-groups are dihedral. Denote by $O(G)$ the largest normal subgroup of odd order in G. Then $G/O(G)$ is either*

(**a**) *a 2-group (which is then isomorphic to a Sylow 2-group of G), or*

(**b**) *the alternating group* A_7 *of order* 2520, *or*
(**c**) *isomorphic to a subgroup of* $P\Gamma L_2(q)$ *containing* $PSL_2(q)$, *for some odd prime power* q.

This is the main result of GORENSTEIN & WALTER 1965. As Γ^* is non-soluble, case (**a**) cannot occur in our situation,[1] and case (**b**) can be excluded as well. Thus Γ^* contains a normal subgroup isomorphic to a $PSL_2(q)$, for odd q. So far, we know of no connection between n and q. But it can be shown that Γ^* must be doubly transitive, except when $n = 9$ and $q = 5$. But in this case **P** is desarguesian, for it is known (see Chapter 5) that there is only one non-desarguesian (A, A)-transitive plane of order 9, and this does not satisfy the hypothesis of **25** that every flag be fixed by a nontrivial elation. In all other cases the following result is applicable:

29. *Suppose that* $PSL_2(q) \subseteqq G \subseteqq P\Gamma L_2(q)$ *for some* q, *and that* G *admits a faithful doubly transitive permutation representation of degree* $m+1$. *Then either* $m = q$, *or we have one of the following cases:*

(**a**) $q = 4$, $m = 5$, $G = PSL_2(4)$ *or* $G = P\Gamma L_2(4)$.
(**b**) $q = 5$, $m = 4$, $G = PSL_2(5)$ *or* $G = PGL_2(5)$.
(**c**) $q = 7$, $m = 6$, $G = PSL_2(7)$.
(**d**) $q = 8$, $m = 27$, $G = P\Gamma L_2(8)$.
(**e**) $q = 9$, $m = 5$, $G = PSL_2(9)$ *or* $G = PSL_2(9)\langle\varphi\rangle$, *with* φ *the involution* $x \to x^3$ *in* $P\Gamma L_2(9)$.
(**f**) $q = 11$, $m = 10$, $G = PSL_2(11)$.

This is Satz 1 of LÜNEBURG 1964c. In our special situation, we have $m = n = p^e$, and **29** shows that $q = n$ except possibly in case (**e**), because both m and q are odd. But in this case **P** is again desarguesian since the order is 5 (cf. **3.2.15**). Finally, if $n = q$, it can be shown that Γ contains $SL_2(n)$, whence result **4.2.13** yields that **P** is desarguesian. This concludes our discussion of the proof of **25**.

We consider now the case of a collineation group Γ of **P** which contains nontrivial homologies. This situation has not been investigated as thoroughly as that where there are elations, but the following analogue of **22** can be proved:

30. *Suppose that every point of* **P** *is the center, or that every line of* **P** *is the axis, of a nontrivial homology in* Γ.

(**a**) *If* $\mathbf{F}(\Gamma) = \emptyset$, *then* **P** *is desarguesian, and* Γ *contains all elations of* **P**.
(**b**) *If* $\mathbf{F}(\Gamma) \neq \emptyset$, *then* $\mathbf{F}(\Gamma)$ *is either a point* c *or a line* A, *and* Γ *is* (c, c)- *or* (A, A)-*transitive, respectively.*

[1] This involves, of course, the solubility of groups of odd order (FEIT and THOMPSON 1963).

Proof: PIPER 1967, Theorem 4; for part (**a**), see also COFMAN 1965, Theorem II. A slightly weaker version of the theorem (with the "or" in the hypothesis replaced by "and") is Theorem 2 of WAGNER 1959. The basic task in the proof of **30** is to find sufficiently many elations, so as to be able to apply **23**. We mention one of the main ideas in solving this, due to WAGNER 1959, p. 117:

31. *If every point is the center of a nontrivial homology in* Γ, *then* Γ *contains nontrivial elations.*

Proof: The result will follow from **2a** if it can be shown that there exists a line A and two points $p, q \,\bar{\mathrm{I}}\, A$ such that $\Gamma(p, A) \neq 1 \neq \Gamma(q, A)$. If this were not so, there would exist a well-defined one-one mapping π from the points onto the lines of **P** such that $p \,\bar{\mathrm{I}}\, p\pi$ and $\Gamma(p) = \Gamma(p, p\pi)$ for every point p of **P**. But it is easily seen that then π would be a polarity without absolute points, contradicting result **3.3.1**.

As in the case of **22** and **23**, the question arises whether **P** must be desarguesian in case (**b**) of **30**. The answer is again negative; examples (to be discussed in Sections 5.1 and 5.2) are found in WAGNER 1959, Section 5.

In the remainder of this section, we shall be concerned with a projective plane of order n, admitting a collineation group Γ which is (p, L)-transitive for certain point-line pairs (p, L). The possible configurations of all such p, L were given in **3.1.20**. Also, it was stated in the context of **3.1.20** (cf. Table I) that many of these cannot occur if Γ is taken as the full collineation group of **P**. We give now some more details, and in particular we shall indicate proofs for the nonexistence of finite planes of Lenz-Barlotti types I.6 and III.2.[1]

First we show that collineation groups of Lenz-Barlotti types I.5 and I.7 are possible only in finite planes. This will imply that there are no planes of these types, and no groups of type I.8 (compare here the remarks after **3.3.2**).

Suppose that **P** is a projective plane which need not be finite, and let Γ be a collineation group of Lenz-Barlotti type I.5 or I.7 of **P**. Thus, for some nonincident point-line pair (o, W),

(5) Γ *is* (x, ox^σ)*-transitive for every* $x \,\mathrm{I}\, W$, *with* σ *a fixed point free involution of the points on* W.

[1]) These and VII.1 are the only Lenz-Barlotti types for which it is known that they cannot occur in the finite case, while the existence problem is either open or settled affirmatively in the infinite case. The corresponding problem for the types I.2, I.3, I.4, II.2 is open and equally interesting, but has not received much attention so far.

(6) *Γ is not (x, Y)-transitive for any other point-line pair (x, Y), except perhaps for $x = o$, $Y = W$.*

Without loss in generality, we can assume that Γ is generated by its subgroups $\Gamma(x, ox^\sigma)$. Let Δ denote the permutation group induced by Γ on W. Then $\sigma\delta = \delta\sigma$ for every $\delta \in \Delta$, and Δ is transitive. It follows that the groups $\Gamma(x, ox^\sigma)$ are conjugate in Γ, and **3.1.8a** shows that each of them contains precisely one involution. Denote the unique involution in $\Gamma(x, ox^\sigma)$ by $\eta(x)$; then $\eta(x)$ fixes x and x^σ and interchanges y and y^σ for all $y \neq x, x^\sigma$ on W.

Now introduce coordinates in **P**, with o as in (5) and $u, v \mathrel{I} W$ with $v = u^\sigma$. Then we have, for any ternary field[1] $\mathfrak{T} = \mathfrak{T}(o, e, u, v)$:

(7) $\mathfrak{T}$ *is linear,*

and

(8) $\mathfrak{T}^\times$ *is a group isomorphic to all* $\Gamma(x, o\, x^\sigma)$.

Hence the multiplicative group $\mathfrak{T}^\times$ contains precisely one involution i which must of course be in the centre:

(9) $i\,x = x\,i$ *for all* $x \in \mathfrak{T}$.

The group $\Gamma(v, ou) = \Gamma(v, ov^\sigma)$ consists of the mappings $(x, y) \to (x, yt)$, $t \in \mathfrak{T}^\times$; the fact that these are collineations implies

(10) $(x + y)\,z = x\,z + y\,z$ *for all* $x, y, z \in \mathfrak{T}$.

Also, the only involution in $\Gamma(v, ou)$ is $\eta(v)$: $(x, y) \to (x, yi)$, mapping the ideal point $(m) \mathrel{I} W$ onto $(mi) = (im)$. Hence

(11) $(m)^\sigma = (im) = (mi)$, *if* $m \neq 0$.

Next, consider the involution $\eta((i))$, with center (i) and axis $(y = x)$. As u and v are interchanged by $\eta((i))$, it follows that

(12) $(x, y)^{\eta((i))} = (y, x)$.

The fact that this is a collineation implies:

(13) *If* $c + c' = 0$, *then* $(x + c) + c' = x$, *for all* $x \in \mathfrak{T}$.

This means that the additive loop $\mathfrak{T}^+$ has the right inverse property.[2] In particular, every element $x \in \mathfrak{T}$ has a unique two-sided additive inverse $-x$. As (10) implies $(-x)\,y = -x\,y$, we can also conclude that $i = -1$. Moreover, (12) shows that that $\eta((i))$ maps the lines with slopes $m \neq 0, \pm 1$ onto those with slope m^{-1}; thus $(m^{-1}) = (m)^\sigma = (im)$ $= (-m)$ by (11), and consequently

(14) $m^2 = i = -1$ *whenever* $0, \pm 1 \neq m \in \mathfrak{T}$.

[1]) For definition and basic properties of this and the following terms, consult Section 3.1.

[2]) Compare this with property (3.1.32) of the multiplicative loop.

But this implies[1]) that $\mathfrak{T}^\times$ is either a cyclic group of order 2 or 4 or the quaternion group[2]) of order 8. Hence we have:

32. *If a projective plane admits a collineation group of Lenz-Barlotti type* I.5 *or* I.7, *then it is finite of order* 3, 5, *or* 9. *A collineation group of Lenz-Barlotti type* I.8 *does not exist.*

This result is due to OSTROM 1957, 1958b. Note that **P** is desarguesian if $n = 3, 5$ and nondesarguesian if $n = 9$ [the multiplicative group of $GF(9)$ is not the quaternion group]. PICKERT 1959a has shown that there are precisely two possibilities, dual to each other, for the case $n = 9$. These planes, i.e. their full collineation groups, are of Lenz-Barlotti types IVa.3 and IVb.3, respectively; they will be exhibited in Section 5.2. — Combining all these results, we have:

33. *There are no projective planes of types* I.5, I.7, *or* I.8. *There exists precisely one projective plane for each of the types* IV*a*.3 *and* IV*b*.3; *these planes are finite of order* 9.

We turn now to the existence problem for groups of Lenz-Barlotti class I.6. Whether these can exist in the infinite case is an unsolved problem; the following considerations will include the main ideas of the nonexistence proof in the finite case.

Suppose that **P** is a projective plane admitting a collineation group Γ of type I.6 or II.3; thus, for some flag (v, W),

(15) Γ *is* (x, x^σ)-*transitive whenever* $v \neq x \mathrel{I} W$; *here* σ *is a one-one mapping of the points* $\neq v$ *on* W *onto the lines* $\neq W$ *through* v.

(16) Γ *is not* (x, Y)-*transitive for any other point-line pair* (x, Y), *except perhaps for* $x = v$, $Y = W$.

Again without loss of generality, assume that Γ is generated by the $\Gamma(x, x^\sigma)$. It is easily checked that σ commutes with every $\gamma \in \Gamma$, and σ is uniquely determined by p^σ and q^σ, for two distinct points $p, q \neq v$ on W; in fact,

$$x^\sigma = [(o\,x)\,(p\,[(o\,q)\,q^\sigma])]\,v \quad \textit{for any } o \neq v \textit{ on } p^\sigma, \textit{ and } x \neq p.$$

Denote by Δ the permutation group induced by Γ on the set $(W) - \{v\}$. Then Δ is doubly transitive unless **P** is of order 2, and Γ is transitive on the points $\not\mathrel{I} W$ unless $n \leqq 4$ (LÜNEBURG 1964a, Satz 2). Also, the

[1]) 1 and i form a normal subgroup H such that $\mathfrak{T}^\times/H$ is elementary abelian and can be identified with the additive group of a vector space over $GF(2)$. If this vector space had rank >2, then $\mathfrak{T}^\times/H$ would contain a subgroup of order 8, corresponding to a subgroup G of order 16 of $\mathfrak{T}^\times$, with exponent 4 and only one involution. But the only finite groups satisfying this condition are of orders $\leqq 8$; cf. HALL 1959, p. 189.

[2]) This is the group of order 8 defined by (4.2.2)—(4.2.5) and (4.2.7) with $r = 4$; see p. 179.

kernel of the canonical epimorphism $\Gamma \to \Delta$ consists precisely of the (v, W)-elations in Γ, and an arbitrary collineation of **P** centralizes Γ if and only if it is a (v, W)-elation. In particular, the kernel of $\Gamma \to \Delta$ is the centre Z of Γ (COFMAN 1966b).

Now we introduce coordinates in **P**. The origin o is chosen arbitrarily $\not{I} W$. Then $ov = u^\sigma$ for a uniquely determined point $u \neq v$ on W. Finally, let w be a point $\neq u, v$ on W and put $e = (ow)\, w^\sigma$, so that $w = (1)$ in the coordinate system with base points o, e, u, v. The ternary field $\mathfrak{T} = \mathfrak{T}(o, e, u, v)$ is linear and $\mathfrak{T}^\times$ associative as in the case of type I.5 above. In fact, $\mathfrak{T}^\times$ is isomorphic to the group $\Gamma(u, ov) = \Gamma(u, u^\sigma)$ which consists of the homologies

$$(17) \qquad \gamma(t)\colon (x, y) \to (x\,t, y) \qquad 0 \neq t \in \mathfrak{T}.$$

Note that all $\Gamma(x, x^\sigma)$ are conjugate in Γ. By arguments similar to, but a little more complicated than those leading to (11) above, it can be proved that

$$(18) \qquad s + (s^{-1} + b) = 1 + b \qquad \textit{whenever } 0{,}1 \neq s \in \mathfrak{T}$$

(PICKERT 1959b, p. 101). Hence

$$(19) \qquad s + s^{-1} = 1 \qquad \textit{for all } s \neq 0{,}1$$

and

$$(20) \qquad s + 1 = 1 + s, \qquad (s + t) + 1 = s + (t + 1) \qquad \textit{for all } s, t \in \mathfrak{T}.$$

The group $\Gamma(w, w^\sigma)$ is sharply transitive on the points $\neq (1, 1)$ of the line $(y = x)$. Let $\varphi(s)$ denote the unique homology in $\Gamma(w, w^\sigma)$ mapping $o = (0, 0)$ onto (s, s), and define a new operation "$\circ$" in $\mathfrak{T}$ as follows:

$$(21) \qquad (x = t)^{\varphi(s)} = (x = t \circ s) \qquad \textit{for } s \neq 1, \textit{ and } t \circ 1 = 1.$$

Then the elements $\neq 1$ of $\mathfrak{T}$ form a group $\mathfrak{T}^\circ \cong \Gamma(w, w^\sigma) \cong \mathfrak{T}^\times$ with neutral element 0, and

$$s \circ s^{-1} = s^{-1} \circ s = 0 \qquad \text{if } s \neq 1,$$

i.e. inverses are generally the same with respect to both multiplications in $\mathfrak{T}$. Also, $\varphi(s^{-1}) = \varphi(s)^{-1}$ and $(m)^{\varphi(s)} = (s^{-1} \circ m)$ whenever $s \neq 0, 1$. Note that $\varphi: s \to \varphi(s)$ is an isomorphism from $\mathfrak{T}^\circ$ onto $\Gamma(w, w^\sigma)$, as $\gamma: t \to \gamma(t)$ [cf. (17)] was an isomorphism from $\mathfrak{T}^\times$ onto $\Gamma(u, u^\sigma)$. If $m \neq 0, 1$, then the mappings

$$\lambda\colon \varphi(s) \to \gamma(m^{-1})\, \varphi(s)\, \gamma(m) \quad \text{and} \quad \mu: \gamma(t) \to \varphi(m^{-1})\, \gamma(t)\, \varphi(m)$$

are isomorphisms from $\Gamma(w, w^\sigma)$ and $\Gamma(u, u^\sigma)$, respectively, onto the same group $\Gamma((m^{-1}), (x = m))$. Consequently, the mapping $\alpha(m) = \gamma\, \mu\, \lambda^{-1}\, \varphi^{-1}$ is an isomorphism from $\mathfrak{T}^\times$ onto $\mathfrak{T}^\circ$. The definitions of γ and φ show that $t^{\alpha(m)} = (m^{-1} t \circ m) m^{-1}$; hence if we define

$$t^{\beta(m)} = (m\, t\, m^{-1})^{\alpha(m)}$$

then

(22) $$t^{\beta(m)} = (t\, m^{-1} \circ m)\, m^{-1} \qquad (m \neq 0),$$

and it turns out that $\beta(m)$ is not only an isomorphism $\mathfrak{T}^{\times} \to \mathfrak{T}^{\circ}$ as $\alpha(m)$, but also an isomorphism $\mathfrak{T}^{\circ} \to \mathfrak{T}^{\times}$:

(23) $$(s\, t)^{\beta(m)} = s^{\beta(m)} \circ t^{\beta(m)} \quad \textit{and} \quad (s \circ t)^{\beta(m)} = s^{\beta(m)}\, t^{\beta(m)}.$$

These results are due to JÓNSSON 1963, Lemma 2.2.3.) It follows from (22) that $x^{\beta(x)} = x^{-1}$ for $x \neq 0$, whence (23) yields

(24) $$(x \circ x)^2 = x^2 \circ x^2 = x \qquad \textit{except when } x \neq 1 = x^2$$

(JÓNSSON 1963, Lemma 2.2.4). From (24) one can now easily derive the following result on the structure of $\mathfrak{T}^{\times}$:

34. *If* $x^2 \neq 1$, *then there exists exactly one square root of* x *in* $\mathfrak{T}^{\times}$, *namely* $x \circ x$. *Consequently, an element of even multiplicative order has order* 2.

Using this and other tools of a similar nature, JÓNSSON 1963 has given a proof that

35. *There exists no plane of Lenz-Barlotti type* II.3,

a theorem originally due to SPENCER 1960. As a matter of fact, the proof shows that a plane which admits a *group* of type II.3 must be of order $\leqq 4$, and hence desarguesian. Thus, in the situation considered here, Γ cannot be (v, W)-transitive. Notice, however, that the mapping

(25) $$\tau: (x, y) \to (x, y + 1)$$

is a nontrivial (v, W)-elation, because of (20); here the easily proved fact is used that the (v, W)-elations are precisely those mappings $(x, y) \to (x, y + r)$ for which r is in the *additive right nucleus*, i.e. satisfies $(x + y) + r = x + (y + r)$ for all $x, y \in \mathfrak{T}$.

Before turning exclusively to the finite case, we note:

36. *If* **P** *is a projective plane with a collineation group of type* I.6, *and if* φ *is a planar collineation of* **P**, *then* $\mathbf{F}(\varphi)$ *is either desarguesian of order* $\leqq 4$ *or of type* I.6.

This is a slight extension of YAQUB 1966b, Lemma 2; the proof rests mainly on **35** and the fact that the only fields, commutative or not, in which (19) holds are the Galois fields $GF(q)$ with $q = 2, 3, 4$. Note that the case $\varphi = 1$ is not excluded, so that **P** itself is of type I.6 if $n > 4$.

We shall now outline some of the main ideas in the proof of the following theorem:

37. *If there exists a projective plane with a collineation group of type* I.6, *then it is either infinite or of order* $\leqq 4$.[1]) *In particular, there is no finite plane of type* I.6.

The result is due to Jónsson 1963, Lüneburg 1964a, Cofman 1966b and Yaqub 1966b. We list first Jónsson's contributions, mostly derived from **34** above:

38. *Suppose that* **P** *is a projective plane of order* n, *and of type* I.6. *Then:*

(a) n *is not a prime.*
(b) $1+1 \neq 0$ *if* $\mathfrak{T}^\times$ *contains involutions (i.e. if* n *is odd).*
(c) *If* $n-1$ *is a power of* 2 *(or, equivalently by* **34**, *if* $\mathfrak{T}^\times$ *is an elementary abelian* 2*-group), then* n *is not a power of a prime. In particular,* $n \neq 9$.
(d) $n \neq 16$.
(e) *If* $n-1$ *is a prime number* p, *then* 2 *is not a primitive root* mod p.

For proofs, see Jónsson 1963, Sections 2 and 3.[2])

Assume now that **37** is false, and consider a minimal counterexample **P**, of order $n > 4$, with a group Γ of type I.6. As the induced group Δ on $(W) - \{v\}$ is doubly transitive, Γ contains involutions. None of these can be planar: otherwise their fixed Baer subplanes would be of order $\leqq 4$, by **36** and the minimality of **P**, whence $n = 9$ or 16, contradicting **38c, d**. Thus all involutions in Γ are central.

Case 1: n *odd*. Then every involution is a homology in one of the $\Gamma(x, x^\sigma)$. From this one derives that $\Gamma(x, x^\sigma)$, considered as permutation group of $(W) - \{v\}$, has odd index in Δ_x (Lüneburg 1964a, p. 78); also it is clear that $\Gamma(x, x^\sigma)$ is normal in Δ_x and sharply transitive on the points $\neq x$ of $(W) - \{v\}$. Hence the following result of Suzuki 1964a, p. 515 is applicable:

39. *Let* Λ *be a doubly transitive permutation group of odd degree* $m+1$ *such that every stabilizer* Λ_x *contains a normal subgroup* Φ_x, *of odd index in* Λ_x, *which is sharply transitive on the* m *letters* $\neq x$. *Then* Λ *contains a normal subgroup* Σ *of odd index which is either*

(a) *sharply doubly transitive,*

---

[1]) **P**(4) actually admits a group of type I.6, namely the stabilizer in $\Gamma_0(\pi)$ of an absolute point with respect to a unitary polarity π; cf. footnote [2]) on p. 123.

[2]) There is an error in Jónsson's Satz 2.3.5. What he actually proves there is the present statement **(e)**.

or one of the following:

(b) $\Sigma \simeq SL_2(m)$ $(m = 2^e)$,

(c) $\Sigma \simeq Sz(\sqrt{m})$ $(m = 2^{2(2r+1)} \geqq 64)$,

(d) $\Sigma \simeq PSU_3(\sqrt[3]{m^2})$ $(m = 2^{3t})$.

The proof in Case 1 then consists in showing that none of the possibilities **(a)**—**(d)** can actually occur under our assumptions. For the details, see LÜNEBURG 1964a.

Case 2: *n even.* Then every involution in Γ is an elation, and **4.1.12** shows that $n \equiv 0 \bmod 4$. Further analysis shows that, in fact, $n \equiv 0 \bmod 8$; this is the contribution of COFMAN 1966b. In view of **38d** we now have $n \geqq 24$. The multiplicative group is of odd order, so that $x^2 \neq 1$ for all $x \neq 1$ in $\mathfrak{T}$. It can then be proved that $1 + 1 = 0$ (YAQUB 1966b, Lemma 4), so that the elation τ of (25) is an involution. Next, if $a \in \mathfrak{T}$ is in the additive right nucleus but $a \neq 0, 1$, then $a + a \neq 0$.[1]) The remarks accompanying (25) now show that there is only one involution in the group of all (v, W)-elations of $\mathbf{P}$, namely τ. In fact, as Γ is of even order, τ is in the centre Z of Γ. By Burnside's Theorem **4.2.2**, the Sylow 2-groups of Γ are cyclic or generalized quaternion groups. But the cyclic case is excluded here by another theorem of BURNSIDE 1911, p. 327:

40. *If the Sylow 2-group Σ of a finite group Γ is cyclic, then there exists a normal subgroup H of Γ such that $\Gamma = \Sigma\mathsf{H}$ and $\Sigma \cap \mathsf{H} = 1$.*

This is impossible in our situation, for H would have odd order, while all normal subgroups $\neq 1$ of the primitive group Δ are transitive of degree n and hence of even order. Thus H would be in the kernel Z of $\Gamma \to \Delta$. But then

$$\Delta \cong \Gamma/\mathsf{Z} = \Sigma\mathsf{H}/\mathsf{Z} \subseteq \Sigma\mathsf{Z}/\mathsf{Z} \cong \Sigma/\Sigma \cap \mathsf{Z},$$

so that Δ would be a 2-group, contrary to the fact that the odd integer $n - 1$ divides $|\Delta|$ (we use the double transitivity of Δ). Hence the Sylow 2-groups of Γ are generalized quaternion groups, and it is not difficult to derive from this that the Sylow 2-groups of Δ must be dihedral. Thus the Gorenstein-Walter Theorem **28** can be applied again, together with Lüneburg's **29**. The analysis of all possibilities shows that $|\mathsf{Z}| = 2t$, with t odd $\geqq 7$.[2]) But it can also be proved, by arguments

[1]) Actually, YAQUB's 1966b Lemma 8 says something much weaker, but the analysis of her arguments shows that this is what she really proves. The main idea in this proof is to consider certain planar collineations, products of the $\varphi(s)$ and $\gamma(t)$ above, and use the fact that their fixed subplanes must be of order $\leqq 4$, by **36**.

[2]) A little less than this is shown in YAQUB 1966b, Lemma 9, but the argument used there, adopted from LÜNEBURG 1964c, Theorem 6, actually proves the present assertion.

for which finiteness is not essential (but rather the minimality of **P**, i.e. the condition that no proper subplane of **P** be of type I.6), that $|Z| \leqq 4$ (YAQUB 1966b, Lemma 6). Hence we have a contradiction, and **37** is proved.

Some of the preceding arguments may be applied also in the case of a collineation group of Lenz-Barlotti type III.1 or III.2; we discuss this now. Let **P** be of order n, with a group Γ such that, for a non-incident point-line pair (u, L),

(26) *Γ is (x, ux)-transitive for every $x \mathrel{I} L$, and*

(27) *Γ is not (x, Y)-transitive for any other (x, Y), except perhaps $x = u$, $Y = L$.*

Again, suppose that Γ is generated by the $\Gamma(x, ux)$, $x \mathrel{I} L$, and let Δ be the permutation group induced on (L). By **4.2.14**, Γ is transitive on the flags (x, X) with $u \neq x \mathrel{I} L$ and $L \neq X \mathrel{I} u$, and as the number of these flags is $n(n^2-1)$, we have

(28) $$|\Gamma| = n(n^2-1)\,|\Gamma_{xX}|, \quad \text{for} \quad u \neq x \mathrel{I} L \neq X \mathrel{I} u, \quad x \mathrel{I} X.$$

Furthermore, $\Gamma(x, ux)$ is, when considered as a permutation group of (L), a normal subgroup of Δ_x, for any $x \mathrel{I} L$, and of course $\Gamma(x, ux)$ is sharply transitive on the points $\neq x$ of L. It follows also that Γ is doubly transitive on (L). The kernel of the canonical epimorphism $\Gamma \to \Delta$ consists of the (u, L)-homologies in Γ, and it coincides with the centre Z of Γ (LÜNEBURG 1964c, p. 441).

Now suppose that n is even, so that the degree $n+1$ of Δ is odd. We assume first that every involution in Γ is an elation. Then $\Gamma(x, ux)$ is of odd index in Δ_x (LÜNEBURG 1964c, p. 441—442); hence all hypotheses of Suzuki's theorem **39** are satisfied again, with $m = n$. Consequently there exists a normal subgroup Σ of odd index in Δ which is either sharply 2-transitive or one of the groups $SL_2(n)$, $Sz(\sqrt{n})$, $PSU_3(\sqrt[3]{n^2})$. But unless $n = 2$, only $SL_2(n)$ can actually occur in the present situation; for the proof of this,[1]) see LÜNEBURG 1964c, Satz 5. It follows now from **4.2.13** that **P** is desarguesian.

Hence if there exists a non-desarguesian plane of even order with a group Γ of type III.1 or III.2, then Γ must contain planar involutions, so that n is a square. The following arguments, due to HERING 1968b, will show, however, that these assumptions lead to a contradiction. Consider a non-desarguesian projective plane of even order which is *minimal* with respect to the property that there exists a collineation group Γ

[1]) This proof is more complicated than the corresponding argument in Case 1 of **37**, one important tool being the results of SCHUR 1904, 1905.

satisfying (26). For every planar involution $\eta \in \Gamma$, consider the set

$$\mathfrak{c}(\eta) = \mathbf{F}(\eta) \frown (L) = \{x \mathrm{I} L : x\eta = x\}.$$

It can be shown that the points of L, together with all the point sets $\mathfrak{c}(\eta)$, where η ranges over all planar involutions in Γ, form an inversive plane of order $\sqrt{n}$. [For the definition of inversive planes, see Section 2.4, in particular the context of (2.4.19)]. From results of DEMBOWSKI 1964b[1]) it then follows that this inversive plane is actually of the form $\mathbf{M}(\sqrt{n}) = \mathbf{E}(2, \sqrt{n})$ [see (2.4.16)], and Δ contains $PSL_2(n) = SL_2(n)$ [n is a power of 2 here]. This implies, finally, that $\mathbf{P}$ admits $SL_2(n)$ as a collineation group again, so that $\mathbf{P}$ is desarguesian by **4.2.13**. This contradicts our initial hypothesis, and therefore we have the following result:

41. *Let* **P** *be a projective plane of even order. If* **P** *admits a collineation group of type* III.1 *or* III.2, *then* **P** *is desarguesian. In particular, there exist no planes of even order of Lenz-Barlotti type* III.1 *or* III.2.

It is worth noting in this context that if Γ is of type III.2, then every involution in Γ must be an elation (LINGENBERG 1962, Satz 12), so that the considerations on inversive planes are not necessary in this case.

We turn now to the case where n is odd. It is very probable that then **P** must again be desarguesian, but so far this has been proved only in special cases.

42. *Let* **P** *be a projective plane of odd order. If* **P** *admits a collineation group* Γ *of type* III.1 *or* III.2 *such that* $\Gamma = \langle \Gamma(x, ux), x \mathrm{I} L\rangle$ *contains at most one involution, then* **P** *is desarguesian.*[2])

This is Satz 6 of LÜNEBURG 1964c. The proof is similar to that of **37**, case 2 above; in particular the Gorenstein-Walter theorem **28** is used once more: the Sylow 2-groups of Γ are generalized quaternion groups (the possibility that they are cyclic is ruled out with **40** as in case 2 of **37**), and $O(\Gamma)$, the largest normal subgroup of odd order in Γ, is contained in the centre Z of Γ which coincides with the kernel of $\Gamma \to \Delta$. If Δ is soluble, then $n = 3$. Otherwise, let σ be the unique involution in Γ, which must clearly be a homology, and N the normal subgroup generated by σ and $O(\Gamma)$. Then $O(\Gamma/\mathsf{N}) = 1$, and Γ/N has dihedral Sylow 2-groups. LÜNEBURG 1964c, p. 445 shows that among all the possibilities of **28** and **29** the only remaining case is that where n is an (odd) prime power and $PSL_2(q) \subseteqq \Gamma/\mathsf{N}$. Finally, an appeal to

[1]) A detailed discussion of these results will be given in Sections 6.2 and 6.3.
[2]) It is a simple exercise to show that the converse of this is also true.

results of SCHUR 1907 yields that Γ must contain $SL_2(n)$, whence $\mathbf{P}$ is desarguesian by **4.2.13**.

The hypothesis in **42** that Γ contain only one involution can actually be proved, provided $n \equiv 3 \bmod 4$ (LÜNEBURG 1964c, Korollar 5) or $n \equiv 5 \bmod 8$ (COFMAN 1966a, Theorem I), or if n is a prime (COFMAN 1966a, Theorem II). With this information and results **41**, **42**, we can therefore say:

43. *If there exists a finite projective plane of Lenz-Barlotti type* III.1, *then its order is a non-prime* $\equiv 1 \bmod 8$.

The conclusion holds also for type III.2, but in this case one can say more:

44. *There exists no finite projective plane of Lenz-Barlotti type* III.2.

This theorem is due to LÜNEBURG 1965d and YAQUB 1967a. Lüneburg's contributions are the following. Let Γ be of type III.2, i.e. we have (u, L)-transitivity in addition to (26) and (27). Here we can assume that Γ is generated by the subgroups $\Gamma(x, ux)$, $x \mathrm{I} L$, and $\Gamma(u, L)$. If (i) all involutions in Γ are homologies, or if (ii) n is either not a square or $n = m^2$ with $m \not\equiv 1 \bmod 4$, then $\mathbf{P}$ is desarguesian.[1]) Hence for the proof of **44**, we know that if $\mathbf{P}$ is a minimal counterexample, then Γ contains planar involutions with desarguesian Baer subplanes, so that the order n of $\mathbf{P}$ is a prime power $q = p^{2e}$ with $p^e \equiv 1 \bmod 4$.

Now we introduce coordinates in $\mathbf{P}$: the points o and $v \neq o$ are chosen arbitrarily on L, so that L has equation $x = 0$, and then e so that o, e, u, v form a quadrangle. The ternary field $\mathfrak{T} = \mathfrak{T}(o, e, u, v)$ is again linear, and both $\mathfrak{T}^+$ and $\mathfrak{T}^\times$ are associative; in fact $\mathfrak{T}^+ \cong \Gamma(v, uv)$ because $\Gamma(v, uv)$ consists of the mappings

$$(29) \qquad \tau(s)\colon (x, y) \to (x, y + s), \qquad s \in \mathfrak{T},$$

and $\mathfrak{T}^\times \cong \Gamma(u, L)$ because $\Gamma(u, L)$ consists of the mappings

$$(30) \qquad \gamma(t)\colon (x, y) \to (x\,t, y), \qquad 0 \neq t \in \mathfrak{T}.$$

Now let σ be a planar involution whose Baer subplane $\mathbf{F}(\sigma) \cong \mathbf{P}(p^e)$ contains o, e, u, v. Then $\mathbf{F}(\sigma)$ is coordinatized by a substructure $\mathfrak{S}$ of $\mathfrak{T}$ which is isomorphic to $GF(p^e)$, and σ can be represented as follows:

$$(31) \qquad \sigma\colon (x, y) \to (a^{-1}\,x\,a,\; a^{-1}\,y\,a),$$

with $a \in \mathfrak{S}^\times$ and a^2, but not a, in the centre $\mathfrak{Z}$ of the multiplicative group $\mathfrak{T}^\times$.[2])

[1]) Condition (ii) implies (i), and (i) can be reduced to **42**; the details are in LÜNEBURG 1965d.

[2]) Conversely, every such collineation is an involution fixing $\mathbf{F} = \mathbf{F}(\sigma)$.

YAQUB 1967a then considers several collineations in Γ for which representations similar to (29)—(31) can be computed. Some of these turn out to be involutions; they must therefore be either of the form (30) with $t^2 = 1$, or of the form (31). This implies[1]) that $-1 \in \mathfrak{T}^\times$ is in $\mathfrak{Z}$ and in fact is the only involution of $\mathfrak{T}^\times$. Hence (cf. **4.2.2**) the Sylow 2-groups of $\mathfrak{T}^\times$ are cyclic or generalized quaternion groups. Let S be a Sylow 2-group of the cyclic group $\mathfrak{S}^\times$, and T a Sylow 2-group of $\mathfrak{T}^\times$ containing S; then $[T:S] = 2$, because $p^e \equiv 1 \bmod 4$. Hence there exists $x \in T$ such that $x \notin \mathfrak{S}$. If T were cyclic, a collineation of the form (31) with $a \in S$ would fix $\mathbf{F} = \mathbf{F}(\sigma)$ and the point $(x, x) \notin \mathbf{F}$; this would imply $\sigma = 1$, a contradiction. Hence T is a generalized quaternion group, and there exists $t \in T - S$ with $t^2 = -1$ and $t^{-1}\,x\,t = x^{-1}$ for all $x \in S$. It follows that, for this t,

$$\varphi : (x, y) \to (t^{-1}\,x\,t,\ t^{-1}\,y\,t) \tag{32}$$

is an involution fixing o, e, u, v and hence a Baer subplane $\mathbf{F}' \cong \mathbf{P}(p^e)$ which is distinct from $\mathbf{F}$ because the only $x \in S$ with $t^{-1}\,x\,t = x$ are $x = \pm 1$. Consider a collineation σ of the form (31) with $a \in S$. It is a simple exercise to show that φ and σ commute, so that φ fixes $\mathbf{F}$ as a whole and thus induces a planar involution in $\mathbf{F}$. This implies that the order p^e of $\mathbf{F}$ is a square, i.e. $e = 2d$. Since $\mathbf{F}$ is desarguesian, the restriction of φ to $\mathbf{F}$ is of the form

$$(x, y) \to (x^{p^d},\ y^{p^d}), \qquad \text{for } x, y \in \mathfrak{S}. \tag{33}$$

Finally, let s be a generator of the (cyclic) Sylow 2-group S of $\mathfrak{S}^\times$. As $|\mathfrak{S}^\times| \mid p^e - 1 = (p^d + 1)(p^d - 1)$ and both $p^d + 1$ and $p^d - 1$ are even, $s^{p^d+1} \neq 1$. But (32) and (33) show that $s^{p^d} = t^{-1}\,s\,t = s^{-1}$, or $s^{p^d+1} = 1$. Hence we have a contradiction, and the proof of **44** is complete.[2])

4.4 Groups with few orbits

The problem to be dealt with in this section is comparable with those treated in Sections 4.2 and 4.3. Here as there we are concerned with a finite projective plane $\mathbf{P}$ and a collineation group Γ of $\mathbf{P}$, and again we wish to draw conclusions on the structure of $\mathbf{P}$ from given information on Γ. In 4.2, this information was on the abstract structure of Γ; and in 4.3, Γ was supposed to contain many central collineations. Here we assume that Γ, when considered as a permutation group on the points or lines of $\mathbf{P}$, has only few orbits, in one sense or another. Again our

[1]) The idea of this proof resembles that indicated in footnote [1]), p. 203. Some results of ANDRÉ 1964 are used here as well. There are some inaccuracies in André's paper [e.g., Satz 3 and case β) on p. 323], but the present proof is not affected.

[2]) For the details of the proof just outlined, the reader is referred to the quoted paper of YAQUB 1967a.

main aim will be to characterize the desarguesian planes $\mathbf{P}(q)$, this time by transitivity properties of their collineation groups.

We begin with the case where Γ is transitive on the points (hence also on the lines, because of **4.2.7**) of $\mathbf{P}$. It has been conjectured that $\mathbf{P}$ must then be desarguesian, and we shall see later how this may be proved under additional assumptions on Γ. The order of Γ is

(1) $$|\Gamma| = (n^2 + n + 1)\, s,$$

where

(2) $$s = |\Gamma_p| = |\Gamma_L|$$

for any point p or line L.

First we consider some aspects of the situation when $s = 1$, i.e. when Γ is sharply transitive. This is true in particular whenever Γ is abelian, and only in this case the situation has been studied extensively.[1]) Hence suppose now that Γ is abelian of order $n^2 + n + 1$, and transitive. Then $\mathbf{P}$ can be represented by a difference set[2]) Δ in Γ. For the following discussion, we fix one such representation; also we identify multipliers of Δ with the collineations they induce on $\mathbf{P}$.

1. *If* $\mathbf{P}$ *is of order* n *and admits a transitive abelian collineation group, then*

(**a**) $\mathbf{P}$ *admits polarities with exactly* $n + 1$ *absolute points, and*
(**b**) *every divisor* t *of* n *is a Hall multiplier of any difference set representing* $\mathbf{P}$, *and either* $\mathbf{F}(t)$ *is a subplane, also admitting a transitive abelian group, or else* $f(t) = 1$ *or* 3.

Proof: (**a**) is a special case of result **1.2.13**, and (**b**) follows from **2.3.30** and **2.3.31**; see also Hall 1947, Theorem 4.6 and Evans & Mann 1951, Theorem 6. We note a corollary:

2. $\mathbf{P}$ *admits an involutorial collineation induced by a Hall multiplier if and only if* n *is a square.*

The necessity follows from **1b** and **4.1.9a**, and for sufficiency notice that the multiplier $n^{3/2}$ induces an involution. It should also be mentioned here that the collineation induced by the Hall multiplier n has order 3.

The preceding results were first proved, by Hall 1947, in the case where Γ is cyclic. In this case $\mathbf{P}$ is called a *cyclic projective plane.*[3]) We mention further results valid for cyclic planes:

---

[1]) Compare, however, result **1.4.17**, of Ellers & Karzel 1964, for the desarguesian case.

[2]) The definitions of difference sets and related concepts are on pp. 87—88.

[3]) This terminology is slightly problematic. Similarly, one could speak of "abelian" and "soluble" planes, etc; and ironically, Rosati 1957b has called a projective plane "non-cyclic" if it admits a non-cyclic sharply transitive group. In view of **1.4.17**, many desarguesian planes would then be cyclic and non-cyclic simultaneously.

3. *If Γ is cyclic and if $t_1, \ldots, t_4$ are Hall multipliers with $t_1 - t_2 = t_3 - t_4$, then $(t_1 - t_2)(t_1 - t_3) \equiv 0 \mod n^2 + n + 1$.*

This useful lemma is due to MANN 1952, Theorem 4. Using it and the preceding results **1**, **2**, EVANS & MANN 1951 have derived the following nonexistence theorems:

4. *Suppose that there exists a cyclic projective plane of order n.*

(a) *If n possesses four prime divisors p_1, p_2, p_3, q such that $p_1, p_2 \neq q$ and $p_i^{k_i} < 2q$ ($i = 1, 2, 3$; k_i positive integers), then $q - p_1^{k_1} \neq p_2^{k_2} - p_3^{k_3}$.*
(b) *If p, q are primes such that $p \mid n$ and $q \mid n^2 + n + 1$, and if $\left(\frac{p}{q}\right) = -1$, then n is a square.*
(c) *Suppose that $n^2 + n + 1$ is a prime. Then the order of a Hall multiplier is odd, and it divides n if $n \equiv 0 \mod 3$ and $n + 1$* if $n \equiv -1 \mod 3$.

For proofs, see EVANS & MANN 1951, Corollaries 2.1, 3.1, 5.1, and 5.2.[1]) For more results in the same direction, the reader is referred to this paper. We mention here only that Evans and Mann have concluded from their investigation that the order of a cyclic projective plane is either a prime power or >1600.[2]) This makes it quite likely that cyclic planes always have prime power order. Clearly this would follow if the conjecture mentioned in the beginning (finite projective planes with transitive groups are desarguesian) is true.

On the other hand, for the proof that cyclic planes are desarguesian, it would not be sufficient to know that they have prime power order; one would also have to prove that there can be only one cyclic projective plane for any given order (SINGER 1938 has shown that all $\mathbf{P}(q)$ are cyclic; cf. p. 105). This problem has not received much attention so far. BRUCK 1960 has shown, by lengthy ad hoc arguments, that the desarguesian planes of orders $n \leqq 9$, which are known to be the only cyclic planes of these orders, can be extended[3]) to a cyclic plane of order n^2 in only one way. Hence:

5. *A cyclic projective plane of square order $\leqq 81$ is desarguesian.*

It is natural to ask whether planes with transitive groups must be desarguesian not only in the projective but also in the affine case. This

---

[1]) The conclusion of **(c)** is given there only under the hypothesis that n is not a square. But this can be proved: if $n = m^2$, then

$$n^2 + n + 1 = (m^2 + m + 1)(m^2 - m + 1)$$

is not a prime.

[2]) This has now been improved up to order 3600 by V. H. Keiser (unpublished).

[3]) We use this word in an informal sense here, not with the technical meaning of Section 2.2.

is false in general (there are nondesarguesian translation planes; see Sections 5.2, 5.3), but the groups involved are almost never cyclic. As a matter of fact, the following result was proved by HOFFMAN 1952, Section 5:

6. *If a finite affine plane* **A** *admits a transitive cyclic collineation group, then* **A** *is of order* 2.

In the same paper, Hoffman investigates *cyclic affine planes,* defined as affine planes **A** with a cyclic collineation group Z, fixing one point and transitive on the others. Clearly, such a group has three orbits when considered as a collineation group of the projective plane corresponding to **A**. As in the projective case, every desarguesian finite affine plane is cyclic (BOSE 1942b), and the converse is probably also true. Some of the arguments supporting this conjecture remain valid also if "cyclic" is replaced by "abelian":

7. *Let* **P** *be a finite projective plane,* Γ *a collineation group of* **P** *fixing the line* W, *and* **A** *the affine plane* $\mathbf{P}^W$.

(**a**) *If, for some point* $p \in \mathbf{A}$, *the stabilizer* Γ_p *contains an abelian subgroup* A *transitive on the points* $\neq p$ *of* **A**, *then* Γ *is* (p, W)*-transitive.*

(**b**) *If there exists more than one such point* p, *then* **P** *is desarguesian.*

For the case that A is cyclic, compare result (2.4) of HOFFMAN 1952. Clearly, (**b**) follows from (**a**) and **3.1.17, 3.1.20.** For the proof of (**a**), observe that the order of A is a multiple of $n^2 - 1$, that A is transitive on the $n+1$ lines through p, and that any $\alpha \in \mathsf{A}$ fixes either all or none of the lines $X \mathrel{I} p$. Hence A_X consists of homologies with center p and has order $n-1$.

Let **A** be a cyclic affine plane, with cyclic group Z fixing p. The incidence structure $\mathbf{A}^p$ of the points $\neq p$ and the lines $\not\mathrel{I} p$ admits Z as a point and line transitive automorphism group; hence it may be described, by result **1.2.12,** by a quotient set in Z. Quotient sets of this kind bear many similarities to difference sets defining projective planes:

8. *Let* Z *be a cyclic group of order* $n^2 - 1$, *and suppose that* Z *contains a subset* Ω *such that*

(D′) *the elements of the form* $\omega' \omega^{-1}$, *with* $\omega, \omega' \in \Omega$, *are all distinct, and they are precisely those elements of* Z *whose orders are* $\equiv 0 \bmod n+1$.

Then the incidence structure $\mathbf{C}(\mathsf{Z}, 1, \Omega)$ *of* **1.2.12** *is of the form* $\mathbf{A}^p$, *with* **A** *a cyclic affine plane of order* n. *Conversely, every cyclic affine plane of order* n *can be represented in this way.*[1]

[1]) It should be noted here that **A** is uniquely determined by $\mathbf{A}^p$. Compare the remarks after **4.2.14** where the situation was essentially the same.

These are results (2.1) and (2.2) of HOFFMAN 1952. Note the resemblance between (D′) and condition (D) of **2.3.29**. We call a quotient set Ω with property (D′) an *affine difference set* (BOSE 1942b). HOFFMAN 1952, Theorem 3.1, has proved the following analogue of the multiplier Theorem **2.3.31**:

9. *If Ω is an affine difference set in the cyclic group* Z *of order* $n^2 - 1$, *then for every divisor t of n, the automorphism $\zeta \to \zeta^t$ of* Z *is a multiplier of* Ω.

Hence there are collineations of the cyclic plane **A** which are not contained in the cyclic group Z. It is then possible to prove nonexistence theorems like **4** for cyclic affine planes; these show that for $n < 212$ every cyclic plane of order n has prime power order. For more details, the reader is referred to Sections 3 and 4 of HOFFMAN 1952.

We return to the case of a projective plane **P** of order n, with a transitive collineation group Γ. The following results will show under what conditions on Γ, besides transitivity, it can be proved that **P** is desarguesian. We shall see that in several cases it can also be concluded that Γ contains all elations, and hence the little projective group $PSL_3(n)$, of $\mathbf{P} = \mathbf{P}(n)$. The basis for all these results is the following theorem of WAGNER 1959, Theorem 3:

10. *If* **P** *has a transitive collineation group Γ containing a nontrivial central collineation, then* **P** *is desarguesian, and Γ contains all elations of* **P**.

The result follows immediately from the transitivity of Γ and **4.3.22**, **4.3.30**. Applying also **4.1.9**, we get the following corollary:

11. *If Γ is transitive of even order, and if n is not a square, then* **P** *is desarguesian, and Γ contains all elations of* **P**.

(WAGNER 1959, p. 122). The following is a complement to **11**:

12. *If Γ is transitive and* $|\Gamma| \equiv 0 \bmod 4$, *and if* $n = m^2$ *with* $m \equiv 2$ *or* $3 \bmod 4$, *then* **P** *is desarguesian, and Γ contains all elations.*

For the proof, see KEISER 1967, lemma; note that the hypothesis on Γ is satisfied in particular when Γ is non-soluble.[1]) In the soluble case, we have the following result:

---

[1]) By theorems of Burnside (cf. HALL 1959, p. 203—204) and FEIT & THOMPSON 1963, the order of a non-soluble finite group is divisible by 8 or 12, hence certainly by 4. Although Keiser proves **12**, he states the much stronger hypothesis that Γ be non-soluble.

13. *If Γ is soluble and point primitive,*[1]) *then $n^2 + n + 1$ is a prime, and Γ contains a (cyclic) normal subgroup of order $n^2 + n + 1$, so that* **P** *is cyclic.*

Proof: The hypotheses on Γ imply that Γ possesses a sharply transitive elementary abelian normal subgroup Z; cf. **1.2.20**. Thus

$$n^2 + n + 1 = p^e \tag{3}$$

for some prime p, and $e \geqq 1$. By a result of LJUNGGREN 1942, p. 11, if $e > 1$ in (3), then $n = 18$, $p = 7$, and $e = 3$. BRUCK 1955a, p. 470, has shown that a plane of order 18, if it exists, cannot admit a sharply transitive abelian collineation group; see also ROTH 1965. Hence $e = 1$ in (3), so that $n^2 + n + 1$ is a prime, and Z is cyclic.

We turn now to the case where Γ is *flag transitive*. By **2.3.5a**, Γ is then also primitive on points and on lines. In this case we have

$$|\Gamma| = (n^2 + n + 1)\,(n + 1)\,t, \tag{4}$$

where

$$t = |\Gamma_{pL}| \quad \text{for} \quad p\,\mathrm{I}\,L; \tag{5}$$

hence $s = (n + 1)\,t$, with s as in (1) and (2). The next results summarize the known conditions under which a plane with such a group must be desarguesian.

14. *Let* **P** *be a projective plane of order n, admitting a flag transitive collineation group Γ. If*
(a) *n is odd and not a fourth power,*
or if
(b) *n is not a square and $n^2 + n + 1$ is not a prime,*
then **P** *is desarguesian, and Γ contains all elations of* **P**.

The proof consists in a reduction to **10**. Part **(a)** is essentially[2]) contained in proposition 10 of HIGMAN & MCLAUGHLIN 1961; an important tool for this part is the following lemma:

15. *The Sylow 2-groups of a flag transitive group on a plane of odd order are non-abelian.*

(HIGMAN & MCLAUGHLIN 1961, Lemma 5.) Part **(b)** of **14** is the corollary on p. 489 of ROTH 1964b. The proof employs **13**, which shows

[1]) It seems to be unknown whether point primitivity implies line primitivity.

[2]) Higman and McLaughlin state the stronger hypothesis that if n is a square, then $n = m^2$ with $m \equiv -1 \bmod 4$. (As remarked by KEISER 1967, this original version is a consequence of **12**: the fact that $e = 1$ in (3), and **19** below, imply that if n is odd, any flag transitive group is non-soluble.) The present stronger version was communicated by J. E. McLaughlin to the author; see also WAGNER 1962a, Theorem 3.

that Γ cannot be soluble; thus $|\Gamma|$ is even (FEIT & THOMPSON 1963), whence **11** yields the desired result.

The hypothesis in **14b** that $n^2 + n + 1$ be not a prime is essential for the conclusion that Γ contains all elations: there are two known cases of flag transitive groups not containing any central collineations $\neq 1$; cf. HIGMAN & MCLAUGHLIN 1961, p. 393. These occur in the desarguesian planes $\mathbf{P}(2)$ and $\mathbf{P}(8)$ [in these cases, $n^2 + n + 1 = 7$ or 73], and they are in fact sharply flag transitive, i.e. $t = 1$ in (4). It is unknown whether there are any other sharply flag transitive groups of projective planes. Such a plane would have to be of even order n such that $n^2 + n + 1$ is a prime, and it would be non-desarguesian, for:

16. *If q is a prime power $\neq 2, 8$, then any flag transitive collineation group of the desarguesian plane $\mathbf{P}(q)$ contains all elations of $\mathbf{P}(q)$.*

This is Theorem 1 of HIGMAN & MCLAUGHLIN 1961, p. 394. Compare here also the context of result **1.4.21**.

We conclude this discussion of flag transitivity on finite projective planes with a result containing a contribution to the excluded case of **14b**, where $n^2 + n + 1$ is a prime:

17. *Suppose that either $n^2 + n + 1$ or $n + 1$ is a prime. Then a flag transitive group is either doubly transitive*[1]) *or it contains a sharply flag transitive subgroup.*

(ROTH 1964b, Theorem 1.) In the doubly transitive case, $\mathbf{P}$ is desarguesian; a proof of this will be outlined below. If Γ is non-soluble, then it is doubly transitive; this is a consequence (both for $n^2 + n + 1$ and $n + 1$ a prime) of the following theorem of BURNSIDE 1911, p. 341:

18. *A non-soluble transitive permutation group of prime degree is doubly transitive.*

For another proof of this, see SCHUR 1908.[2]) If Γ is soluble, then the following theorem of Galois is employed:

19. *A transitive permutation group of prime degree is soluble if and only if it is either regular or a Frobenius group.*

A proof of this appears in WIELANDT 1964, 11.6. It is then not too difficult to find a sharply flag transitive subgroup of Γ. Note that every sharply flag transitive group must itself be a Frobenius group,

---

[1]) On points or on lines. This equivalence is a simple consequence of **4.2.7**; see also **2.3.4**.

[2]) Schur's proof appears also in CARMICHAEL 1937, Section 60.

with Frobenius kernel of prime order; this result is contained in Proposition 4 of HIGMAN & MCLAUGHLIN 1961.

The difficulties in the case where n is a square can be overcome if Γ is assumed to be *doubly transitive:*

20. *If a finite projective plane* **P** *admits a doubly transitive collineation group* Γ, *then* **P** *is desarguesian and* Γ *contains all elations of* **P**.

This is the celebrated Theorem of OSTROM & WAGNER 1959. Since Γ, being doubly transitive, must have even order, **20** follows immediately from **11** except in the case when n is a square. Before turning to the proof in the general case, we mention two supplementary results:

21. *Let* **P** *be of order* n, *and suppose that the collineation group* Γ *of* **P** *is either*

(**a**) *transitive on non-incident point-line pairs, or*

(**b**) *2-homogeneous on points (i.e. transitive on unordered point pairs; see Section* 2.4), *and* $n > 2$.

Then Γ *is doubly transitive, so that (by* **20**) **P** *is desarguesian and* Γ *contains all elations of* **P**.

For the proof of (**a**), see OSTROM 1958a. Result (**b**), due to A. Wagner, is contained in **2.4.14**; we sketch a direct proof. Assume that Γ is soluble. Since 2-homogeneity implies primitivity, we get from **13** and **19** that Γ is a Frobenius group on the points. For any line L, the stabilizer Γ_L is a soluble primitive Frobenius group[1]) on (L); hence its Frobenius kernel Φ is elementary abelian. It follows that Φ fixes a unique point $o \mathrel{I} L$ and permutes the remaining points $\mathrel{I} L$ in orbits of length $n+1$. As Φ is normal in Γ_L, the point o is also fixed by Γ_L. But this implies that Γ_L contains non-trivial collineations with two fixed points, a contradiction. This shows that Γ is non-soluble and hence (FEIT & THOMPSON 1963) of even order. But 2-homogeneous groups of even order are doubly transitive.

We turn now to the proof of **20**. Actually, we shall outline a proof of the following more general theorem:

22. *Let* **A** *be a finite affine plane and suppose that* **A** *admits a collineation group* Π *transitive on the lines of* **A**. *Then* **A** *is a translation plane, and* Π *contains all translations of* **A**.

(WAGNER 1965, Theorem 4.) In the situation of **20**, the stabilizer Γ_L of a line L in **P** can be considered as the Π of **22**, with $\mathbf{A} = \mathbf{P}^L$; hence it is clear that **22** implies **20**. For the proof of **22**, note first that

23. *The following statements concerning the collineation group* Π *of a finite affine plane* **A** *are equivalent:*

---

[1]) Here the hypothesis $n > 2$ is needed. If $n = 2$, then Γ_L may be regular; the sharply flag transitive groups of **P**(2) actually are 2-homogeneous.

(**a**) *Π is transitive on the lines.*
(**b**) *Π is flag transitive.*
(**c**) *Π is transitive on the points as well as on the parallel classes.*

Note that (**b**) implies primitivity of Π on the points; cf. **2.3.7a**. Also, (**c**) shows that Π, when considered as acting on the projective plane corresponding to **A**, has 2 orbits. **23** is proved by simple counting arguments.[1]) If k denotes the order of a stabilizer Π_{pL} of a flag (p, L) in **A**, and if $\Pi_{\mathfrak{P}}$ is the stabilizer of the parallel class $\mathfrak{P}$ (i.e. $\mathfrak{P}$ is fixed as a whole, not elementwise), then

(6) $$|\Pi| = n^2(n+1)\,k, \quad |\Pi_{\mathfrak{P}}| = n^2\,k, \quad |\Pi_L| = n\,k,$$
$$|\Pi_p| = (n+1)\,k, \quad |\Pi_{p\mathfrak{P}}| = k.$$

These relations follow at once from **23**; see WAGNER 1965, Theorem 2 and Corollary. Next, **4.2.7** implies:

24. *If Γ is any collineation group of* **A**, *then Γ is point transitive if and only if, for any parallel class $\mathfrak{P}$, the group $\Gamma_{\mathfrak{P}}$ is transitive on the lines of $\mathfrak{P}$.*

(WAGNER 1965, Theorem 3.) Hence in our case, if $\mathfrak{P}$, $\mathfrak{Q}$ are distinct parallel classes and $L, M \in \mathfrak{P}$, we get

(7) $$|\Pi_{\mathfrak{P}\mathfrak{Q}L}| = |\Pi_{\mathfrak{P}\mathfrak{Q}M}|.$$

Now suppose first that n is *even*. Let 2^u and 2^v be the highest powers of 2 which divide n and k, respectively, so that a Sylow 2-group Σ of Π has order $2^{2u+v} \geqq 4$. An involution σ in the centre of Σ is either a translation or a shear (cf. **3.1.29**) or planar; assume the second or third of these alternatives, so that σ has exactly n fixed points. If p is one of these, and $\mathfrak{p} = p\Sigma$ the Σ-orbit containing p, then on the one hand $|\Sigma_p| \mid 2^v$ since $\Sigma_p \subseteqq \Pi_p$; here (6) is used. On the other hand, $n|\Sigma_p| = |\mathfrak{p}|\,|\Sigma_p| = |\Sigma| = 2^{2u+v}$, whence $n \equiv 0 \bmod 2^{2u}$, a contradiction. This shows that σ must be a translation, whence **22** follows from **23c** and **4.3.4c**.

The case where n is *odd* is much more complicated and can only be outlined here. The main difficulty lies in showing that n must be a prime power.

To show this, it suffices (because of **4.1.10**) to consider the case where no involution in Π is planar. Also, the case where Π contains involutorial dilatations is easily settled by **4.3.2a**. Thus there remains the case that every involution in Π is a strain, i.e. a homology with finite axis and ideal center, and further analysis shows that we need only consider the following special situation: There exists a point o

[1]) Alternatively, one may use result **2.2.7**.

of **A** such that, for any line L I o and any ideal point (parallel class) $\mathfrak{P}$ not incident with L, the group Π contains a strain $\neq 1$ with center $\mathfrak{P}$ and axis L. The group Γ generated by these strains induces a permutation group Λ of the ideal points which satisfies the hypothesis of the following theorem.

25. *Let Λ be a doubly transitive permutation group of a set W with $|W| = n + 1$ even. Suppose that every stabilizer Λ_x, $x \in W$, contains a subgroup Φ_x of even order which is a Frobenius group on $W - \{x\}$. Then Λ_x is primitive on $W - \{x\}$.*

The proof of this is in WAGNER 1964; compare the hypotheses with those of **4.3.39**. In our case, Φ_x is the subgroup generated by the strains with axis ox in Γ; this is a Frobenius group on the ideal line W, by **4.3.2a**. Moreover, Φ_x is a normal subgroup of Λ_x in this case; hence the Frobenius kernel K_x of Φ_x, which is characteristic in Φ_x, is also normal and therefore, by **25**, transitive on the n points $\neq x$ of W. On the other hand, $|\Phi_x| \equiv 0 \bmod 2$ implies that K_x is abelian (BURNSIDE 1911, p. 172) and hence even elementary abelian [by a theorem of Galois; (cf. WAGNER 1964, Lemma 2 and Corollary (i)]. Thus the degree n of K_x is a power of a prime.

The remainder of the proof of **22** is then similar to that in the even case: a collineation of order p in the centre of a Sylow p-group of Π must be a translation. For the details, the reader is referred to the paper of WAGNER 1965.

The hypotheses of **22** are not sufficient to guarantee that **A** is desarguesian; examples showing this are again provided by the planes of LÜNEBURG 1965c, Section 6, to be defined in Section 5.2 below.

The following is known about the structure of Π in **22**:

26. *Let Π be a soluble collineation group of the translation plane* **A** *of order $q = p^e$, and suppose that Π is flag transitive.*[1]) *Then, with precisely 16 exceptions, Π is isomorphic to a group of permutations of $GF(q^2)$ of the form $x \to x^\alpha c + a$, where $a, c \in GF(q^2)$, $c \neq 0$, and $\alpha \in \mathrm{Aut}\, GF(q^2)$.*

This is the main result of FOULSER 1964b. The 16 exceptions are all listed there; 13 of these had been previously exhibited by HUPPERT 1957b as the only exceptional soluble doubly transitive permutation groups. A corollary of **26** is:

27. *Let* **A** *be a translation plane of order q. If Π is a sharply flag transitive soluble collineation group of* **A**, *then the desarguesian affine plane* **A**(q) *possesses a sharply flag transitive group similar to Π as a permutation group on the points.*

---

[1]) This hypothesis is, because of **23**, equivalent to saying that Π is transitive on lines.

(Foulser 1964b, Corollary 2.2.) The hypothesis of sharp flag transitivity appears to be much more restrictive than that of flag transitivity in general, for the projective as well as the affine case. The following result of Lüneburg 1965c, Korollar 2, supports the conjecture that, at least in the affine case, all examples are desarguesian:

28. *Suppose that Π is sharply flag transitive on the finite affine plane* **A**. *If, for some point $p \in$* **A**,
(**a**) Π_p *is cyclic, and*
(**b**) *every point orbit $\neq \{p\}$ of Π_p is an oval,*
then **A** *is desarguesian.*

This should be compared with result **7** above: if "sharply flag transitive" is replaced by "doubly transitive", then (**b**) is superfluous, even when "cyclic" in (**a**) is replaced by "abelian".

There are examples of point transitive collineation groups of finite affine planes which are not translation planes; see Section 5.4. Hence the hypothesis of line transitivity in **22** cannot be replaced by point transitivity. The following result of Keiser 1966 shows, however, that point primitivity is often sufficient:

29. *Let* **A** *be an affine plane whose order is either not a square or of the form m^2 with $m \not\equiv 1 \bmod 4$. If* **A** *admits a point primitive collineation group Γ, then* **A** *is a translation plane and Γ contains all translations of* **A**.

Result **22** can be stated in the following equivalent projective form: If the group Γ of the finite projective plane **P** fixes the line L and is transitive on the other lines, then **P** is (L, L)-transitive, and Γ contains all elations with axis L. It has been conjectured[1]) that the same conclusion holds also if the permutation group Δ induced by Γ on the points of L is doubly transitive. (The planes of Lüneburg 1965c, Section 6, show that one cannot expect **P** to be desarguesian, at least when n is even.) The situation seems to be different when n is odd:

30. *Let* **P** *be a projective plane of odd order $n \not\equiv 1 \bmod 8$. If* **P** *admits a collineation group Γ which fixes a line L and induces a doubly transitive permutation group on the points of L, then* **P** *is desarguesian.*

The proof of this result (Cofman 1966a) involves another application of the Gorenstein-Walter Theorem **4.3.28** and Lüneburg's **4.2.13**.

In conclusion, we list three further results of J. Cofman which generalize some of the earlier theorems.

31. *Let $\mathfrak{o}$ be a point set in the projective plane* **P** *of order n, such that*
(8) *not all points of $\mathfrak{o}$ are collinear, and*
(9) *some line of* **P** *carries more than two points of $\mathfrak{o}$.*

[1]) By D. R. Hughes.

If **P** *admits a collineation group* Γ *whose involutions are central collineations, such that*

(10) $\mathfrak{o}\Gamma = \mathfrak{o}$, *and*

(11) Γ *is transitive on the ordered triangles in* $\mathfrak{o}$,

then either

(a) $\mathfrak{o}$ *is the point set of a desarguesian subplane* **Q**, *and the group induced by* Γ *on* **Q** *contains all elations of* **Q**, *or*

(b) $\mathfrak{o}$ *is the point set of a translation plane* $\mathbf{Q}^W$, *where* **Q** *is a projective subplane of* **P** *and* W *a line fixed by* Γ, *and the group induced by* Γ *on* **Q** *is* (W, W)*-transitive.*

(Cofman 1968) In the affine case **(b)**, when $|\mathfrak{o}|$ is not too small, (11) can be relaxed to double transitivity:

32. *Let* $\mathfrak{o}$ *be a point set in the affine plane* **A** *of order* n, *such that*

$$|\mathfrak{o}| > n + 1. \tag{12}$$

If **A** *admits a collineation group* Γ *whose involutions are central collineations in the projective plane corresponding to* **A**, *such that* (10) *holds and*

(13) Γ *is doubly transitive on* $\mathfrak{o}$,

then **A** *is a translation plane. Furthermore,* Γ *contains all translations of* **A** *provided that* $n > 8$; *hence in this case* $\mathfrak{o}$ *consists of all points of* **A**.

This is essentially Theorem 1 of Cofman 1967b.[1]) Finally:

33. *Let* $\mathfrak{o}$ *be an oval in the projective plane* **P** *of odd order* n. *If* **P** *admits a collineation group* Γ *whose involutions are central collineations, such that* (10) *and* (13) *hold, then* **P** *is desarguesian, hence* $\mathfrak{o}$ *is a conic* (*cf.* **1.4.50**), *and* Γ *contains all collineations in the little projective group of* **P** *which leave* $\mathfrak{o}$ *invariant.*

(Cofman 1967c.) The condition in **31—33** that all involutions in Γ be central collineations is probably superfluous; cf. the footnotes to **4.2.13, 4.2.15, 4.2.16**.

The group Γ in **33** has three point orbits but no fixed elements. There is only one known class of finite projective planes whose full collineation group has no fixed elements but more than one orbit. These planes were discovered by Veblen & Wedderburn 1907 and Hughes 1957b; they will be properly defined in Section 5.4. Their full collineation groups each have two orbits; a characterization by transitivity properties will be given in 5.4 also (see **5.4.3**).

---

[1]) Cofman leaves out the essential hypothesis $n > 8$. If $n = 4$ or $n = 8$, then $\mathfrak{o}$ may also consist of 6 or 28 points, respectively. These are the only counterexamples.

5. Construction of finite planes

The only finite projective and affine planes that we have actually defined so far (cf. Section 1.4) are the desarguesian planes $\mathbf{P}(q)$ and $\mathbf{A}(q)$. We have mentioned repeatedly that there exist non-desarguesian finite planes as well, and in this chapter we shall present all known such planes. The known construction techniques for finite planes all use a finite vector space in a more or less obvious, but always essential way. This is the reason that these constructions always lead to planes of prime-power order. It is one of the major unsolved problems of the theory whether or not there also exist planes of non-prime-power order.

Section 5.1 is concerned with general methods of construction and representation for finite planes; we shall not give actual examples there. We discuss representations of spreads and quasifields, which in 3.1 were seen to be intimately connected with the construction of translation planes. Also, we introduce the concepts of derivation and planar function; these may be used to construct finite affine planes which are neither translation nor shears planes.

In Sections 5.2 and 5.3 we list all known finite non-desarguesian translation planes, those of Lenz-Barlotti types IV (coordinatized by quasifields which are not semifields) being collected in 5.2, and those of type V (coordinatizable by semifields which are not fields) in 5.3.

Finally, in Section 5.4 all known finite projective planes are listed which are not (L, L)- or (c, c)-transitive for any line L or point c. This means that no affine plane of these is a translation or a shears plane. An important tool for the construction of such planes is the derivation procedure mentioned above.

5.1 Algebraic representations

A *t-spread*, in an arbitrary projective geometry $\mathcal{P}$, was defined in Section 1.4 as a collection $\mathcal{S}$ of t-dimensional subspaces such that every point of $\mathcal{P}$ is on exactly one member of $\mathcal{S}$. It was pointed out in Section 3.1 that there exists a canonical correspondence between translation planes of order q^{t+1}, with kernel $GF(q)$, and t-spreads in $\mathcal{P}(2t+1, q)$. Hence these spreads are of fundamental importance for the construction of translation planes; we begin our discussion with an algebraic representation for them.

As in 1.4, let $V(n, q)$ denote the vector space of rank n over $GF(q)$. We represent $V(n, q)$ as the set of all n-tuples of elements in $GF(q)$, and we call a collection $\mathfrak{C}$ of $(t+1, t+1)$-matrices over $GF(q)$ a *t-spread set* if it satisfies the following conditions:

(1) $$|\mathfrak{C}| = q^{t+1},$$

(2) $$O \in \mathfrak{C} \quad \textit{and} \quad I \in \mathfrak{C},$$

(3) *If* $X, Y \in \mathfrak{C}$ *and* $X \neq Y$, *then* $\det(X - Y) \neq 0$.

Here O and I are the zero and identity $(t+1, t+1)$-matrices, respectively. Note that (2) and (3) imply the nonsingularity of every $C \in \mathfrak{C}$. The connection between spreads and spread sets is the following:

1. *Let* $\mathfrak{C}$ *be a t-spread set and* $(a_0, \ldots, a_t, b_0, \ldots, b_t)$ *a basis of* $W = V(2t+2, q)$. *Let* S *be the subspace spanned by* $a_0, \ldots, a_t$ *and, for each* $C \in \mathfrak{C}$, *let* $S(C)$ *be the subspace spanned by all* $a_i C + b_i$, $i = 0, \ldots, t$. *Then the subspaces* S *and* $S(C)$ *for* $C \in \mathfrak{C}$ *form a partition of* W, *and therefore they represent the subspaces of a t-spread* $\mathscr{S} = \mathscr{S}(\mathfrak{C})$ *in* $\mathscr{P}(2t+1, q)$. *Conversely, every such spread may be represented in this fashion.*

For the proof, see BRUCK & BOSE 1964, Section 5.[1]) Let e be a nonzero vector in $V = V(t+1, q)$ and $\mathfrak{C}$ a t-spread set. Then every vector $y \in V$ can be written in a unique way as $y = eC$, with $C = C(y) \in \mathfrak{C}$. We define a multiplication in V by

(4) $$xy = xC(y);$$

t is not difficult to verify (BRUCK & BOSE 1964, Section 6) that with this multiplication and the original addition, V becomes a quasifield $\mathfrak{Q} = \mathfrak{Q}(\mathfrak{C})$ of order q^{t+1}, coordinatizing the translation plane $\mathbf{A} = \mathbf{A}(\mathfrak{C})$ defined by the spread $\mathscr{S} = \mathscr{S}(\mathfrak{C})$. A simple corollary of this observation is:

2. $\mathfrak{Q}(\mathfrak{C})$ *is a nearfield if and only if* $\mathfrak{C}$ *is closed under multiplication, and* $\mathfrak{Q}(\mathfrak{C})$ *is a semifield if and only if* $\mathfrak{C}$ *is closed under addition. Consequently,* $\mathbf{A}(\mathfrak{C})$ *is desarguesian if and only if* $\mathfrak{C}$ *is a ring.*

(BRUCK & BOSE 1966, Section 11.) In fact if $\mathfrak{C}$ is a ring, then $\mathfrak{C}$ is the Galois field $GF(q^{t+1})$.

The spreads connected with the desarguesian planes can be characterized synthetically as follows. A *t-regulus* in $\mathscr{P}(2t+1, q)$ is a set $\mathscr{R}$ of

---

[1]) Condition (3) guarantees that $S(C) \cap S(D) = 0$ if $C \neq D$, and a counting argument then shows that the $S(C)$, $C \in \mathfrak{C}$, represent a spread. In the infinite case, (1) has to be replaced by the following condition: if $a \neq 0$ and b are in $V(t+1, \mathfrak{K})$, then there exists $X \in \mathfrak{C}$ such that $aX = b$. Such an X is necessarily unique.

t-subspaces such that

(5) $|\mathscr{R}| = q + 1$,

(6) *If* $X, Y \in \mathscr{R}$ *and* $X \neq Y$, *then* $X \cap Y = \emptyset$,

(7) *If a line* L *meets three distinct members of* $\mathscr{R}$, *then* $L \cap X \neq \emptyset$ *for all* $X \in \mathscr{R}$.

Such a line L is called a *transversal* of $\mathscr{R}$. It is clear that a transversal meets every member of $\mathscr{R}$ in a unique point, and conversely that every point of a transversal belongs to a unique member of $\mathscr{R}$. Also, it is not difficult to prove that there is a unique transversal through each point of an arbitrary member of $\mathscr{R}$, so that in particular all transversals of $\mathscr{R}$ are mutually skew. The existence of reguli is well known; in fact the nondegenerate quadrics of (maximal) index $t+1$ in $\mathscr{P}(2t+1, q)$ are always covered by reguli (cf. **1.4.44, 1.4.45**). One can say even more: given any three mutually disjoint t-subspaces A, B, C in $\mathscr{P}(2t+1, q)$, there is a unique t-regulus $\mathscr{R} = \mathscr{R}(A, B, C)$ containing A, B, C. We say now that the t-spread $\mathscr{S}$ of $\mathscr{P}(2t+1, q)$ is *regular* provided that

(8) $$A, B, C \in \mathscr{S} \quad \textit{implies} \quad \mathscr{R}(A, B, C) \subset \mathscr{S}.$$

The basic fact relevant in the present context then is:

3. *For* $q > 2$, *a* t*-spread* $\mathscr{S}$ *in* $\mathscr{P}(2t+1, q)$ *is regular if and only if the translation plane defined by* $\mathscr{S}$ *is desarguesian.*

For the proof, see BRUCK & BOSE 1966, Theorem 12.1.[1]) The case $q = 2$ is truly exceptional: every t-spread in $\mathscr{P}(2t+1, 2)$ is regular.

Results **1** and **2** should be useful in constructing translation planes, but this possibility has found little interest so far. On the other hand, translation planes can also be represented as coordinate planes, in the sense of (3.1.25), over quasifields, and many construction techniques for quasifields have been found. We postpone special constructions of this kind to Sections 5.2 and 5.3; here we shall give a general method for the representation of finite quasifields within Galois fields.

The additive group $\mathfrak{Q}^+$ of a finite quasifield $\mathfrak{Q}$, being isomorphic to a group of translations of the plane over $\mathfrak{Q}$, is elementary abelian[2]) and hence isomorphic to the additive group $\mathfrak{F}^+$ of a Galois field $\mathfrak{F}$ with $|\mathfrak{F}| = |\mathfrak{Q}|$. Let $|\mathfrak{Q}| = q = p^e$; then the order of the kernel $\mathfrak{K}(\mathfrak{Q})$

[1]) The result is valid, mutatis mutandis, also in the infinite case; in fact it is well known in special cases — e.g., for projective 3-space over the real numbers (cf. WEISS 1936; ANDRÉ 1954a, PICKERT 1955, p. 205—206). In the infinite case, "desarguesian" must be replaced by "pappian".

[2]) A more direct proof that $\mathfrak{Q}^+$ is abelian, valid also in the infinite case, is in PICKERT 1952; see also PICKERT 1955, p. 91.

is p^d with $d \mid e$, say $e = m\,d$. Hence $\mathfrak{F}$ contains a subfield $\mathfrak{K} \cong \mathfrak{K}(\mathfrak{Q})$, and both $\mathfrak{F}$ and $\mathfrak{Q}$ can be regarded as (left) vector spaces of rank m over $\mathfrak{K}$. Let ω be a linear transformation from $\mathfrak{Q}$ onto $\mathfrak{F}$ over $\mathfrak{K}$, so that

$$(k \cdot x)\,\omega = k\,(x\,\omega),$$

where "$\cdot$" denotes multiplication in $\mathfrak{Q}$. Then a new multiplication "$\circ$" is defined in $\mathfrak{F}$ by

$$x \circ y = (x\,\omega^{-1} \cdot y\,\omega^{-1})\,\omega,$$

and with respect to the original addition and this multiplication, $\mathfrak{F}$ is a quasifield $\mathfrak{F}^\circ$ isomorphic to $\mathfrak{Q}$. Also, $k \circ x = k\,x$ for $k \in \mathfrak{K}$, so that $\mathfrak{K}$ is the kernel of $\mathfrak{F}^\circ$.

Now define, for every $t \neq 0$ in $\mathfrak{F}$, a mapping $\lambda(t)$ by

$$x^{\lambda(t)} = (x \circ t)\,t^{-1},$$

where t^{-1} denotes the inverse with respect to the original field multiplication. It is easily verified that $\lambda(t)$ is a nonsingular linear automorphism of the vector space $\mathfrak{F}$ over $\mathfrak{K}$. These linear automorphisms satisfy

(9) $$\lambda(1) = 1$$

and

(10) $$\textit{if} \quad x \neq 0 \quad \textit{and} \quad x^{\lambda(s)}\,s = x^{\lambda(t)}\,t, \quad \textit{then} \quad s = t.$$

The significance of these considerations lies in the fact that, conversely, (9) and (10) always determine a quasifield:

4. *Let $\mathfrak{F} = GF(p^{md})$ and $\mathfrak{K} = GF(p^d) \subseteq \mathfrak{F}$. Suppose that to every $t \neq 0$ in $\mathfrak{F}$ there is assigned a linear automorphism $\lambda(t)$ of $\mathfrak{F}$, considered as a vector space over $\mathfrak{K}$, such that* (9) *and* (10) *are satisfied. Then, with the original addition and a new multiplication "$\circ$" defined by $x \circ 0 = 0$ and*

(11) $$x \circ t = x^{\lambda(t)}\,t \quad \textit{for} \quad t \neq 0,$$

$\mathfrak{F}$ becomes a quasifield with kernel containing $\mathfrak{K}$. Conversely, every finite quasifield can be represented in this way.

We note an immediate corollary:

5. *Let $\mathfrak{F}_\lambda$ be the quasifield defined by the family $\{\lambda(t), t \neq 0\}$ of linear automorphisms satisfying* (9) *and* (10). *Then $\mathfrak{F}_\lambda$ is a nearfield if and only if*

(12) $$[x^{\lambda(s)}\,s]^{\lambda(t)} = x^{\lambda(s^{\lambda(t)}t)}\,s^{\lambda(t)},$$

and it is a semifield if and only if

(13) $$x^{\lambda(s+t)}\,(s+t) = x^{\lambda(s)}\,s + x^{\lambda(t)}\,t,$$

for all $x, s, t \in \mathfrak{F}$.

Results **4** and **5** ought to be compared with **1** and **2**; there is of course an obvious relationship. Note, however, that the matrices C of **1** do not represent the linear automorphisms $\lambda(t)$ of **4**: comparison of (4) and (11) shows rather that they represent the mappings $\lambda(t)\,\mu(t)$, where $\mu(t)\colon x \to x\,t$. Note also that **2** provides much more satisfactory criteria for associativity and distributivity than **5**.

It follows from **4** that the problem of finding all finite quasifields is equivalent to that of constructing all possible sets $\{\lambda(t), t \neq 0\}$ of linear automorphisms satisfying (9) and (10). This has so far been investigated mainly in the case where the $\lambda(t)$ are subject to the further restriction of actually being field automorphisms of $\mathfrak{F}$, fixing the kernel $\mathfrak{K}$ elementwise. The group A of such automorphisms is cyclic of order $m = e/d$, and the multiplicative group of $\mathfrak{F}$ is cyclic of order $p^e - 1$. Let α and c be generating elements of these groups; then $\lambda(c^i) = \alpha^{f(i)}$ for some mapping f from $Z(p^e - 1)$ into $Z(m)$, where $Z(j)$ denotes the ring of integers mod j. The conditions (9) and (10) for the $\lambda(t)$ are equivalent to the following for f:

$$f(0) = 0 \tag{14}$$

and

$$\text{if} \quad i \equiv j \bmod p^{rd} - 1, \quad \text{where} \quad r = (m, f(i) - f(j)), \quad \text{then} \quad i \equiv j \bmod p^{md} - 1. \tag{15}$$

This was proved by Foulser 1962, p. 138; see also Foulser 1967a. We shall see in Section 5.2 how such functions f can actually be constructed.

Next, we discuss a construction technique, due to Ostrom 1964a, for finite planes from given planes of the same (square) order. Let $\mathbf{P}$ be a projective plane of order $n = m^2$. A *derivation set* in $\mathbf{P}$ is a set $\mathfrak{d}$ of $m + 1$ points, all on the same line W, such that for any two distinct points $x, y \,\mathrm{I}\!\!\!/\, W$ with $(x\,y)\,W \in \mathfrak{d}$, the subplane $\langle\{x, y\} \cup \mathfrak{d}\rangle$ has order m (and is, therefore, a Baer subplane of $\mathbf{P}$). We call $\mathbf{P}$ *derivable* if there exists a derivation set in $\mathbf{P}$. We can then define a new projective plane $\mathfrak{d}\mathbf{P}$ as follows: Consider the points of the affine plane $\mathbf{A} = \mathbf{P}^W$, together with the point sets (X), where X is a line of $\mathbf{A}$ with $X\,W \notin \mathfrak{d}$, and the point sets of the affine Baer subplanes of $\mathbf{A}$ whose ideal points are precisely those of $\mathfrak{d}$. This incidence structure is denoted by $\mathfrak{d}\mathbf{A}$, and it is easily verified that conditions (3.2.5), (3.2.7), (3.2.9) are satisfied for $\mathfrak{d}\mathbf{A}$, so that result **3.2.4b** yields:

6. $\mathfrak{d}\mathbf{A}$ *is an affine plane of order* n.

This affine plane will be said to be *derived* from $\mathbf{A}$ with respect to $\mathfrak{d}$; and the projective plane corresponding to $\mathfrak{d}\mathbf{A}$ is the projective plane $\mathfrak{d}\mathbf{P}$ *derived* from the derivable plane $\mathbf{P}$.

7. *If* $\mathbf{P}$ *is derivable, and* $\mathfrak{d}\mathbf{P}$ *derived from* $\mathbf{P}$, *then* $\mathfrak{d}\mathbf{P}$ *is derivable, and* $\mathbf{P}$ *can be derived from* $\mathfrak{d}\mathbf{P}$.

For if L is a line $\neq W$ of $\mathbf{P}$ with $LW \in \mathfrak{d}$, then the set $(L) - \{LW\}$ of the affine points of L is easily seen to be the point set of an affine Baer subplane of $\mathfrak{d}\mathbf{A}$. All these Baer subplanes have the same ideal points, which form a derivation set $\mathfrak{d}'$ of $\mathfrak{d}\mathbf{P}$ such that $\mathfrak{d}'(\mathfrak{d}\mathbf{P}) \cong \mathbf{P}$.

It is clear from the definitions that

8. *Any collineation of a derivable plane* $\mathbf{P}$ *which keeps the derivation set* $\mathfrak{d}$ *invariant may be interpreted also as a collineation of the derived plane* $\mathfrak{d}\mathbf{P}$.

Consequently, for any finite affine plane $\mathbf{A}$:

9. *The translation groups of* $\mathbf{A}$ *and* $\mathfrak{d}\mathbf{A}$ *are the same. In particular,* $\mathfrak{d}\mathbf{A}$ *is a translation plane if and only if* $\mathbf{A}$ *is.*

The following result shows that the derivable planes constitute a rather large class.

10. *Let* $\mathfrak{T}$ *be a ternary field of order* $n = m^2$ *whose additive loop is an abelian group. Suppose that* $\mathfrak{T}$ *contains a ternary subfield* $\mathfrak{K}$ *of order* m *such that the following conditions hold:*

(**a**) $\mathfrak{T}$ *is* $\mathfrak{K}$*-linear*[1]), *i.e.* $x \cdot a \circ y = xa + y$ *whenever* $a \in \mathfrak{K}$, *for all* $x, y \in \mathfrak{T}$.
(**b**) *If* $0 \neq x \in \mathfrak{T}$, *if either* $xa = xb + xc$ *or* $xa = (xb)c$, *and if two of the elements* a, b, c *are in* $\mathfrak{K}$, *then so is the third.*
(**c**) *If* $a, x, y \in \mathfrak{T}$ *and* $x + y \neq 0$, *then there is a unique* $b \in \mathfrak{T}$ *such that* $(x + y)a = xb + yb$, *and* $a \in \mathfrak{K}$ *if and only if* $b \in \mathfrak{K}$.

Then the projective plane coordinatized by $\mathfrak{T}$ *is derivable. In fact, the ideal points of the Baer subplane coordinatized by* $\mathfrak{K}$ *are those of a derivation set, and the affine point sets of the subplanes which become lines in the derived plane are*

$$R(a, b, c) = \{(ax + b, ay + c) : x, y \in \mathfrak{K}\}, \text{ with } a, b, c \in \mathfrak{T} \text{ and } a \neq 0.$$

For the proof, see OSTROM 1964a, Theorem 9. Note that $R(a, b, c) = R(a', b', c')$ does not imply $(a, b, c) = (a', b', c')$.

We shall now restrict our attention to the case of a translation plane $\mathbf{A}$ of square order q^2 which possesses a Baer subplane $\mathbf{B}$ of order q. It is clear from **3.1.5** that $\mathbf{B}$ is also a translation plane, coordinatizable by a sub-quasifield $\mathfrak{S}$ of the quasifield $\mathfrak{Q}$ coordinatizing $\mathbf{A}$. We need the following additional condition:

---

[1]) Compare here (3.1.55) and **3.1.37**, on p. 137.

(16) *For any two points x, y of* $\mathbf{A}$, *there exists a collineation σ of* $\mathbf{A}$ *such that $x\sigma$ and $y\sigma$ are both in* $\mathbf{B}$, *and the set of ideal points of* $\mathbf{B}$ *is invariant under σ.*

Note that this is not only satisfied if $\mathbf{A}$ is desarguesian, but also, for example, if $\mathfrak{S}$ is the kernel $\mathfrak{K}(\mathfrak{Q})$, in which case $\mathbf{B}$ is desarguesian and the dilatations of $\mathbf{B}$ extend to all of $\mathbf{A}$. It is not difficult to prove that

11. *If* (16) *holds, then the ideal points of* $\mathbf{B}$ *form a derivation set.*

The derived plane $\mathfrak{d}\mathbf{A}$ is then also a translation plane, by **9**, but in general the planes $\mathbf{A}$ and $\mathfrak{d}\mathbf{A}$ are not isomorphic. For the case that $\mathbf{A} = \mathbf{A}(q^2)$, $q > 2$, this non-isomorphy is easily proved as follows: Put $q = p^e$, let L be a line of $\mathbf{B}$, and let Γ be the group of those dilatations fixing L which leave $\mathbf{B}$ invariant. Then $|\Gamma| = q(q-1)$, and **8** implies that Γ is also a collineation group of $\mathfrak{d}\mathbf{A}$. Now (L) becomes a Baer subplane in $\mathfrak{d}\mathbf{A}$, and the number of collineations fixing a Baer subplane of $\mathbf{A}(q^2)$ is $2e$. But $2e < q(q-1)$ unless $q = 2$, so that $\mathfrak{d}\mathbf{A}(q^2) \not\cong \mathbf{A}(q^2)$.[1])

Except for special cases, the relations between the quasifields of $\mathbf{A}$ and $\mathfrak{d}\mathbf{A}$ seem to be undetermined. In the case mentioned above, where $\mathfrak{S} = \mathfrak{K}(\mathfrak{Q})$, the plane $\mathfrak{d}\mathbf{A}$ can be represented as follows: Points are the ordered pairs (x, y) of elements in $\mathfrak{Q}$, and lines are the following two types of point sets:

$$\text{(17)}\quad \begin{aligned} Q(a,b) &= \{(x,y) \vdots\, y = xa + b\}; \quad a \notin \mathfrak{S}; \quad b \in \mathfrak{Q}, \\ R(a,b,c) &= \{(x,y) \vdots\, x - b \in a\mathfrak{S},\, y - c \in a\mathfrak{S}\}; \quad 0 \neq a \in \mathfrak{S};\ b, c \in \mathfrak{Q}. \end{aligned}$$

This again is proved in Ostrom 1964a.

The translation planes of order q^2 with kernel of order q are remarkable also from another point of view. Such a plane can be represented by a 1-spread $\mathscr{S}$ in $\mathscr{P}(3, q)$, and this is an exceptional situation for the following reason: Suppose that $\mathscr{S}$ contains a regulus $\mathscr{R}$; then the set of all transversals of $\mathscr{R}$ is itself a regulus $\mathscr{R}'$. (Clearly this is false for t-reguli with $t > 1$.) It is obvious that $\mathscr{R}$ and $\mathscr{R}'$ cover the same points of $\mathscr{P}(3, q)$, whence

$$\text{(18)}\qquad \mathscr{S}' = (\mathscr{S} - \mathscr{R}) \cup \mathscr{R}'$$

is again a 1-spread in $\mathscr{P}(3, q)$. In general, $\mathscr{S}$ and $\mathscr{S}'$ define non-isomorphic translation planes. In fact if $\mathscr{S}$ is regular, i.e. if $\mathbf{A} = \mathbf{A}(\mathscr{S})$ is desarguesian by **3**, then $\mathscr{S}'$ cannot be regular, so that $\mathbf{A}' = \mathbf{A}(\mathscr{S}')$

[1]) The actual structure of $\mathfrak{d}\mathbf{A}(q^2)$ was determined by Albert 1961, § 6. These planes can be coordinatized by a class of quasifields first exhibited by Hall 1943, which will be defined in the next section.

is non-desarguesian.[1]) The transformation $\mathcal{S} \to \mathcal{S}'$ defined by (18) can, of course, be repeated; the interesting question as to how the resulting translation planes are related seems to have received little attention so far.

Let $\mathbf{N} = (\mathfrak{p}, \mathfrak{L}, \mathrm{I})$ be a *net* in the sense of (3.2.12)—(3.2.14). We call $\mathbf{N}$ *replaceable* (with Ostrom 1966b) if there exists a collection $\mathfrak{L}^*$ of subsets of $\mathfrak{p}$ such that the following conditions are satisfied:

(19) *The incidence structure* $\mathbf{N}^* = (\mathfrak{p}, \mathfrak{L}^*, \mathrm{I}^*)$, *with* $p\,\mathrm{I}^*\,\mathfrak{l}$ *defined by* $p \in \mathfrak{l}$ (*for* $p \in \mathfrak{p}$ *and* $\mathfrak{l} \in \mathfrak{L}^*$) *is a net.*

(20) *Two points of* $\mathfrak{p}$ *are joined by a line of* $\mathbf{N}$ *if and only if they are joined by a line of* $\mathbf{N}^*$.

(21) *There exists an* $\mathfrak{l} \in \mathfrak{L}^*$ *such that* $\mathfrak{l} \neq (X)$ *for any* $X \in \mathfrak{L}$.

The notation is that of (1.1.5) again. It is easily verified that if the replaceable net $\mathbf{N}$ is finite, then the parameters of $\mathbf{N}$ and $\mathbf{N}^*$ are the same. Thus if $\mathbf{N}$ is an r-net of order n (cf. **3.2.8**), then so is $\mathbf{N}^*$.

If $\mathfrak{d}$ is a derivation set of the projective plane $\mathbf{P}$ of order m^2, contained in the line W, then the points of the affine plane $\mathbf{A} = \mathbf{P}^W$, together with the lines $\neq W$ meeting W in points of $\mathfrak{d}$, form a replaceable $(m+1)$-net $\mathbf{N}$ of order m^2, the point sets of $\mathfrak{L}^*$ being those of the Baer subplanes having precisely $\mathfrak{d}$ in common with W. The affine plane $\mathfrak{d}\mathbf{A}$ of **6** is, therefore, obtained from $\mathbf{A}$ by replacing the lines of $\mathbf{N}$ by those of $\mathbf{N}^*$.

This suggests a more general construction procedure. Let $\mathbf{A}$ be an affine plane of order n and suppose that the r-net $\mathbf{N}$ of order n, formed by the points of $\mathbf{A}$ and the lines of r distinct parallel classes, is replaceable by $\mathbf{N}^*$. Let $\mathbf{A}^*$ be the incidence structure whose points are those of $\mathbf{A}$ and whose lines are (i) the lines of $\mathbf{A}$ which are not in $\mathbf{N}$ and (ii) the lines of $\mathbf{N}^*$. It is fairly obvious that $\mathbf{A}^*$ is again an affine plane of order n; in general $\mathbf{A}$ and $\mathbf{A}^*$ are not isomorphic.

We have remarked in the introduction to this chapter that all known finite affine planes have prime-power order q. Hence for any such plane $\mathbf{A}$, there exists a one-one correspondence between its points and those of the desarguesian affine plane $\mathbf{A}(q)$. Consequently, $\mathbf{A}(q)$ may be regarded as a replaceable net, and $\mathbf{A} \cong \mathbf{A}(q)^*$. For a more detailed discussion of this point of view, and for representations of the known finite planes in terms of replaceable nets, the reader is referred to Ostrom 1960, 1962a, and 1968.[2])

---

[1]) For this special case, Bruck & Bose 1964, Section 9, have shown that $\mathbf{A}' = \mathfrak{d}\mathbf{A}$, the plane represented by (17) with $\mathfrak{Q} = GF(q^2)$. That $\mathbf{A}'$ may be coordinatized by one of the quasifields of Hall 1943 [cf. footnote [1]) on p. 225], was proved earlier by Panella 1959.

[2]) In particular, Ostrom gives certain general constructions of sets $\{\lambda(t)\}$ satisfying (9) and (10) by means of replaceable nets.

In the remainder of this section, we discuss a general construction principle for a certain type of (v, W)-transitive planes, with $v \mathrm{I} W$, and the corresponding cartesian groups. Let G and H be two finite groups of the same order n, both written additively but not necessarily commutative. For any function f from G into H, single-valued but not necessarily one-one, we define an incidence structure $\mathbf{S} = \mathbf{S}(G, H, f)$ as follows: Points are the elements of the direct product $G \times H$, blocks are the point sets

$$L(a, b) = \{(x, f(x - a) + b) \vdots x \in G\}, \quad a \in G, \quad b \in H \tag{22}$$

and

$$L(c) = \{(c, y) \vdots y \in H\}, \quad c \in G, \tag{23}$$

and incidence is set theoretic inclusion. For any $g \neq 0$ in G, let the mappings $\lambda(g)$ and $\varrho(g)$ from G into H be defined by

$$x^{\lambda(g)} = f(g + x) - f(x) \tag{24}$$

and

$$x^{\varrho(g)} = -f(x) + f(x + g). \tag{25}$$

Then:

12. *The following conditions on G, H, and f are equivalent:*

(**a**) *$\lambda(g)$ is one-one and onto for every $g \neq 0$,*
(**b**) *$\varrho(g)$ is one-one and onto for every $g \neq 0$,*
(**c**) *$\mathbf{S}(G, H, f)$ is an affine plane.*

For the simple proof, see Dembowski & Ostrom 1968, Lemma 12; result **3.2.4e** is used here. If f satisfies one, and hence both, of the conditions (**a**), (**b**), then it is called a *planar function.* Thus every planar function between two groups of order n gives rise to an affine plane of order n. The mappings

$$\sigma(g, h): (x, y) \to (x + g, y + h)$$

are automorphisms of $\mathbf{S}(G, H, f)$; obviously they form a group, isomorphic to $G \times H$ and transitive on the points of $\mathbf{S}$ as well as on each of the two classes of lines given by (22) and (23). Furthermore, the $\sigma(0, h)$ fix every line (23), so that they form a (v, W)-transitive group of translations whenever $\mathbf{S}$ is an affine plane. These properties are not only necessary, but also sufficient for the representability by planar functions:

13. *An affine plane of order n can be represented in the form $\mathbf{S}(G,H,f)$ by a planar function f from G to H if, and only if, the corresponding projective plane possesses a collineation group Γ of order n^2 such that*

(**a**) *Γ is (c, A)-transitive for some flag (c, A).*
(**b**) *The (c, A)-elations in Γ form a direct factor Π of Γ, and*
(**c**) *Γ has three orbits.*

If this is the case, then n is odd, $\Gamma \cong G \times H$, and $\Pi \cong H$.

Proof: DEMBOWSKI & OSTROM 1968, Lemma 9 and Theorem 5. Note that **13**, together with **4.2.10**, has the following consequence:

14. *If a projective plane* **P** *of order* 2^e *admits an elementary abelian collineation group* Γ *of order* 2^{2e}, *then* Γ *consists entirely of elations, so that* **P** *is* (L, L)- *or* (c, c)-*transitive for some line* L *or point* c.

(DEMBOWSKI & OSTROM 1968, Theorem 4.)

Result **13** shows that an affine plane of the form $\mathbf{S}(G, H, f)$ can be coordinatized by a cartesian group (cf. **3.1.22a**). In conclusion of this section, we give a direct description of such a cartesian group, by means of the planar function f. Passing, if necessary, from $f(x)$ to $f(x-a)+b$ for suitable $a \in G$, $b \in H$, we can assume that f is a *normed* planar function in the sense that $f(0) = f(e) = 0$ for some $e \neq 0$ in G. Then:

15. *Let* f *be a normed planar function from* G *to* H, *and define a multiplication in* H *by the rule*

$$x \cdot y = -f(x^{\lambda(e)^{-1}}) + f(x^{\lambda(e)^{-1}} + y^{\varrho(e)^{-1}}) - f(y^{\varrho(e)^{-1}}); \tag{26}$$

cf. (24), (25). *Then, with the original addition and this multiplication,* H *becomes a cartesian group coordinatizing* $\mathbf{S}(G, H, f)$.

This is Theorem 6 of DEMBOWSKI & OSTROM 1968. In the same paper, there are also necessary and sufficient conditions for $\mathbf{S}(G, H, f)$ to be a translation plane, in terms of the multiplication (26). Examples of planar functions, all with $G \cong H$ the elementary abelian group of order n, but some of them defining non-desarguesian planes, will be given in Section 5.3.

5.2 Planes of type IV

In this section, we begin to list the known quasifields. We restrict ourselves here to quasifields which coordinatize planes of one of the Lenz-Barlotti types IVa; by reasons of duality, this will be sufficient also for type IVb.[1]) Result **3.1.22** shows that planes of type IVa are coordinatized by non-distributive quasifields; hence our task will be to construct finite quasifields which are not semifields. Note, however, that the plane coordinatized by such a quasifield may well be of type V.1;[2]) thus additional considerations will be necessary to ensure in each case that the plane under consideration is really of type IV.

---

[1]) Note that planes of types IVa or IVb cannot be self-dual.

[2]) A contrary statement concerning type IVa, in PICKERT 1955, p. 108, is incorrect. In order that **P** be of type IVa it is necessary that not only one but all coordinatizing quasifields are non-distributive.

We begin with types IV a.2 and IV a.3; these planes are coordinatized by planar[1]) nearfields. We have already defined an infinite class of finite nearfields, the *regular* nearfields, in Section 1.4; we repeat the definition here in a slightly different form. As in Section 5.1, let $\mathfrak{F} = GF(p^{md}) = GF(q^m) \supseteqq GF(q) = \mathfrak{K}$, and suppose that every prime divisor of m divides $q-1$. Also, let $m \not\equiv 0 \bmod 4$ if $q \equiv -1 \bmod 4$. Define the function f from $Z(q^m-1)$ into $Z(m)$ by

$$(1) \qquad (q^{f(i)}-1)(q-1)^{-1} \equiv i \bmod m;$$

then f satisfies conditions (5.1.14) and (5.1.15) and thus defines a quasi-field which can be shown to be the regular nearfield $N(m,q)$ as defined on pp. 33–34.

The following results will provide some information about the multiplicative structure of the regular nearfields.

1. *A finite nearfield $\mathfrak{N}$ is regular, i.e. of the form $N(m,q)$, if and only if its multiplicative group $\mathfrak{N}^\times$ contains a cyclic normal subgroup C such that $\mathfrak{N}^\times/C$ is also cyclic. If this is the case, and if $\mathfrak{K}^\times$ and $\mathfrak{F}^\times$ denote the multiplicative groups of $\mathfrak{K} = GF(q)$ and $\mathfrak{F} = GF(q^m)$, respectively, then*

(**a**) *The centre of $\mathfrak{N}^\times$ is the cyclic group $\mathfrak{K}^\times$, and*
(**b**) *The elements of a subgroup of $\mathfrak{F}^\times$ also form a subgroup of $\mathfrak{N}^\times$.*

These results are contained in ZASSENHAUS 1935a; see also HALL 1959, Theorem 20.7.2. Note that one can actually prove a little more than (**b**): If S is a subgroup of $\mathfrak{F}^\times$, then S is characteristic since $\mathfrak{F}^\times$ is cyclic. As multiplication in $\mathfrak{N} = N(m,q)$ is defined by (5.1.11) with $\lambda(t) \in \operatorname{Aut}\mathfrak{F}$ [see also (1.4.14)], we get

$$S \circ t = S^{\lambda(t)}\, t = S\, t \qquad \text{for all } t \in \mathfrak{N}.$$

Hence the right cosets of S are the same in $\mathfrak{F}^\times$ and $\mathfrak{N}^\times$.

2. *If $p^{md} \neq 9$, then the automorphism group of $N(m, p^d)$ is cyclic of order dividing $m\,d$. The automorphism group of $N(2,3)$ is the non-abelian group S_3 of order 6; it is sharply transitive on the elements not in the kernel $GF(3)$ of $N(2,3)$.*

This result, due to ZASSENHAUS 1935a, is a first reason for calling $N(2,3)$ the *exceptional nearfield*. Another more geometric reason for this will be encountered below.

We mention two more theorems on finite nearfields; these are not restricted to regular nearfields.

[1]) This word may be omitted in our context: every finite nearfield is easily proved to be planar.

3. *The centre* $\mathfrak{Z}$ *of a finite nearfield* $\mathfrak{N}$ *is the intersection* $\mathfrak{D}$ *of all conjugates of its kernel* $\mathfrak{K}$.

Proof.[1] That $\mathfrak{Z} \subseteqq \mathfrak{D}$ is obvious, hence it suffices to prove $\mathfrak{D} \subseteqq \mathfrak{Z}$. Since finite division rings are commutative (WEDDERBURN 1905), there is nothing to prove if $\mathfrak{K} = \mathfrak{N}$; hence suppose $0 \neq t \in \mathfrak{N} - \mathfrak{K}$. For arbitrary $d \in \mathfrak{D}$ there exist $d_1, d_2 \in \mathfrak{D}$ such that $t\,d = d_1 t$ and $(t+1)\,d = d_2(t+1)$. Clearly $d_2 \in \mathfrak{K}$; hence

$$d_1 t + d = t\,d + d = (t+1)\,d = d_2(t+1) = d_2 t + d_2$$

and therefore $(d_1 - d_2)\,t = d_2 - d$. But $d_1 - d_2 \in \mathfrak{K}$ and $d_2 - d \in \mathfrak{K}$, whereas $t \notin \mathfrak{K}$; this implies $d_1 = d_2 = d$, so that $t\,d = dt$. Hence $\mathfrak{N} - \mathfrak{D}$ is in the centralizer of $\mathfrak{D}$, and this implies $\mathfrak{D} \subseteqq \mathfrak{Z}$.

4. *If* G *is a subgroup of order* $p_1 p_2$ *of the multiplicative group* $\mathfrak{N}^\times$ *of a finite nearfield* $\mathfrak{N}$, *where* p_1 *and* p_2 *are (not necessarily distinct) prime numbers, then* G *is cyclic. Consequently, the Sylow* p*-groups of* $\mathfrak{N}^\times$ *are cyclic if* $p > 2$, *and cyclic or generalized quaternion groups if* $p = 2$.

For the proof of this (also contained in ZASSENHAUS 1935a), see HALL 1959, p. 389—390.[2] The multiplicative group of $N(2,3)$ is of order 8 and non-abelian, hence it is the quaternion group. The unique involution in this group is the element $-1 \in N(2,3)$; it satisfies

(2) $\quad x^2 = -1$ *for every* $x \neq 0, \pm 1$ *in* $N(2,3)$.

This property is not shared by any other proper nearfield. Comparing (2) with (4.3.14) and using **4.3.32**, we see that

5. *There is only one plane of Lenz-Barlotti type* IV a.3, *namely that over the exceptional nearfield* $N(2,3)$.

This result is due to ANDRÉ 1955, Hilfssatz 10.

We shall now describe seven finite nearfields, discovered by DICKSON 1905, which are not regular and will be called here the *irregular nearfields*. Each of these is of order p^2, with p an odd prime. By results **5.1.1** and **5.1.2**, their structure will be completely determined when we present a group G of (2, 2)-matrices over $GF(p)$, isomorphic to the multiplicative group of the irregular nearfield in question, which is sharply transitive on the non-zero elements of the vector space $V(2,p)$. We shall define this group G, in each of the seven cases which follow, by giving a set of generators. One of these generators will always be the matrix $A = \begin{pmatrix} 0 & -1 \\ -1 & 0 \end{pmatrix}$, over the respective $GF(p)$. We shall also

[1] The following argument, which does not depend essentially on finiteness, is due to R. BRAUER 1949; another proof was given by ANDRÉ 1963a.

[2] The result is actually true in the more general situation where G is in the stabilizer of a Frobenius group.

give information about the automorphism groups of the irregular nearfields (these were determined by FOULSER 1962, p. 176). The following order is that of ZASSENHAUS 1935a.

I. $p=5$, $G=\langle A, B\rangle$, with $B=\begin{pmatrix} 1 & -2 \\ -1 & -2 \end{pmatrix}$. The automorphism group is cyclic of order 4.

II. $p=11$, $G=\langle A, B, C\rangle$, with $B=\begin{pmatrix} 1 & 5 \\ -5 & -2 \end{pmatrix}$ and $C=\begin{pmatrix} 4 & 0 \\ 0 & 4 \end{pmatrix}$. The group $\langle A, B\rangle$ is of order 24, isomorphic to G in case I. Hence G in case II is the direct product of a group of order 24 and the cyclic group $\langle C\rangle$ of order 5; thus G is soluble of order 120. The automorphism group has order 2.

III. $p=7$, $G=\langle A, B\rangle$, with $B=\begin{pmatrix} 1 & 3 \\ -1 & -2 \end{pmatrix}$. The automorphism group is of order 3.

IV. $p=23$, $G=\langle A, B, C\rangle$, with $B=\begin{pmatrix} 1 & -6 \\ 12 & -2 \end{pmatrix}$ and $C=\begin{pmatrix} 2 & 0 \\ 0 & 2 \end{pmatrix}$. Here $\langle A, B\rangle$ is of order 48, isomorphic to G in case III, so that G in case IV is the direct product of a group of order 48 with the cyclic group $\langle C\rangle$ of order 11. Hence G is soluble of order 528. The automorphism group is trivial.

V. $p=11$, as in case II, and $G=\langle A, B\rangle$, with $B=\begin{pmatrix} 2 & 4 \\ 1 & -3 \end{pmatrix}$. Here $G\cong SL_2(5)$, with $G/\mathfrak{Z}G$ isomorphic to the alternating group on 5 symbols. Hence G is non-soluble in this case, which shows that II and IV are non-isomorphic. The automorphism group is of order 5.

VI. $p=29$, $G=\langle A, B, C\rangle$, with $B=\begin{pmatrix} 1 & -7 \\ -12 & -2 \end{pmatrix}$ and $C=\begin{pmatrix} 16 & 0 \\ 0 & 16 \end{pmatrix}$. The subgroup $\langle A, B\rangle$ is isomorphic to the non-soluble group G of case V, and G in case VI is the direct product of $\langle A, B\rangle$ with $\langle C\rangle$ of order 7. The automorphism group is of order 2.

VII. $p=59$, $G=\langle A, B, C\rangle$, with $B=\begin{pmatrix} 9 & 15 \\ -10 & -10 \end{pmatrix}$ and C as in case II. Again, $\langle A, B\rangle$ is the non-soluble group G of case V, and G is the direct product of $\langle A, B\rangle$ with $\langle C\rangle$ of order 29. The automorphism group is trivial.

The following major theorem is due to ZASSENHAUS 1935a:

6. *Every finite nearfield is either regular or of one of the types* I—VII *just described.*

For the proof, see PASSMAN 1967, §§ 18—20, also WOLF 1967, Chapters 5, 6. Result **6** shows that we now have all finite planes of type IVa.2:

7. *The finite planes of type* IV a.2 *are in a one-one-correspondence*[1]) *with the non-exceptional finite nearfields, i.e. the regular nearfields* $N(m, q) \neq N(2, 3)$ *and the seven irregular nearfields* I—VII.

The collineation groups of the planes over finite nearfields were determined by ANDRÉ 1955. We note here only the following:

8. *The complete collineation group of a plane over a finite nearfield is soluble, except in four cases where the only non-abelian composition factor is the alternating group of order* 60.

These cases are (i) the planes over the three irregular nearfields V—VII, whose homology groups $\Gamma(u, ov)$ are non-soluble, and (ii) the exceptional nearfield plane of order 9 and type IV a.3: here the collineation group induces all possible permutations of the 5 point pairs on the ideal line which define the canonical involution characteristic for type IV a.3 (ANDRÉ 1955, p. 158—159). Hence in this case the symmetric group of degree 5 is a factor group of the collineation group.

This concludes our discussion of finite nearfields and the planes coordinatized by them. For the remainder of this section, we will be concerned with collecting the known planes of type IV a.1; but we shall have no completeness theorem like **5** or **7** here.

We begin with a construction, due to ANDRÉ 1954a, p. 185, of a family $\lambda(t)$ of linear automorphisms satisfying the conditions (5.1.9) and (5.1.10), and thus defining a quasifield. As above, let $\mathfrak{F} = GF(p^{md}) = GF(q^m) \supseteqq \mathfrak{K} = GF(p^d) = GF(q)$ and let A be the (cyclic) group of those automorphisms of $\mathfrak{F}$ which fix $\mathfrak{K}$ elementwise. For every $t \neq 0$ in $\mathfrak{F}$, define

$$N(t) = \prod_{\alpha \in \mathsf{A}} t^{\alpha}.$$

It is easily checked that $t \to N(t)$ is an endomorphism of $\mathfrak{F}^{\times}$. Let μ be an arbitrary mapping of the elements of the form $N(t)$ into A such that $1^{\mu} = 1$, the identity automorphism. Then the automorphisms $\lambda(t)$ defined by

$$\lambda(t) = N(t)^{\mu}, \quad 0 \neq t \in \mathfrak{F}, \tag{3}$$

satisfy (5.1.9) and (5.1.10), so that $x \circ y = x^{\lambda(y)} y$ is a quasifield multiplication. We denote the quasifield so defined by $A(m, q, \mu)$.

9. *Unless* μ *maps every* $N(t)$ *onto the identical automorphism,* $A(m, q, \mu)$ *is a quasifield but not a semifield. The plane coordinatized by* $A(m, q, \mu)$ *is of one of the types* IV a.

That such a plane is not of type V can be proved by showing that every ideal point is moved by some collineation.[2]) The plane may be

[1]) Recall here result **3.1.34**.

[2]) This is an unpublished result of A. A. Albert and D. R. Hughes.

of type IV a.1, 2, or 3, for many of the regular nearfields $N(m, q)$ discussed above, but not all of them[1]), are among the $A(m, q, \mu)$: if μ is taken to be a homomorphism from $N(\mathfrak{F}^\times)$ into A, then (5.1.12) holds, and $A(m, q, \mu) \cong N(m, q)$.

We remark here that the duals of certain of the projective planes under **9** have been defined in another way and investigated by PIERCE 1961, under the name of "finite Moulton planes". A geometric characterization of these planes was given by PICKERT 1964.

Next, we define a class of quasifields constructed by FOULSER 1962, p. 154, 1967a; this may be considered a generalization of the above construction of ANDRÉ 1954a. Let $\mathfrak{F}$, $\mathfrak{K}$, and A be as above, and choose a divisor $g > 1$ of m such that every prime divisor of g divides $q - 1$. Furthermore, let r and s be integers with $(r, g) = (s, m) = 1$. If ν is the map from $Z(q^m - 1)$ onto $Z(g)$ defined by $\nu(i) \equiv i \bmod g$, then

$$f(i) \equiv s\,\nu(r\,i) \bmod m \tag{4}$$

defines a map f from $Z(q^m - 1)$ into $Z(m)$ with properties (5.1.14) and (5.1.15), and thus a quasifield which we will denote by $F(m, q, g)$.

10. *The plane over $F(m, q, g)$ is of one of the types* IV a, *and it can also be coordinatized by $A(m, q, \mu)$ as defined by* (3) *if, and only if, $F(m, q, g) \cong A(m, q, \mu)$.*

For the proof, see FOULSER 1962, Chapter 4.3.2, or FOULSER 1967a, b.

The next class of quasifields to be discussed here is due to HALL 1943.[2]) We do not describe these by their respective sets $\{\lambda(t) : t \neq 0\}$, but in a slightly modified form of Hall's original fashion. Let $\mathfrak{K}$ be a Galois field of order $q > 2$ and suppose that the quadratic polynomial $f(x) = x^2 - r\,x - s$ is irreducible over $\mathfrak{K}$. In the vector space $V(2, q)$ of ordered pairs of elements in $\mathfrak{K}$, define a multiplication as follows:

$$(x, y)\,(u, v) = \begin{cases} (x\,u,\, y\,u) & \text{if } v = 0, \\ (x\,u - y\,v^{-1} f(u),\, x\,v - y\,u + y\,r) & \text{if } v \neq 0. \end{cases} \tag{5}$$

With this multiplication and vector space addition, $V(2, q)$ becomes a quasifield which we denote by $H(q, f)$; its kernel is $\{(x, 0) : x \in \mathfrak{K}\} \cong \mathfrak{K}$. Note that

$$f((x, y)) = (x, y)\,(x, y) - (r, 0)\,(x, y) - (s, 0) = 0$$

whenever $y \neq 0$.

---

[1]) The statement in HUGHES 1959b, p. 49, that the regular nearfields are precisely the associative systems $A(m, q, \mu)$, is incorrect: $N(m, q)$ can be of the form $A(m, q, \mu)$ only if m divides $q - 1$.

[2]) See also HALL 1959, Theorem 20.4.7, where these quasifields are called "Hall systems".

The planes over the quasifields $H(q, f)$ constitute one of the most thoroughly investigated classes of non-desarguesian finite planes. We list a few results of these investigations.

11. *$H(q, f)$ is the exceptional nearfield if $q = 3$ and $f(x) = x^2 + 1$; otherwise $H(q, f)$ is non-associative and non-distributive. $H(q_1, f_1)$ and $H(q_2, f_2)$, not necessarily isomorphic, coordinatize isomorphic planes if and only if $q_1 = q_2$.*

For proofs see HALL 1943; HUGHES 1959a; PANELLA 1959, 1960a, b. In the case $H(q, f) \neq N(2, 3)$, the ideal points of these planes are permuted in two orbits of lengths > 1 by the collineation group; hence type V cannot occur. The planes over $H(q, f)$ can also be coordinatized by certain $A(2, q, \mu)$, and they are characterized within this larger class of planes as follows:

12. *A plane over $A(2, q, \mu)$ can be coordinatized by an $H(q, f)$ if, and only if, its full collineation group is non-soluble.*

This was proved by HUGHES 1959a, Theorem 3.1.[1]) The collineation groups of the planes over $H(q, f)$ involve the simple groups $PSL_2(q)$. PANELLA 1960a has investigated the other quasifields coordinatizing the planes over $H(q, f)$.

The following are some results on the subplane structure of the planes over $H(q, f)$.

13. *Let **P** be a projective plane over a quasifield $H(q, f)$, of order $q^2 = p^{2d}$. Then:*

(a) ***P** contains desarguesian Baer subplanes.*

(b) ***P** contains subplanes isomorphic to **P**(2) as well as quadrangles with non-collinear diagonal points.*

(c) *If p is odd, then **P** is generated by a quadrangle.*

(d) *If $p = 2$, then **P** contains a quadrangle generating a non-desarguesian subplane **Q** which contains a **P**(2).*

Result (**a**) is clear: the subplane over the kernel of $H(q, f)$ serves. Note that this implies, by **5.1.11**, that the plane **A** over $H(q, f)$ is derivable. In fact:

14. *𝔡**A** is the desarguesian plane **A**(q^2), and conversely 𝔡**A**(q^2) is the plane over $H(q, f)$.*

(ALBERT 1961b). This was mentioned briefly in Section 5.1, as was the next result, due to PANELLA 1959:

---

[1]) See also FOULSER 1967a. We remark here that the determination by HUGHES 1959a of the full collineation group of the planes over $H(q, f)$ is incomplete; for a complete account, see PANELLA 1960b.

15. *Let $\mathcal{S}$ be a regular 1-spread in $\mathcal{P}(3, q)$, $q > 2$, and let $\mathcal{S}'$ be the spread defined by* (5.1.18). *Then the plane defined by $\mathcal{S}'$ can be coordinatized by $H(q, f)$.*

We return to **13**. The proof of (**b**) was given by H. NEUMANN 1954; note that GLEASON's 1956 theorem **3.4.23** implies that not all quadrangles can have collinear diagonal points. The proof of (**c**) is in COFMAN 1964; a weaker form of this, as well as (**d**), were proved earlier by WAGNER 1956. In this context, compare also LENZ 1953 and KILLGROVE 1964, 1965.

Next, we discuss a class of quasifields introduced by LÜNEBURG 1965c, Section 6; the corresponding translation planes were mentioned repeatedly in Chapter 4. Let $\mathfrak{K} = GF(q)$ with $q = 2^d$, d odd and > 1. Also, let σ be the unique automorphism of $\mathfrak{K}$ with

$$x^{\sigma^2} = x^2 \quad \text{for all} \quad x \in \mathfrak{K},$$

viz., $x \to x^{2^{(d+1)/2}}$. In $V(2, q)$, we then define a multiplication by[1])

$$(x, y)(u, v) = (x u + y(u^\sigma + v^{\sigma+1}), x v + u y). \tag{6}$$

This turns $V(2, q)$ into a quasifield which we denote by $L(q)$. As in the case of the $H(q, f)$, the kernel is $\{(x, 0) : x \in \mathfrak{K}\} \cong \mathfrak{K}$. The planes over these quasifields can be characterized as follows:

16. *A translation plane* **A** *can be coordinatized by a quasifield of the form $L(q)$ if, and only if,*

(**a**) **A** *is finite of order $q^2 = 2^{2d}$, with odd $d > 1$,*
(**b**) *The kernel $\mathfrak{K}$(**A**) is $GF(q)$, and*
(**c**) **A** *admits a collineation group isomorphic to $Sz(q)$.*

This is Satz 10 of LÜNEBURG 1965c. In particular, the collineation group is always non-soluble. Using **4.2.16**, we can further conclude that the collineation group is doubly transitive on the ideal points; no other non-desarguesian finite plane with this property is known. Also, (6) shows that the distributor of $L(q)$, as defined by (3.1.42), coincides with $\mathfrak{K}$ and is therefore non-trivial. Thus **3.1.30** implies:

17. *The group of shears with fixed axis, in the affine plane over $L(q)$, has order q.*

LÜNEBURG 1965c, Section 6, has determined the full collineation groups Γ of these planes. If Δ is the (soluble) group of all dilatations and Π a semi-direct product of Δ with $Sz(q)$, then there is a normal chain

$$1 \lhd \Delta \lhd \Pi \lhd \Gamma$$

with $\Gamma/\Pi \cong \operatorname{Aut} GF(q)$.

---

1) Lüneburg's formula is slightly more complicated than (6). Another equivalent definition appears in OSTROM 1966a, p. 203.

As $L(q)$ has rank 2 over its kernel, the translation planes over $L(q)$ can be represented by a 1-spread in $\mathscr{P}(3, q)$, $q = 2^d$. These spreads first appeared in TITS 1962b; we have, in fact, mentioned them in **1.4.56b**: They consist of the ψ-absolute lines with respect to a polarity ψ of the configuration $\mathbf{W} = \mathbf{W}(q)$ formed by the absolute points and totally isotropic lines of a symplectic polarity in $\mathscr{P}(3, q)$.

We conclude by mentioning briefly the few other known non-desarguesian finite planes whose collineation groups are transitive on the ideal points. They are all translation planes, and besides the planes of Lüneburg just discussed, there are just four more. One of them was already mentioned: the plane over the exceptional nearfield of order 9, of type IVa.3. The other three are of type IVa.1 again. Two of them are of order 25; they were discovered by FOULSER 1964b, Section 6. The groups of these planes are actually flag transitive; cf. **4.4.23**.

The last plane to be discussed here, of order 27, was found by C. Hering, and as Hering's result is so far unpublished, we present the definition of his plane here, in terms of a 2-spread in $\mathscr{P}(5, 3)$. Let U_i, V_i $(i = 1, 2)$ be the subspaces of $V(6, 3)$, defined by basis vectors as follows:

$$U_1 = \langle (0,0,0,1,0,0),\ (0,0,0,0,1,0),\ (0,0,0,0,0,1)\rangle,$$

$$U_2 = \langle (1,0,0,1,0,-1),\ (0,1,0,0,-1,0),\ (0,0,1,-1,0,0)\rangle,$$

$$V_1 = \langle (1,0,0,0,0,0),\ (0,1,0,0,0,0),\ (0,0,1,0,0,0)\rangle,$$

$$V_2 = \langle (1,0,0,1,0,1),\ (0,1,0,0,1,0),\ (0,0,1,1,0,0)\rangle.$$

Also, let α be the linear automorphism of $V(6, 3)$ defined by the matrix

$$\begin{pmatrix} 0 & -1 & 0 & 0 & 0 & 0 \\ -1 & 0 & -1 & 0 & 0 & 0 \\ 1 & 0 & 0 & 0 & 0 & 0 \\ 0 & 0 & 0 & -1 & -1 & 0 \\ 0 & 0 & 0 & 0 & 0 & -1 \\ 0 & 0 & 0 & 1 & 0 & 0 \end{pmatrix}.$$

Then the subspaces U_1, V_1, $U_2\alpha^i$, $V_2\alpha^i$ $(0 \leqq i \leqq 12)$ form a congruence partition of $V(6, 3)$, and the corresponding 2-spread in $\mathscr{P}(5, 3)$ defines the desired translation plane of order 27. The full collineation group of this plane is doubly transitive on the affine points.

5.3 Planes of type V

In this section we complete our survey of the known finite quasifields, by listing finite semifields. Geometrically, this means that we collect all known finite planes of Lenz-Barlotti type V.1: these are precisely the planes coordinatizable by semifields which are not fields. Apart from just listing special construction techniques, we present also some general theory on finite semifields and their planes.[1])

By a *pre-semifield* we mean (with KNUTH 1965b, p. 185) a finite[2]) set $\mathfrak{S}$, together with addition and multiplication, such that

(1) *$\mathfrak{S}$ is a group with respect to addition,*[3])

(2) $x(y+z) = xy + xz$ *and* $(x+y)z = xz + yz$ *for all* $x, y, z \in \mathfrak{S}$, and

(3) *if* $xy = 0$, *then* $x = 0$ *or* $y = 0$.

It is clear that the finite semifields[4]) are precisely those pre-semifields which have a multiplicative identity element, in other words for which the following holds:

(4) *There exists* $1 \in \mathfrak{S}$ *such that* $1x = x1 = x$ *for all* $x \in \mathfrak{S}$.

In what follows, it will sometimes be more convenient to work with pre-semifields than semifields. It should be noted, however, that pre-semifields are not really more general than semifields as far as the construction of projective planes is concerned:

1. *Let $\mathfrak{S}$ be a pre-semifield and $0 \neq e \in \mathfrak{S}$. Define a new multiplication "$\circ$" among the elements of $\mathfrak{S}$ by*

(5) $$ae \circ eb = ab.$$

Then with this multiplication and the given addition, $\mathfrak{S}$ becomes a semifield whose identity element is ee.

(KNUTH 1965b, p. 204.) This semifield is isotopic to the given pre-semifield, in the sense of Section 3.1. By considering isotopy, one can in fact find many semifields from a given pre-semifield, but by **3.1.32** these all coordinatize the same plane. For a more thorough discussion of results and problems related to isotopy, see ALBERT 1960.

The *left, middle,* and *right nucleus* of a pre-semifield $\mathfrak{S}$ are defined, respectively, as follows:

$$\mathfrak{L} = \mathfrak{L}(\mathfrak{S}) = \{l \in \mathfrak{S} : l(xy) = (lx)y \text{ for all } x, y \in \mathfrak{S}\},$$
$$\mathfrak{M} = \mathfrak{M}(\mathfrak{S}) = \{m \in \mathfrak{S} : x(my) = (xm)y \text{ for all } x, y \in \mathfrak{S}\},$$
$$\mathfrak{R} = \mathfrak{R}(\mathfrak{S}) = \{r \in \mathfrak{S} : x(yr) = (xy)r \text{ for all } x, y \in \mathfrak{S}\}.$$

[1]) Some such results, independent of finiteness, were already given in Section 3.1.

[2]) Finiteness is essential here, because of condition (3) below. The notion of pre-semifield can, however, be extended to the infinite case by strengthening (3) to say that $ax = b$ and $ya = b$ have unique solutions x, y whenever $a \neq 0$.

[3]) This group is commutative, as follows from (2) and (3) below.

[4]) These were defined on p. 129.

Also, the *centre* of $\mathfrak{S}$ is

$$\mathfrak{Z} = \mathfrak{Z}(\mathfrak{S}) = \{z \in \mathfrak{S} : z\,x = x\,z \text{ for all } x \in \mathfrak{S}\}.$$

Each of $\mathfrak{L}, \mathfrak{M}, \mathfrak{N}$ is a Galois field (but the identity elements of these do not, in general, act as an identity for $\mathfrak{S}$). The centre $\mathfrak{Z}$ need not be a field (there exist commutative semifields which are not associative; see **6** below), but each of the intersections $\mathfrak{Z} \cap \mathfrak{L}$, $\mathfrak{Z} \cap \mathfrak{M}$, $\mathfrak{Z} \cap \mathfrak{N}$ is a Galois field. All these subfields of $\mathfrak{S}$ have the same characteristic p, and $\mathfrak{S}$ can be regarded as a (not necessarily associative) algebra over $GF(p)$. This means that $\mathfrak{S}$ is a vector space over $GF(p)$, and

$$(6) \qquad t(x\,y) = (t\,x)\,y = x(t\,y) \quad \text{for} \quad t \in GF(p),\; x, y \in \mathfrak{S}.$$

Suppose now that the pre-semifield $\mathfrak{S}$ is an algebra of rank m over $\mathfrak{K} = GF(q)$. (In addition to the case $q = p$ just mentioned we also have this situation whenever $\mathfrak{K} \subseteq \mathfrak{L} \cap \mathfrak{M} \cap \mathfrak{N} \cap \mathfrak{Z}$.) Let $(e_1, \ldots, e_m)$ be a basis of $\mathfrak{S}$ over $\mathfrak{K}$, and put

$$(7) \qquad e_i\,e_j = \sum_{k=1}^{m} a_{ijk}\,e_k \qquad (i, j = 1, \ldots, m;\; a_{ijk} \in \mathfrak{K}).$$

It follows from (2) and (6) that the *cubical array* A of the m^3 elements $a_{ijk} \in \mathfrak{K}$ determines the multiplication in $\mathfrak{S}$ uniquely:

$$(8) \qquad \left(\sum_{i=1}^{m} x_i\,e_i\right)\left(\sum_{j=1}^{m} y_j\,e_j\right) = \sum_{k=1}^{m}\left(\sum_{i,j=1}^{m} x_i\,y_j\,a_{ijk}\right) e_k.$$

Also, condition (3) for pre-semifields implies that $A = (a_{ijk})$ is *nonsingular* in the sense that

(9) *for any m-vector $(x_1, \ldots, x_m) \neq 0$, the matrix (c_{jk}) defined by $c_{jk} = \sum_{i=1}^{m} x_i\,a_{ijk}$ is nonsingular.*[1]

The converse is likewise true:

2. *Let $A = (a_{ijk})$, $i, j, k = 1, \ldots, m$, be a cubical array of elements $a_{ijk} \in GF(q)$, and define a multiplication in the vector space $V(m, q)$ by (8). This turns $V(m, q)$ into a pre-semifield if and only if A is nonsingular.*

Proof: Knuth 1965b, Theorem 4.4.1. In the same paper, Knuth shows that the special role of the first subscript i of a_{ijk} in (9) is inessential: Let σ denote any permutation of $\{1, 2, 3\}$, and let A^σ be the cubical array obtained from A by applying σ to the subscripts [for example, if $\sigma = (123)$, then $a^\sigma_{ijk} = a_{jki}$]. Then A^σ is nonsingular if

[1] This implies, of course, that for every i the matrix $A_i = (a_{ijk})$ is nonsingular in the usual sense. On the other hand, even if all these matrices are nonsingular, the cubical array A may be singular.

and only if A is nonsingular.[1]) Hence, by applying all six permutations in the symmetric group S_3 to the subscripts of a nonsingular cubical array A, we get six such arrays. Each of these defines, by **2**, a pre-semifield, and from **1** we then get six semifields, and hence six planes of type V. If **P** is one of these projective planes, then the dual $\overline{\mathbf{P}}$ is also among them: the permutation (12) replaces a_{ijk} by a_{jik}, so that, by (8), the pre-semifields defined by A and $A^{(12)}$ have their respective multiplications "$\cdot$" and "$*$" related by

$$x * y = y \cdot x.$$

This carries over to the corresponding semifields; but then the planes coordinatized by these are dual to each other (cf. **3.1.36** above).

A similar construction, with less obvious geometric significance, was investigated by Knuth 1965b, Section 5: Let A be a nonsingular cubical array, $\mathfrak{S}$ the corresponding pre-semifield, and **P** the corresponding projective plane. Then consider the permutation $\tau = (23) \in S_3$. The cubical array A^τ defines (uniquely) another plane $\mathbf{P}^\tau$, called by Knuth the *transpose* of **P**. Since $\sigma = (12)$ and $\tau = (23)$ generate all of S_3, we get:

3. *Let* **P** *be a finite projective plane of type* V.1. *Then a series of at most six nonisomorphic planes of type* V.1 *can be generated from* **P** *by repeated dualization and transposition. The collineation groups of these planes all have the same order.*

For this, see Knuth 1965b, Theorems 5.2.1 and 5.3.1. The collineation groups mentioned in **3** need not be all isomorphic.

The affine plane **A** over a semifield $\mathfrak{S}$ admits the translations

$$\tau(a, b)\colon (x, y) \to (x + a, y + b)$$

and the shears with axis $(x = 0)$:

$$\sigma(c)\colon (x, y) \to (x, xc + y).$$

These collineations together generate a metabelian normal subgroup Π of the full collineation group Γ of **A**, and in fact Π has a complement A in Γ:

$$\Pi \mathsf{A} = \Gamma, \quad \Pi \cap \mathsf{A} = 1.$$

The nature of Π was investigated by Cronheim 1965. The complement A is isomorphic to the stabilizer of the points $(0, 0)$, (0), and v, which is essentially identical with the autotopism group of $\mathfrak{S}$; cf. **3.1.33**. It

[1]) This is a special case of Theorem 4.3.1 of Knuth 1965b, where arrays of higher dimension d are also investigated. Ordinary matrix theory appears here as the special case $d = 2$; for example, the matrix analogue of the result stated in the text is the fact that a matrix is nonsingular if and only if its transpose is nonsingular.

has been conjectured that the autotopism group of a finite semifield is always soluble. For most of the semifields to be listed below, this conjecture has been verified, but in the general case the best known result is:

4. *Let* **P** *be a finite projective plane of type* V.1, *coordinatized by the semifield* $\mathfrak{S}$ *of order* n. *Suppose that either*
(**a**) n *is even, and* **P** *has no Baer subplane, or*
(**b**) n *is odd, and* $\mathfrak{S}$ *has odd rank over one of its nuclei.*

Then the autotopism group of $\mathfrak{S}$, *and hence the full collineation group of* **P**, *is soluble.*

This result is due to BURMESTER 1964a and BURMESTER & HUGHES 1965, Theorem 1.[1]) In particular, Aut**P** is soluble if n is not a square.

Another simple fact about collineations in semifield planes is:

5. *Let* **P** *be finite of type* V.1 *and let* Γ *be a group of homologies of* **P** *which all have the same axis (and, therefore, the same center). Then* Γ *is cyclic.*

For such a group must be isomorphic to a subgroup of the multiplicative group of one of the nuclei of any coordinatizing semifield (cf. **3.1.28**–**3.1.30**).

Before proceeding to actual construction techniques, we discuss briefly the case where $\mathfrak{S}$ is a special type of quadratic extension of a Galois field. We say that the subfield $\mathfrak{F}$ of the semifield $\mathfrak{S}$ is a *weak nucleus* provided that

(10) $(x\,y)\,z = x\,(y\,z)$ *whenever at least two of* x, y, z *are in* $\mathfrak{F}$.

It is clear that $\mathfrak{S}$ must then be a vector space over $\mathfrak{F}$. However, $\mathfrak{S}$ need not be an algebra over $\mathfrak{F}$: (6) may be violated when $GF(p)$ is replaced by $\mathfrak{F}$. Note that the intersection of any two of the nuclei $\mathfrak{L}$, $\mathfrak{M}$, $\mathfrak{R}$ are weak nuclei. On the other hand, there exist examples (all of order 16; we shall see later that this is the least possible order of a proper semifield) where $\mathfrak{F}$ is not contained in any of $\mathfrak{L}$, $\mathfrak{M}$, $\mathfrak{R}$, and others where $\mathfrak{L}$ is not a weaknucleus (KNUTH 1965b, p. 212).

We consider now the situation in which the finite[2]) semifield $\mathfrak{S}$ is of rank 2 over its weak nucleus $\mathfrak{F}$. If $e \notin \mathfrak{F}$, then $\{1, e\}$ is a basis of $\mathfrak{S}$ over $\mathfrak{F}$, so that every element of $\mathfrak{S}$ is of the form $x + y\,e$, with $x, y \in \mathfrak{F}$. It can then be proved (KNUTH 1965, Theorem 7.2.1) that, for suitable

---

[1]) It is, in fact, a special case of a more general theorem on the solubility of the autotopism group of a quasifield which need not be a semifield (BURMESTER & HUGHES 1965, Theorem 2); but in this more general case, the full group Aut**P** may be non-soluble even when the autotopism group is soluble. The main tool for the Burmester-Hughes theorem is the solubility of groups of odd order (FEIT & THOMPSON 1963).

[2]) Finiteness is quite essential for the proof of (11) below.

choice of e,

(11) $\quad e\,t = t^\sigma e$ *for all* $t \in \mathfrak{F}$ *and fixed* $\sigma \in \operatorname{Aut} \mathfrak{F}$.

This implies

(12) $\quad (x + y\,e)\,(u + v\,e) = x\,u + (x\,v + y\,u^\sigma)\,e + (y\,e)\,(v\,e)$,

so that multiplication in $\mathfrak{S}$ is determined by σ and the products of the form $(y\,e)\,(v\,e)$. KNUTH 1965b, Section 7.3, has investigated under what circumstances the definition

(13) $$(y\,e)\,(v\,e) = y^\alpha v^\beta f + (y^\gamma v^\delta g)\,e,$$

with fixed $f, g \in \mathfrak{F}$ and $\alpha, \beta, \gamma, \delta \in \operatorname{Aut} \mathfrak{F}$, satisfies the critical axiom (3) and hence defines a semifield multiplication. The necessary and sufficient condition for this is clearly the following:

(14) *If* $x\,u + y^\alpha v^\beta f = 0$ *and* $x\,v + y\,u^\sigma + y^\gamma v^\delta g = 0$, *then either* $x = y = 0$ *or* $u = v = 0$.

This leads to our first actual constructions:

6. *Let q be an odd prime power and f a non-square in $GF(q)$. Also suppose that $\alpha, \beta, \sigma \in \operatorname{Aut} GF(q)$ are not all the identity. Then the multiplication defined by*

(15) $\quad (x + y\,e)\,(u + v\,e) = x\,u + y^\alpha v^\beta f + (x\,v + y\,u^\sigma)\,e$

turns $V(2, q)$ into a semifield of order q^2 which is not a field.

This follows from (11)—(14) with $g = 0$. The condition that at least one of α, β, σ is not the identity is necessary to assure the non-associativity of the multiplication (15). In particular, we must have $q = p^d$ with $d \geqq 2$; hence all semifields of this form are of order $p^{2d} \geqq p^4 \geqq 81$. The case $\alpha = \beta \neq 1 = \sigma$ was first exhibited by DICKSON 1905; these are precisely the commutative ones among the semifields given by **6**. The autotopism groups of Dickson's semifields were determined and shown to be soluble by BURMESTER 1962 and SANDLER 1962c.

The case $g \neq 0$ is more complicated. Here the following has been proved by KNUTH 1965b, Section 7.4:

7. *Let f and g be non-zero elements in $\mathfrak{F} = GF(q)$, and $1 \neq \sigma \in \operatorname{Aut} \mathfrak{F}$ such that the polynomial*

(16) $$x^{\sigma+1} + g\,x - f$$

is irreducible in $\mathfrak{F}$. Then each of the following defines a semifield multiplication in $V(2, q)$:

(17) $\quad (x + y\,e)\,(u + v\,e) = x\,u + y\,v^\sigma f + (x\,v + y\,u^\sigma + y\,v^\sigma g)\,e$;

(18) $\quad (x + y\,e)\,(u + v\,e) = x\,u + y^{\sigma^{-2}} v^{\sigma^{-1}} f + (x\,v + y\,u^\sigma + y^{\sigma^{-1}} v\,g)\,e$;

(19) $\quad (x + y\,e)\,(u + v\,e) = x\,u + y\,v^{\sigma^{-1}} f + (x\,v + y\,u^\sigma + y\,v\,g)\,e$.

Moreover, the semifields defined by (17), (18), (19) *are precisely those of rank* 2 *over* $\mathfrak{L} = \mathfrak{M}$, $\mathfrak{R} = \mathfrak{M}$, $\mathfrak{L} = \mathfrak{R}$, *respectively*.

That (18) characterizes the case $\mathfrak{F} = \mathfrak{R} = \mathfrak{M}$ had been previously shown by HUGHES & KLEINFELD 1960.[1]) Besides (17)—(19), Knuth gives one more possibility, namely

(20) $$(x + y\,e)\,(u + v\,e) = x\,u + y^{\sigma^{-2}}\,v^{\sigma}\,f + (x\,v + y\,u^{\sigma} + y^{\sigma^{-1}}\,v^{\sigma}\,g)\,e;$$

there seems to be no characterization of these semifields comparable to those given in **7**. The semifields defined by (17)—(20) are again of order $p^{2d} \geqq p^4$, but in contrast to **6** the case $p = 2$ is not excluded here. Thus these constructions also yield semifields of order 16.[2]) For the types (17) and (18), HUGHES 1960a has proved that the autotopism groups, and hence the collineation groups of the corresponding planes, are soluble.[3]) For (19) and (20), the autotopism groups do not appear to have been investigated so far.

We turn now to constructions of semifields which are not necessarily of rank 2 over some subfield. First, we give two constructions of ALBERT 1952, 1958, 1960. Consider the prime power $q = p^d$, and let $m > d$. In $\mathfrak{F} = GF(p^m)$, consider an element c which is not a $(q-1)$th power. Such an element exists if and only if $q - 1$ and $p^m - 1$ have a common factor >1; this is trivially the case if $p > 2$, but for $p = 2$ we need the additional condition $(d, m) > 1$. The multiplication "$\circ$" defined by

(21) $$x \circ y = x\,y^q - c\,x^q\,y$$

then turns $\mathfrak{F}$ into a pre-semifield, and with the help of **1** we obtain a semifield $T(p^d, p^m, c)$. These semifields are also called *twisted fields*.

8. *If* $m = k\,d$ *and* $k > 2$, *then* $T(p^d, p^m, c)$ *is non-associative. If in addition* $c \neq -1$, *then* $T(p^d, p^m, c)$ *is non-commutative.*

For the proof, see ALBERT 1958. The planes coordinatized by twisted fields again have soluble collineation groups; this was shown by ALBERT 1963.

---

[1]) The multiplication formula given by Hughes and Kleinfeld differs from (18) but is easily proved to be equivalent to (18). The difference is due to the fact that they write the elements of $\mathfrak{S}$ as $x + e\,y$ rather than $x + y\,e$.

[2]) This is the least possible order for a proper semifield; see **10** below. The semifields of order 16 are all known (KLEINFELD 1960); there are 23 of them, falling into two isotopy classes (see also KNUTH 1965b, Section 6.2). Hence there are two non-isomorphic projective planes of order 16 and type V.1; each of these is self-dual and self-transpose.

[3]) For more details or special cases, see also ALBERT 1959 and SANDLER 1962a.

Generalized twisted fields were introduced and investigated by ALBERT 1961 a, c. Here the multiplication is not (21) but rather

$$x \circ y = x\,y - c\,x^{\alpha}\,y^{\beta}, \tag{22}$$

with $\alpha, \beta \in \mathrm{Aut}\,\mathfrak{F}$ and $c \neq x^{\alpha-1}\,y^{\beta-1}$ for all $x, y \in \mathfrak{F}$. The isotopy problem for these semifields is investigated in ALBERT 1961 c.

The next class of semifields was found by SANDLER 1962b. Let $q = p^d$ and $m > 1$, and consider the Galois fields $\mathfrak{F} = GF(q^m)$ and $\mathfrak{K} = GF(q) \subset \mathfrak{F}$. Also, let c denote an element of $\mathfrak{F}$ whose defining irreducible polynomial in $\mathfrak{K}$ has degree $\geqq m$, so that $f(c) \neq 0$ for all polynomials f of degree $< m$ in $\mathfrak{K}$. For example, a primitive element of $\mathfrak{F}$ satisfies this condition. Finally, let $\alpha : x \to x^q$; this is the generating automorphism of the subgroup of $\mathrm{Aut}\,\mathfrak{F}$ fixing $\mathfrak{K}$ elementwise. Then a multiplication "$\circ$" is introduced on the vector space $V(m, q^m)$ as follows: Let $a_0, \ldots, a_{m-1}$ be a basis of $V(m, q^m)$, and define

$$\begin{aligned} a_i \circ x &= x^{\alpha^i}\,a_i && \text{for all } x \in \mathfrak{F}, \\ a_i \circ a_j &= \begin{cases} a_{i+j} & \text{if } i + j < m, \\ c\,a_{i+j-m} & \text{if } i + j \geqq m. \end{cases} \end{aligned} \tag{23}$$

Also, we require the distributive laws (2) for the multiplication "$\circ$". It can then be verified that (23) determines a unique semifield structure on $V(m, q^m)$, with identity element $a_0 = 1$. We denote this semifield by $S(m, q, c)$; it has q^{m^2} elements, and its kernel (left nucleus) is $\mathfrak{K} = GF(q)$. The autotopism groups for $S(m, q, c)$ were determined in case $m = 2$ by SANDLER 1962b; they are again soluble. Sandler's paper also contains information about the autotopism groups in case $m > 2$.

The last class of semifields to be exhibited here is due to KNUTH 1965 a; these semifields are all of characteristic 2. Let m be odd and $q = 2^d$, and consider an arbitrary linear function f from $\mathfrak{F} = GF(q^m)$ into $\mathfrak{K} = GF(q)$. Then the new multiplication "$\circ$" in $\mathfrak{F}$ defined by

$$x \circ y = x\,y + \big(f(x)\,y + f(y)\,x\big)^2 \tag{24}$$

turns $\mathfrak{F}$ into a pre-semifield, and with the help of **1** we obtain a commutative semifield $K(d, m, f)$ of order 2^{dm}. In special situations, for example if $f = 0$ or $dm = 3$, we have $K(d, m, f) \cong GF(2^{dm})$. In general, however, $K(d, m, f)$ is a proper semifield:

9. *If f and g are linear functions from $\mathfrak{F} = GF(2^{dm})$ onto (not merely into) $\mathfrak{K} = GF(2^d)$, then the semifields $K(d, m, f)$ and $K(d, m, g)$ are isotopic; hence they coordinatize the same plane* **P**. *Moreover, if $dm > 3$, these semifields are non-associative, so that* **P** *is of type* V.1.

For the proof of these statements, see KNUTH 1965a, Section 2.[1]) In Theorem 10 of the same paper, Knuth shows furthermore that if $dm = d'm'$ but $d \neq d'$, then the semifields $K(d, m, f)$ and $K(d', m', f')$, with f from $GF(2^{dm})$ onto $GF(2^d)$ and f' from $GF(2^{dm})$ onto $GF(2^{d'})$, coordinatize non-isomorphic planes.

The constructions given so far establish the "if" part of the following theorem:

10. *Let p^e be an arbitrary prime power. There exists a projective plane of type* V.1 *and of order p^e if, and only if, $e \geqq 3$ for $p > 2$ and $e \geqq 4$ for $p = 2$.*

For the simple proof of the converse, see KNUTH 1965b, Section 6.1.

In the remainder of this section, we make some remarks on representations of certain planes of type V.1 by means of planar functions. Let **A** be the affine plane over the semifield $\mathfrak{S}$ of order $n = p^e$, and let a be a non-zero element of $\mathfrak{S}$ satisfying

$$x(y\,a) = y(x\,a) \qquad \text{for all } x, y \in \mathfrak{S}. \tag{25}$$

It is clear that if $\mathfrak{S}$ is commutative, then every $a \in \mathfrak{N}(\mathfrak{S})$ satisfies (25); but in the noncommutative case the existence problem for such an a seems to be open. It is straightforward to verify that the permutations $\varphi(u, v)$ of **A** defined by

$$(x, y)^{\varphi(u,v)} = (x + u, x(u\,a) + y + v) \tag{26}$$

form an abelian collineation group Γ of order n^2 which, when considered as a group on the corresponding projective plane **P**, satisfies conditions (**a**) and (**c**) of **5.1.13**. Conversely, every abelian collineation group satisfying (**a**) and (**c**) can be represented in this fashion (see DEMBOWSKI 1965c, (4.1) for the non-associative case, and DEMBOWSKI & OSTROM 1968, Corollary 1 for the associative case). Furthermore,

$$\varphi(u, v)^i = \varphi\left(i\,u, \frac{i(i-1)}{2}\,u(u\,a) + i\,v\right) \qquad \text{for } i = 1, 2, \ldots; \tag{27}$$

this implies that Γ is elementary abelian if and only if n is odd, i.e. $\mathfrak{S}$ of characteristic $\neq 2$. But in this case where Γ is even elementary

[1]) Since m is odd and $dm > 3$, we have $dm \geqq 5$, so that $|K(d, m, f)| = 2^{dm} \geqq 32$. The semifields of order 32 were tabulated by WALKER 1963. There are five distinct projective planes of order 32 and type V.1; some of these may be coordinatized by as many as 31^2 distinct (isotopic but not isomorphic) semifields of order 32. The first such semifield was discovered by D. E. Knuth on a digital computer, and from this example the $K(d, m, f)$ arose as a generalization. Compare also ALBERT 1962.

abelian, condition (**b**) of **5.1.13** is clearly also satisfied, so that **A** can be represented by a planar function $f\colon \mathfrak{S}^+ \to \mathfrak{S}^+$.

We present some planar functions of this kind, in the case where $\mathfrak{S}$ is commutative. As we are assuming n odd, the element $a = 1 + 1$ satisfies condition (25) in this case, and the corresponding group Γ is the direct product of $\Phi = \{\varphi(u, u^2) \vdots u \in \mathfrak{S}\}$ and $\Pi = \{\varphi(0, v) \vdots v \in \mathfrak{S}\}$. It is not difficult to derive[1]) from this observation that the plane over $\mathfrak{S}$ is then represented by the planar function

$$f(x) = x^2, \tag{28}$$

where x^2 denotes the square with respect to the semifield multiplication in $\mathfrak{S}$.

We conclude with two special cases of (28). Firstly, let $\mathfrak{K} = GF(p^d)$, $p \neq 2$, c a non-square in $\mathfrak{K}$, and $1 \neq \alpha \in \operatorname{Aut}\mathfrak{K}$. Then

$$f((x, y)) = (x^2 + y^{2\alpha} c,\ 2xy) \tag{29}$$

defines a planar function from $G = \mathfrak{K}^+ \oplus \mathfrak{K}^+$ into G, representing the plane over Dickson's semifield of order p^{2d}, defined by (15) with $\alpha = \beta \neq 1 = \sigma$.

The second special situation is the following. Let $\mathfrak{F} = GF(p^e)$, $p \neq 2$, and suppose that

$$x \circ y = \sum_{i,j=0}^{e-1} a_{ij}\,(x^{p^i} y^{p^j} + x^{p^j} y^{p^i}) \tag{30}$$

turns $\mathfrak{F}$ into a pre-semifield.[2]) Then

$$f(x) = \sum_{i,j=0}^{e-1} a_{ij}\, x^{p^i + p^j} \tag{31}$$

is a planar function representing the plane determined by this pre-semifield. Finally, we observe that if $k = 0$ or $(k, e) = 1$, and if $a_{k0} = 1$ and $a_{ij} = 0$ for $(i, j) \neq (k, 0)$, then (30) becomes $x \circ y = x^{p^k} y + y^{p^k} x$, which is (21) with $q = p^k$ and $c = -1$. The corresponding planar function is

$$f(x) = x^{p^k + 1} \tag{32}$$

with either $k = 0$ (in which case we have (28) again, but here with Galois field multiplication) or $(k, e) = 1$; these conditions on k are necessary and sufficient that the automorphism $\sigma\colon x \to x^{p^k}$ of $\mathfrak{F}$ have 0 as the only solution of $x + x^\sigma = 0$.

---

[1]) For more details, see the proof of Theorem 5 in DEMBOWSKI & OSTROM 1968.

[2]) It is clear that (30) always satisfies the distributive laws (2); thus (3) is the only property that must be stipulated.

5.4 Planes of types I and II

There are only a few classes of known finite projective planes which are not (L, L)- or (p, p)-transitive for some line L or point p. We shall list these in this section; they are all of Lenz-Barlotti types I.1 and II.1.

The first class of planes to be discussed here was discovered by HUGHES 1957b.[1] The construction is based on a nearfield $\mathfrak{N}$ of order q^2 whose kernel $\mathfrak{K}$ is of order q and coincides with the centre of $\mathfrak{N}$. From the classification of finite nearfields (see Section 5.2) it follows that q must be an odd prime power. Consider the additive group $V = \mathfrak{N}^+ \oplus \mathfrak{N}^+ \oplus \mathfrak{N}^+$, with a right scalar multiplication by elements of $\mathfrak{N}$, defined by

(1) $$xa = (x_1, x_2, x_3)\, a = (x_1 a, x_2 a, x_3 a),$$

for $x = (x_1, x_2, x_3) \in V$ and $a \in \mathfrak{N}$. We call *points* the subgroups $x\mathfrak{N} = \{xa : a \in \mathfrak{N}\}$, where $(0, 0, 0) \neq x \in V$. Let Γ be the group $GL_3(q)$ of all nonsingular $(3, 3)$-matrices with elements in $\mathfrak{K} = GF(q)$. For any $A \in \Gamma$, the mapping

(2) $$(x_1, x_2, x_3) \to \left(\sum_{i=1}^{3} x_i\, a_{i1},\ \sum_{i=1}^{3} x_i\, a_{i2},\ \sum_{i=1}^{3} x_i\, a_{i3}\right)$$

is an automorphism of V, mapping points onto points. It is not difficult to verify (see, e.g., ROSATI 1958b, Lemma 1), that the points fall into precisely two Γ-orbits $\mathfrak{p}$ and $\mathfrak{q}$, with

(3) $$\mathfrak{p} = \{(k_1, k_2, k_3)\, \mathfrak{N} : k_i \in \mathfrak{K},\ i = 1, 2, 3\}$$

and $\mathfrak{q}$ the set of all other points.

Let t be an arbitrary element of $\mathfrak{N}$. Then the set of all those $(x_1, x_2, x_3) \in V$ for which the equation

(4) $$x_1 + t\, x_2 + x_3 = 0$$

is satisfied, consists of full points; this point set we denote by $L(t)$. Now we define the *lines* as the point sets

(5) $$L(t)\, A = \{(xA)\, \mathfrak{N} : x\mathfrak{N} \subset L(t)\},$$

with $t = 1$ or $t \notin \mathfrak{K}$, and $A \in \Gamma$. Note that $L(t)\, A = L(s)\, B$ does not imply $t = s$ and $A = B$.

With *incidence* defined in the obvious way by set theoretic inclusion, we obtain an incidence structure of points and lines, which will be called the *Hughes plane* over $\mathfrak{N}$, and denoted by $\mathbf{H} = \mathbf{H}(\mathfrak{N})$. We summarize the main properties of $\mathbf{H}$ as follows:

---

[1] The smallest member of this class, of order 9, had actually been found by VEBLEN & WEDDERBURN 1907. — The following description differs from that of Hughes; it is mainly due to ZAPPA 1957 and ROSATI 1958b.

1. **(a)** $\mathbf{H}$ *is a projective plane of odd square prime power order* q^2, *and it is of Lenz-Barlotti type* I.1.
(b) *The points in the set* $\mathfrak{p}$ *defined by* (3), *together with the lines joining them, form a desarguesian Baer subplane* $\mathbf{H}_0$ *of* $\mathbf{H}$.
(c) *The projective group of* $\mathbf{H}_0$ *is faithfully induced by the collineation group* Γ *of* $\mathbf{H}$. *Every central collineation of* $\mathbf{H}_0$ *extends to a central collineation of* $\mathbf{H}$ *in* Γ.
(d) *The full collineation group* Φ *of* $\mathbf{H}$ *has two point, two line, and four flag orbits, one point orbit being the set* $\mathfrak{p}$ *defined by* (3). *Also,* Φ *is a semidirect product* $\Gamma \cdot \operatorname{Aut} \mathfrak{N}$; *this product is direct if and only if* q *is a prime.*
(e) $\mathbf{H}$ *is self-dual. In fact, every orthogonal polarity of* $\mathbf{H}_0$ *can be extended to a polarity* π *of* $\mathbf{H}$, *but in general* π *is not regular in the sense of Section* 3.3.

These results are in HUGHES 1957b; ROSATI 1958b, 1960a; OSTROM 1964a. We outline some of the proofs. That $\mathbf{H}$ is a projective plane, obviously of order q^2, is more or less a direct verification,[1]) and **(b)** is also clear. Furthermore, the first part of **(c)** is obvious from the construction. The second part of **(c)**, that central collineations of $\mathbf{H}_0$ are induced by central collineations of $\mathbf{H}$, is also not difficult; cf. ZAPPA 1957, Lemma I; ROSATI 1958b, Lemma 2.

For the remaining proofs, we consider a ternary field $\mathfrak{T} = \mathfrak{T}(o, e, u, v)$ for $\mathbf{H}$, with o, e, u, v points in $\mathbf{H}_0$. It follows from (4) and (5) that an arbitrary line of $\mathbf{H}$ is represented by an equation of the form

$$(6)\quad \sum_{i=1}^{3} a_i x_i + s \sum_{i=1}^{3} b_i x_i = 0, \quad \text{with } a_i, b_i \in \mathfrak{K}, \text{ not all } 0, \text{ and } s \notin \mathfrak{K}.$$

In particular, the line $W = L(1)\begin{pmatrix} 1 & 0 & 1 \\ 0 & 1 & 1 \\ 0 & 0 & 1 \end{pmatrix}$ has equation $x_3 = 0$; hence the points of the affine plane $\mathbf{H}^W = \mathbf{A}$ are characterized by $x_3 \neq 0$. Putting

$$x = x_1 x_3^{-1} \quad \text{and} \quad y = x_2 x_3^{-1},$$

we see that every point of $\mathbf{H}^W$ can be represented by an element $(x, y, 1) \in V$, and the correspondence between the pairs $(x, y) \in \mathfrak{N} \times \mathfrak{N}$ and the points $(x, y, 1)$ of $\mathbf{H}^W$ is one-one. From (6) it now follows that the equations of lines in $\mathbf{H}^W$, with respect to the inhomogeneous coordinates (x, y), are

$$a_1 x + a_2 y + a_3 + s(b_1 x + b_2 y + b_3) = 0,$$

---

[1]) Finiteness is not essential for this part. For the infinite analogues of the Hughes planes, see ROSATI 1960b.

with a_i, b_i $(i = 1, 2)$ not all 0; $a_j, b_j \in \mathfrak{K}$ $(j = 1, 2, 3)$; and $s \notin \mathfrak{K}$. It can be shown (HUGHES 1957b) that every equation of this form is equivalent to one of the following: $x = c$ with $c \in \mathfrak{N}$; $y = kx + b$ with $k \in \mathfrak{K}$ and $b \in \mathfrak{N}$; or $y = s(x + k) + k'$ with $s \in \mathfrak{N} - \mathfrak{K}$ and $k, k' \in \mathfrak{K}$. It follows that

$$(7)\quad x \cdot s \circ t = \begin{cases} s x + t & \text{if } s \in \mathfrak{K} \\ s(x + k) + k' & \text{if } s \in \mathfrak{N} - \mathfrak{K} \text{ and } t = ks + k', \text{ with } k, k' \in \mathfrak{K}, \end{cases}$$

defines a ternary operation on $\mathfrak{N}$, yielding the ternary field $\mathfrak{T} = \mathfrak{T}(o, e, u, v)$ of $\mathbf{H}$, with $o = (0, 0, 1)$, $e = (1, 1, 1)$, $ou = (y = 0)$, $o\,v = (x = 0)$, and $u\,v = W$.[1])

Note that $x \cdot 1 \circ y = x + y$ and $x \cdot y \circ 0 = yx$ for all $x, y \in \mathfrak{N}$, but in general $x \cdot s \circ t \neq sx + t$ or $xs + t$. Thus $\mathfrak{T}$ is not linear, and $\mathbf{H}$ is non-desarguesian. Result **4.4.10** then shows that the orbits of Γ must be those of the full collineation group of $\mathbf{H}$. This proves the first claim of (**d**). It is a trivial consequence that $\mathbf{H}$ is not (p, L)-transitive for any point-line pair (p, L); hence $\mathbf{H}$ is of Lenz-Barlotti type I.1. For the remainder of (**d**), see ROSATI 1958b, Lemma 5 and Teorema 4.

Finally, that $\mathbf{H}$ is self-dual was shown in special cases by HUGHES 1957b and in general by ROSATI 1960a. The additional statements in (**e**) were proved by OSTROM 1964a, Theorem 15; that $\mathbf{H}$ can admit non-regular polarities was observed by HUGHES 1957b.

There are two more reasons why we have exhibited the definition (7) for the ternary field $\mathfrak{T} = \mathfrak{T}(o, e, u, v)$ of $\mathbf{H}$. The first of these is the following: Let $\mathfrak{N}^*$ be the algebraic structure anti-isomorphic to $\mathfrak{N}$, defined by (3.1.49). Then $\mathfrak{N}^*$ satisfies the requirements of Result **3.1.37**, and comparison of (7) with (3.1.56) shows:

2. *For any Hughes plane* $\mathbf{H}$ *and any line* W *in* $\mathbf{H}_0$, *the affine plane* $\mathbf{H}^W$ *is a semi-translation plane.*

It is conceivable that this is a characteristic property of the Hughes planes, in other words that if a projective plane $\mathbf{P}$ has a Baer subplane $\mathbf{B}$ such that, for every choice of W in $\mathbf{B}$, the affine plane $\mathbf{P}^W$ is a semi-translation plane, then $\mathbf{P}$ is either desarguesian or a Hughes plane. So far, this has been proved only under more restrictive hypotheses:

3. *Let* $\mathbf{P}$ *be a non-desarguesian projective plane of square order* $n = m^2$, *containing a Baer subplane* $\mathbf{B}$ *of order* m. *If* $\mathbf{B}$ *contains a line* W *such that, for any point* c *of* $\mathbf{B}$ *on* W *and any line* A *of* $\mathbf{B}$, *the* (c, A)*-collineations of* $\mathbf{P}$ *which leave* $\mathbf{B}$ *invariant induce a* (c, A)*-transitive group on* $\mathbf{B}$, *then* $\mathbf{P}$ *is a Hughes plane* $\mathbf{H}$, *and* $\mathbf{B} = \mathbf{H}_0$.

---

[1]) For the affine description of the Hughes planes, see also PICKERT 1965b.

For the proof, see OSTROM 1965c.[1]) Note that under the hypotheses of **3**, m must be an odd prime power.

The second reason for the importance of (7) is result **5.1.10**, which shows:

4. *If W is a line contained in the subplane $\mathbf{H}_0$ of the Hughes plane $\mathbf{H}$, then the set $\mathfrak{d}$ of points in $\mathbf{H}_0$ on W is a derivation set for $\mathbf{H}$.*

Thus the Hughes planes are derivable. The planes derived from the Hughes planes have been investigated by OSTROM 1964b and ROSATI 1963, 1964. These planes can be described as follows. Call *points* the ordered pairs (x, y) of elements in the nearfield $\mathfrak{N}$ of order q^2, with centre equal to the kernel $\mathfrak{K}$ of order q. *Lines* are the following two types of point sets:[2])

$$(8)\quad \begin{aligned} &Q(s, ks+k') = \{(x, y) \vdots y = s(x+k)+k'\}, \quad s \in \mathfrak{N}-\mathfrak{K} \text{ and } k, k' \in \mathfrak{K}, \\ &R(a, b, c) = \{(x, y) \vdots x-b \in a\mathfrak{K},\ y-c \in a\mathfrak{K}\}, \quad a, b, c \in \mathfrak{N},\ a \neq 0. \end{aligned}$$

According to **5.1.10**, this defines an affine plane, and the corresponding projective plane is the derived Hughes plane $\mathfrak{d}\mathbf{H}(\mathfrak{N})$.

5. *Unless $\mathfrak{N}$ is the exceptional nearfield $N(2, 3)$ of order 9, in which case $\mathfrak{d}\mathbf{H}(N(2, 3))$ is the shears plane over $N(2, 3)$, of type* IVb.3, *the derived Hughes plane $\mathfrak{d}\mathbf{H}(\mathfrak{N})$ is of Lenz-Barlotti type* II.1.

That $\mathfrak{d}\mathbf{H}(\mathfrak{N})$ must be (p, L)-transitive for some flag (p, L) follows from **2**, **4** and the following more general result:

6. *Let $\mathbf{A}$ be a semi-translation plane, with distinguished Baer subplane $\mathbf{B}$, as defined by* (3.1.11) *and* (3.1.12). *In the corresponding projective plane $\mathbf{P}$, let $\mathfrak{d}$ denote the set of ideal points of $\mathbf{B}$. If $\mathfrak{d}$ is a derivation set for $\mathbf{P}$, then $\mathfrak{d}\mathbf{P}$ is (p, W)-transitive, for the carrier line W of $\mathfrak{d}$ and the intersection point p of W with the line $\mathbf{B}$ of $\mathfrak{d}\mathbf{P}$.*

Proof: The translation group T of $\mathbf{B}$ extends to $\mathbf{A}$ and fixes $\mathfrak{d}$, hence it is also a translation group of $\mathfrak{d}\mathbf{P}$. But T fixes $\mathbf{B}$ and is transitive on the points of $\mathbf{B}$; thus the result follows from the fact that $\mathbf{B}$ is a line of $\mathfrak{d}\mathbf{P}$.

For the more difficult part of **5**, that there are no other (p, L)-transitivities in $\mathfrak{d}\mathbf{H}(\mathfrak{N})$ except when $\mathfrak{N} = N(2, 3)$, see OSTROM 1964b, 1966c, ROSATI 1964.[3]) Cartesian groups for the planes $\mathfrak{d}\mathbf{H}$, for different

---

[1]) Ostrom also proves a similar result for the infinite analogues of the Hughes planes. Note that the hypothesis in **3** is a little stronger than necessary: if $p, q \in \mathbf{B}$ such that $p \neq c \neq q \mathrel{I} W \mathrel{I} p$, and $cp = cq$, then a (c, A)-collineation of $\mathbf{P}$ mapping p onto q automatically leaves $\mathbf{B}$ invariant (cf. **3.1.5**).

[2]) This should be compared with (5.1.17).

[3]) The proof of Theorem 3, case (3), in OSTROM 1964b contains an error which, however, does not affect the validity of **5**.

choices of o, e, u, v, were exhibited by FRYXELL 1964; PANELLA 1965; PICKERT 1967.

The process of derivation can be used to construct other classes of planes of types I and II, in a way similar to that of constructing the planes $\mathfrak{d}\mathbf{H}$. We know from **5.1.9** that a derived plane of a translation plane is again a translation plane; so in order to construct non-translation planes, the planes to be derived must also be non-translation planes. One way to achieve this is to dualize before deriving, in other words to consider derived planes of shears planes. This is the main idea in the following construction of OSTROM 1962a.[1])

Let $\mathfrak{Q}$ be an arbitrary quasifield which is of rank 2 over its kernel $\mathfrak{K}$, i.e. $|\mathfrak{K}| = q$ and $|\mathfrak{Q}| = q^2$, for some prime power q. We denote by $\mathbf{A} = \mathbf{A}(\mathfrak{Q})$ the affine plane over $\mathfrak{Q}$, and by $\mathbf{P} = \mathbf{P}(\mathfrak{Q})$ the corresponding projective plane. Also, consider the system $\mathfrak{Q}^*$ anti-isomorphic to $\mathfrak{Q}$, defined by (3.1.49), and let $\mathbf{A}^* = \mathbf{A}(\mathfrak{Q}^*)$ and $\mathbf{P}^* = \mathbf{P}(\mathfrak{Q}^*)$. Note that $\mathbf{P}$ and $\mathbf{P}^*$ are dual to each other, by result **3.1.36**. The affine plane $\mathbf{A}^*$ is a shears plane and hence (v, W)-transitive; let T be the group of (v, W)-translations $(x, y) \to (x, y + t)$, $t \in \mathfrak{Q}^*$, of order q^2. It is easily verified that $\mathfrak{Q}^*$ and $\mathfrak{K}^* \subset \mathfrak{Q}^*$ satisfy the derivability conditions of **5.1.10**. Thus $\mathbf{A}^*$ is derivable, and T is, by **5.1.8**, also a collineation group of $\mathfrak{d}\mathbf{A}^*$. But a line through v in $\mathbf{A}^*$ becomes a subplane in $\mathfrak{d}\mathbf{A}^*$; hence it now follows that

7. *For any quasifield $\mathfrak{Q}$ of rank* 2 *over its kernel, the derived plane $\mathfrak{d}\mathbf{A}^*$ of the shears plane $\mathbf{A}^*$ over $\mathfrak{Q}^*$ is a semi-translation plane.*

Now if $\mathfrak{Q}$ is a semifield, $\mathbf{A}^*$ is a translation plane, and hence $\mathfrak{d}\mathbf{A}^*$ is also a translation plane. But if $\mathfrak{Q}$ is non-distributive, then neither $\mathbf{A}^*$ nor $\mathfrak{d}\mathbf{A}^*$ are translation planes (OSTROM 1962a). Furthermore, $\mathfrak{d}\mathbf{A}^*$ is not a shears plane. Also, the following holds:

8. *Let $\mathfrak{d}\mathbf{P}^*$ denote the projective plane corresponding to the affine plane $\mathfrak{d}\mathbf{A}^*$ derived from the shears plane $\mathbf{A}^*$ over $\mathfrak{Q}^*$, where $\mathfrak{Q}$ is a quasifield of rank* 2 *over its kernel. If $|\mathfrak{Q}| = q^2 > 16$, then every collineation φ of $\mathfrak{d}\mathbf{P}^*$ fixes the ideal line of $\mathfrak{d}\mathbf{A}^*$. Moreover, when φ is considered as a permutation of the points of $\mathfrak{d}\mathbf{A}^*$, then φ is also a collineation of $\mathbf{A}^*$.*

This was proved by OSTROM 1965a. The case $q \leqq 4$ seems to have received no special attention in this respect.

[1]) The idea has been followed further by FRYXELL 1964, who considered the sequence of planes obtained by alternating dualizations and derivations, beginning with either a Hughes plane or a translation plane. Fryxell has found that these sequences repeat themselves after at most eight steps, and that no new planes arise after the second step.

The planes $\mathfrak{d}\mathbf{A}^*$ can be represented algebraically in the form (5.1.17), but with $\mathfrak{Q}^*$ instead of $\mathfrak{Q}$. Thus the points are the pairs (x, y) of elements of $\mathfrak{Q}$, and the lines are the point sets

$$(9)\qquad \begin{aligned} Q(a, b) &= \{(x, y) \vdots\, y = a\,x + b\}, && a \in \mathfrak{Q} - \mathfrak{K}, \\ R(a, b, c) &= \{(x, y) \vdots\, x - b \in \mathfrak{K}a,\; y - c \in \mathfrak{K}a\}, && 0 \neq a \in \mathfrak{K}. \end{aligned}$$

The semi-translation planes so defined seem to be mostly of Lenz-Barlotti type I.1. In particular, this is true if $\mathfrak{Q}$ is either a nearfield[1]) or one of the Hall quasifields defined by (5.2.5). One cannot expect, however, to prove this in general:

9. *Let* $\mathfrak{Q} = L(q)$, $q = 2^{m+1}$, *be the Lüneburg quasifield defined by* (5.2.6), *and* $\mathbf{Q}$ *the projective plane* $\mathfrak{d}\mathbf{P}^*$ *corresponding to the derived plane* $\mathfrak{d}\mathbf{A}^*$ *of the shears plane over* $\mathfrak{Q}^*$. *Then*

(**a**) $\mathbf{Q}$ *is of Lenz-Barlotti type* II.1; *the only flag* (v, W) *for which* $\mathbf{Q}$ *is* (v, W)*-transitive consists of the ideal point* v *of* $R(1, 0, 0)$ *and the ideal line* W *of* $d\,\mathbf{A}^*$.

(**b**) *The full translation group of* $\mathfrak{d}\mathbf{A}^*$ *is of order* q^3; *every ideal point* $\neq v$ *is the center of exactly* q *translations.*

(**c**) *The group of all collineations of* $\mathbf{Q}$ *has at most four point orbits. One of these consists of all points of* $\mathfrak{d}\mathbf{A}^*$, *another of* v *alone.*

These are the main results of OSTROM 1966a. The derived Hughes planes $\mathfrak{d}\mathbf{H}$ and the planes $\mathbf{Q}$ of **9** are the only known finite planes of Lenz-Barlotti type II.1. Their orders are all square prime powers; those of $\mathfrak{d}\mathbf{H}$ are odd and those of $\mathbf{Q}$ even.

---

[1]) In this case one might first expect that $\mathfrak{d}\mathbf{A}^*$ is an affine plane of a Hughes plane constructed over the same nearfield. That this is not the case follows from **8**.

6. Inversive planes

It is the object of this chapter to collect the known facts on finite inversive planes, which were defined in Section 2.4 as designs of type (1, 3) satisfying (2.4.19). We shall give a more geometric definition here, which is meaningful also in the infinite case. In fact, Section 6.1 (which may be compared with Section 3.1 on projective and affine planes) is concerned with inversive planes in general; finiteness will usually not be essential there.

Sections 6.2—6.4 follow the same general pattern as our earlier discussions of designs and planes: In 6.2 we will be concerned with purely combinatorial properties of finite inversive planes, 6.3 deals with automorphisms and automorphism groups, and 6.4 with the two known classes of finite inversive planes. These classes were actually defined in Section 2.4; here we shall mostly be interested in characterizing them intrinsically, by combinatorial features as well as by properties of their automorphism groups.

The notion of "egglike" inversive planes, to be defined in Section 6.1, plays a major role throughout this chapter. In fact, all known finite inversive planes are egglike. Since egglike inversive planes are intimately related to ovoids in 3-dimensional projective geometries, some of the results in Section 1.4 concerning ovoids are significant for inversive planes as well. However, we shall only point out the more important of these connections here.

6.1 General definitions and results

An *inversive plane*[1]) is an incidence structure $\mathbf{I} = (\mathfrak{p}, \mathfrak{C}, \mathrm{I})$, whose blocks are called *circles*, such that the following axioms are satisfied:

(1) $\qquad [o, p, q] = 1$ *for distinct* $o, p, q \in \mathfrak{p}$.

[1]) Also "Möbius-Ebene" or "Möbius-plane" (Ewald 1960, 1961a, 1962, 1967; Dembowski 1964b, 1965a; Lüneburg 1964b, 1966c; Hering 1965), "Möbiusebene im engeren Sinne" (Benz 1960a), or "conformal plane". Related geometrical objects in higher dimensions have been studied by Hoffman 1951a (some of these cannot exist in the finite case; see Section 6.2) and Benz 1960a; note that Benz uses the term "Möbiusebene" in a sense more general than that of the present "inversive plane". For an extensive bibliography on the classical inversive plane, see Coxeter 1966.

(2) *If* $p, q \in \mathfrak{p}$ *and* $C \in \mathfrak{C}$ *such that* $p \mathrm{I} C \not\mathrm{I} q$, *then there is a unique* $D \in \mathfrak{C}$ *with* $p \mathrm{I} D \mathrm{I} q$ *and* $(C, D) = \{p\}$.

(3) $|\mathfrak{p}| \geqq 4$, *there exist* $p \in \mathfrak{p}$ *and* $C \in \mathfrak{C}$ *with* $p \not\mathrm{I} C$, *and* $[X] > 0$ *for every* $X \in \mathfrak{C}$.

We are using here again the notations (1.1.5) and (1.1.6); thus $(\mathfrak{x})$ is the set of all elements incident with each $x \in \mathfrak{x}$, and $[\mathfrak{x}] = |(\mathfrak{x})|$. A shorter way of expressing (1)—(3) is the following:

(4) *For every* $p \in \mathfrak{p}$, *the internal structure* $\mathbf{I}_p$ *is an affine plane.*

For the definition of internal structures, see Sections 1.1 and 2.2. The equivalence of (1)—(3) with (4) is easily proved; for example, (2) is equivalent to the axiom of parallels (Euclidean axiom) in $\mathbf{I}_p$.

Because of (1), there is a unique circle through any three distinct points. It follows that

(5) $$[B, C] = 0, 1 \text{ or } 2, \qquad \text{for distinct } B, C \in \mathfrak{C}.$$

According to these possibilities we say that B and C are *disjoint, tangent,* or *intersecting*, respectively. We also say that two circles *meet* if they are not disjoint. A set $\mathfrak{c}$ of points is called *concircular*[1]) if there exists a circle C with $\mathfrak{c} \subseteqq (C)$.

We shall be concerned with some particularly important types of sets of circles later on; we introduce these now. A *bundle*[1]) of circles is any set of the form (p, q) with $p \neq q$, i.e. the set of all circles through two distinct points. These points p, q are called the *carriers* of the bundle (p, q); the set of all bundles in the inversive plane $\mathbf{I}$ will be denoted by $\mathcal{B} = \mathcal{B}(\mathbf{I})$. A *pencil*[1]) is any maximal set of mutually tangent circles through a common point p, called the *carrier* of the pencil. Thus the pencils with given carrier p are essentially identical with the parallel classes of lines in the affine plane $\mathbf{I}_p$. Any point-circle flag $p \mathrm{I} C$, or alternatively any pair of tangent circles C, D, determines a unique pencil, denoted by $\langle p, C\rangle$ or $\langle C, D\rangle$, respectively.[2]) We denote the set of all pencils in the inversive plane $\mathbf{I}$ by $\mathcal{P} = \mathcal{P}(\mathbf{I})$. The axioms (1) and (2) imply:

1. *Let* $\mathfrak{S}$ *be a set of circles which is either a bundle or a pencil. Then every point of* $\mathbf{I}$ *which is not a carrier of* $\mathfrak{S}$ *is on precisely one circle of* $\mathfrak{S}$.

---

[1]) Other customary terminologies are "concyclic" instead of "concircular", and "hyperbolic", "parabolic", resp. "elliptic pencil" instead of "bundle", "pencil", or "flock". In a special situation (for egglike inversive planes, see p. 255 below) these terms would indeed be preferable, but for general inversive planes it is more convenient to have the present shorter terms available.

[2]) In Section 3.1, the symbol $\langle \mathfrak{x} \rangle$ was used for the closed configuration generated by the subset $\mathfrak{x}$ of a projective plane. Clearly, $\langle \ldots \rangle$ is used with a different meaning here, but no confusion ought to arise from this.

There is a third type of circle sets, with a property similar to **1**: A *flock*[1]) is a set $\mathfrak{F}$ of mutually disjoint circles in **I** such that, with the exception of precisely two points p, q, every point of **I** is on a (necessarily unique) circle of $\mathfrak{F}$. These points are again called the *carriers* of the flock. The existence of flocks is not as clear as that of bundles and pencils, and a flock need not be determined by its carriers. In special situations, however, this can be proved, for example for the finite inversive planes in the class be defined next.

Let $\mathfrak{o}$ be an *ovoid* in a 3-dimensional projective geometry $\mathcal{G}$, i.e. [cf. (1.4.27) and (1.4.28′)] a set of points in $\mathcal{G}$ such that

(6) $|\mathfrak{o} \cap L| \leqq 2$ *for any line L of $\mathcal{G}$*

and

(7) *For any $x \in \mathfrak{o}$, the set* $T(x) = \bigcup_{\mathfrak{o} \cap L = \{x\}} L$ *is a plane of $\mathcal{G}$.*

Then an incidence structure $\mathbf{I} = \mathbf{I}(\mathfrak{o})$ is defined as follows. Points are the elements of $\mathfrak{o}$, circles are the planes P with $|\mathfrak{o} \cap P| > 1$ in $\mathcal{G}$, and incidence is that of $\mathcal{G}$. It is straightforward to prove that

2. $\mathbf{I} = \mathbf{I}(\mathfrak{o})$ *is an inversive plane.*

Furthermore, if $\mathfrak{F}$ denotes the (not necessarily commutative) field underlying the geometry $\mathcal{G}$, then

3. *For any point p of the inversive plane $\mathbf{I} = \mathbf{I}(\mathfrak{o})$ defined by the ovoid $\mathfrak{o}$, the affine plane $\mathbf{I}_p$ is the (desarguesian) plane over $\mathfrak{F}$.*

For the proof of **3**, one has to observe only that the projective plane corresponding to $\mathbf{I}_p$ is essentially the same as the quotient geometry $\mathcal{G}/p$ of $\mathcal{G}$, as defined in Section 1.4. See also Dembowski 1964b, (4.1).[2])

We call an inversive plane *egglike*[3]) if it is isomorphic to an $\mathbf{I}(\mathfrak{o})$, for some ovoid $\mathfrak{o}$. Among the egglike inversive planes is the classical model $\mathbf{I}(\mathfrak{s})$, where $\mathfrak{s}$ is a sphere in euclidean 3-space; this is isomorphic to $\mathbf{I}(\mathfrak{q})$, where $\mathfrak{q}$ is any *non-ruled quadric*, i.e. a quadric of index 1 as defined in 1.4, in three-dimensional projective space over the real numbers. Also, the two classes of finite inversive planes $\mathbf{M}(q)$ and $\mathbf{S}(q)$ defined in Section 2.4 are egglike, the corresponding ovoids being the non-ruled quadrics in $\mathcal{P}(3, q)$ and the ovoids $\mathfrak{t}(\psi)$ of **1.4.56**, respectively; these two classes will be discussed in detail in Section 6.4. Also, there

[1]) See footnote [1]) on p. 253.

[2]) One should also notice the connection with the familiar process of "stereographic projection" here.

[3]) This is the terminology of Dembowski & Hughes 1965. Dembowski 1964b uses the term "ovoidal" in German.

will be various results on finite egglike inversive planes in Sections 6.2 and 6.3. At this point, we mention only that there exist non-egglike inversive planes: EWALD 1962 has constructed inversive planes **I** for which the affine planes $\mathbf{I}_p$ are not all isomorphic; it follows from **3** that such an **I** cannot be egglike. Ewald's examples are infinite, and in fact there are no finite non-egglike inversive planes known. It is an interesting question whether or not finite inversive planes are necessarily egglike; this problem will be dealt with in Section 6.2.

Let $\mathfrak{o}$ be an ovoid in the 3-dimensional projective geometry $\mathcal{G}$, and $\mathbf{I} = \mathbf{I}(\mathfrak{o})$. If L is any line in $\mathcal{G}$, then the planes through L which are circles of **I** form a bundle, pencil, or flock of **I**, according as L and $\mathfrak{o}$ have two, one, or no point in common. Conversely, every bundle or pencil of **I** can be interpreted in this way, but for flocks this is not true in general.[1] It has not yet been investigated which finite egglike inversive planes possess flocks that cannot be interpreted as pencils of planes in $\mathcal{G}$; compare here **1.4.54** above and **6.2.12b**, **6.2.13** below.

We formulate next two configurational propositions known as the "Bundle Theorem" and the "Theorem of Miquel". Let us call a *4-chain* any quadruple $C_1, \ldots, C_4$ of circles such that no three of the C_i have a common point, but $(C_i, C_{i+1}) \neq \emptyset$, with the subscripts taken mod 4. We put

(8) $$(C_i, C_{i+1}) = \{a_i, b_i\},$$

where the possibility $a_i = b_i$ is not excluded; but $a_1, \ldots, a_4$ are, of course, four distinct points. We say that **I** satisfies the *Bundle Theorem* provided that

(B) *For any 4-chain $C_1, \ldots, C_4$, the bundles (or pencils) determined by C_1, C_2 and C_3, C_4 have a common circle if, and only if, the bundles (or pencils) determined by C_1, C_3 and C_2, C_4 have a common circle.*

Note that both intersections in question contain at most one circle; if $a_i \neq b_i$ $(i = 1, \ldots, 4)$, the condition says that a_1, b_1, a_3, b_3 are concircular if, and only if, a_2, b_2, a_4, b_4 are concircular (cf. Fig. 8, p. 256).

An inversive plane is said to be *miquelian* if it satisfies the following condition, called the *Theorem of Miquel:*[2]

(M) *For any 4-chain with the convention* (8), *the points $a_1, \ldots, a_4$ are concircular if, and only if, the points $b_1, \ldots, b_4$ are concircular* (cf. Fig. 9).

---

[1]) For example, let S be the ordinary sphere and E its equatorial plane, and consider two distinct lines L, M in E which do not meet S. The planes through L intersecting S in the northern hemisphere, together with the planes through M intersecting S in the southern hemisphere, give rise to a flock not corresponding to a single line.

[2]) Cf. BAKER 1930, p. 72; MORLEY & MORLEY 1933; PECZAR 1950; HOFFMAN 1951b; BENZ 1958b.

The significance of the conditions (**B**) and (**M**) with respect to egg-likeness lies in the following results.

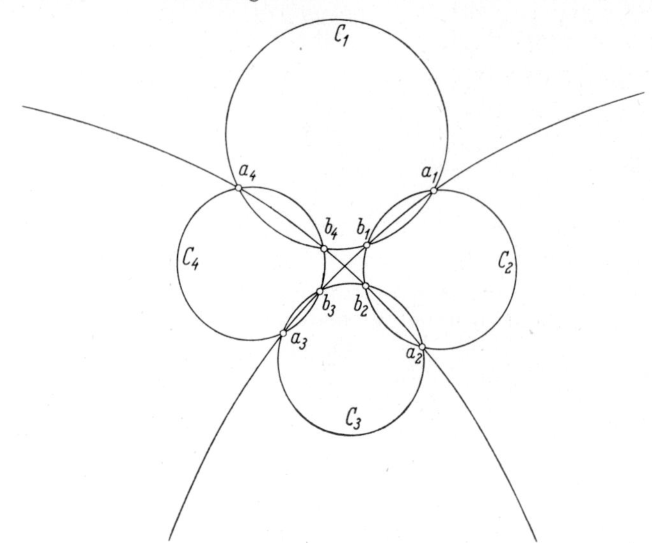

Fig. 8. Bundle Theorem

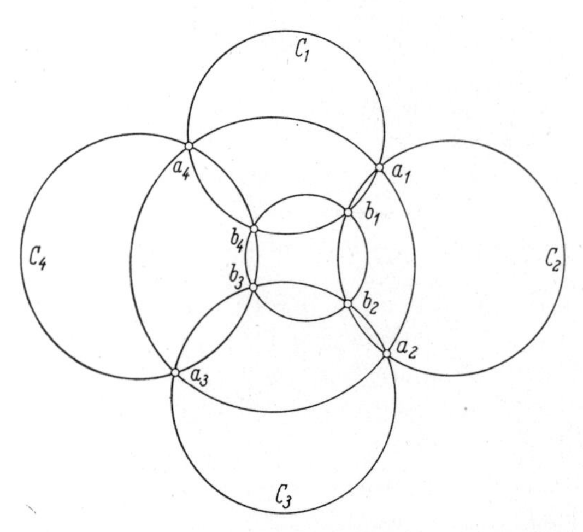

Fig. 9. Theorem of Miquel

4. *Every egglike inversive plane satisfies the Bundle Theorem* (**B**).

For the proof, see VAN DER WAERDEN & SMID 1935, or DEMBOWSKI 1964b, (4.2). In fact, (**B**) is an immediate consequence of the dual of axiom (1.4.3) for projective 3-space. The converse of **4** is conceivably

For proofs of **12** and **13**, see HERING 1965, Hilfssätze 2.1—2.3. The hypothesis in **13a** that **I** possesses more than 5 points is essential; in fact there is only one inversive plane with only 5 points, namely the miquelian plane $\mathbf{M}(2)$, its circles are precisely its 3-element point sets, and (p, q)-transitivity for $p \neq q$ means no restriction for this plane.

Using **12** and **13**, HERING 1965 has classified all possibilities for the set

(11) $\qquad \mathcal{T} = \mathcal{T}(\Gamma) = \{\mathfrak{X} \in \mathcal{B}(\mathbf{I}) \cup \mathcal{P}(\mathbf{I}) : \Gamma \text{ is } \mathfrak{X}\text{-transitive}\}.$

The result is the following analogue to **3.1.20**:

14. *For any automorphism group* Γ *of an inversive plane* $\mathbf{I} = (\mathfrak{p}, \mathfrak{C}, \mathrm{I})$ *with more than* 5 *points, the set* $\mathcal{T}(\Gamma)$ *is of one of the following types:*

I.1. $\mathcal{T} = \emptyset$.
I.2. $\mathcal{T}$ *consists of a single bundle.*
I.3. $\mathcal{T} = \{(x, x^\sigma) : x \,\mathrm{I}\, C\}$, *with* σ *a fixed point free involution of the points of the circle* C.
I.4. $\mathcal{T}$ *consists of bundles only*; if $\mathfrak{X} = (x, x')$ *and* $\mathfrak{Y} = (y, y')$ *are in* $\mathcal{T}$ *and* $\mathfrak{X} \neq \mathfrak{Y}$, *then* x, x', y, y' *are four distinct concircular points, and* (x, x', y, y') *is the only circle in* (x, y) *each of whose points is a carrier of some bundle in* $\mathcal{T}$; *there exist three bundles in* $\mathcal{T}$ *which have no circle in common.*
I.5. $\mathcal{T} = \{(x, x^\tau) : x \in \mathfrak{p}\}$, *with* τ *a fixed point free involution of* $\mathfrak{p}$.

II.1. $\mathcal{T}$ *consists of a single pencil.*
II.2. $\mathcal{T} = \{\mathfrak{P}\} \cup \{(p, x) : p \neq x \,\mathrm{I}\, C\}$, *where* $\mathfrak{P}$ *is a pencil with carrier* p, *and* $C \in \mathfrak{P}$.

III.1. $\mathcal{T}$ *consists of all pencils with the same fixed carrier* p.
III.2. $\mathcal{T} = \{\mathfrak{X} \in \mathcal{P}(\mathbf{I}) : \mathfrak{X} \text{ has carrier } p\} \cup \{(p, x) : x \neq p\}$.

IV.1. $\mathcal{T}$ *consists of all pencils containing the same fixed circle* C.
IV.2. $\mathcal{T} = \{\mathfrak{X} \in \mathcal{P}(\mathbf{I}) : C \in \mathfrak{X}\} \cup \{(x, y) : x \neq y \text{ and } x \,\mathrm{I}\, C \,\mathrm{I}\, y\}$.
IV.3. $\mathcal{T} = \{\mathfrak{X} \in \mathcal{P}(\mathbf{I}) : C \in \mathfrak{X}\} \cup \{(z, z^\varrho) : z \,\not\mathrm{I}\, C\}$, *with* ϱ *a fixed point free involution of* $\mathfrak{p} - (C)$.
IV.4. $\mathcal{T} = \{\mathfrak{X} \in \mathcal{P}(\mathbf{I}) : C \in \mathfrak{X}\} \cup \{(x, y) : y \neq x \,\mathrm{I}\, C \,\mathrm{I}\, Y\} \cup \{(z, z^\varrho) : z \,\not\mathrm{I}\, C\}$, *with* ϱ *as in* IV.3.

V. $\mathcal{T}$ *consists of pencils only; if* $\mathfrak{X}$ *and* $\mathfrak{Y}$, *with carriers* x, y, *respectively, are distinct pencils in* $\mathcal{T}$, *then* $x \neq y$, *and there is a (necessarily unique) circle* C *common to* $\mathfrak{X}$ *and* $\mathfrak{Y}$, *which is the only circle in* (x, y) *each of whose points is the carrier for some pencil in* $\mathcal{T}$; *there exist three pencils without common circle in* $\mathcal{T}$.
VI.1. *Every point* $x \in \mathfrak{p}$ *is the carrier for one and only one pencil* $\mathfrak{P} = \mathfrak{P}(x) \in \mathcal{T}$.

VI.2. $\mathscr{T} = \{\mathfrak{P}(x) \vdots x \in p\} \cap \{(x, y) \vdots x \neq y \mathrm{I} C(x)\}$, *where* $\mathfrak{P}(x)$ *is as in* VI.1. *and* $C(x)$ *a distinguished circle in* $\mathfrak{P}(x)$.

VII.1. $\mathscr{T} = \mathscr{P}(\mathbf{I})$.

VII.2. $\mathscr{T} = \mathscr{P}(\mathbf{I}) \cup \mathfrak{B}(\mathbf{I})$.

This is the main result of HERING 1965. Which of these types can actually occur has not yet been completely clarified; the present state of our knowledge is contained in Table II, which is the analogue of

Table II: *Hering types*

Type	Finite Case		Infinite Case	
	Group exists	Plane exists	Group exists	Plane exists
I.1.	yes	?	yes	yes
I.2.	yes	?	yes	?
I.3.	yes	?	?	?
I.4.	yes	?	?	?
I.5.	yes	?	?	?
II.1.	yes	?	yes	yes
II.2.	yes	?	yes	yes
III.1.	yes	?	yes	yes
III.2.	yes	?	yes	yes
IV.1.	yes	?	?	?
IV.2.	yes	?	yes	?
IV.3.	Only in $\mathbf{M}(3)$	no	no	no
IV.4.	Only in $\mathbf{M}(3)$	no	no	no
V.	no	no	?	?
VI.1.	yes	yes, $\mathbf{S}(q)$	yes	yes
VI.2.	Only in $\mathbf{M}(3)$	no	?	?
VII.1.	no	no	?	?
VII.2.	yes	yes, only $\mathbf{M}(q)$	yes	yes

Table I for Lenz-Barlotti classes of projective planes and their groups; see Section 3.1. Proofs for the assertions in the first two columns of table II will be indicated in Sections 6.3 and 6.4. The entries "yes" in the last column, except those for VI.1. and VII.2., are due to EWALD 1967.

6.2 Combinatorics of finite inversive planes

We have defined a finite inversive plane in Section 2.4 as a design of type (1, 3), with parameters of the form

$$v = n^2 + 1, \quad k = n + 1, \quad b = n(n^2 + 1),$$
$$r = n(n + 1), \quad \lambda = n + 1, \quad b_3 = 1, \tag{1}$$

for some integer n. It is a simple exercise to verify that this definition is equivalent to the finite case of either (6.1.1)—(6.1.3) or (6.1.4). In particular, the integer n of (1) is the common order of each of the affine planes $\mathbf{I}_p$, for the given inversive plane $\mathbf{I}$. We call n the *order*[1]) of $\mathbf{I}$.

We begin our discussion of finite inversive planes with two results showing that the definitions (6.1.1)—(6.1.3) and (6.1.4) can be relaxed in the finite case.

1. *Let* $\mathbf{I} = (\mathfrak{p}, \mathfrak{C}, \mathrm{I})$ *be a finite incidence structure with the following properties:*

(a) $[o, p, q] \geqq 1$ *for distinct* $o, p, q \in \mathfrak{p}$.

(b) *If* $A, B, C \in \mathfrak{C}$ *are distinct and incident with* $p, q \in \mathfrak{p}$, *where* $p \neq q$, *then*

$$(A, B) = (B, C) = (C, A) = (A, B, C).$$

(c) $[C] = k = \text{const.}$, *for all* $C \in \mathfrak{C}$.

(d) $[B, C] = 0, 1,$ *or* t, *whenever* $B, C \in \mathfrak{C}$ *are distinct, for some fixed integer* t.

If $\mathbf{I}$ *also satisfies axioms* (6.1.2) *and* (6.1.3), *then* $\mathbf{I}$ *is an inversive plane.*[2])

Proof: Let $p \neq q$. By (**a**), (**b**), (**d**), precisely t points of $\mathbf{I}$ are incident with every $X \in (p, q)$, while each of the remaining $v - t$ points is on exactly one $X \in (p, q)$. Hence (**c**) gives $[p, q] = (v - t)\,(k - t)^{-1} = \lambda > 1$, independent of p, q, so that $\mathbf{I}$ is a design. With $r = [x]$ for any $x \in \mathfrak{p}$, we get from (2.1.5):

$$\text{(i)} \qquad r(k-1) = \lambda(v-1) = \lambda[t + \lambda(k-t) - 1].$$

Consider $C \in \mathfrak{C}$ such that $p \,\mathrm{I}\, C \,\not\mathrm{I}\, q$. By (6.1.2), there are exactly $\lambda - 1$ circles in (p, q) not tangent to C. Each of these meets C in $t - 1$ points $\neq p$, by (**d**). Also, (**b**) implies that any two of these circles have only one point of C, namely p, in common. On the other hand, every $x \,\mathrm{I}\, C$ is on one of these $\lambda - 1$ circles, by (**a**). Therefore,

$$\text{(ii)} \qquad k - 1 = (\lambda - 1)\,(t - 1).$$

Eliminating k from (i) and (ii), we get $r = \lambda(\lambda - 1)$, whence $v - 1 = (\lambda - 1)^2\,(t - 1)$. The first equation (2.1.5) then gives

$$b[1 + (\lambda - 1)\,(t - 1)] = b\,k = r\,v = \lambda(\lambda - 1)\,[1 + (\lambda - 1)^2\,(t - 1)],$$

---

[1]) This is an inconsistency similar to that pointed out in footnote [2]) on p. 259: The "order" as defined here is not the integer $r - \lambda$ of (2.1.9), which would be $n^2 - 1$ for an inversive plane. The present definition, however, is quite natural, and there ought to arise no confusion.

[2]) Clearly, (**a**)—(**d**) hold if (6.1.1) holds; thus the converse of **1** is also true. Finiteness is an essential hypothesis here: there exist infinite incidence structures satisfying (6.1.2), (6.1.3), and (**a**)—(**d**), with k and t suitable infinite cardinals, but not (6.1.1); see EWALD 1957b.

so that $b = |\mathfrak{C}|$ must be of the form $(\lambda - 1)[(\lambda - 1)c + 1]$, for some integer c. Substituting this, we get after obvious simplifications

$$\text{(iii)} \qquad (t-1)[(\lambda-1)(\lambda-c)-1] = c-1.$$

The right side of (iii) is non-negative; hence $\lambda - c > 0$ or $\lambda \geqq c + 1$. Now (iii) gives

$$t - 1 = 1 - \frac{\lambda(\lambda - c - 1)}{\lambda(\lambda - c - 1) + c - 1} \leqq 1,$$

so that $t \leqq 2$. But then $t = 2$, so that (6.1.1) holds and $\mathbf{I}$ is an inversive plane.

2. *A finite incidence structure* $\mathbf{I}$ *is an inversive plane if, and only if, for each point* p, *the internal structure* $\mathbf{I}_p$ *is an affine design.*[1])

This is a consequence of result **2.2.20** which says that $\mathbf{I}_p$ must actually be an affine plane.

We list now some simple numerical properties of finite inversive planes, which supplement (1).

3. *Let* $\mathbf{I}$ *be an inversive plane of order* n. *Then*

(a) *The number of points of* $\mathbf{I}$ *is* $n^2 + 1$.
(b) *The number of circles of* $\mathbf{I}$ *is* $n(n^2 + 1)$.
(c) *Each circle is incident with* $n + 1$ *points.*
(d) *Each point is incident with* $n(n + 1)$ *circles.*
(e) *Each bundle consists of* $n + 1$ *circles.*
(f) *Each pencil consists of* n *circles.*
(g) *Each flock consists of* $n - 1$ *circles.*
(h) *Each circle is tangent to* $n^2 - 1$ *other circles.*
(i) *Each circle is disjoint from* $n(n - 1)(n - 2)/2$ *other circles.*

For proofs, see DEMBOWSKI 1964b, § 2.

The next result is fundamental for many investigations in finite inversive planes.

4. *Let* p, C *be a non-incident point-circle pair in an inversive plane of order* n, *and denote by* $\mathfrak{T}(p, C)$ *the set of all circles through* p *which are tangent to* C. *Then* $|\mathfrak{T}(p, C)| = n + 1$, *and*

(a) *If* n *is even,* $\mathfrak{T}(p, C)$ *is a bundle* (p, q), *with a uniquely determined point* $q \,\mathrm{I}\, C$.
(b) *If* n *is odd, and* $\mathfrak{X}$ *a bundle or pencil with carrier* p, *then* $|\mathfrak{X} \cap \mathfrak{T}(p, C)| = 0$ *or* 2.

---

[1]) Finiteness is again essential here: The points and non-trivial hyperplane sections of an m-sphere in euclidean $(m + 1)$-space form an incidence structure $\mathbf{S}$ such that all $\mathbf{S}_p$ are isomorphic to the system of points and hyperplanes of the m-dimensional affine geometry over the real numbers.

Proof: That $|\mathfrak{T}(p, C)| = n + 1$ follows immediately from (6.1.2). In the projective plane $\mathbf{I}_p^*$ corresponding to the affine plane $\mathbf{I}_p$ of (6.1.4), the points of C form an oval $\mathfrak{o}$ such that the ideal line of $\mathbf{I}_p$ is exterior to $\mathfrak{o}$. But then a direct application of result **3.2.23** proves the claims of **4**. A more direct proof of (**a**), without appeal to QVIST's 1952 result **3.2.23**, appears in DEMBOWSKI 1964b, Satz 2. We note an immediate corollary of **4**:

5. *In an inversive plane* **I** *of order n, let $\mathfrak{P}$ be a pencil and C a circle not incident with the carrier of $\mathfrak{P}$.*

(**a**) *If n is even, C is tangent to precisely one circle of $\mathfrak{P}$.*
(**b**) *If n is odd, C is tangent to either none or precisely two circles of $\mathfrak{P}$.*[1])

Next, we note the following results:

6. *If an inversive plane of order n has a proper subplane of order m, then*
(**a**) $m \equiv n \bmod 2$
and
(**b**) $m^2 + m \leqq n$.

This is Satz 1.3 of DEMBOWSKI 1965a. Note that **6** is a rather stronger result than the analogous **3.2.18** for finite projective planes.

For the remainder of this section we shall be concerned with an intrinsic characterization of egglikeness in finite inversive planes. Among other things, this investigation will yield the result that inversive planes of even order are egglike; cf. **14** below.

An *orthogonality*[2]) of an inversive plane $\mathbf{I} = (\mathfrak{p}, \mathfrak{C}, \mathrm{I})$ is a binary relation, denoted by $\perp$, in $\mathfrak{C}$ which satisfies the following axioms:

(2) $\perp$ *is symmetric: $B \perp C$ if and only if $C \perp B$.*

(3) *If $C \mathrel{\mathrm{I}} p \neq q$, then there exists a unique $D \perp C$ such that $p \mathrel{\mathrm{I}} D \mathrel{\mathrm{I}} q$.*

(4) *Let A, B be distinct circles in a bundle $\mathfrak{B} = (p, q)$. If $A \perp C \perp B$, then $C \perp X$ for every $X \in \mathfrak{B}$.*

Suppose that the inversive plane **I** possesses an orthogonality $\perp$. We define, for an arbitrary point-circle pair p, C:

(5) $$\mathfrak{O}(p, C) = \{X \in \mathfrak{C} \colon p \mathrel{\mathrm{I}} X \perp C\}.$$

---

[1]) This implies that, in the language of BENZ 1958b, c, 1960a, every finite inversive plane is an "(F)-plane". This is not true for infinite inversive planes.

[2]) This concept, defined here as in DEMBOWSKI & HUGHES 1965, is also meaningful in infinite inversive planes, but it does not in general have the same significance there (cf. **13** below). Other notions of orthogonality, more suitable for certain infinite situations, were considered by EWALD 1956b and BENZ 1958c, 1960a, p. 16; see also COXETER 1966, p. 219.

Then (3) implies immediately that

7. *If $p \mathrm{I} C$, then $\mathfrak{D}(p, C)$ is a pencil with carrier p.*

Here **I** need not be finite. In the finite case, more can be said:

8. *If* **I** *is finite and $p \not\mathrm{I} C$, then $\mathfrak{D}(p, C)$ is a bundle (p, q) with $q \not\mathrm{I} C$.*

For the proof, see DEMBOWSKI & HUGHES 1965, (2.16). The following results show that an orthogonality in a finite inversive plane, if it exists, can be described by means of the given incidence alone.

9. *Let* **I** *be an inversive plane of order n and suppose that there exists an orthogonality relation $\perp$ in* **I**.

(**a**) *If n is even, then*

$$C \perp D \text{ if and only if } C = D \text{ or } [C, D] = 1. \tag{6}$$

(**b**) *If n is odd, then $C \perp D$ implies $C \neq D$ and $[C, D] = 2$ or 0. Moreover, if $(C, D) = \{p, q\}$, then $C \perp D$ if, and only if, for any $x \neq p, q$ on D, the unique circles through x which are tangent to C in p and q are tangent to each other; and if $(C, D) = \emptyset$, then $C \perp D$ if and only if there exist distinct points $p, q \mathrm{I} D$ as well as distinct circles $A, B \in (p, q)$ such that $[A, C] = [B, C] = 2$ and $A \perp C \perp B$.*

For the proof, see DEMBOWSKI & HUGHES 1965, pp. 175—176. A circle C is called *isotropic* if $C \perp C$. It follows from **9** that there exist isotropic circles if and only if n is even, and that in this case every circle is isotropic. Another corollary of **9** is:

10. *A finite inversive plane admits at most one orthogonality.*

(DEMBOWSKI & HUGHES 1965, Theorem 2). Result **9** describes the only possibilities for defining an orthogonality in a finite inversive plane by means of incidence; but it is not clear whether, conversely, these must actually yield orthogonalities. In the even case, this can be proved:

11. *Every finite inversive plane of even order admits precisely one orthogonality, namely that defined by* (6).

This is a simple consequence of (6.1.2) and **4a** above. It is conceivable that an analogous result holds in the odd case as well, in other words that the description of **9b**, when interpreted as definition of $\perp$, must satisfy conditions (2)—(4) in every inversive plane of odd order. This is an unsolved problem.

12. *Let* $\mathbf{I}$ *be an inversive plane of order* n *with an orthogonality. Then:*

(**a**) *Each circle is orthogonal to* n^2 *circles.*
(**b**) *If* o, p, q *are distinct points, then there is a unique circle* $C \mathrm{I} o$ *such that* $C \perp X$ *for all* $X \in (p, q)$.[1])
(**c**) *For any two disjoint circles* C, D, *there exists a unique unordered point pair* p, q *such that* $C \perp X \perp D$ *for all* $X \in (p, q)$.

The proofs are again in Dembowski & Hughes 1965, pp. 177—179.

We shall now outline a proof of the following theorem, which is the principal reason for introducing the concept of orthogonality in the present context:

13. *A finite inversive plane is egglike if and only if it admits an orthogonality.*

(Dembowski & Hughes 1965, Theorem 1). That egglike finite inversive planes possess an orthogonality is a consequence of result **1.4.54** which guarantees that to any ovoid $\mathfrak{o}$ in $\mathscr{P}(3, q)$ there exists a polarity π such that $x\pi = T(x)$ for every $x \in \mathfrak{o}$; cf. (6.1.6), (6.1.7). It is a simple consequence of the definitions that, with such a polarity, we get an orthogonality of the egglike inversive plane $\mathbf{I}(\mathfrak{o})$ by defining

$$C \perp D \quad \textit{if and only if} \quad C\pi \mathrm{I} D. \tag{7}$$

The converse is more difficult; only the main ideas can be outlined here. Let $\mathbf{I} = (\mathfrak{p}, \mathfrak{C}, \mathrm{I})$ be a finite inversive plane of order n, with an orthogonality $\perp$. We define an incidence structure $\mathbf{S} = \mathbf{S}(\mathbf{I})$, whose blocks will be called "planes", as follows. Points of $\mathbf{S}$ are, firstly, the points of $\mathbf{I}$; these are called *real* points. Secondly, the symbols C^*, where $C \in \mathfrak{C}$, are points of $\mathbf{S}$; these are called *ideal* points. Similarly, the planes of $\mathbf{S}$ are firstly the circles of $\mathbf{I}$, called *real* planes, and secondly the symbols p^*, where $p \in \mathfrak{p}$; these are the *ideal* planes. Incidence is defined as follows:

$p \mathrm{I} C$ *in* $\mathbf{S}$ *if and only if* $p \mathrm{I} C$ *in* $\mathbf{I}$.
$p \mathrm{I} q^*$ *in* $\mathbf{S}$ *if and only if* $p = q$.
$C^* \mathrm{I} p^*$ *in* $\mathbf{S}$ *if and only if* $C \mathrm{I} p$ *in* $\mathbf{I}$.
$C^* \mathrm{I} D$ *in* $\mathbf{S}$ *if and only if* $C \perp D$ *in* $\mathbf{I}$.

With the help of **12**, it may now be shown that $\mathbf{S}$ has the following properties:

(8) *The number of points, and that of planes, of* $\mathbf{S}$ *is* $(n+1)(n^2+1)$.

(9) *Every plane of* $\mathbf{S}$ *is incident with* n^2+n+1 *points.*

[1]) Another way of saying this is: the set of all circles orthogonal to every $X \in (p, q)$ is a flock with carriers p and q.

(10) *Any two points of* $\mathbf{S}$ *are connected by exactly* $n+1$ *planes.*

(11) *The* $n+1$ *planes through two distinct points have exactly* $n+1$ *common points.*

By (8)—(10), $\mathbf{S}$ is a projective design, and (11) shows, because of **2.1.23**, that $\mathbf{S}$ is isomorphic to $\mathbf{P}_2(3, n)$. In particular, n must be a prime power. Finally, the real points of $\mathbf{S}$ form an ovoid $\mathfrak{o}$ in $\mathscr{P}(3, n)$, and $\mathbf{I}$ is essentially identical with the egglike inversive plane $\mathbf{I}(\mathfrak{o})$. Furthermore, it is fairly obvious that the correspondence $p \longleftrightarrow p^*$, $C \longleftrightarrow C^*$ is essentially the same as the polarity π of (7). Note that this yields a proof of result **1.4.54**; cf. DEMBOWSKI 1964b, Zusatz 4.

We conclude with two consequences of **13**.

14. *Every inversive plane of even order* n *is egglike. Consequently, it satisfies the Bundle Theorem, and* n *is a power of* 2.

(DEMBOWSKI 1963b, 1964b). Besides **13**, one needs **11** and **6.1.4** for this result.

15. *An inversive plane of odd order admits an orthogonality if, and only if, it is miquelian.*

This follows from a combination of **13** with results **6.1.5** (VAN DER WAERDEN & SMID 1935) and **1.4.50** (BARLOTTI 1955, PANELLA 1955).

6.3 Automorphisms

In this section we discuss general facts concerning automorphisms of finite inversive planes. Some of these will be applied in 6.4, where the known finite inversive planes $\mathbf{M}(q)$ and $\mathbf{S}(q)$, defined in 2.4, will be characterized in various ways. As in the preceding sections, we denote by $\mathbf{I}_p$ the affine plane of (6.1.4) and by $\mathbf{I}_p^*$ the projective plane corresponding to it, for an arbitrary point p of the inversive plane $\mathbf{I}$. If p is a fixed point of the automorphism $\varphi \in \operatorname{Aut}\mathbf{I}$, then the induced collineations in $\mathbf{I}_p$ and $\mathbf{I}_p^*$ are denoted by φ_p and φ_p^*, respectively. Also, we call $f = f(\varphi)$ the number of fixed points of φ, and $F_p = F_p(\varphi)$ the number of fixed pencils[1]) with carrier p. As these fixed pencils correspond to the fixed ideal points of φ_p in $\mathbf{I}_p$, application of **4.1.2** to φ_p^* yields:

1. *For any point* p *fixed by* $\varphi \in \operatorname{Aut}\mathbf{I}$, *the number of* φ*-fixed circles through* p *is* $f(\varphi) + F_p(\varphi) - 2$.

(DEMBOWSKI 1965a, Lemma 3.1.) It follows in particular that this integer is non-negative.[2]) We have seen earlier (cf. **6.1.7** above) that

[1]) As before, this means fixed as a whole, not elementwise.

[2]) This was left open in DEMBOWSKI 1965a, Lemma 3.1.

if φ fixes four non-concircular points (i.e. if φ is planar), then the fixed points of φ are those of a subplane $\mathbf{F}(\varphi)$. It is natural to ask for the relations between the orders of $\mathbf{I}$, φ, and $\mathbf{F}(\varphi)$. The following can be proved:

2. *Let φ be a nontrivial planar automorphism of an inversive plane of order n, and suppose that $\mathbf{F}(\varphi)$ is of order m. Then*

(**a**) *φ is of odd order,*
and
(**b**) $m^3 \leqq n$.

Equality holds in (**b**) *if and only if every non-fixed point is on some circle of $\mathbf{F}(\varphi)$.*

For proofs, see DEMBOWSKI 1965a, Sätze 2.2 and 2.5. The main reasons for (**a**) are **4.1.9** and **6.2.6b**: no power of φ can be an involution. On the other hand, (**b**) is analogous to **4.1.6** in a similar way as **6.2.6b** is to **3.2.18**; again we have a stronger result for inversive than for projective planes. The condition for equality in (**b**) is satisfied, for example, if $o(\varphi) = 3$.[1]) If $\mathbf{I}$ is miquelian, then conversely every planar automorphism φ with $n = m^3$ has order 3 (this will follow from the more general result **6.4.2** below). In the general case, the following weaker result can be proved:

3. *If φ is planar and $\mathbf{F}(\varphi)$ of order m such that $m^3 = n$, then $o(\varphi)$ is (odd and) a divisor of $m(m^2 - 1)$.*

(DEMBOWSKI 1965a, Zusatz 2.6.)

Result **2a** leads to the following classification of involutions in finite inversive planes:

4. *Let φ be an involutorial automorphism of an inversive plane of order n. If φ is not an inversion (cf.* **6.1.8a**), *then φ is either*

(**a**) *a translation, in which case n must be even,*
or else
(**b**) *fixed point free or a dilatation which is not a translation; in this case n must be odd.*

This is Satz 2.3 of DEMBOWSKI 1965a. Note that fixed point free involutions occur in every finite miquelian plane of odd order: Let c be a non-square in $GF(q)$, where q is odd; then the mapping

$$x \to x^{-1}c \tag{1}$$

[1]) This remark is comparable to **3.1.6**, that involutorial collineations of projective planes are quasicentral, and facilitates a classification analogous to **4.1.9** of automorphisms of order 3 in finite inversive planes. We shall not give such a classification here; note, however, that every automorphism of order 3 has a fixed point, because -1 is not a quadratic residue mod 3.

can be interpreted as an automorphism of $\mathbf{M}(q)$ [cf. (6.1.9)], and it is obviously fixed point free and of order 2. Result **4a** has the following converse:

5. *In an inversive plane of even order, every non-trivial translation is involutorial.*[1])

This is a simple consequence of **6.2.14**, **6.1.3**, and **4.3.4**. The result can also be inferred from a more general result on automorphisms of egglike inversive planes; this is of considerable interest in its own right:

6. *Let* $\mathfrak{o}$ *be an ovoid in* $\mathscr{P}(3, q)$. *Then the group* $\mathrm{Aut}\,\mathbf{I}(\mathfrak{o})$ *is essentially identical with the group* $\Gamma(\mathfrak{o})$ *of those collineations of* $\mathscr{P}(3, q)$ *which leave* $\mathfrak{o}$ *invariant.*

(Dembowski 1964b, Zusatz 3.) We list some consequences of **6** for inversive planes of even order, which are egglike by **6.2.14**.

7. *Let* $\mathbf{I}$ *be an inversive plane of even order* $q = 2^e$. *Then* $\mathrm{Aut}\,\mathbf{I}$ *is a subgroup of* $P\Gamma L_4(q)$; *let* $\Lambda = \mathrm{Aut}\,\mathbf{I} \cap PGL_4(q) \lhd \mathrm{Aut}\,\mathbf{I}$. *Then:*

(**a**) *Every involution of* $\mathbf{I}$ *is in* Λ.
(**b**) *Every* $\lambda \in \Lambda$ *with* $f(\lambda) > 2$ *is an inversion.*
(**c**) *If* $\sigma \in \mathrm{Aut}\,\mathbf{I}$ *has* $f(\sigma) = 1$, *then the centralizer of* σ *in* Λ *is a* 2*-group.*

For proofs, see Lüneburg 1966c, Lemmas 3—5. Results analogous to (**a**) and (**b**) hold also in egglike inversive planes of odd order, but as these are miquelian by **6.2.13** and **6.2.15**, this is not so interesting. Result **7** will be useful in Section 6.4.

We return to automorphisms of arbitrary finite inversive planes $\mathbf{I}$. Let us call $\varphi \in \mathrm{Aut}\,\mathbf{I}$ *circular* if $\mathbf{F}(\varphi)$ contains exactly one circle. This means[2]) that $f(\varphi) \geqq 3$, and that all φ-fixed points are incident with the same circle $A = A(\varphi)$. Clearly, all inversions are circular. The following results are generalizations of **6.1.8** for the finite case:

8. *Let* φ *be a circular automorphism of an inversive plane of even order* n. *Then:*

(**a**) φ *is of even order, and the highest power of* 2 *dividing* $o(\varphi)$ *divides* n.[3])
(**b**) *The number of* φ*-fixed circles through an arbitrary fixed point* p *is* $f(\varphi) - 1$; *these circles all belong to the pencil* $\langle p, A(\varphi)\rangle$.
(**c**) *The total number of* φ*-fixed circles is* $(f(\varphi) - 1)^2$.
(**d**) *If* φ *is not an inversion, i.e. if* $f(\varphi) < n + 1$, *then* $(f(\varphi) - 1)^2 \leqq n$, *and even* $(f(\varphi) - 1)^3 \leqq n$ *except when* $o(\varphi)$ *is a power of* 2.

---

[1]) The analogous converse of **4b** is clearly false: miquelian inversive planes of odd order $n > 3$ admit non-involutorial dilatations; cf. (6.1.9).

[2]) See the definition of $\mathbf{F}(\varphi)$ in the context of **6.1.7**.

[3]) Note that n itself must be a power of 2, by **6.2.14**.

Proofs are in DEMBOWSKI 1965a, Sätze 3.4 and 3.7. In the miquelian inversive planes $\mathbf{M}(2^e)$, the cases $(f-1)^2 = n$ and $(f-1)^3 = n$ can actually occur; this will again follow from **6.4.2** below.—Much less than **8** is known for odd orders:

9. *Let φ be a circular automorphism of an inversive plane of odd order n. Then:*

(**a**) *There are exactly $f(\varphi)$ fixed circles through each fixed point, and $f(\varphi) - 1$ of these belong to the same pencil, not containing $A(\varphi)$.*
(**b**) *The total number of fixed circles which are distinct but not disjoint from $A(\varphi)$ is $f(\varphi)[f(\varphi)-1]/2$.*

This is also contained in Satz 3.4 of DEMBOWSKI 1965a. Result **6.4.2** below will show that **9** can be improved in the miquelian case; in particular one can show that circular automorphisms in $\mathbf{M}(q)$ have even order, also for odd q. It seems plausible to conjecture that this is true in general, i.e. that the order of a circular automorphism must always be even.[1]) The following result on inversions is an obvious consequence of **8b** and **9a**:

10. *Let φ be an inversion of an inversive plane of order n, and x a point on the axis A of φ. As in* **6.1.8b**, *let $\mathfrak{P}_x$ denote the unique pencil with carrier x that is fixed elementwise by φ. Then*

(**a**) *If n is even, $A \in \mathfrak{P}_x$;*
(**b**) *If n is odd, $A \notin \mathfrak{P}_x$.*

This provides a proof for a remark made after **6.1.8** above; actually it is easier to prove **10** directly than by way of **8** and **9**.

The remainder of this section will be concerned with automorphism groups of finite inversive planes. We begin with some results on 2-groups of such automorphisms.

11. *Let* **I** *be an inversive plane of order $n \equiv -1 \bmod 4$ and Γ an automorphism group of* **I** *which, when considered as a permutation group of the points, consists of even permutations only. Then all involutions in Γ are dilatations, and the Sylow 2-group of Γ is cyclic, dihedral, semi-dihedral, or a generalized quaternion group.*

This is Satz 3 of HERING 1967b; the proof relies on **4.2.2** and **4.2.3** and is similar to that of **4.2.5**. That all involutions in Γ are dilatations follows from **4**: fixed point free involutions are odd permutations, and so are inversions if $n \equiv -1 \bmod 4$. A simple consequence of **11** is that in case $n \equiv -1 \bmod 4$ the Sylow 2-group of Aut**I** has a subgroup of

[1]) It can be proved, for example, that if φ is circular of odd prime order p, and if every point is on some φ-fixed circle (this is satisfied, for example, if $p = 3$), then $n = f(\varphi)[f(\varphi)-1]/2$ which is not a prime power for $f(\varphi) > 3$.

index $\leqq 2$ which is cyclic, dihedral, semi-dihedral or generalized quaternion.

Under the hypotheses of **11**, the product of two commuting involutions in Γ, which must also be an involution, is in fact a dilatation. This is a special case of the following more general result:

12. *Let* $\mathbf{I}$ *be an inversive plane of order* n, *and* α, β *two involutions in* $\mathrm{Aut}\,\mathbf{I}$ *such that* $\alpha\beta = \beta\alpha = \gamma \neq 1$. *Then* γ *is also an involution, and*

(**a**) *If* α *and* β *are inversions, then* γ *is a dilatation, and in fact a translation if and only if* n *is even.*
(**b**) *If* α *and* β *are translations, then* n *is even and* γ *is also a translation.*
(**c**) *If* α *and* β *are dilatations but not translations, then* n *is odd and* γ *is also such a dilatation.*

Result (**a**) is Satz 4.2 of Dembowski 1965a; it is also proved there that the center(s) of γ is (are) either the common point(s) of the axes of α and β or, if these are disjoint (which happens only when n is odd), the points of the unique point orbit of length 2 not on $A(\alpha)$ or $A(\beta)$, under the (noncyclic) group of order 4 generated by α and β. In case (**b**), it is clear that α and β have the same center; the result then follows from the corresponding fact for affine planes. Finally, result (**c**) is essentially Hilfssatz 2.7 of Hering 1965; the only non-trivial case here is that where the centers of α and β are four distinct points. We note an immediate corollary of (**a**) and **4.2.2**:

13. *If all involutions in a 2-group* $\Gamma \subseteqq \mathrm{Aut}\,\mathbf{I}$ *are inversions, then* Γ *is cyclic or a generalized quaternion group.*

This will be useful in Section 6.4.

In the remainder of this section, we report on what little is known about the existence problem for finite inversive planes (and automorphism groups) of the types I.1—VII.2 of Hering 1965, listed in **6.1.14**.

14. *Let* C *be a circle in a finite inversive plane* $\mathbf{I}$, *and suppose that* $\Gamma \subseteqq \mathrm{Aut}\,\mathbf{I}$ *is* $\langle x, C\rangle$*-transitive*[1]) *for every* $x \mathrel{\mathrm{I}} C$. *Then* Γ_C *is transitive on the points non-incident with* C.

(Hering 1965, Hilfssatz 2.6.) In fact, a Γ_C-orbit of points not on C is easily seen to contain at least $n^2/2$ points, while the total number of points not on C is $n(n-1) < n^2$. An immediate corollary of **14** is:

[1]) Recall that $\langle x, C\rangle$ is the pencil with carrier x containing C.

15. *No finite inversive plane admits an automorphism group of Hering type* V.

For the next result, the following theorem on permutation groups is essential:

16. *Let G be a transitive permutation group of degree $n^2 + 1$, and suppose that the stabilizer G_x of the letter x permutes the n^2 letters $\neq x$ in two orbits of length n and $n(n-1)$. Then $n = 2, 3, 7$, or 57.*

Proof: D. G. HIGMAN 1964, Theorem 1. Now observe that if an inversive plane $\mathbf{I}$ of order n admits an automorphism group Γ of Hering type VI.2, then Γ satisfies the hypotheses of **16** when considered as permutation group of the points of $\mathbf{I}$. But $n = 57$ is excluded by the Bruck-Ryser Theorem **3.2.13**, and $n = 7$ is also impossible, by another result of HIGMAN 1966. Since (x, y)-transitivity with $x \neq y$ is a vacuous condition for the inversive plane $\mathbf{M}(2)$ of order 2, it follows that

17. *The only finite inversive plane admitting an automorphism group of Hering type* VI.2 *is the miquelian plane* $\mathbf{M}(3)$ *of order* 3.[1])

In particular, there exist no finite inversive planes of type VI.2.— An analogous result holds for the types IV.3 and IV.4:

18. *The only finite inversive plane admitting an automorphism group of Hering type* IV.3 *or* IV.4 *is the miquelian plane* $\mathbf{M}(3)$ *of order* 3.

This was proved by HERING 1965, pp. 260—261; **6.1.10** is an essential tool. Again it follows that there are no finite inversive planes of Hering type IV.3 or IV.4.

6.4 The known finite models

We have mentioned repeatedly that the only known finite inversive planes are the $\mathbf{M}(q)$ and $\mathbf{S}(q)$ defined in Section 2.4. We shall now list some basic properties of these and give intrinsic characterizations of $\mathbf{M}(q)$ and $\mathbf{S}(q)$, mostly by properties of their automorphism groups.

We begin with the finite miquelian inversive planes $\mathbf{M}(q)$. As pointed out in Section 6.1, these may be defined in two different ways: Firstly, $\mathbf{M}(q)$ is the egglike inversive plane $\mathbf{I}(\mathfrak{q})$, with $\mathfrak{q}$ a non-ruled quadric (i.e. a quadric of index 1) in $\mathcal{P}(3, q)$; secondly, the points of $\mathbf{M}(q)$ can be represented by the elements of $GF(q^2) \cup \{\infty\}$ and the circles by the images of the point set $K = GF(q) \cup \{\infty\}$ under the group $P\Gamma L_2(q^2)$ of permutations

$$(1) \qquad x \to \frac{x^\alpha a + c}{x^\alpha b + d}, \qquad a d \neq b c, \qquad \alpha \in \operatorname{Aut} GF(q^2);$$

[1]) That there is only one inversive plane of order 3, namely $\mathbf{M}(3)$, was proved by WITT 1938b, Satz 3.

cf. (6.1.9). The permutations (1) with $\alpha = 1$ form a normal subgroup $PGL_2(q^2)$ of Aut $\mathbf{M}(q)$, and in fact it suffices[1]) to define the circles as the images of K under $PGL_2(q^2)$: It is well known (and easily proved) that $PGL_2(q^2)$ is sharply 3-transitive on the points of $\mathbf{M}(q)$; from axiom (6.1.1) it then follows that $PGL_2(q^2)$ is transitive on the circles of $\mathbf{M}(q)$. Also, the stabilizer of the point ∞ contains the group of all automorphisms

(2) $$x \to x + c, \quad c \in GF(q),$$

which is $\mathfrak{P}$-transitive for the pencil $\mathfrak{P}$ with carrier ∞ containing K, and the group of all

(3) $$x \to xa, \quad 0 \neq a \in GF(q),$$

which is $\mathfrak{B}$-transitive for the bundle $\mathfrak{B} = (0, \infty)$. Therefore:

1. *For the miquelian inversive plane* $\mathbf{I} = \mathbf{M}(q)$, *with* $q = p^e$, *the group* Aut$\,\mathbf{I} \cong P\Gamma L_2(q^2)$ *has order* $2eq^2(q^4 - 1)$. *It is triply transitive on points and transitive on circles, and it is of Hering type* VII.2.

By triple transitivity on points, the stabilizers of any three points are isomorphic. But (1) shows that the stabilizer of $0, 1, \infty$ consists of the permutations

(4) $$x \to x^\alpha, \quad \alpha \in \text{Aut}\, GF(q^2);$$

hence it is isomorphic to the automorphism group of $GF(q^2) = GF(p^{2e})$ which is cyclic of order $2e$.

2. *The automorphism* (4) *of* $\mathbf{M}(q)$ *has precisely* $1 + p^{2eo(\alpha)^{-1}}$ *fixed points, and it is planar or circular according as* $o(\alpha)$ *is odd or even.*

This is easily verified; cf. Dembowski 1965a, 6.2. In particular, it follows that there exist circular automorphisms φ with $(f(\varphi) - 1)^2 = q$ if (and, trivially, only if) q is a square, and then $o(\varphi) = 4$. Similarly, there exist planar automorphisms ψ with $\mathbf{F}(\psi)$ of order $\sqrt[3]{q}$ whenever q is a cube, and then $o(\psi) = 3$. (Compare here the remarks after **6.3.2** and **6.3.8**.) The cyclic group Aut $GF(p^{2e})$ contains a unique involution, namely

(5) $$x \to x^q.$$

This fixes every point of the circle $K = GF(q) \cup \{\infty\}$ and is, therefore, an inversion.

3. *Every circle of* $\mathbf{M}(q)$ *is the axis of an inversion. The group* Π *generated by all inversions contains all mappings* (1) *with* $\alpha = 1$ *and* $ad - bc$ *a non-zero square in* $GF(q^2)$; *these form a subgroup* $\Delta \cong PSL_2(q^2)$ *which is doubly transitive on the points, with* $\Delta_{xyz} = 1$. *Moreover,*

[1]) Compare here footnote [2]) of p. 257.

$\Pi = \Delta Z$, *where* Z *is the group of order* 2 *generated by the inversion* (5). *For any circle* C, *the stabilizer* Π_C *is triply transitive on the points of* C, *and*

(**a**) *If* q *is even, then* $|\Pi| = 2q^2(q^4 - 1)$, *and* Π *is transitive on the circles, hence* 3*-transitive on the points of* $\mathbf{M}(q)$.
(**b**) *If* q *is odd, then* $|\Pi| = q^2(q^4 - 1)$, *and* Π *permutes the circles in two orbits of equal length* $q(q^2 - 1)/2$.

For the proof of these results, see DEMBOWSKI 1965a, Section 6.[1])
We shall see later that several of the properties of the $\mathbf{M}(q)$ listed in **1**—**3** are in fact characteristic for these inversive planes. Others, however, are shared by the $\mathbf{S}(q)$ which we will now discuss briefly.

By definition, $\mathbf{S}(q)$ is the egglike inversive plane $\mathbf{I}(\mathfrak{t})$, where $\mathfrak{t} = \mathfrak{t}(\psi)$ is the ovoid of **1.4.56a**, represented in $\mathscr{A}(3, q)$, $q = 2^e$, e odd and > 1, by the equation

$$z = xy + x^{\sigma+2} + y^\sigma, \tag{6}$$

where $\sigma\colon x \to x^{2^{(e+1)/2}}$ is the unique automorphism of $GF(q)$ satisfying $x^{\sigma^2} = x^2$ for all $x \in GF(q)$.

4. *For the inversive plane* $\mathbf{S}(q)$, *with* $q = 2^e$, e *odd* > 1, *the group* Aut $\mathbf{S}(q)$ *is the semidirect product* $Sz(q) \cdot$ Aut $GF(q)$, *of order* $eq^2(q^2 + 1)(q - 1)$. *It is of Hering type* VI.1, *doubly transitive on points and transitive on circles, but not flag transitive and hence not triply transitive on points.*

This follows from the corresponding results **1.4.57** and **1.4.58** on the ovoids $\mathfrak{t}(\psi)$. Incidentally, these results also show that some of the properties of **4** are characteristic for the $\mathbf{S}(q)$; this will be given in more detail further below. A simple corollary of **4** is:

5. $\mathbf{S}(q)$ *admits no inversions, and all dilatations are translations.*

For the stabilizer of $(0, 0, 0)$ and the ideal point of the z-axis [both these points are in $\mathfrak{t}(\psi)$] consists of the mappings

$$(x, y, xy + x^{\sigma+2} + y^\sigma) \to (tx^\alpha, t^{\sigma+1}y^\alpha, t^{\sigma+2}(xy + x^{\sigma+2} + y^\sigma)^\alpha)$$

with $0 \neq t \in GF(q)$ and $\alpha \in$ Aut $GF(q)$; all these are of odd order, and only the identity fixes all planes $ax + by = 0$; hence **5** follows from **6.1.8a** and **6.1.9b.**

[1]) The reader will notice that Satz 6.8 of DEMBOWSKI 1965a differs essentially from the present result. As a matter of fact, Satz 6.8 as stated there is false; the proof contains an error on the bottom of p. 135. The correct statement is that the circle K can be mapped onto Kc within the stabilizer of 0 and ∞ if and only if c is a square in $GF(q^2)$. Hence this stabilizer decomposes the circles of the bundle $(0, \infty)$ into two orbits of length $(n + 1)/2$. The other arguments concerning Satz 6.8, or obvious modifications of them, are correct.

We turn now to the question which of the properties in **1**—**5** are characteristic for the inversive planes $\mathbf{M}(q)$ and $\mathbf{S}(q)$. We begin with the following results of LÜNEBURG 1964b.

6. *Let* $\mathbf{I}$ *be an inverse plane of order* $n > 2$, *and suppose that* $\mathrm{Aut}\,\mathbf{I}$ *contains a group* Γ *satisfying the following conditions:*

(7) Γ *is doubly transitive on the points of* $\mathbf{I}$.

(8) $\Gamma_{xyz} = 1$, *for distinct points* x, y, z.

Then n *is a prime power* p^e, *and either*

(**a**) Γ *contains a subgroup isomorphic to* $PSL_2(n^2)$,
or
(**b**) $p = 2$ *and odd* $e > 1$, *and* $\Gamma \cong Sz(n)$.

The result will follow from **4.3.27** if it can be shown that Γ is not of one of the types (**a**) or (**b**) of **4.3.27**. If Γ were of one of these types, then Γ would contain a sharply 2-transitive subgroup Δ. As the degree of a sharply 2-transitive permutation group is a prime power, we would have

(9) $$n^2 + 1 = p^r$$

for some prime p. But

7. *If* (9) *holds for some integer* n *and a prime* p, *then* $r = 1$.

This follows from a theorem of LEBESGUE 1850. Hence in our case Γ and Δ are of prime degree, whence the stabilizer Δ_x of a point x in $\mathbf{I}$ is cyclic. On the other hand, Δ_x induces a cyclic transitive collineation group of the affine plane $\mathbf{I}_x$, which by **4.4.6** is possible only if $n = 2$. Since this was excluded, **6** follows.

8. *Let* $\mathbf{I}$ *be an inversive plane of prime power order* q, *and* $\Gamma \subseteqq \mathrm{Aut}\,\mathbf{I}$.

(**a**) *If* $\Gamma \cong PSL_2(q^2)$, *then* $\mathbf{I} \cong \mathbf{M}(q)$.
(**b**) *If* $q = 2^e$, *with odd* $e > 1$, *and* $\Gamma \cong Sz(q)$, *then* $\mathbf{I} \cong \mathbf{S}(q)$.

These are Theorems 1 and 2 of LÜNEBURG 1964b; we give here the basic idea of the proof. If q is even and $\Gamma \cong PSL_2(q^2) = PGL_2(q^2)$, then it can be shown that Γ is flag transitive on $\mathbf{I}$, so that $\mathbf{I}$ can be represented in the form $\mathbf{K}(\Gamma, \Gamma_p, \Gamma_C)$, for some incident point-circle pair p, C [cf. **1.2.17**]. But $\mathbf{M}(q)$ also possesses a representation in this form, and it is then not difficult to verify that these two representations are the same. This proves (**a**) for even q. If q is odd, or if we are in case (**b**), then Γ is not flag transitive, so that the situation is more difficult. It can be shown, however, that in these cases the representations $\mathbf{C}(\Gamma, \Gamma_p, \Delta)$, with suitable quotient sets Δ, are essentially unique, so that $\mathbf{I} \cong \mathbf{M}(q)$ or $\mathbf{S}(q)$, according as $\Gamma \cong PSL(q^2)$ or $Sz(q)$. For more details, the reader is referred to the quoted paper by Lüneburg.

Combining **6** and **8**, we see that the existence of an automorphism group Γ satisfying (7) and (8) forces an inversive plane to be either $\mathbf{M}(q)$ or $\mathbf{S}(q)$. This result will be improved further below; cf. **10**.

9. *If every circle of the finite inversive plane* $\mathbf{I}$ *is the axis of an inversion, then* $\mathbf{I}$ *is miquelian.*

This is Satz 5.5 of DEMBOWSKI 1965a; note that the converse is also true, by **3**. Again we give only the main idea of the proof. Let Π denote the group generated by all inversions; then Π is doubly transitive on the points of $\mathbf{I}$. (In fact, Π_x contains a transitive translation group of $\mathbf{I}_x$ for every point x, because of **6.3.12a, 4.3.4c** if n is even, and **6.3.12a, 3.1.8a, 4.3.2a** if n is odd.) In case n even, Π is even triply transitive on the points (DEMBOWSKI 1964b, Lemma 2); the result then follows from **6.2.14** and **1.4.51b**. Hence suppose that n is odd, and denote by $\pi(A)$ the inversion with axis A. Define

(10) $\quad A \perp B$ *if and only if* $A \neq B$ *and* $\pi(A)\,\pi(B) = \pi(B)\,\pi(A)$;

then this relation $\perp$ can be proved to be an orthogonality in the sense of (6.2.2)—(6.2.4), an essential tool in this proof being **6.3.12a**. That $\mathbf{I}$ is miquelian then follows from **6.2.15**.

In **8** and **9**, the double point transitivity of the groups involved played an important role, so that it is natural to ask whether this property alone guarantees that $\mathbf{I}$ is either $\mathbf{M}(q)$ or $\mathbf{S}(q)$. This is indeed the case:

10. *Let* $\mathbf{I}$ *be an inversive plane of order* n, *admitting an automorphism group* Γ *which is doubly transitive on the points of* $\mathbf{I}$. *Then* n *is a prime power, and* $\mathbf{I}$ *is either the miquelian plane* $\mathbf{M}(n)$, *in which case* Γ *contains* $PSL_2(n^2)$, *or else* $n = 2^e$ *with odd* $e > 1$, $\mathbf{I} = \mathbf{S}(n)$, *and* $\Gamma \supseteq Sz(n)$.

For even n, this is a simple consequence of **6.2.14** and **1.4.51b, 1.4.57b**. We shall now outline a proof of **10** for odd n; this is due to HERING 1967b, where more details can be found. It clearly suffices to consider the case where Γ is minimal, i.e. no proper subgroup of Γ is 2-transitive. Let Γ_0 denote the normal subgroup of those automorphisms in Γ which induce even permutations on the points of $\mathbf{I}$. It is easily checked that involutions in Γ_0 cannot be free of fixed points; in fact they are dilatations if $n \equiv -1 \bmod 4$. (cf. **6.3.11**).

Let Δ be a non-trivial normal subgroup of Γ which is contained in Γ_0. Then Δ is transitive of even degree $n^2 + 1$, hence contains an involution, and it can be inferred from **6.3.13** that there actually exists an involutorial dilatation in Δ. As $\Delta \lhd \Gamma$ and Γ is doubly transitive, every pair of points in $\mathbf{I}$ consists of the centers of an involutorial dilatation in Δ. Hence if p is an arbitrary point, the stabilizer Δ_p induces a collineation group of $\mathbf{I}_p$ which, for every point q in $\mathbf{I}_p$, contains an involutorial dilatation with center q. Result **4.3.2a** then implies that $\mathbf{I}_p$ is a translation plane, and Δ_p contains the full translation group of $\mathbf{I}_p$. This shows that Δ is at least of Hering type VII.1. In particular, Δ is doubly transitive, so that the minimality of Γ implies $\Delta = \Gamma_0 = \Gamma$. Furthermore, Γ must be a simple group.

For the remainder of the proof, the following theorem on permutation groups is needed:

11. *Let Π be a doubly transitive permutation group of degree $m+1$, such that*

(11) $\quad |\Pi_{xy}| \equiv 0 \not\equiv |\Pi_{xyz}| \bmod 2,$ *for distinct points x, y, z.*

Then m is an odd prime power, and:

(**a**) *The Sylow 2-groups of Π are dihedral or semi-dihedral;*
(**b**) *If $m \neq 5$, then $PSL_2(m) \lhd \Pi \subseteqq P\Gamma L_2(m)$.*

For the proof, which uses **4.3.28**, the reader is referred to HERING 1968a. We return to the proof of **10** for odd n and show next that

(*) $\qquad \Gamma$ *contains no inversions.*

If every circle were the axis of an inversion in Γ, then $\mathbf{I}$ would be miquelian by **9**. But then **3** shows that Γ would contain a doubly transitive proper subgroup, contradicting the minimality of Γ. Hence if (*) is false, there exists a Γ-orbit $\mathfrak{X}$ of circles which are axes of inversions in Γ, as well as a circle orbit $\mathfrak{Y}$ whose circles do not admit such inversions. If $\mathfrak{Z}$ is any circle orbit, then its circles, together with all points of $\mathbf{I}$, form a design with parameters $v = n^2+1$, $k = n+1$, and $\lambda < n+1$; hence the basic equations (2.1.5) yield

$$(n+1)\,|\mathfrak{Z}| = n(n^2+1)\,\lambda.$$

But $(n(n^2+1), n+1) = 2$, so that $|\mathfrak{Z}| = n(n^2+1)$ or $n(n^2+1)/2$. The first alternative would mean that Γ is transitive on circles which is impossible in our situation. Hence

$$|\mathfrak{X}| = |\mathfrak{Y}| = n(n^2+1)/2,$$

and there are no circle orbits besides $\mathfrak{X}$ and $\mathfrak{Y}$.

Now let $C \in \mathfrak{Y}$. Then the stabilizer Γ_C acts faithfully and 2-transitively on the $n+1$ points of C, since Γ is of Hering type at least VII.1. Also, Γ_C satisfies condition (11) when regarded as permutation group on (C). Hence **11a** shows that a Sylow 2-group Σ of Γ_C is dihedral or semi-dihedral. On the other hand, since $|\Gamma : \Gamma_C| = |\mathfrak{Y}| = n(n^2+1)/2$ is odd, Σ is actually a Sylow 2-group of Γ, and since

$$|\Gamma| = n^2(n^2+1)\,|\Gamma_{pq}| = \tfrac{1}{2} n^2(n^2+1)\,|\Gamma_{\{pq\}}|,$$

where $\Gamma_{\{pq\}}$ denotes the group of all automorphisms in Γ which either fix or interchange the distinct points p and q, we can assume $\Sigma \subseteqq \Gamma_{\{pq\}}$.

It can then be shown that Σ contains inversions σ, τ such that $p\sigma = q = q\tau$ and involutorial dilatations α, β with $p\alpha = q = q\beta$. It follows that $\alpha, \beta, \sigma, \tau$ belong to 4 different conjugate classes of Σ; but dihedral and semi-dihedral groups have only at most 3 conjugate classes of involutions. This contradiction proves (*).

Since there are no fixed point free involutions in Γ, it follows now from **6.3.4** that every involution in Γ is a dilatation, with exactly 2 fixed points. Hence Γ satisfies the hypotheses of **11b** because $n^2 \neq 5$, and we can conclude that Γ contains $PSL_2(n^2)$ as a normal subgroup. But then **8a** shows that $\mathbf{I} \cong \mathbf{M}(n)$, and **10** is proved.

We list some consequences of **10**.

12. *A finite inversive plane* **I** *is isomorphic to an* $\mathbf{M}(q)$ *if and only if* Aut**I** *is of Hering type* VII.2. *There are no finite inversive planes of type* VII.1.[1])

For a group Γ of type VII is easily seen to be doubly transitive on the points, so that **12** follows from **10**, **1**, and **4**.

13. *A finite inversive plane is miquelian if, and only if, it admits a flag transitive automorphism group.*

Proof: That Aut $\mathbf{M}(q)$ is flag transitive is clear from **1**. Suppose conversely that Γ is flag transitive. Then for any point p, the stabilizer Γ_p induces a collineation group in $\mathbf{I}_p$ which is transitive on the lines. It follows from **4.4.22** that $\mathbf{I}_p$ is a translation plane, and Γ_p contains the full translation group of $\mathbf{I}_p$. This implies that Γ is at least of type VII.1, so that **13** follows from **12**.[2])

In the remainder of this section, we give some characterizations of the $\mathbf{M}(2^e)$ and $\mathbf{S}(2^e)$ among the inversive planes of even order. The first is a rather isolated result.

14. *Suppose that the inversive plane* **I** *of even order satisfies condition* (**M**) *of Section* 6.1 *whenever the circles of the four-chain involved have a common tangent circle. Then* (**M**) *is satisfied universally in* **I**, *so that* **I** *is miquelian.*

For the proof it is shown (Dembowski 1964b, Lemma 1) that if (**M**) holds for all four-chains with common tangent circle C, then there exists an inversion with axis C. The result is then a consequence of **9**.

---

[1]) The preceding arguments actually show that there does not even exist a finite group of type VII.1.

[2]) The proof actually shows that flag transitivity implies point 2-transitivity in finite inversive planes; the converse is not true since $\mathbf{S}(q)$ has a doubly point transitive, but no flag transitive group. Note that in projective planes we have precisely the opposite logical relationship: here double point transitivity obviously implies flag transitivity, but not conversely; cf. **1.4.21**.

Now let the inversive plane **I** of even order n be represented in the form $\mathbf{I}(\mathfrak{o})$, with $\mathfrak{o}$ an ovoid in $\mathscr{P}(3, n)$; this is possible by **6.2.14**. By **6.3.6**, we then have $\mathrm{Aut}\,\mathbf{I} = \Gamma(\mathfrak{o})$, the group of the ovoid $\mathfrak{o}$; cf. **1.4.51**. As in **6.3.7**, put $\Lambda = \Gamma(\mathfrak{o}) \cap PGL_4(n)$; i.e. Λ is the group of those automorphisms of **I** which are induced by projective collineations of $\mathscr{P}(3, n)$.

15. *Let* **I** *be of even order* n *and* $\Gamma \subseteqq \Lambda$. *If*
(**a**) $|\Gamma| = n^2 + 1$,
or if
(**b**) Γ *is soluble and transitive,*
then **I** *is miquelian.*

Result (**a**) is an immediate consequence of **1.4.51f**, and (**b**) can be derived from (**a**) as follows (LÜNEBURG 1966c, Theorem 1): Γ is of order $(n^2+1)\,|\Gamma_p|$, and since $\Gamma \subseteqq PGL_4(n)$, we have

$$|\Gamma_p| \,\Big|\, n^6 (n-1)^3 (n+1)^2,$$

whence $(|\Gamma_p|, n^2+1) = 1$. But then a theorem of P. HALL 1928 (see also M. HALL 1959, Theorem 9.3.1) implies that Γ has a subgroup of order $n^2 + 1$.

The following is an improvement of **10** for the case of even order.

16. *Let* **I** *be an inversive plane of even order* n *and suppose that* Aut **I** *contains a subgroup* Γ *of even order which is transitive on the points. Then* $\mathbf{I} = \mathbf{M}(n)$ *or* $\mathbf{I} = \mathbf{S}(n)$.

This is Theorem 2 of LÜNEBURG 1966c. Certain parts of the proof resemble that of **10**. Again the main difficulty lies in dealing with the case when Γ contains inversions; here **9** is used again. If there are no inversions, **6.3.7c** is used to show that the hypotheses of **4.3.26** are satisfied. In case (**a**) of that theorem, Γ is doubly transitive on points, so that **10** gives the desired result; in case (**b**), on the other hand, Γ must be soluble,[1] whence **16** follows from **15b**. More details are in the quoted paper by Lüneburg.

We conclude with a result which is not restricted to the case of even order:

17. *A finite inversive plane is of one of the types* $\mathbf{M}(q)$ *or* $\mathbf{S}(q)$ *if, and only if, it admits an automorphism group transitive on the circles.*

Such a group is also transitive on the points, by **2.3.2**. Since the number of circles is even (cf. **6.2.3b**), the result follows from **16** in case **I** is of even order (LÜNEBURG 1966c, Corollary). In the case of odd order, it can actually be shown that the group must be doubly transitive on the points (LÜNEBURG 1967c), so that the result follows from **10**.

[1]) This is immediate from the results of THOMPSON 1960a, 1964 and FEIT & THOMPSON 1963, but the solubility of Γ may also be proved without these deep theorems.

7. Appendices

In this last chapter we will be concerned with four topics which, for one reason or another, do not fit naturally into the general setup of the earlier parts of the book. There are nevertheless enough connections to justify the inclusion of these subjects here.

The central concept of the chapter is that of "partial design", a generalization of the notion of "design" dealt with in Chapter 2. In 7.1, some of the combinatorial properties of partial designs will be developed. In 7.2 and 7.3, we deal with two special types of geometries ("Hjelmslev planes" and "Generalized polygons") which give rise to partial designs in a natural fashion. In 7.4, finally, we consider a class of partial planes (which will be called semi-planes here), and we will be concerned with the problem of embedding these into finite projective planes of specified orders. The most homogeneous semi-planes (the "elliptic" ones, see **7.4.3** below) are again special partial designs.

7.1 Association schemes and partial designs

Let $\mathfrak{p}$ be a finite set, and put $|\mathfrak{p}| = v$. We denote by $\mathfrak{p}(2)$ the set of all $v(v-1)/2$ subsets $\mathfrak{x}$ with $|\mathfrak{x}| = 2$ of $\mathfrak{p}$. Consider a *partition*

$$\mathscr{A} = \{A_1, \ldots, A_m\}$$

of $\mathfrak{p}(2)$; this means that the A_i are non-empty sets of unordered pairs of different elements in $\mathfrak{p}$, and that for any two distinct $x, y \in \mathfrak{p}$ there is precisely one $i \in \{1, \ldots, m\}$ such that $\{x, y\} \in A_i$. We call $\mathscr{A}$ an *association scheme*[1]) on $\mathfrak{p}$ if the following condition is satisfied:

(1) *Given* $\{x, y\} \in A_h$, *the number of* $z \in \mathfrak{p}$ *for which* $\{x, z\} \in A_i$ *and* $\{y, z\} \in A_j$ *depends only on* $h, i, j \in \{1, \ldots, m\}$, *not on* x *and* y.

This number will be denoted by p_{ij}^h throughout; the definition implies that $p_{ij}^h = p_{ji}^h$ for all $h, i, j \in \{1, \ldots, m\}$. The sets $A_1, \ldots, A_m$ of the association scheme $\mathscr{A}$ are called its *classes*, and the fact that $\{x, y\} \in A_i$ will also be expressed by saying that x and y are i^{th} *associates*

---

[1]) This term is due to Bose & Shimamoto 1952, but the concept is inherent in Bose & Nair 1939. These authors include the first part of Result **1** below, which follows easily from (1), in their definition.

$(i = 1, \ldots, m)$. The integer m is called the *class number* of $\mathcal{A}$. Note that the unique partitions of $\mathfrak{p}(2)$ with $m = 1$ and $m = v(v-1)/2$ are association schemes on $\mathfrak{p}$. Since these are rather uninteresting, we shall usually assume

$$1 < m < v(v-1)/2$$

for the class numbers m of the association schemes to be considered later on.

1. *Let $\mathcal{A}$ be an association scheme of class number m on $\mathfrak{p}$. Given $x \in \mathfrak{p}$, the number of $y \in \mathfrak{p}$ with $\{x, y\} \in A_i$ depends only on $i \in \{1, \ldots, m\}$, not on x. In fact, if this number is denoted by n_i, then*

$$n_i = \delta_{hi} + \sum_{j=1}^{m} p_{ij}^h \qquad \text{for } i, h = 1, \ldots, m, \tag{2}$$

where δ_{hi} is the Kronecker symbol, and

$$n_h p_{ij}^h = n_i p_{jh}^i = n_j p_{hi}^j, \qquad \text{for } h, i, j = 1, \ldots, m. \tag{3}$$

The proofs, which are essentially contained in Bose & Nair 1939, are effected by simple counting arguments. An obvious consequence of the definition of the n_i is

$$\sum_{i=1}^{m} n_i = v - 1. \tag{4}$$

A *partial design*[1]) is a tactical configuration $\mathbf{S} = (\mathfrak{p}, \mathfrak{B}, \mathrm{I})$, together with an association scheme $\mathcal{A}$ on $\mathfrak{p}$, such that

(5) *Given $\{x, y\} \in A_i$, the number $[x, y]$ of blocks through x and y depends only on $i \in \{1, \ldots, m\}$, not on x, y.*

This number will be denoted by λ_i; note that $i \neq j$ does not imply $\lambda_i \neq \lambda_j$.[2]) Any design, in the sense of Chapter 2, is a partial design with respect to the trivial association scheme of class number $m = 1$; hence the notion of "partial design" may be considered a generalization of that of "design". It should be noted that non-isomorphic partial designs may have isomorphic association schemes, and that the same tactical configuration may be a partial design with respect to non-isomorphic association schemes. The *class number* of a partial design is that of the

---

[1]) This is essentially what the statisticians call "partially balanced incomplete block designs" (PBIBD); cf. Bose & Nair 1939, Bose & Shimamoto 1952, and the other references quoted further below. The concept is very general; we note here that, for example, the Desargues and Pappus configurations are partial designs.

[2]) Some authors use the more restrictive definition that the partition of $\mathfrak{p}(2)$ given by the equivalence relation $\{x, x'\} \sim \{y, y'\}$ if and only if $[x, x'] = [y, y']$ should define an association scheme on $\mathfrak{p}$. In this case, $\lambda_i = \lambda_j$ implies $i = j$. The present wider definition is that of Bose & Shimamoto 1952.

accompanying association scheme, and its *parameters* are, besides the usual v, r, b, k, the integers p_{ij}^h and n_i of (1) and **1**, for the accompanying association scheme, as well as the λ_i $(i = 1, \ldots, m)$ of (5).

2. *The parameters of a partial design satisfy, in addition to* (2)—(4), *the equations*

$$v\,r = b\,k \tag{6}$$

and

$$\sum_{i=1}^{m} \lambda_i\, n_i = r(k-1). \tag{7}$$

This is rather obvious; cf. Bose & Nair 1939. There are, however, also some less trivial relations between the parameters of an association scheme or a partial design, at least for $m > 2$. These will follow from the following matrix arguments due to Bose & Mesner 1959.[1])

Given an association scheme $\mathcal{A} = \{A_1, \ldots, A_m\}$ on the set $\mathfrak{p} = \{x_1, \ldots, x_v\}$, define the (v, v)-matrices $B_h = (b_{ih}^j)$, with $i, j \in \{1, \ldots, v\}$ and $h \in \{1, \ldots, m\}$, as follows:

$$b_{ih}^j = \begin{cases} 1 & \textit{if } \{x_i, x_j\} \in A_h, \\ 0 & \textit{if } \{x_i, x_j\} \notin A_h. \end{cases} \tag{8}$$

In particular, $b_{ih}^i = 0$ for all i, h. Furthermore, we put

$$B_0 = I = I_v, \tag{9}$$

the (v, v)-identity matrix. These matrices $B_0, \ldots, B_m$ are called the *association matrices* of $\mathcal{A}$, relative to the chosen numbering of the elements in $\mathfrak{p}$. The definitions (8) and (9) imply

$$\sum_{h=0}^{m} B_h = J = J_v, \tag{10}$$

the (v, v)-matrix all of whose entries are 1. A simple consequence of (10) is that the matrices $B_0, \ldots, B_m$ are linearly independent over any field. The equation (10) only reflects that $\mathcal{A}$ is a partition; that $\mathcal{A}$ is in fact an association scheme leads to the following non-trivial formula:

$$B_i\, B_h = \sum_{j=0}^{m} p_{ih}^j\, B_j \qquad \textit{for } i, h = 0, \ldots, m; \tag{11}$$

(W. A. Thompson 1958), where

$$p_{ih}^0 = n_i\, \delta_{ih}, \quad p_{i0}^j = p_{0i}^j = \delta_{ij}, \quad \text{and} \quad n_0 = 1.$$

Also, it follows from **1** that

$$J\, B_h = n_h\, J \qquad \textit{for } h = 0, \ldots, m. \tag{12}$$

---

[1]) See also Connor & Clatworthy 1954.

Result (11) shows that the B_h $(h = 0, \ldots, m)$ form the basis of a commutative (since $p_{ih}^j = p_{hi}^j$) matrix algebra of rank $m+1$ over any field.[1] Computing the product $B_i B_j B_h$ in two different ways with (11), one gets the formula

$$\sum_{s=0}^{m} p_{ij}^s p_{sh}^t = \sum_{s=0}^{m} p_{jh}^s p_{is}^t \tag{13}$$

for $i, j, h, t = 0, \ldots, m$. Next, consider the $(m+1, m+1)$-matrices

$$P_h = (p_{ih}^j), \qquad \text{for } i, j, h = 0, \ldots, m;$$

then (13) and $p_{ij}^h = p_{ji}^h$ show that

$$P_i P_h = \sum_{j=0}^{m} p_{ih}^j P_j \qquad \textit{for } i, h = 0, \ldots, m, \tag{14}$$

whence the P_i multiply in the same manner as do the B_i. It is easily seen that the P_i are also linearly independent; hence they form the basis of an algebra isomorphic to that spanned by the (v, v)-matrices B_i, and in particular $P_i P_h = P_h P_i$. Bose & Mesner 1959 have shown that, furthermore, the matrices $B = \sum_{i=0}^{m} c_i B_i$ and $P = \sum_{i=0}^{m} c_i P_i$ have the same eigenvalues; hence B has at most $m+1$ distinct eigenvalues.

Suppose now that **S** is a partial design. Then, in addition to the numbering of the points already used in (8), we number the blocks in an arbitrary way as $X_1, \ldots, X_b$ and consider the (v, b) incidence matrix C thus defined. It is a simple consequence of (5) that

$$CC^T = \sum_{i=0}^{m} \lambda_i B_i, \tag{15}$$

where we use the convention $\lambda_0 = r =$ number of blocks per point in **S**. It follows from the preceding remarks that CC^T has at most $m+1$ distinct eigenvalues. Moreover, Connor & Clatworthy 1954, Sections 3, 4, have shown:

3. *Let C be an incidence matrix of a partial design with class number m. Then*[2])

$$\det CC^T = rk \prod_{i=1}^{m'} (r - z_i)^{\alpha_i}, \tag{16}$$

---

[1]) The preceding matrix formulas provide in turn simple proofs of (2)—(4). For example, comparing row sums in (10) gives (4), and (2) follows from

$$n_h \sum_{i=0}^{m} B_i = n_h J = J B_h = \sum_{j=0}^{m} B_j B_h = \sum_{i=0}^{m} \left(\sum_{j=0}^{m} p_{jh}^i \right) B_i.$$

[2]) For the cases $m = 1$ or $\lambda_1 = \cdots = \lambda_m = 1$, compare **2.1.17**. Also, the more general investigations of A. Brauer 1952 on doubly stochastic matrices should be compared in this context.

with $m' \leqq m$ *and* $\sum_{i=1}^{m'} \alpha_i = v - 1$, *and rational* $z_i \leqq r$; $i = 1, \ldots, m'$. *Furthermore,*

$$b \geqq v - \alpha, \tag{17}$$

where α *is the least* α_i *for which* $z_i = r$, *in case such an* i *exists, and* $\alpha = 0$ *if* $z_i < r$ *for* $i = 1, \ldots, m'$.

The inequality (17) can be considered a generalization of FISHER's 1940 inequality $b \geqq v$ for designs (cf. Section 2.1); it follows immediately from the obvious fact that CC^T has rank $\leqq b$. We note here that there exist partial designs with $v > b$: SHRIKHANDE 1952 has shown that the duals of certain non-projective designs are partial designs of class number 2.[1]) The fact that $v > b$ implies a non-trivial relation $z_i = r$ among the parameters of a partial design was first noticed by NAIR 1943; for the special case $m = 2$ see result **6** below.

The remarks after (14) show that the matrix $\sum_{i=0}^{m} \lambda_i P_i$ has the same eigenvalues as CC^T; in particular, rk is one of these. Applications of this observation to existence problems for partial designs are given in Sections 4—6 of BOSE & MESNER 1959. Among other things, the following is proved there:

4. *Let* $B_0 = I_v, B_1, \ldots, B_m$ *be* (v, v)*-matrices, and* C *a* (v, b)*-matrix, with entries* 0 *and* 1. *If these matrices satisfy the equations* (10), (11), *and* (15), *then they are association matrices and incidence matrix, respectively, of a partial design with parameters* $v, b, k, r, \lambda_i, p_{ij}^h$ $(h, i, j = 0, \ldots, m)$.

For most of the remainder of this section, we restrict our attention to association schemes and partial designs of class number 2. The following are useful supplements to **3**:

5. *Let* $\mathbf{S}$ *be a partial design of class number* 2, *and put*

$$\beta = p_{12}^1 + p_{12}^2, \quad \gamma = p_{12}^1 - p_{12}^2, \quad \delta = 1 + 2\beta + \gamma^2. \tag{18}$$

Then the z_i *and* α_i *of* **3** *are given by*

$$\left.\begin{aligned} z_i &= \left[(\lambda_1 - \lambda_2)\left(\gamma + (-1)^i \sqrt{\delta}\right) + \lambda_1 + \lambda_2\right]/2 \quad &(19)\\ \alpha_i &= \left[(v-1)\left((-1)^{i+1}\gamma + \sqrt{\delta} + 1\right) - 2n_i\right]/2\sqrt{\delta} \quad &(20)\end{aligned}\right\} \quad i = 1, 2,$$

Proof: CONNOR & CLATWORTHY 1954, Section 5. The fact that the multiplicities α_1 and α_2 must be integers can be used to exclude various

[1]) See also HOFFMAN 1963.

sets of parameters compatible with (2)—(4), (6), (7). Also, if $\lambda_1 \neq \lambda_2$, then $\delta > 0$ and $z_1 \neq z_2$; moreover $z_1 < z_2$ if and only if $\lambda_1 > \lambda_2$.

6. *If the partial design* **S** *of class number 2 has more points than blocks, i.e. if $v > b$, and if $\lambda_1 > \lambda_2$, then δ is a square,*

$$(2b - v - 1)\sqrt{\delta} \geqq (v-1)(\gamma - 1) + 2n_2, \tag{21}$$

and

$$2r = (\lambda_1 - \lambda_2)(\gamma + \sqrt{\delta}) + \lambda_1 + \lambda_2, \tag{22}$$

where γ and δ are as in (18). *Furthermore,*

$$(r - \lambda_1)(r - \lambda_2) = (\lambda_1 - \lambda_2)\left[p_{12}^1 (r - \lambda_2) - p_{12}^2 (r - \lambda_1)\right]. \tag{23}$$

Clearly, (21) and (22) follow from (17), (19), and (20). The equation (23) was proved in a similar fashion by NAIR 1943. We note another consequence of **5**:

7. *Let* **S** *be a partial design of class number 2, with $\lambda_1 \neq \lambda_2$. If the multiplicities α_1 and α_2 of* (16) *are equal, then*

$$n_1 = n_2 = \alpha_1 = \alpha_2 = (v-1)/2 \tag{24}$$

and

$$p_{12}^1 = p_{12}^2 = (v-1)/4. \tag{25}$$

This is essentially Theorem 5.2 of CONNOR & CLATWORTHY 1954. Clearly, $v \equiv 1 \bmod 4$ under the hypothesis of **7**. For further results in this direction, also for $m > 2$ and for association schemes rather than designs, see CONNOR & CLATWORTHY 1954, Section 6; also BOSE & MESNER 1959.

An interesting special class of association schemes of class number 2 is that with $p_{12}^1 = 0$ or $p_{12}^2 = 0$. Clearly there is no loss in generality to assume

$$p_{12}^1 = 0, \tag{26}$$

which means that $\{x, y\} \in A_1$ and $\{y, z\} \in A_1$ implies $\{x, z\} \in A_1$, i.e. the relation of being first associates is transitive. We consider here partial designs of class number 2 satisfying (26). The observation just made shows that these may be defined in another way:

8. *Let* $\mathbf{S} = (\mathfrak{p}, \mathfrak{B}, \mathrm{I})$ *be a tactical configuration which is not a design.* **S** *is a partial design satisfying* (26) *if, and only if,*

(**a**) *There are two distinct integers λ_1 and λ_2 such that $[x, y] = \lambda_1$ or λ_2, for distinct $x, y \in \mathfrak{p}$, and*

(**b**) *There exists a partition of $\mathfrak{p}$ into subsets of equal size s such that $[x, y] = \lambda_1$ if and only if x and y are in the same subset.*

If these conditions hold, then the parameters of **S** *satisfy*

$$\begin{aligned} & p_{11}^1 = s-2, \quad p_{12}^1 = p_{11}^2 = 0, \quad p_{22}^1 = s(t-1), \\ & p_{12}^2 = s-1, \quad p_{22}^2 = s(t-2) \\ & (s-1)\lambda_1 + s(t-1)\lambda_2 = r(k-1), \quad n_1 = s-1, \\ & n_2 = s(t-1), \quad v = st, \end{aligned} \tag{27}$$

where t *denotes the number of subsets in* (**b**).

For proofs, see Bose & Connor 1952 and Bose & Shimamoto 1952. We call partial designs of this kind *divisible,*[1]) and the subsets of points given in (**b**) its *cosets.*[1])

Combination of **3** and **5** with (27) gives most of the following:

9. *An incidence matrix* C *of a divisible partial design satisfies*

$$\det CC^T = rk(rk - \lambda_2 v)^{t-1}(r-\lambda_1)^{t(s-1)}, \tag{28}$$

where s *and* t *have the same significance as in* **8**. *Moreover, the factors on the right of* (28) *are all non-negative, so that in particular*

$$rk \geqq \lambda_2 v. \tag{29}$$

Equality in (29) *holds if, and only if, every block carries equally many points of every coset. Hence if* (29) *holds, then* t *divides* k, *and every block carries* kt^{-1} *points of every coset.*

For a direct proof of **9**, see Bose & Connor 1952, (3.7) and Theorems 3, 4.[2]) The equation (28) shows that CC^T is a non-singular matrix if and only if $rk > \lambda_2 v$ and $r > \lambda_1$. If these conditions are satisfied, then the divisible partial design in question is called *regular*. A non-regular divisible partial design is called *semi-regular* if $rk = \lambda_2 v$ and

---

[1]) The statisticians use the term "group divisible" instead of "divisible", and "group" instead of "coset". We avoid here the use of the word "group", since there is no connection whatsoever with groups in the usual mathematical sense. Our use of the term "coset" is motivated by the fact that in certain cases these are the point cosets of a naturally defined epimorphism of the given divisible partial design onto a design; cf. **10** and **7.2.5** below.

[2]) Their proof uses matrix arguments. The claims concerning (29), however, can be proved in a more elementary way; we indicate this proof here, as a simple application of result **1.1.2**: Let $\mathfrak{c}$ be one of the cosets, and $p \notin \mathfrak{c}$. Let c_p denote the number of blocks through p which carry points of $\mathfrak{c}$, and d_p the average number of points in $\mathfrak{c}$ on these blocks. Counting flags $x \mathrm{I} X$ with $p \mathrm{I} X \mathrm{I} x \in \mathfrak{c}$ in two ways, we get $c_p d_p = s\lambda_2$. Let **C** be the substructure of all points in $\mathfrak{c}$ and all blocks carrying points of $\mathfrak{c}$, and let v_1 denote, as in (1.1.9), the average number of points in $\mathfrak{c}$ on a block of **C**. Choosing $p \notin \mathfrak{c}$ so that d_p is minimal, we have $d_p \leqq v_1$; also, clearly $r \geqq c_p$. Finally, (1.1.9) with $m = 1$ gives $r(v_1 - 1) \leqq \lambda_1(s-1)$ since $v_0 = s$, $b_1 = r$, $b_2 = \lambda_1$. Combining these facts with (27), we get

$$rk - \lambda_2 v \geqq v_1 r - s\lambda_2 = v_1 r - c_p d_p \geqq c_p(v_1 - d_p) \geqq 0,$$

i.e. (29). The claims about equality follow from the corresponding facts in **1.1.2**.

$r > \lambda_1$, and *singular* if $r = \lambda_1$. BOSE & CONNOR 1952, Theorem 2, have shown:

10. *The singular divisible partial designs are precisely those finite incidence structures* **S** *which can be obtained from a design* **D** *as follows: Every point x of* **D** *is replaced by s new points $x_1, \ldots, x_s$, and each of these x_i is incident with precisely those blocks that are incident in* **D** *with x.*

Hence if φ is defined by $x_i\varphi = x$ $(i = 1, \ldots, s)$ and $X\varphi = X$ for all blocks X, then φ is a natural epimorphism of the singular partial design **S** onto the design **D**, and the point cosets of φ are the cosets of **S**.

In view of **10**, we can restrict our attention to non-singular divisible partial designs. Note that (17) and (28) give:

11. *For every non-singular divisible partial design* **S**,

(**a**) $b \geqq v$ *if* **S** *is regular,*
(**b**) $b \geqq v - t + 1$ *if* **S** *is semi-regular.*

Little more is known about non-singular divisible partial designs in general. The following are some results for the symmetric case $v = b$.

12. *Let* **S** *be a non-singular divisible partial design with equally many points and blocks, i.e. with $b = v$ and $r = k$.*

(**a**) *If* **S** *is regular and $\lambda_1 > \lambda_2$, then*

$$\lambda_1 \geqq [X, Y] \geqq \lambda_2 (r - \lambda_1)(r^2 - \lambda_2 v)^{-1}$$

for any two distinct blocks X, Y. If $\lambda_1 < \lambda_2$, the reversed inequalities hold.
(**b**) *If* **S** *is regular and $(r^2 - \lambda_2 v, \lambda_1 - \lambda_2) = 1$, then the dual of* **S** *is also a regular divisible partial design, with the same parameters as* **S**.
(**c**) *If* **S** *is semi-regular, then $\lambda_1 < \lambda_2$, and*

$$\lambda_1 \leqq [X, Y] \leqq 2\lambda_2 r^2 (r + \lambda_2 v - \lambda_1)^{-1} - \lambda_1$$

for any two distinct blocks X, Y.

For proofs, see CONNOR 1952b, Theorems 5.1, 6.2, 7.1, and Lemma 7.1.

13. *Let* **S** *be a regular divisible partial design with $v = b$ and $r = k$. Also, let s and t have the same significance as in* **8**.

(**a**) *If t is even, then $r^2 - \lambda_2 v$ is a square. Furthermore, if $t \equiv 2 \bmod 4$ and s even, then $r - \lambda_1$ is a sum of two squares.*
(**b**) *If t is odd and s even, then $r - \lambda_1$ is a square, and*

$$(r^2 - \lambda_2 v)\, x^2 + (-1)^{t(t-1)/2}\, s\, \lambda_2\, y^2 = z^2 \tag{30}$$

has a non-trivial solution in integers.

(**c**) *If* s *and* t *are both odd, then* (30) *has a non-trivial solution if and only if*

$$(r - \lambda_1) x^2 + (-1)^{s(s-1)/2} s y^2 = z^2$$

has such a solution.

This is Theorem 9 of BOSE & CONNOR 1952. The proof consists in an application of the Minkowski-Hasse theory of quadratic forms, as outlined in Section 1.3. The result is comparable to the Bruck-Ryser Theorem **3.2.13** for finite projective planes; it excludes many sets $s, t, r, \lambda_1, \lambda_2$ of integers satisfying (27) from the possible parameters of regular divisible partial designs.

Next, we discuss briefly two other types of association schemes and partial designs of class number 2. The first of these is the class of so-called *triangular*[1]) association schemes, which are defined as follows: Let $\mathfrak{q}$ be a finite set of size $|\mathfrak{q}| = N > 3$, and put $\mathfrak{p} = \mathfrak{q}(2)$. The elements of $\mathfrak{p}(2)$ are then of the form $\{\{x, x'\}, \{y, y'\}\}$, with $x, x', y, y' \in \mathfrak{q}$, $x \neq x'$, $y \neq y'$, and $\{x, x'\} \neq \{y, y'\}$. Let

$$A_1 = \{\{\{x, x'\}, \{y, y'\}\} : \{x, x'\} \cap \{y, y'\} \neq \emptyset\}$$

and $A_2 = \mathfrak{p}(2) - A_1$; then $\mathscr{A} = \{A_1, A_2\}$ is an association scheme of class number 2, namely, the triangular association scheme to be defined. (BOSE & SHIMAMOTO 1952, Section IV.) The parameters of $\mathscr{A}$ are

(31) $$p_{11}^1 = N - 2, \quad p_{11}^2 = 4, \quad n_1 = 2(N-2);$$

the remaining parameters can be computed from this. It is an interesting fact that these association schemes are, in general, characterized by (31):

14. *If an association scheme of class number* 2 *has parameters* (31), *with* $N \neq 8$, *then it is triangular.*

Proof: CONNOR 1958, SHRIKHANDE 1959. The condition $N \neq 8$ is essential: CHANG 1959, 1960 has shown that for $N = 8$ there are precisely four non-isomorphic association schemes with parameters (31). See also HOFFMAN 1960a, b.

The other topic to be discussed here is that of partial designs of class number 2 with

(32) $$\lambda_1 = 0 \quad \text{and} \quad \lambda_2 = 1.$$

These are all partial planes, and some of them occur in a natural fashion as substructures of finite projective planes. For example, let **P** be a

[1]) This term was coined by BOSE & SHIMAMOTO 1952 who gave a thorough discussion of partial designs of class number 2. Besides the divisible and triangular types, they also investigate "singly linked", "Latin square type", and "cyclic" association schemes. These construction techniques cannot be discussed here.

projective plane of square order q^2 which contains a Baer subplane $\mathbf{Q}$ of order q. Consider the incidence structure $\mathbf{S} = \mathbf{P} - \mathbf{Q}$ of all points and lines of $\mathbf{P}$ which do not belong to $\mathbf{Q}$. Then $\mathbf{S}$ is a partial design of class number 2 satisfying (32). In fact, the parameters of $\mathbf{S}$ satisfy

$$(33) \qquad v = b = q(q^3 - 1), \quad r = k = q^2, \quad p_{22}^1 = n_2,$$

from which the remaining parameters may be computed with (2)—(4) and (7). In particular, $p_{12}^1 = n_2 - p_{22}^1 = 0$, so that $\mathbf{S}$ is even a regular divisible partial design.[1])

There is a converse similar to **14**:

15. *If a partial design of class number* 2 *has parameters satisfying* (32) *and* (33), *then it is isomorphic to a projective plane of order* q^2 *from which the points and lines of a subplane of order* q *are removed.*

This follows from the more general result **7.4.12c** below; the proof of **15** will be indicated at the end of Section 7.4.

Not much is known in general about partial designs with (32). The following parameter relations can be proved:

16. *Let* $\mathbf{S}$ *be a partial design of class number* 2, *satisfying* (32). *Then*

(**a**) $n_2 = r(k-1)$,
(**b**) $k - 2 \leqq p_{22}^2 \leqq k - 2 + (r-1)^2$, *and*
(**c**) *If* $k > r$, *then* $r\,p_{12}^2 - (r-1)\,p_{12}^1 = r(r-1)$.

Proofs: (**a**) follows immediately from (32) and (7). Result (**b**) is Lemma 4.1 of Bose & Clatworthy 1955,[2]) and (**c**) is the obvious special case of Nair's 1943 equation (23). Bose & Clatworthy 1955 have determined the possible parameters of $\mathbf{S}$ for the case $k > r$. In the limiting case $p_{22}^2 = k - 2$ of (**b**), it is not difficult to see that $\mathbf{S}$ must be "free of triangles" in the sense that if x, y, z are three points not on a common block (line), then $[x, y] = [y, z] = [z, x] = 1$ is impossible. Similarly, $p_{22}^2 = k - 2 + (r-1)^2$ if and only if $[x, y] = 1 = [y, z]$ implies $[x, z] = 1$.

[1]) Clearly, $\mathbf{S}$ is self-dual if $\mathbf{P}$ and $\mathbf{Q}$ are. This is the case, for example, whenever $\mathbf{P}$ is desarguesian; hence we have here an infinite class of self-dual regular divisible partial designs. This shows that partial designs may possess correlations, but so far there seem to exist no papers dealing with them in general. However, Hoffman, Newman, Straus, Taussky 1956, in their investigation of absolute points of correlations in finite projective planes, mention the possibility of extending their methods to divisible partial designs.

[2]) The additional hypothesis $k > r$ made by these authors is superfluous, as their own proof shows. Note also that Bose and Clatworthy assume $\lambda_1 = 1$ and $\lambda_2 = 0$ rather than (32); this results in an interchange of the subscripts 1 and 2 in their formulas as compared with ours.

Finally, we mention briefly some examples of partial designs. It would lead much too far to attempt to be complete here; hence we concentrate on examples which are rather closely related to finite projective geometries.[1])

As in Section 1.4, let $\mathscr{P}(d, q)$ be the d-dimensional projective geometry over $GF(q)$. Let $0 \leqq s < t \leqq d - 1$, call "points" the s-dimensional, and "blocks" the t-dimensional subspaces of $\mathscr{P}(d, q)$, and define incidence by set theoretic inclusion. The resulting incidence structure $\mathbf{P}_{s,t}(d, q)$ is a partial design of class number $m = s + 1$, the corresponding association scheme $\mathscr{A} = \{A_0, \ldots, A_s\}$ being defined as follows: The "points" S and T are i^{th} associates (i.e. $\{S, T\} \in A_i$) if and only if $S \cap T$ is of rank i (dimension $i - 1$), for $i = 0, \ldots, s$.[2]) Note that the λ_i need not be all distinct. For example, $\lambda_i = 0$ for $i < 2s - t + 1$; cf. (1.4.5). The parameters of $\mathbf{P}_{s,t}(d, q)$ can easily be computed by counting arguments using (1.4.6). Clearly, we get nothing new for $s = 0$: the design $\mathbf{P}_t(d, q)$ of Section 1.4 is the same as $\mathbf{P}_{0,t}(d, q)$ in our present terminology.

Many partial designs arise as substructures of the $\mathbf{P}_{s,t}(d, q)$ just defined. For example, BRUCK & BOSE 1966 have shown that every finite projective plane is isomorphic to such a substructure. Other models consist of isotropic subspaces with respect to a polarity of $\mathscr{P}(d, q)$. For example, all absolute points together with all totally isotropic t-subspaces, of a polarity π of $\mathscr{P}(d, q)$, form a partial design of class number 2; the parameters of these were computed by WAN 1964, WAN & YANG 1964, DAI & FENG 1964a, b. For the unitary case, see also BOSE & CHAKRAVARTI 1965. Other constructions along similar lines were given by WAN 1965.

7.2 Hjelmslev planes

A *projective Hjelmslev plane*[3]) is an incidence structure $\mathbf{H} = (\mathfrak{p}, \mathfrak{B}, \mathrm{I})$ such that

(1) $[x, y] \geqq 1$ *for* $x, y \in \mathfrak{p}$, *and* $[X, Y] \geqq 1$ *for* $X, Y \in \mathfrak{B}$.

[1]) Similar models from affine geometries may also be constructed. Many other references to construction techniques are in the bibliography. We mention here the survey of GUÉRIN 1965, where further references can be found.

[2]) Here we use the convention that the empty subspace has dimension -1. Note also that here "0th associates" has a meaning different from that in (11)—(14) above.

[3]) After HJELMSLEV 1929a, b, 1942, 1945a, b, 1949. These structures were called "Projektive Ebenen mit Nachbarelementen" by KLINGENBERG 1954, 1955a. The present definition (1)—(3) is that of KLINGENBERG 1955a. For a generalization of the concept, see KLINGENBERG 1956.

(2) *There exists a canonical epimorphism* ν *of* $\mathbf{H}$ *onto a projective plane* $\mathbf{H}^*$ *such that* $x\nu = y\nu$ *if and only if* $[x, y] > 1$ *and* $X\nu = Y\nu$ *if and only if* $[X, Y] > 1$.

It follows easily from (1) and (2) that

(3) $[x, y] > 1 < [y, z]$ *implies* $[x, z] > 1$, *and* $[X, Y] > 1 < [Y, Z]$ *implies* $[X, Z] > 1$.

Every projective plane is a projective Hjelmslev plane; in fact the (ordinary) projective planes are precisely those projective Hjelmslev planes which satisfy $[x, y] = 1$ for any two distinct points x, y; for these the epimorphism of (2) is the identical automorphism. A projective Hjelmslev plane will be called *proper* if it is not a projective plane.

In an arbitrary projective Hjelmslev plane, two points x, y are said to be *neighbors* if either $x = y$ or $x \neq y$ and $[x, y] > 1$, and a neighbor relation among the blocks[1]) is defined dually. We denote the neighbor relation by the symbol $\bigcirc$, i.e.

$$(4)\quad \begin{cases} x \bigcirc y \quad \textit{if and only if} \quad x = y \quad \textit{or} \quad x \neq y \quad \textit{and} \quad [x, y] > 1, \\ \qquad\qquad\qquad\qquad\qquad\qquad\qquad\qquad \textit{for} \quad x, y \in \mathfrak{p}, \\ X \bigcirc Y \quad \textit{if and only if} \quad X = Y \quad \textit{or} \quad X \neq Y \quad \textit{and} \quad [X, Y] > 1, \\ \qquad\qquad\qquad\qquad\qquad\qquad\qquad\qquad \textit{for} \quad X, Y \in \mathfrak{B}. \end{cases}$$

It is an immediate consequence of this definition and (3) that

1. *The neighbor relation is an equivalence relation, both on points and on blocks, for every projective Hjelmslev plane.*

Also, the equivalence classes of $\bigcirc$ are, by (2), the cosets of the canonical epimorphism ν of $\mathbf{H}$ onto $\mathbf{H}^*$. In fact, $\mathbf{H}^*$ can be identified with the following incidence structure $(\mathfrak{p}^*, \mathfrak{B}^*, \mathrm{I}^*)$: The elements of $\mathfrak{p}^*$ and $\mathfrak{B}^*$ are the $\bigcirc$-classes of points and blocks, respectively, of $\mathbf{H}$, and $\mathfrak{x}\, \mathrm{I}^*\, \mathfrak{X}$ if and only if the class $\mathfrak{x} \in \mathfrak{p}^*$ contains a point x incident with some block X of the class $\mathfrak{X} \in \mathfrak{B}^*$.

It is straightforward to verify that (2) and (3) imply the following properties of the neighbor relation:

(5) *If* $X, Y, Z\, \mathrm{I}\, p$ *for some* $p \in \mathfrak{p}$, *and if* $X \bigcirc Y$ *and* $Y \varnothing Z$, *then* $X \varnothing Z$.

(6) *If* $X \bigcirc Y$, $Y \varnothing Z$, $X\, \mathrm{I}\, x\, \mathrm{I}\, Z$ *and* $Y\, \mathrm{I}\, y\, \mathrm{I}\, Z$, *then* $x \bigcirc y$.

(7) *If* $x \bigcirc y$, $y \varnothing z$, $x\, \mathrm{I}\, X\, \mathrm{I}\, z$ *and* $y\, \mathrm{I}\, Y\, \mathrm{I}\, z$, *then* $X \bigcirc Y$.

---

[1]) Although it is customary to call the blocks of a Hjelmslev plane "lines", we adhere to the term "block", for consistency and because we shall have occasion to use the term "line", in the same general sense as in Section 2.1, for the proof of result **5** below.

(8) *There exist four points of which no three are on one block, and of which no two are neighbors, such that no two of the six blocks joining any two of these points are neighbors.*

KLINGENBERG 1954, Section 1.3, has shown that, conversely, (1) together with (5)—(8) implies (2) and hence (3). Therefore:

2. *An incidence structure satisfying* (1) *is a projective Hjelmslev plane if, and only if, the relation* $\bigcirc$ *defined by* (4) *satisfies conditions* (5)—(8).

The definition (1)—(2) clearly shows that the dual of a projective Hjelmslev plane is likewise a projective Hjelmslev plane. Hence (5)—(8) are, in the presence of (1), also equivalent to their duals. Note that (6) and (7) are dual to each other.

Now we restrict our attention to finite projective Hjelmslev planes.

3. *Let* $\mathbf{H} = (\mathfrak{p}, \mathfrak{B}, \mathrm{I})$ *be a finite projective Hjelmslev plane. Then there exists an integer* j *such that:*

(**a**) *Given any flag* $p \mathrm{I} B$ *in* $\mathbf{H}$, *there are exactly* j *points* x *such that* $p \bigcirc x \mathrm{I} B$ *and exactly* j *blocks* X *such that* $B \bigcirc X \mathrm{I} p$.

(**b**) $\mathbf{H}$ *is a tactical configuration with* $r = k$ *and* $v = k^2 - jk + j^2$.

(**c**) *The projective plane* $\mathbf{H}^*$ *of* (3) *is of order* $n = (k - j)j^{-1}$, *whence* $k \equiv 0 \bmod j$.

(**d**) *Every coset* (*of points or blocks*) *of the epimorphism* $\nu : \mathbf{H} \to \mathbf{H}^*$ *of* (2) *consists of precisely* j^2 *elements.*

(**e**) *If* $\mathbf{H}$ *is proper, or equivalently if* $j > 1$, *then* $k \leqq j^2 + j$.

This is essentially Theorem 1 of KLEINFELD 1959; the proof is elementary. Comparison of (**b**) and (**d**) with result **7.1.8** shows that a proper finite projective Hjelmslev plane is a divisible partial design (whose cosets are the neighbor classes of points) if, and only if, any two distinct neighbor points are joined by equally many blocks. This observation may be improved as follows.

We call a (not necessarily finite) projective Hjelmslev plane $\mathbf{H}$ *uniform*[1]) if

(9) $\quad B \mathrm{I} p \mathrm{I} C \mathrm{I} q, \quad B \bigcirc C$ *and* $p \bigcirc q$ *imply* $B \mathrm{I} q$.

This property may be expressed in several other ways if $\mathbf{H}$ is finite: Let p be an arbitrary point of $\mathbf{H}$, and call $\lambda_1 = \lambda_1(p)$ the average number of blocks joining p with its neighbors $\neq p$. Counting the flags $x \mathrm{I} X$ with $X \mathrm{I} p \neq x \bigcirc p$ in two ways, one obtains

(10) $$k = \lambda_1(j + 1)$$

[1]) This is definition 2.4 of LÜNEBURG 1962. Lüneburg is concerned with the affine case only (see below), but the same definition may be used here also. A more restrictive definition, valid in the finite projective case only, was given earlier by KLEINFELD 1959, p. 405.

with j as in **3**. Hence λ_1 is independent of the choice of p. After this, it is not difficult to derive the following from **3e**:

4. *The following properties of a proper finite projective Hjelmslev plane are equivalent:*

(**a**) **H** *is uniform,*
(**b**) λ_1 *is an integer,*
(**c**) $\lambda_1 = j$,
(**d**) $k = j(j+1)$,
(**e**) $\mathbf{H}^*$ *is of order* j.

For the proof, see KLEINFELD 1959, Theorem 2; also LÜNEBURG 1962, Satz 2.13. It is now obvious that **H** is a divisible partial design if uniform, λ_1 having the same meaning as in **7.1.8**. Moreover:

5. *The finite proper uniform projective Hjelmslev planes are precisely those regular divisible partial designs* **H** *whose parameters satisfy*

$$(11)\quad s = j^2, \quad t = j^2 + j + 1, \quad r = k = j(j+1), \quad \lambda_1 = j, \quad \lambda_2 = 1,$$

and whose duals $\overline{\mathbf{H}}$ *are likewise divisible partial designs with these parameters.*

One of the two claims of **5** is obvious from the preceding (see KLEINFELD 1959, Theorem 2). For the converse, suppose that **H** and $\overline{\mathbf{H}}$ are divisible partial designs with parameters (11). As condition (1) is trivially satisfied, we have to prove only (2).

A counting argument similar to that leading to (10) shows that a block of **H** meets a point coset of **H** either not at all or in precisely j points. Also, the j common points of two blocks in the same block coset (point coset of $\overline{\mathbf{H}}$) all belong to the same point coset. It then follows that, *on the average*, there exist $j+1$ block cosets with blocks containing points of an arbitrary given point coset. Next, define the *lines* of **H**, as in Section 2.1, as the point sets $((x, y))$ for distinct points x, y. It follows that the j^2 points of any point coset $\mathfrak{c}$, together with the lines joining them, form a design with parameters $v = j^2$, $k = j$ and $\lambda = 1$, i.e. an affine plane $\mathbf{A} = \mathbf{A}(\mathfrak{c})$ of order j (cf. **3.2.4b**). If two lines $\mathfrak{l}$ and $\mathfrak{m}$ of **A** are not parallel, then their unique common point is also the unique common point of any block L containing $\mathfrak{l}$ with any block M containing $\mathfrak{m}$; hence L and M belong to different block cosets. This shows that there are *at least* $j+1$ block cosets with blocks containing points of $\mathfrak{c}$. Consequently, if $\mathfrak{l}$ and $\mathfrak{m}$ are parallel in $\mathbf{A}(\mathfrak{c})$, then the blocks containing $\mathfrak{l}$ and those containing $\mathfrak{m}$ all belong to the same block coset. It follows that if two blocks B, C with $[B, C] = 1$ both contain points of $\mathfrak{c}$, then their unique common point is also in $\mathfrak{c}$. Thus, if $\mathfrak{c}, \mathfrak{c}'$ are distinct point cosets, then all blocks joining points of $\mathfrak{c}$ with points of $\mathfrak{c}'$ are in the same block coset. This and a dual argument show that condition (2) is satisfied, so that $\mathbf{H}$ is a projective Hjelmslev plane.

The result that $\mathbf{A} = \mathbf{A}(\mathfrak{c})$ is an affine plane for every point coset $\mathfrak{c}$ is of independent interest. In fact, this may be extended as follows. For any class $\mathfrak{c}$ of neighbor points in the projective Hjelmslev plane $\mathbf{H} = (\mathfrak{p}, \mathfrak{B}, \mathrm{I})$, let $\mathbf{A}(\mathfrak{c})$ be the incidence structure whose points are those of $\mathfrak{c}$, whose blocks are the point sets $(X) \cap \mathfrak{c}$, $X \in \mathfrak{B}$, and whose incidence is set theoretic inclusion. Then:

6. *The following properties of the finite proper projective Hjelmslev plane* $\mathbf{H}$ *are equivalent:*

(**a**) $\mathbf{H}$ *is uniform.*
(**b**) *There exists a class* $\mathfrak{c}$ *of neighbor points for which* $\mathbf{A}(\mathfrak{c})$ *is an affine plane.*
(**c**) *For every class* $\mathfrak{c}$ *of neighbor points,* $\mathbf{A}(\mathfrak{c})$ *is an affine plane.*

For the proof of this, in a more general context, see LÜNEBURG 1962, Satz 2.12.

There exist uniform and non-uniform proper finite projective Hjelmslev planes; some constructions will be indicated further below. CRAIG 1964 has shown that

7. *Every finite projective plane is the canonical epimorphic image, in the sense of* (2), *of some proper uniform finite projective Hjelmslev plane.*

On the other hand, it is an open question which k and j (cf. **3**) can occur as the parameters of a finite projective Hjelmslev plane. **7.1.13** yields that, in the uniform case,

$$j x^2 + (-1)^{j(j+1)/2} y^2 = z^2 \tag{12}$$

must have a nontrivial solution in integers; this follows also from the Bruck-Ryser Theorem **3.2.13** for $\mathbf{H}^*$.

Little is known beyond this. KLEINFELD 1959, Section 3, has proved by an elementary argument that

8. *A finite projective Hjelmslev plane with parameters* j *and* k (cf. **3**), *uniform or not, can exist only if* $k - j \neq 6$.

We shall now introduce affine Hjelmslev planes. As a preparation for this, we define a relation $\underset{a}{\bigcirc}$ among the blocks of an arbitrary incidence structure:

(13) $B \underset{a}{\bigcirc} C$ *if, and only if, to every point* x *on one of* B, C *there exists a point* y *on the other such that* $[x, y] > 1$.

An *affine Hjelmslev plane*[1]) is an incidence structure $\mathbf{J} = (\mathfrak{p}, \mathfrak{B}, \mathrm{I})$ satisfying the following axioms:

(14) $[x, y] \geqq 1$, *for* $x, y \in \mathfrak{p}$.

(15) *There is a binary relation* $\|$ *("parallel") on* $\mathfrak{B}$ *such that, given any point-line pair* p, B, *there exists a unique block* C *such that* $p \mathrm{I} C$ *and* $B \| C$.

(16) *If* $B \neq C$ *and* $B \underset{a}{\bigcirc} C$ *in the sense of* (13), *then* $[B, C] \neq 1$.

(17) *There exists a canonical epimorphism* μ *of* $\mathbf{J}$ *onto an affine plane* $\mathbf{J}^*$ *such that* $x\mu = y\mu$ *if and only if* $[x, y] > 1$, *for* $x, y \in \mathfrak{p}$, *and* $X\mu = Y\mu$ *if and only if* $X \underset{a}{\bigcirc} Y$, *in the sense of* (13), *for* $X, Y \in \mathfrak{B}$.

The following is easily verified:

9. *Let* $\mathbf{H} = (\mathfrak{p}, \mathfrak{B}, \mathrm{I})$ *be a projective Hjelmslev plane,* W *a block of* $\mathbf{H}$, *and* $\mathfrak{W}$ *the class of all blocks which are neighbors of* W. *Then the incidence structure* $\mathbf{J} = (\mathfrak{q}, \mathfrak{C}, \mathrm{J})$, *where*

$$\mathfrak{q} = \mathfrak{p} - \bigcup_{X \in \mathfrak{W}} (X), \quad \mathfrak{C} = \mathfrak{B} - \mathfrak{W}, \quad \textit{and} \quad \mathrm{J} = \mathrm{I} \cap (\mathfrak{q} \times \mathfrak{C}),$$

is an affine Hjelmslev plane, with $B \| C$ *if and only if* $(B, C) \cap (W) \neq \emptyset$ *in* $\mathbf{H}$.

(Cf. KLINGENBERG 1954, Satz 3.6; LÜNEBURG 1962, Section 2.) The converse of **9**, i.e. the question as to whether or not every affine Hjelmslev plane can be interpreted in this fashion, is an open problem. We shall see, however, that in certain cases such an interpretation is indeed possible; see **18** below.

As $x = y$ is not excluded in (13), every affine plane is an affine Hjelmslev plane. Again, we call an affine Hjelmslev plane *proper* if it is not an ordinary affine plane.

Result **3** has the following analogue for finite affine Hjelmslev planes:

10. *Let* $\mathbf{J}$ *be a finite affine Hjelmslev plane. Then there exists an integer* j *such that:*

(**a**) *To any flag* $p \mathrm{I} B$ *of* $\mathbf{J}$, *there are precisely* j *points* $x \mathrm{I} B$ *such that* $[p, x] > 1$.

(**b**) $\mathbf{J}$ *is a tactical configuration with* $r = k + j$ *and* $v = k^2$, *whence* $b = k(k + j)$ *and* $|\mathfrak{X}| = k$ *for every parallel class* $\mathfrak{X}$ *of blocks.*

(**c**) *The affine plane* $\mathbf{J}^*$ *of* (17) *has order* kj^{-1}, *whence* $k \equiv 0 \bmod j$.

---

[1]) Introduced by KLINGENBERG 1954, under the name "Affine Ebene mit Nachbarelementen". The present definition is essentially that of LÜNEBURG 1962, Satz 2.6. As Lüneburg remarks in his Introduction, these structures are rather artificial (cf. result **9** below, and the special role of the line W); but the results below will show that their investigation is nevertheless interesting.

(d) *Every coset (of points or blocks) of the epimorphism* $\mu : \mathbf{J} \to \mathbf{J}^*$ *of* (17) *consists of* j^2 *elements.*
(e) *If* **J** *is proper, then* $k \leq j^2$.

Proof: LÜNEBURG 1962, Satz 2.11. Note that for the case where **J** can be extended to a projective Hjelmslev plane, **10** is an almost immediate consequence of **3**.

An affine Hjelmslev plane is called *uniform* if it satisfies condition (9) above. Analogously to **4**, LÜNEBURG 1962, Satz 2.13 has shown:

11. *A proper finite affine Hjelmslev plane is uniform if and only if* $k = j^2$ *or, equivalently, if* $\mathbf{J}^*$ *has order* j.

An automorphism τ of the affine Hjelmslev plane **J** is called a *translation* if some parallel class of blocks is fixed elementwise by τ, and if τ is either the identity or free of fixed points. We denote the set of all translations of **J** by $\mathsf{T}(\mathbf{J})$ or simply T. It is not known whether or not T must always be a group. If $\mathsf{T}(\mathbf{J})$ is a group and transitive on the points of **J**, then **J** is called a *Hjelmslev translation plane.* The following discussion will include, among other things, a complete description of the finite uniform Hjelmslev translation planes.

Let G be an arbitrary group and $\mathscr{C}$ a *covering* of G, i.e. a set of subgroups such that every $g \in G$ is in some $C \in \mathscr{C}$. Consider the incidence structure $\mathbf{J}(G, \mathscr{C})$, defined in Section 1.2, whose points are the elements of G, whose blocks are the cosets Cx, with $C \in \mathscr{C}$, and whose incidence is set theoretic inclusion. $\mathbf{J}(G, \mathscr{C})$ possesses a natural parallelism satisfying condition (15), namely that defined by

$$(18) \qquad Cx \parallel Dy \quad \text{if and only if} \quad C = D.$$

Furthermore, every $g \in G$ determines an automorphism $x \to xg$ of $\mathbf{J}(G, \mathscr{C})$, mapping every block onto a parallel one. Also, it is clear that if G is abelian, then this automorphism fixes every block Cx, $C \in \mathscr{C}$, for which $g \in C$.[1])

12. *The incidence structure* $\mathbf{J}(G, \mathscr{C})$ *is an affine Hjelmslev plane if and only if*

(a) *there exist* $C, D \in \mathscr{C}$ *with* $C \cap D = 1$, *and* $CD = G$ *for any such* C, D,
and
(b) *the set* $N = N(C)$ *of all those elements in* G *which are contained in more than one member of* $\mathscr{C}$ *is a normal subgroup satisfying*
$C \nsubseteq N$ *for all* $C \in \mathscr{C}$,

---

[1]) Compare here the context of **1.2.21**, also **2.4.29**. For more details concerning the incidence structures $\mathbf{J}(G, \mathscr{C})$ in general, see also ANDRÉ 1961b and LÜNEBURG 1967a, Section 1.

$NC \cap ND = N$ for all $C, D \in \mathscr{C}$ with $C \cap D = 1$, and
$NC = ND$ for all $C, D \in \mathscr{C}$ with $C \cap D \neq 1$.

Furthermore, if these conditions are satisfied, then $\mathbf{J}(G, \mathscr{C})$ *is uniform if and only if*
(**c**) *the groups $N \cap C$, for $C \in \mathscr{C}$, form a partition*[1]) *of N.*
Proof: Lüneburg 1962, Satz 4.2 and Satz 7.1.

13. *Let* $\mathbf{J} = \mathbf{J}(G, \mathscr{C})$ *be an affine Hjelmslev plane.*
(**a**) *G acts as a group of translations on* $\mathbf{J}$ *if, and only if, G is abelian.*
(**b**) *If G is abelian and contains an element $\neq 1$ of finite order, then G/N is elementary abelian.*
(**c**) *If* $\mathbf{J}$ *is uniform and G abelian, with an element $\neq 1$ of finite order, then G is either elementary abelian or a direct product of cyclic groups of order p^2, for some prime number p.*

Proof: Lüneburg 1962, Sätze 4.5, 8.2, 8.3.[2]) These results now permit the following complete description of the finite uniform Hjelmslev translation planes:

14. *The finite uniform Hjelmslev translation planes are precisely those incidence structures* $\mathbf{J}(G, \mathscr{C})$ *for which G is of order p^{4e}, with p a prime number, and either elementary abelian or a direct product of cyclic groups of order p^2, and for which $\mathscr{C}$ satisfies conditions* (**a**)—(**c**) *of* **12**.

Proof: Lüneburg 1962, Section 9. The following improvement of **7** is also proved there:

15. *Every finite affine translation plane is the canonical epimorphic image, in the sense of* (17), *of two non-isomorphic proper uniform Hjelmslev translation planes.*

Namely, both the elementary abelian type $(p, \ldots, p)$ and the type $(p^2, \ldots, p^2)$ can occur as the structure of the translation group of the pre-image (Lüneburg 1962, Corollaries 9.1 and 9.2).

Let $\mathsf{T} = \mathsf{T}(\mathbf{J})$ be, as before, the translation group of the Hjelmslev translation plane $\mathbf{J}$. For an arbitrary parallel class $\mathfrak{X}$ of blocks in $\mathbf{J}$, denote by $\mathsf{T}(\mathfrak{X})$ the subgroup of those $\tau \in \mathsf{T}$ which fix every block in $\mathfrak{X}$. In analogy to the Definition (3.1.37), we define the *kernel* of $\mathbf{J}$ as the set $\mathfrak{H} = \mathfrak{H}(\mathbf{J})$ of those endomorphisms α of T for which

(19) $\mathsf{T}(\mathfrak{X})^{\alpha} \subseteqq \mathsf{T}(\mathfrak{X})$, *for all parallel classes* $\mathfrak{X}$.

---

[1]) This means here that $N \cap C \cap D \neq 1$ implies $N \cap C = N \cap D$. See also the context of **1.2.21** and **3.4.18**.

[2]) It should be mentioned that there also exist finite non-abelian groups G with a covering $\mathscr{C}$ such that $\mathbf{J}(G, \mathscr{C})$ is an affine Hjelmslev plane; cf. Lüneburg 1962, p. 282.

As T is abelian by **13a** and **14**, the set $\mathfrak{H}$ is a ring with 1, and it can be shown that

16. *For every Hjelmslev translation plane* $\mathbf{J}$, *the ring* $\mathfrak{H} = \mathfrak{H}(\mathbf{J})$ *has the following properties:*

(**a**) *The zero divisors of* $\mathfrak{H}$ *form a unique maximal ideal* $\mathfrak{N}$.
(**b**) *The group of units of* $\mathfrak{H}$ *consists precisely of the non-zero divisors.*

(Klingenberg 1955a, Section 4; Lüneburg 1962, Section 5.)

We call the affine Hjelmslev plane $\mathbf{J}$ *desarguesian* if it is a Hjelmslev translation plane and if $\mathfrak{H}(\mathbf{J})$ is transitive on the translations in $\mathsf{T}(\mathfrak{X})$, for all parallel classes $\mathfrak{X}$.[1] Unlike arbitrary Hjelmslev translation planes, desarguesian affine Hjelmslev planes are uniquely determined by their kernels:

17. *Let* $\mathfrak{H}$ *be a ring satisfying conditions* (**a**) *and* (**b**) *of* **16**. *Define points as the ordered pairs* (x, y) *of elements in* $\mathfrak{H}$, *and blocks as the point sets*

$$L(s, t, u, v) = \{(x s + u, x t + v), x \in \mathfrak{H}\}, \text{ with } s \notin \mathfrak{N} \text{ or } t \notin \mathfrak{N}.$$

Then the incidence structure so defined is a desarguesian affine Hjelmslev plane $\mathbf{J}$, *with* $\mathfrak{H} \cong \mathfrak{H}(\mathbf{J})$. *Conversely, every desarguesian affine Hjelmslev plane can be represented in this fashion.*

(Klingenberg 1955a). Furthermore, the quotient ring $\mathfrak{H}/\mathfrak{N}$ is a field isomorphic to the ternary field of the affine plane $\mathbf{J}^*$. Using **17**, one can show:

18. *Let* $\mathfrak{H}$ *be as in* **17**. *Define points as the sets*

$$p(x_0, x_1, x_2) = \{(s x_0, s x_1, s x_2) \vdots s \in \mathfrak{H}\} \text{ with at least one } x_i \notin \mathfrak{N},$$

blocks as the sets

$$B(u_0, u_1, u_2) = \{(u_0 t, u_1 t, u_2 t) \vdots t \in \mathfrak{H}\} \text{ with at least one } u_i \notin \mathfrak{N},$$

and incidence of $p(x_0, x_1, x_2)$ *with* $B(u_0, u_1, u_2)$ *by*

$$\sum_{i=0}^{2} x_i u_i = 0.$$

Then the incidence structure so defined is a projective Hjelmslev plane $\mathbf{H}$. *Moreover, for every line* W *of* $\mathbf{H}$, *the corresponding affine Hjelmslev plane (as described in* **9**) *is desarguesian, and in fact isomorphic to the*

---

[1]) Actually, it suffices to postulate this for only one $\mathfrak{X}$. The definition is that of Klingenberg 1955a; the term "desarguesian" is, of course, motivated by result **3.1.25**. Also, we shall see below that if $\mathbf{J}$ is desarguesian, then $\mathbf{J}^*$ is desarguesian in the usual sense. Desarguesian Hjelmslev planes need not be uniform. In fact, we have uniformity if and only if the ideal $\mathfrak{N}$ of **17** satisfies $\mathfrak{N}^2 = 0$.

desarguesian affine Hjelmslev plane over $\mathfrak{H}$ as given in **17**. *Consequently, every desarguesian affine Hjelmslev plane is of the form* **J** *described in* **9**.

(KLINGENBERG 1955a). The ring $Z(p^e)$ of residue classes mod p^e, where p is a prime and $e > 1$, satisfies conditions (**a**) and (**b**) of **16**, and hence gives rise to a finite desarguesian proper affine Hjelmslev plane.

Finally, we mention another construction for finite rings $\mathfrak{H}$ satisfying conditions (**a**), (**b**) of **16**; this is due to KLEINFELD 1959, p. 407. Let $\mathfrak{F} = GF(q)$ and $\alpha \in \operatorname{Aut}\mathfrak{F}$. In the direct sum $\mathfrak{H} = \mathfrak{F} \oplus \mathfrak{F}$, define a multiplication by

$$(x, y)(u, v) = (xu, xv + yu^\alpha).$$

This turns $\mathfrak{H}$ into a ring of the desired kind, and $\mathfrak{H}$ is commutative if and only if $\alpha = 1$. Furthermore, the (projective or affine) Hjelmslev plane over $\mathfrak{H}$ is always uniform. Similar constructions also yield rings with properties (**a**), (**b**) of **16** which give rise to non-uniform desarguesian Hjelmslev planes.

7.3 Generalized polygons

The purpose of this section is to discuss briefly a class of incidence structures introduced by TITS 1959, in connection with certain group theoretical problems. There is not enough space here to make these connections explicit, but the purely combinatorial aspect of the incidence structures in question is interesting enough to deserve being mentioned here.

Let $\mathbf{S} = (\mathfrak{p}, \mathfrak{B}, \mathrm{I})$ be an arbitrary incidence structure. A *chain* in $\mathbf{S}$ is a finite sequence $\mathfrak{C} = (x_0, \ldots, x_h)$ of elements[1]) in $\mathbf{S}$ such that

(1) $$x_{i-1} \,\mathrm{I}\, x_i \quad \text{for} \quad i = 1, \ldots, h.$$

The integer h is the *length* of the chain $\mathfrak{C}$. The case $h = 0$ is not excluded; such a chain of just one element will be called *trivial*. The chain $\mathfrak{C} = (x_0, \ldots, x_h)$ is said to *join* the elements x_0 and x_h of $\mathbf{S}$. Since the x_i must alternatingly belong to $\mathfrak{p}$ and $\mathfrak{B}$, we have

1. *The length of a chain is even if it joins two points or two blocks, and odd if it joins a point and a block.*

The chain $\mathfrak{C} = (x_0, \ldots, x_h)$ is *closed* if $x_0 = x_h$. Clearly, closed chains have even length. $\mathfrak{C}$ is *irreducible* if

(2) $x_{i-1} \neq x_{i+1}$ $(i = 1, \ldots, h-1)$, *and* $x_1 \neq x_{h-1}$ *in case* $\mathfrak{C}$ *is closed.*

[1]) The elements of $\mathbf{S}$ are the points and the blocks. Thus it follows from (1) that, although we usually use lower case letters only for points, about half of the x_i are blocks.

Clearly, all chains of length $\leqq 1$, and in particular the trivial chains, are irreducible. On the other hand, no closed chain of length 2 is irreducible. A *subchain* of $\mathfrak{C} = (x_0, \ldots, x_h)$ is a subsequence $\mathfrak{C}' = (x_{i_0}, \ldots, x_{i_{h'}})$ with

(3) $\quad h' \leqq h, \quad i_{j-1} < i_j, \quad$ *and* $\quad x_{i_{j-1}} \,\mathrm{I}\, x_{i_j}, \quad$ *for* $\quad j = 1, \ldots, h'.$

This implies, of course, that $\mathfrak{C}'$ is a chain in its own right, of length h'. Obviously:

2. *Every chain joining two elements x and y of* **S** *contains an irreducible subchain which also joins x and y.*

If **S** is *connected*, in the obvious sense that any two of its elements can be joined by some chain[1]), then

(4) $$\varrho(x, y) = \min\{h \,\vdots\, \textit{some chain of length } h \textit{ joins } x \textit{ and } y\}$$

is a well-defined non-negative integer, for all $x, y \in \mathfrak{p} \cup \mathfrak{B}$. The function ϱ so defined satisfies the usual rules for a *metric*:

(5) $$\varrho(x, y) = \varrho(y, x) \geqq 0, \quad \varrho(x, y) = 0 \textit{ if and only if } x = y,$$
$$\textit{and} \quad \varrho(x, y) + \varrho(y, z) \geqq \varrho(x, z).$$

Also, it is clear that

3. *If $\varrho(x, y) = h$, then every chain of length h joining x and y is irreducible.*

We define now a *generalized n-gon*[2]) as a connected incidence structure $\mathbf{G} = (\mathfrak{p}, \mathfrak{L}, \mathrm{I})$ satisfying the following conditions:

(6) $$\varrho(x, y) \leqq n, \quad \textit{for all} \quad x, y \in \mathfrak{p} \cup \mathfrak{L}.$$

(7) *If $\varrho(x, y) = h < n$, then there is only one chain of length h joining x and y.*

(8) *Given $x \in \mathfrak{p} \cup \mathfrak{L}$, there exists $y \in \mathfrak{p} \cup \mathfrak{L}$ such that $\varrho(x, y) = n$.*

A *generalized polygon* is an incidence structure which is a generalized n-gon for some integer n. Condition (7) implies that there are no irreducible closed chains of length $< 2n$, while (6) and (8) together guarantee that every element of **G** is contained in some irreducible closed chain of length $2n$. The generalized polygons with $[x] = [X] = 2$ for all $x \in \mathfrak{p}$ and $X \in \mathfrak{B}$ are precisely the polygons in the usual sense of the word.

Generalized n-gons for $n = 0, 1, 2$ are easily classified; they are quite uninteresting. Therefore, we shall be concerned here only with the cases $n > 2$. A simple consequence of (7), with $h = 1$, is

---

[1]) This definition was also given in Section 1.2.

[2]) Feit & Higman 1964 call these objects "nondegenerate generalized n-gons", the "nondegenerate" referring to condition (8).

4. *For $n \neq 2$, every generalized n-gon is a partial plane.*

This means (cf. Section 1.2) that $[x, y] \leqq 1$ whenever x, y are distinct points. In view of **4**, we call the blocks of a generalized polygon *lines*. It follows readily from (6)—(8) that

5. *The generalized 3-gons are precisely the incidence structures with $[x, y] = 1 = [X, Y]$ for distinct points x, y and distinct blocks X, Y.*[1])

From now on we restrict our attention to finite generalized polygons. The most homogeneous among these are those which are also tactical configurations; we call these *polygonal configurations*. More specifically, an *n-gonal configuration* is a tactical configuration satisfying conditions (6)—(8) above. It is obvious from **5** that

6. *The 3-gonal configurations are precisely the ordinary triangle and the finite projective planes.*

Because of **6**, we can now restrict ourselves further to the cases $n \geqq 4$. The main result on finite generalized polygons is Theorem 1 of FEIT & HIGMAN 1964. We present this now, with a supplement not explicitly stated by these authors.

7. *Let **G** be an n-gonal configuration, with $n \geqq 4$, and with parameters v, b, k, r. Put $c = (r-1)(k-1)$, and let c^* denote the square-free part of c. Also assume, without loss in generality, that $k \leqq r$. Unless $r = k = 2$, i.e. unless **G** is an ordinary n-gon, there are only the following possibilities:*

(a) $n = 4$;
(b) $n = 6$ *and* $c^* = 1$;
(c) $n = 8$ *and* $c^* = 2$;
(d) $n = 12$ *and* $k = 2$.

*In each of these four cases, the parameters of **G** satisfy*

$$v = kt \quad \text{and} \quad b = rt, \quad \text{where} \quad t = (c^{n/2} - 1)(c-1)^{-1}. \tag{9}$$

The proof that **(a)**—**(d)** are the only possibilities is rather complicated and can only be sketched here. The basic tool is:

8. *Let ω be a simple root of the polynomial f, so that*

$$f(x) = (x - \omega) f_0(x) \quad \text{and} \quad f_0(\omega) \neq 0.$$

Then, if the square matrix M satisfies $f(M) = 0$, the multiplicity of ω as an eigenvalue of M is

$$\alpha = \operatorname{tr}\left(f_0(M)/f_0(\omega)\right). \tag{10}$$

---

[1]) This includes, of course, all projective planes, but also some degenerate types, like the ordinary triangle. The statement on p. 116 of FEIT & HIGMAN 1964, that generalized 3-gons are precisely the projective planes, is not correct for the definition of "projective plane" used in this book.

(FEIT & HIGMAN 1964, Lemma 3.4.) The point is that $\operatorname{tr}(f_0(M)/f_0(\omega))$ must be a rational integer. On the other hand, let C be an incidence matrix of the n-gonal configuration **G**, and put $M = CC^T$ and $f(x) = (x - rk)\, g(x)$, where $g(x)$ is defined in different ways according as n is even or odd.[1]) Then $\operatorname{tr}(g(M)/g(rk))$ can be computed in other ways, and the fact that the result must be an integer, by **8**, excludes all cases except (**a**)—(**d**). For this computation, the following fact is essential:

9. *Let C be an incidence matrix of a polygonal configuration* **G**, *and h an integer. Then the (i, j)-entry of the matrix $(CC^T)^h$ is precisely the number of chains of length $2h$ joining the ith point and the jth point of* **G**.

(FEIT & HIGMAN 1964, Lemma 3.1.) This number can be computed precisely from n, h and the parameters of **G**; it depends only on $\varrho(p_i, p_j)$, not on the choice of the points p_i and p_j (cf. FEIT & HIGMAN 1964, Lemma 2.4). For more details of the proof of the unicity of cases (**a**)—(**d**), the reader is referred to the quoted paper of Feit and Higman.

The proof of (9) is much easier: Let L be a line of **G** and denote by $d_j(L)$ the number of elements x with $\delta(L, x) = j$. A simple inductive argument gives:[2])

$$(11)\qquad \begin{aligned} d_{2i+1}(L) &= k\, c^i && \text{for } 1 \leqq 2i + 1 \leqq n - 1, \\ d_{2i}(L) &= k(r - 1)\, c^{i-1} && \text{for } 2 \leqq 2i \leqq n - 2, \end{aligned}$$

with c as in **7**. Hence

$$v = \sum_{i=0}^{(n-2)/2} d_{2i+1}(L) = k \sum_{i=0}^{(n-2)/2} c^i = k\, t,$$

as claimed in (9). A dual argument yields the second equation (9).

Automorphisms of generalized polygons and, in particular, of polygonal configurations, are of high group theoretical interest, but it is beyond the scope of this book to deal with them explicitly here. We mention only the following references: TITS 1959, 1962c, 1963, 1964a, 1967, and again FEIT & HIGMAN 1964.

Anti-automorphisms of polygonal configurations seem to have received little attention so far. We mention here the following result:

---

[1]) For details, see Lemmas 4.1, 5.1, 6.1 of FEIT & HIGMAN 1964. If n is odd, the investigation is facilitated by the fact that then $r = k$. This is quite elementary, cf. FEIT & HIGMAN 1964, Lemma 2.2.

[2]) Note that n even and $\geqq 4$ is not needed here. It follows from (11) that $d_j(L)$ depends on j only, not on L. Furthermore, it can be shown along similar lines that the points as well as the lines of **G** form an association scheme of class number $n/2$. This is also implicitly contained in FEIT & HIGMAN 1964; the parameters of these association schemes depend in a rather complicated way on n, r, k, v, and b.

10. *No hexagonal configuration with* $c > 1$ *(cf.* **7***) is self-polar.*

This is Theorem 1 of PAYNE 1969. In view of **1.3.15**, it suffices to prove that, in case $n = 6$, there cannot exist a symmetric incidence matrix. Payne also proves some restrictions for symmetric incidence matrices of tetragonal ($n = 4$) configurations.

We conclude this section by surveying all known finite generalized n-gons with $n \geqq 4$.[1] We shall see in particular that all four cases of **7** can actually occur.

For $n = 4$, let $\mathfrak{q}$ be either a quadric or the set of absolute points of a unitary polarity π in the projective geometry $\mathscr{P}(d, q)$, and suppose that the index[2] of $\mathfrak{q}$ or π, respectively, is 2. If $d = 3, 4$, or 5 in the orthogonal and $d = 3$ or 4 in the unitary case, then (and only then) the points of $\mathfrak{q}$ together with the lines in $\mathfrak{q}$ (which are then precisely the maximal totally isotropic subspaces of $\mathfrak{q}$ resp. π) form a generalized 4-gon. [It should be noted that the configuration $\mathbf{W}(q)$ of **1.4.55** is contained in the construction just mentioned: $\mathbf{W}(q)$ is the dual of the 4-gon arising from a quadric of index 2 in $\mathscr{P}(3, q)$.]

Next, let $d = 2$ or 3 and consider an ovoid[3] $\mathfrak{o}$ in $\mathscr{P}(d, q)$. Let $\mathscr{P}(d, q)$ be embedded as a hyperplane H in $\mathscr{P}(d+1, q) = \mathscr{G}$. Define *points* as (i) the points of $\mathscr{G} - H$, (ii) the hyperplanes X of $\mathscr{G}$ for which $|X \cap \mathfrak{o}| = 1$, and (iii) one new symbol u. *Lines* are (a) the lines of $\mathscr{G}$ which are not contained in H and meet $\mathfrak{o}$ (necessarily in a unique point), and (b) the points of $\mathfrak{o}$. *Incidence* is defined as follows: Points of type (i) are incident only with lines of type (a); here the incidence is that of $\mathscr{G}$. A point X of type (ii) is incident with all lines $\subseteqq X$ of type (a) and precisely one line of type (b), namely the one represented by the unique point of $\mathfrak{o}$ in X. Finally, the unique point u of type (iii) is incident with no line of type (a) and all lines of type (b). It is a simple exercise to verify that the incidence structure so defined is a 4-gonal configuration with parameters

$$(12) \qquad r = q^{d-1} + 1, \quad k = q + 1, \quad v = (q+1)(q^d + 1), \quad b = (q^{d-1} + 1)(q^d + 1).$$

Examples of 6- and 8-gonal configurations will not be given here explicitly. The known configurations of this kind are closely connected with certain finite simple groups (among others, the groups of REE 1961 a, b), and their explicit description requires a rather intimate knowledge of these groups. For this the reader is referred to TITS 1959, 1960, 1963. See also SCHELLEKENS 1962.

---

[1]) For the following, the author is indebted to J. Tits.

[2]) This was defined in Section 1.4 as the common rank of the maximal totally isotropic subspaces of $\mathfrak{q}$ or π.

[3]) For the definition of ovoids, see (1.4.27) and (1.4.28′).

Finally, let $\mathbf{G} = (\mathfrak{p}, \mathfrak{L}, \mathrm{I})$ be a generalized n-gon (not necessarily an n-gonal configuration), with $n \geqq 3$. Define new points as the elements of $\mathfrak{p} \cup \mathfrak{L}$, new lines as the elements of I (flags of $\mathbf{G}$), and incidence between new points and lines in the obvious fashion. The incidence structure $\mathbf{G}^*$ so defined is a generalized $2n$-gon with $[X] = 2$ for every line X, and it is a $2n$-gonal configuration (with $k = 2$) if and only if the given $\mathbf{G}$ is an n-gonal configuration with $r = k$. Applying the construction to the examples mentioned above, we obtain other generalized 6- and 8-gons, and moreover generalized n-gons with $n = 12, 16, 24, \ldots$ (The generalized $2n$-gons so constructed cannot be tactical configurations whenever $n > 6$, by **7**; hence for these there exist points x and y with $[x] \neq [y]$). Conversely, it can be shown[1]) that every generalized $2n$-gon with $n \geqq 6$ is of the form $\mathbf{G}^*$ just described.

7.4 Finite semi-planes

The incidence structures to be considered in this section are all partial planes, i.e. they satisfy

(1) $\qquad [x, y] \leqq 1$ *for distinct points* x, y.

As before, we call the blocks of such an incidence structure *lines*. In addition to (1), we require:

(2) *Given a non-incident point-line pair* p, L, *there exists at most one line* $M \mathrm{I} p$ *such that* $[L, M] = 0$, *and at most one point* $q \mathrm{I} L$ *such that* $[p, q] = 0$.

This is obviously a simultaneous generalization and dualization of the axiom of parallels for an affine plane.[2]) We call an incidence structure $\mathbf{S} = (\mathfrak{p}, \mathfrak{L}, \mathrm{I})$ a *semi-plane*[3]) if it satisfies (1), (2), and the following non-degeneracy condition:[4])

(3) $\mathfrak{p} \neq \emptyset$ *and* $[x] \geqq 3$ *for all* $x \in \mathfrak{p}$; *dually* $\mathfrak{L} \neq \emptyset$ *and* $[X] \geqq 3$ *for all* $X \in \mathfrak{L}$.

---

[1]) This is an unpublished result of J. Tits.

[2]) Cf. (2.2.5). In fact, the affine planes are precisely those partial planes which satisfy (apart from a non-degeneracy condition) the stronger version of (2) obtained by replacing the first "at most" by "exactly" and the second by "no".

[3]) A more restricted class of partial planes, namely that satisfying (2) with the second "at most" replaced by "no", was called "semi-affine planes" by DEMBOWSKI 1962a. Other special cases of (2) were considered by LÜNEBURG 1964c, p. 441, and CRONHEIM 1965, p. 2. In all these situations, embeddability into a finite projective plane was proved; here we pursue the same objective under more general conditions.

[4]) This is a little stronger than usual; for example the affine plane $\mathbf{A}(2)$ is excluded. However, (3) is quite convenient for the investigations below, and it is not difficult to classify the partial planes satisfying (2) but not (3).

Clearly, projective and affine planes (the latter of order > 2) are semi-planes. More generally, it is easily verified that

1. *If* $\mathbf{P}$ *is a projective plane of order* > 2 *and* $\mathbf{C}$ *a closed subset*[1]) *of* $\mathbf{P}$, *then the substructure* $\mathbf{S} = \mathbf{P} - \mathbf{C}$ *of* $\mathbf{P}$ *is a semi-plane.*

For example, if $\mathbf{C}$ consists of one line and all points on it, then $\mathbf{S}$ is an affine plane, and if $\mathbf{C} = \emptyset$, then $\mathbf{S}$ is a projective plane.

We shall be concerned here with the converse question, i.e. the problem as to when a given semi-plane can be interpreted as $\mathbf{P} - \mathbf{C}$, for some projective plane $\mathbf{P}$ and some closed subset $\mathbf{C}$ of $\mathbf{P}$. We shall see that this is not always the case, not even when the attention is restricted to the finite case, as will be done here.[2]) But the counterexamples are rare, and for large classes of finite semi-planes the converse of **1** is true; cf. results **8**, **12**, **13** below.—As most of the following results have not been published before,[3]) we shall at least indicate all necessary proofs.

Let $\mathbf{S}$ be an arbitrary semi-plane. As for an affine plane, we define two lines L, M of $\mathbf{S}$ to be *parallel*, written $L \parallel M$, if either $L = M$ or $[L, M] = 0$. Dually, we call two points p, q of $\mathbf{S}$ *parallel*, written $p \parallel q$, if either $p = q$ or $[p, q] = 0$. It follows easily from (2) that

2. *Parallelism is an equivalence relation.*

The equivalence classes of $\parallel$, which obviously consist either entirely of points or entirely of lines, will be called the *parallel classes*, or simply the *classes*, of $\mathbf{S}$. A class $\mathfrak{x}$ of points is an *ideal line* if $|\mathfrak{x}| > 1$, and a class $\mathfrak{X}$ of lines is an *ideal point* if $|\mathfrak{X}| > 1$.

Let $\mathscr{P}$ denote the set of all ideal points and $\mathscr{l}$ that of all ideal lines. Also, denote by $\mathscr{I}$ the set of all pairs $(X, \mathfrak{X})$ with $X \in \mathfrak{X} \in \mathscr{P}$, all pairs $(x, \mathfrak{x})$ with $x \in \mathfrak{x} \in \mathscr{l}$, and all pairs $(\mathfrak{X}, \mathfrak{x})$ with $\mathfrak{X} \in \mathscr{P}$, $\mathfrak{x} \in \mathscr{l}$, and $x \mathrm{I} X$ whenever $x \in \mathfrak{x}$ and $X \in \mathfrak{X}$. We define an extended incidence structure $\mathbf{S}^+ = (\mathfrak{p}^+, \mathfrak{L}^+, \mathrm{I}^+)$ of the semi-plane $\mathbf{S} = (\mathfrak{p}, \mathfrak{L}, \mathrm{I})$ by

$$\mathfrak{p}^+ = \mathfrak{p} \cup \mathscr{P}, \quad \mathfrak{L}^+ = \mathfrak{L} \cup \mathscr{l}, \quad \mathrm{I}^+ = \mathrm{I} \cup \mathscr{I}. \tag{4}$$

---

[1]) A closed subset of a projective plane was defined in Section 3.1 as a subset $\mathbf{C}$ such that if two distinct points (lines) are in $\mathbf{C}$, then so is the line joining them (their intersection point).

[2]) See result **11** below. A class of examples in DEMBOWSKI 1962a, Section 3, shows that there is no hope of classifying the counterexamples in the infinite case. These examples are "free semi-plane extensions", modeled after HALL's 1943 complete free extensions of partial planes; see also PICKERT 1955, Section 1.3.

[3]) The author has reported on the material in this section at the International Congress of Mathematicians in Stockholm 1962. Some of the results can be considered as special cases of more general theorems of possibly infinite semi-planes (cf. DEMBOWSKI 1962a), but this will not be pursued here.

In many situations to be encountered later,[1] $\mathbf{S}^+$ will again be a semi-plane, so that $\mathbf{S}^{++}$ can be formed. We shall see that if $\mathbf{S}$ is finite and $\mathbf{S}^+$ a semi-plane, then $\mathbf{S}^{++}$ is usually a projective plane $\mathbf{P}$, and $\mathbf{S} = \mathbf{P} - \mathbf{U}$, for some well-defined subset $\mathbf{U}$ of $\mathbf{P}$; see **8, 11, 12, 13** below.

For all that follows, $\mathbf{S} = (\mathfrak{p}, \mathfrak{L}, \mathrm{I})$ is a finite semi-plane. The *degree* of a point $p \in \mathfrak{p}$, or a line $L \in \mathfrak{L}$, is the number $[p]$ of lines through p, or the number $[L]$ of points on L, respectively. We call $n + 1$ the largest of the degrees of the elements of $\mathbf{S}$; the integer n (> 1) so defined is the *order*[2] of $\mathbf{S}$.

3. *Let* $\mathbf{S}$ *be a semi-plane of order* n, *and let* D *denote the set of integers which occur as degrees of elements in* $\mathbf{S}$. *Then there are only three possibilities for* D, *namely*

(a) $D = \{n - 1, n, n + 1\}$,
(b) $D = \{n, n + 1\}$, *or*
(c) $D = \{n + 1\}$.

We shall call $\mathbf{S}$ *hyperbolic, parabolic,* or *elliptic,* according as D is of form **(a)**, **(b)**, or **(c)**.

For the proof of **3**, we note first the following obvious lemma.

4. *Let* p, L *be a non-incident point-line pair in a finite semi-plane. Then there are precisely four possibilities:*

(a) *There is no parallel to* p *on* L *and no parallel to* L *through* p. *In this case* $[p] = [L]$.
(b) *There is no parallel to* p *on* L *but a parallel to* L *through* p. *In this case* $[p] = [L] + 1$.
($\bar{\mathbf{b}}$) *There is a parallel to* p *on* L *but no parallel to* L *through* p. *In this case* $[L] = [p] + 1$.
(c) *There is a parallel to* p *on* L *and a parallel to* L *through* p. *In this case* $[L] = [p]$.

According to these cases, we shall refer to the non-incident point-line pair p, L as being of *type* **(a)**, **(b)**, **($\bar{\mathbf{b}}$)**, or **(c)**, respectively.

As a corollary of **4**, we have:

(5) $$p \,\not\mathrm{I}\, L \quad \text{implies} \quad |[p] - [L]| \leqq 1.$$

Similarly:

(6) $$p \,\|\, q \quad \text{implies} \quad |[p] - [q]| \leqq 1, \quad \text{and, dually,}$$
$$L \,\|\, M \quad \text{implies} \quad |[L] - [M]| \leqq 1.$$

---

[1]) An exceptional situation is that of **13c** below.

[2]) This definition agrees with that in Section 3.2 in case $\mathbf{S}$ is a projective or affine plane.

For by (3), there exists $L \mathrel{I} p, q$, and by **2**, any $x \mathrel{I} L$ is parallel to either both or none of p, q. Hence p and q are incident with equally many lines $\nparallel L$. The first claim of (6) now follows from (2), and the second is dual to it.

Next, if p, q are any two points, then

$$|[p] - [q]| \leqq |[p] - [L]| + |[L] - [q]|$$

for any line $L \mathrel{I} p, q$. Hence (5) gives

$$|[p] - [q]| \leqq 2 \quad \textit{for} \quad p, q \in \mathfrak{p}, \quad \textit{and, dually,} \tag{7}$$
$$|[L] - [M]| \leqq 2 \quad \textit{for} \quad L, M \in \mathfrak{L}.$$

Now assume that **3** is false. Then, without loss in generality, $[q] = n + 1$ for some $q \in \mathfrak{p}$, and $[L] < n - 1$ for some $L \mathrel{I} q$. Let $p \mathrel{I} L$; then $[p] = n - 1$ by (5) and (7), and $[L] = n - 2$ by (5). Hence $[p] > [L]$, so that (p, L) is of type (**b**); cf. **4**. But then there exists $M \parallel L$ through p, and (6) gives $[M] \leqq n - 1$ while (5) and $q \mathrel{I} M$ give $[M] \geqq n$. This contradiction proves **3**.

For an arbitrary line L of **S**, define

$$\left.\begin{matrix} \alpha(L) \\ \beta(L) \\ \gamma(L) \end{matrix}\right\} = \text{number of } x \mathrel{I} L \text{ such that} \left\{\begin{matrix} [x] = n - 1 \\ [x] = n \\ [x] = n + 1, \end{matrix}\right.$$

and

$$\pi(L) = \text{number of lines parallel to } L.$$

Obviously, $\alpha(L), \beta(L), \gamma(L) \geqq 0 < \pi(L) \leqq n + 1$, and

$$\alpha(L) + \beta(L) + \gamma(L) = [L]. \tag{8}$$

Furthermore, simple counting shows that

$$\pi(L) + (n-2)\,\alpha(L) + (n-1)\,\beta(L) + n\,\gamma(L) = |\mathfrak{L}|,$$

whence elimination of $\gamma(L)$ gives

$$\pi(L) - 2\alpha(L) - \beta(L) = |\mathfrak{L}| - n[L]. \tag{9}$$

Obvious analogues to (8) and (9) hold for the numbers $\alpha(p), \ldots, \pi(p)$, defined for points p in the dual fashion.

5. *If the semi-plane* **S** *of order n contains two distinct parallel elements one of which has degree $n + 1$, then* **S** *is elliptic.*

Proof. Without loss in generality, suppose $[L] = n + 1$ and $L \parallel M \neq L$. For any $p \mathrel{I} M$, the pair (p, L) is of type (**c**), whence $[p] = [L] = n + 1$, or

$$\alpha(M) = \beta(M) = 0.$$

If there were a point z of degree $n-1$, then $z \mathrm{I} L$, by (5). On the other hand, (6) gives $[M] \geqq n > [z]$, whence (z, M) would be of type $(\bar{\mathbf{b}})$, contradicting $z \mathrm{I} L \parallel M$. Hence no such z exists, and

$$\alpha(X) = 0, \quad \textit{for all lines } X.$$

Next, we prove $[M] = n+1$. The alternative to this is $[M] = n$, which would imply $\beta(L) = n$. But then there would be, to every point $p \mathrm{I} M$, a unique parallel $q \mathrm{I} L$, and $[p] = n+1$, $[q] = n$. Arguments dual to the above then give $\beta(p) = n$ for every $p \mathrm{I} M$, and $\alpha(q) = \beta(q) = 0$ for precisely n points $q \mathrm{I} L$. Of the $n+1$ lines through $p \mathrm{I} M$, precisely $n-1$ intersect L in points q with $\alpha(q) = \beta(q) = 0$, and as $n-1 > 2$ by (3), we get $\gamma(p) \geqq 2$. But then $\beta(p) + \gamma(p) > n+1$, contradicting the dual of (8) and thus proving $[M] = n+1$.

Assume $[w] < n+1$ for some point w. As $\alpha(X) = 0$ for all X, we have $[w] = n$. We can assume $w \not\mathrm{I} M$; as $[M] > [w]$, there exists $v \parallel w$ on M; cf. **4$\bar{\mathbf{b}}$**. We have already proved that an element parallel to an element of degree $n+1$ also has degree $n+1$; thus $[v] \leqq n$. But this would mean $\alpha(M) + \beta(M) > 0$, a contradiction.

Hence if there exists an element of degree $< n+1$, it must be a line W. If $[W] = n-1$, then W would be incident with every point of degree $n+1$, by (5), hence with all points of $\mathbf{S}$, which is impossible. Thus the only remaining possibility is $[W] = n$. But then (9) would give

$$\begin{aligned} \pi(W) &= \pi(W) - 2\alpha(W) - \beta(W) = |\mathfrak{L}| - n^2 \\ &= \pi(L) - 2\alpha(L) - \beta(L) + n = \pi(L) + n, \end{aligned}$$

or $\pi(L) \leqq 1$, contradicting $L \parallel M \neq L$. This proves **5**.

6. *In a semi-plane of order n, any class of parallels contains at most one element of degree $n-1$.*

Proof. Otherwise, there would be (without loss in generality) two distinct parallel lines E, F of degree $n-1$. By (5), there could be no points of degree $n+1$; hence there is a line H with $[H] = n+1$ which, by **5**, intersects all other lines. Let e and f be the intersection points of H with E and F, respectively. If $f \neq x \mathrm{I} E$, then there is no point parallel to x on F, but some $x' \parallel x$ on H. Analogously, if $e \neq y \mathrm{I} F$, no point on E is parallel to y, but there exists $y' \parallel y$ on H. Hence the $n-2$ points $x' \mathrm{I} H$ are distinct from the $n-2$ points $y' \mathrm{I} H$, and $n-1 = [H] - 2 \geqq 2(n-2)$ or $n \leqq 3$. On the other hand, (3) gives $n-1 = [E] \geqq 3$, or $n \geqq 4$. This contradiction proves **6**.

7. *If* $\mathbf{S}$ *is non-elliptic of order n and $\mathfrak{P}$ an ideal point in* $\mathbf{S}$, *then*

(**a**) *every point of degree $n+1$ is incident with a line of $\mathfrak{P}$,*
(**b**) *no point of degree $n-1$ is incident with a line of $\mathfrak{P}$, and*

(**c**) *to $p \not\mathrm{I}\, X$ for all $X \in \mathfrak{P}$, there exists a parallel q on a line of $\mathfrak{P}$ if, and only if, $[p] = n - 1$. In this case, precisely the $X \in \mathfrak{P}$ with $[X] = n$ carry parallels of p.*

Dual statements hold for ideal lines of $\mathbf{S}$.

Proof. By **5** and **6**, $\mathfrak{P}$ contains a line L of degree n. For a point p which is on no line of $\mathfrak{P}$, the pair (p, L) is of type (**a**) or ($\bar{\mathbf{b}}$); cf. **4**. Hence $[p] \leqq n$, proving (**a**). Furthermore, there exists $q \,||\, p$ on L if and only if (p, L) is of type ($\bar{\mathbf{b}}$), which means $[p] = n - 1$. Hence to prove (**c**) it remains to show that if $E \in \mathfrak{P}$ and $[E] = n - 1$, then E contains no point $||\,p$. But if this were the case, then $[p] < [E] = n - 1$, contradicting **3**.

For the proof of (**b**), assume that the point e satisfies $[e] = n - 1$ and $e \,\mathrm{I}\, X$ for some $X \in \mathfrak{P}$. Consider the line L of degree n in P. If $e \not\mathrm{I}\, L$, then $[e] \geqq [L]$ by **4b,c**, a contradiction. Hence $e \,\mathrm{I}\, L$, and L is the only line of degree n in $\mathfrak{P}$. Since $|\mathfrak{P}| > 1$, there exists $E \neq L$ in $\mathfrak{P}$. But $[E] \neq n + 1$ by **5**, so $[E] = n - 1$, and E is the only line $\neq L$ in $\mathfrak{P}$, by **6**. Thus we have shown:

$$\mathfrak{P} = \{L, E\}, \quad [L] = n, \quad [E] = n - 1, \quad \text{and} \quad e \,\mathrm{I}\, L.$$

The points of degree $n + 1$ are all on E, by (5); hence $\gamma(L) = 0$ and $\alpha(L) + \beta(L) = [L] = n$. Also, $\alpha(E) = 0$ and $\pi(L) = \pi(E) = 2$; hence (9) gives

$$\alpha(L) = n^2 - n + 2 - |\mathfrak{L}| = \beta(E).$$

If $x \not\mathrm{I}\, L, E$, then (x, E) is of type (**a**) or ($\bar{\mathbf{b}}$), whence $[x] \leqq [E]$ and $[x] = n - 1$. It follows that (x, L) is of type ($\bar{\mathbf{b}}$) and that there exists $x' \,||\, x$ on L, with $[x'] = n$. Furthermore, two distinct points $x, y \not\mathrm{I}\, E, L$ give rise to distinct parallels, by **6**. Thus the number of points not on L or E is $\leqq \beta(L)$. It follows that $\beta(L) \geqq 2$ and therefore $\alpha(L) = n - \beta(L) \leqq n - 2$. But then

$$\beta(E) \leqq n - 2 < [E],$$

so that there exists $w \,\mathrm{I}\, E$ with $[w] = n + 1$. Hence there are n lines $\neq E$ through w, and we conclude from (3) that there are at least n points not incident with either of E, L. By the above, this implies $\beta(L) \geqq n$. But then $e \,\mathrm{I}\, L$ and $[e] = n - 1$ give $\alpha(L) + \beta(L) \geqq n + 1 > [L]$, contradicting (8). This completes the proof of **7**.

8. *Let* $\mathbf{S}$ *be a semi-plane of order n which possesses no ideal lines, i.e. in which any two distinct points are joined by a line. Then* $\mathbf{S} = \mathbf{P} - \mathbf{C}$, *where* $\mathbf{P}$ *is a projective plane and* $\mathbf{C}$ *a closed subset in* $\mathbf{P}$, *of one of the following four types:*

(**a**) $\mathbf{C} = \emptyset$;

(**b**) $\mathbf{C}$ *consists of one point and no line*;

(**c**) **C** *consists of one line and* n *points on it;*
(**d**) **C** *consists of one line and all its points.*

In case (**a**), **S** *is elliptic, and in cases* (**b**)—(**d**), **S** *is parabolic. Dual statements hold if* **S** *possesses no ideal points.*

For the proof of this, see DEMBOWSKI 1962a, Satz 3.[1]) Using **7** and **8**, we can now show that there are at most one point and one line of degree $n-1$ in any finite semi-plane. This will follow from

9. *If the semi-plane* **S** *of order* n *contains a point* e *of degree* $n-1$, *then the lines through* e *are precisely the lines of degree* $n+1$ *in* **S**. *A dual statement holds if* **S** *contains a line of degree* $n-1$.

Proof. That all lines of degree $n+1$ pass through e follows from (5); hence we need only show the converse. Thus, let H be an arbitrary line through e. By **7b**, every line of **S** intersects H; hence if $x \not{I} H$, then (x, H) is of one of the types (**a**), ($\bar{\mathbf{b}}$) of **4**.

Now, since **S** is hyperbolic, **8** implies that **S** possesses an ideal point $\mathfrak{P}$, and by **5** and **6** there is a line $L \in \mathfrak{P}$ with $[L] = n$. Also, $e \not{I} L$ by **7b**. Next, **7c** implies the existence of $p \parallel e$ on L, and it follows that (p, H) is of type ($\bar{\mathbf{b}}$). Hence $n = [p] < [H]$ or $[H] = n+1$, proving **9**.

We call a parallel class *pure* if all its elements have the same degree; this is then also called the *degree* of the pure class in question.

10. *Let* $\mathbf{S} = (\mathfrak{p}, \mathfrak{L}, \mathrm{I})$ *be a semi-plane of order* n, *and suppose that* **S** *contains two parallel classes* $\mathfrak{G}$ *and* $\mathfrak{H}$ *of lines, such that the numbers* g, h *of lines of degree* n *in* $\mathfrak{G}$ *and* $\mathfrak{H}$, *respectively, satisfy*

$$0 < g < h.$$

Then:

(**a**) **S** *is not hyperbolic.*
(**b**) *All parallel classes of* **S** *are pure.*
(**c**) *The points of degree* $n+1$ *are precisely the* gh *intersection points of the lines in* $\mathfrak{G}$ *and* $\mathfrak{H}$.
(**d**) $\mathfrak{G}$ *is the only line class of degree* n *which has size* g; *all other line classes of degree* n *have size* h.
(**e**) *Either* $h = n-1$ *or* $h = n$. *Let* r *denote the number of line classes of size* h; *then* $|\mathfrak{p}| = n(n+1) - r$. *If* $h = n-1$, *then* $r = n+1-g$ *and* $|\mathfrak{p}| = |\mathfrak{L}| = n^2 + g - 1$; *if* $h = n$, *then* $|\mathfrak{L}| = n^2 + g$, *and the number of lines of degree* $n+1$ *is* $n(n-r)$.

Dual statements hold for finite semi-planes possessing two point classes containing different numbers of points of degree n.

---

[1]) The result had been previously proved, but not published, by N. H. Kuiper. The finiteness assumption is essential, as was shown by means of examples in DEMBOWSKI 1962a, Satz 2.

Proof. Assume that there is a point e of degree $n-1$, and let L be a line of degree n. Then $e \,\bar{\mathrm{I}}\, L$ by **9**, and $[e] < [L]$ implies that L carries a point $p \parallel e$; cf. **4$\bar{\mathbf{b}}$**. Conversely, let $x \parallel e \neq x$; then $[x] = n$ by **5** and **6**. If $x \neq p$, then $x \,\bar{\mathrm{I}}\, L$, and now $x \parallel p \,\mathrm{I}\, L$ and $[x] = [p] = n$ imply the existence of $L' \parallel L$ through x, with $[L'] = n$. Thus the number of lines of degree n in an arbitrary parallel class is either 0 or $\pi(e) - 1$. This contradicts the hypothesis on $\mathfrak{G}$ and $\mathfrak{H}$; hence in order to prove (**a**) we must show that there is no line of degree $n-1$. Assume $[E] = n-1$. By **8**, there exists an ideal line $\mathfrak{q}$, consisting of points of degree n. Arguments dual to the above show that $|\mathfrak{q}| = \pi(E) - 1$, independently of the choice of $\mathfrak{q}$. Let $[X] = n$; then **9** shows that there is a point q of degree n on X, and we can assume $q \in \mathfrak{q}$. For any line $Y \parallel X \neq Y$, the pair (q, Y) is of type (**c**), and it follows that the class $\mathfrak{X}$ containing X consists of $|\mathfrak{q}| = \pi(E) - 1$ lines. As X was arbitrary, we have the same contradiction as above, so that (**a**) is proved. That (**b**) also holds is now an immediate consequence of **5**.

Next, consider the gh intersection points of the lines in $\mathfrak{G}$ and $\mathfrak{H}$. If such a point p were of degree n, then, for any $L \in \mathfrak{G}$ with $p \,\bar{\mathrm{I}}\, L$, the pair (p, L) would be of type (**c**), whence $\pi(p) = |\mathfrak{G}|$ as above. But in the same way one could also show $\pi(p) = |\mathfrak{H}| = h > g = |\mathfrak{G}|$. This contradiction shows that the intersection points of $\mathfrak{G}$- and $\mathfrak{H}$-lines have degree $n+1$. Conversely, assume $[p] = n+1$ and $p \,\bar{\mathrm{I}}\, L$ for $[L] = n$. Then (p, L) is of type (**b**), and there exists a parallel to L through p. This shows a little more than (**c**), namely:

(10) *If* $[p] = n+1$, *then every line class of degree* n *contains a line through* p.

Now assume that some line class of degree n has size $j \neq g, h$. Then the same arguments as above show that the number of points of degree $n+1$ is $gj \neq gh$. Hence g and h are the only possible sizes for line classes of degree n.

If there were a line class $\mathfrak{W} \neq \mathfrak{G}$ of degree n with $|\mathfrak{W}| = |\mathfrak{G}| = g$, then the points of degree $n+1$ would also be the intersection points of $\mathfrak{W}$ and $\mathfrak{H}$; hence every $W \in \mathfrak{W}$ would contain precisely h points of degree $n+1$. These h points are also on the g lines of $\mathfrak{G}$, and $g < h$ implies that some $L \in \mathfrak{G}$ contains more than one of these points. This would mean $[L, W] > 1$, contradicting (1). This proves (**d**), and it remains to prove (**e**).

Note first that $h = |\mathfrak{H}| \leqq n$, since otherwise $|\mathfrak{H}| = n+1$, implying that any $X \in \mathfrak{G}$ intersects the lines of $\mathfrak{H}$ in $n+1$ distinct points, or $[X] = n+1$. Suppose now that $h < n$; we prove $h = n-1$.

Since $h - g > n - h > 0$, there exist $a \,\mathrm{I}\, G \in \mathfrak{G}$ and $b \,\mathrm{I}\, H \in \mathfrak{H}$ with $[a] = [b] = n$; let $\mathfrak{a}$, $\mathfrak{b}$ be the point classes containing a and b, respectively. We have

$$|\mathfrak{a}| = |\mathfrak{G}| = g < h = |\mathfrak{H}| = |\mathfrak{b}|;$$

hence $\mathfrak{a}$ and $\mathfrak{b}$ have different sizes, and dualizing (**c**) we get that the lines of degree $n+1$ are precisely those joining points of $\mathfrak{a}$ and $\mathfrak{b}$.

With r the number of line classes $\mathfrak{X}$ with $|\mathfrak{X}| = h$ [cf. (**e**)], we have

$$|\mathfrak{L}| = rh + g + gh,$$

for there is only one class of size g and degree n, and gh lines of degree $n+1$. Let z be a point of degree $n+1$; then $\pi(z) = 1$ by **5**, $\alpha(z) = 0$ by (**a**), and $\beta(z) = r + 1$ by (10). Hence the dual of (9) gives

$$(11) \qquad |\mathfrak{p}| = n(n+1) - r.$$

Next, if $L \in \mathfrak{G}$, then $\pi(L) = g$, $\alpha(L) = 0$, and $\beta(L) = n - h$. This and a dual consideration, together with (9), give

$$|\mathfrak{L}| = n(n-1) + g + h = |\mathfrak{p}|.$$

Combination of these equations now yields, after elementary calculations, the quadratic equation

$$(2n - h)(h + 1) = n(n + 1)$$

for h, whose solutions are $h = n$, $n - 1$. But $h = n$ was excluded above, so that $h = n - 1$, as claimed.

If $h = n - 1$, the last equation can also be used to show that $r = n + 1 - g$ and $|\mathfrak{p}| = |\mathfrak{L}| = n^2 + g - 1$, proving (**e**) in this case.

Finally, let $h = n$. Then the gn points on the lines of $\mathfrak{G}$ are precisely the points of degree $n+1$, by (**c**). Call i the number of points of degree n, then

$$i = r(n - g) + (n - r)(n + 1 - g),$$

and since $|\mathfrak{p}| = gn + i$, we get again (11). Let j denote the number of lines of degree $n+1$; counting the flags (x, Y) with $[x] = n$ and $Y \notin \mathfrak{G}$ in two ways gives

$$in = rn(n - g) + j(n + 1 - g),$$

and we can conclude $j = n(n - r)$. Finally, it now follows that

$$|\mathfrak{L}| = j + g + rn = n^2 + g.$$

This concludes the proof of **10.**

We are now ready to classify completely all hyperbolic and parabolic finite semi-planes. For the next two theorems (**11** and **12**) we omit some of the details of the proofs, indicating only the basic idea which is the same in all cases.

11. *The hyperbolic semi-planes of order n are precisely the incidence structures of the form* $\mathbf{P} - \mathbf{U}$, *where* $\mathbf{P}$ *is a projective plane of order n and* $\mathbf{U}$ *a subset of* $\mathbf{P}$ *of one of the following types:*

(**a**) $\mathbf{U}$ *consists of two distinct points u, v and all lines $\neq uv$ through u.*[1])

($\overline{\mathbf{a}}$) $\mathbf{U}$ *consists of two distinct lines U, V and all points $\neq UV$ on U.*[1])

(**b**) $\mathbf{U}$ *consists of a non-incident point-line pair u, U, all points on U, and n lines through u.*

($\overline{\mathbf{b}}$) $\mathbf{U}$ *consists of a non-incident point-line pair u, U, all lines through u, and n points on U.*

(**c**) $\mathbf{U}$ *consists of a non-incident point-line pair u, U, of n points on U, and of n lines through u. Furthermore, the unique point $\notin \mathbf{U}$ on U is not incident with the unique line $\notin \mathbf{U}$ through u.*

Note that in none of the cases (**a**)—(**c**) is $\mathbf{U}$ a closed subset of $\mathbf{P}$.

That every incidence structure of the form $\mathbf{P} - \mathbf{U}$ under discussion is a semi-plane (also in the infinite case), and that this semi-plane is hyperbolic if finite, is straightforward. For the converse, one verifies that, for any hyperbolic semi-plane $\mathbf{S}$ of order n, the incidence structure $\mathbf{S}^+$ defined earlier is again a semi-plane, no longer hyperbolic, and that $\mathbf{S}^{++}$ is a projective plane $\mathbf{P}$ of order n. The new elements added in the process of forming $\mathbf{S}^+$ and $\mathbf{S}^{++}$ form the set $\mathbf{U}$ of **11**. In fact, (**a**) arises if $\mathbf{S}$ contains a line but no point of degree $n - 1$ and if there are lines of degree $n + 1$, ($\overline{\mathbf{a}}$) in the situation dual to this; (**b**) arises if $\mathbf{S}$ contains a line but no point of degree $n - 1$ and if there are no lines of degree $n + 1$, ($\overline{\mathbf{b}}$) in the dual situation; and (**c**) arises if $\mathbf{S}$ contains a point as well as a line of degree $n - 1$.

The verification of these claims is quite straightforward, although not brief, and will be omitted. We mention only that results **3**—**10** above, and in particular **9** and **10**, are essential tools in this verification.

12. *The parabolic semi-planes of order n are precisely the incidence structures of the form* $\mathbf{P} - \mathbf{C}$, *where* $\mathbf{P}$ *is a projective plane of order n and* $\mathbf{C}$ *a closed subset of* $\mathbf{P}$ *which is not a Baer subset.*[2])

Note that, except for the trivial case of a projective plane, **12** may be considered a generalization of **8**.

That incidence structures of the form $\mathbf{P} - \mathbf{C}$ in question are semi-planes, and parabolic if finite, is again straightforward. For the converse, one verifies again that $\mathbf{S}^{++}$ is a projective plane of order n. In fact, in certain cases (see below), $\mathbf{S}^+$ is already a projective plane. The

[1]) uv means, of course, the line joining the points u and v in $\mathbf{P}$, and UV has the dual meaning.

[2]) A Baer subset of a projective plane $\mathbf{P}$ was defined in Section 3.1 as a closed subset $\mathbf{B} \neq \mathbf{P}$ of $\mathbf{P}$ such that every element of $\mathbf{P}$ is incident with an element of $\mathbf{B}$.

verification depends again on the previous results **3—10**, and it is a little more complicated than in the case of **11**. Without giving details, we say here only how the various possibilities for **C** can arise.

As **S** is parabolic, **5** implies that all parallel classes are pure, and that the ideal points (lines) of **S** consist of lines (points) of degree n. Also, it can be assumed, by **8**, that there actually exist ideal points and lines.

Consider first the possibility that **S** possesses two ideal points $\mathfrak{G}$, $\mathfrak{H}$ of different sizes: $|\mathfrak{G}| = g < h = |\mathfrak{H}|$. Then **10** is applicable, and it turns out that *either* $h = n - 1$ (cf. **10e**) in which case $\mathbf{S} = \mathbf{P} - \mathbf{C}$, where **P** is a projective plane of order n and **C** a closed configuration consisting of a non-incident point-line pair u, U, of r points on U and the r lines connecting them with u (here $2 < r < n + 1$ with r having the same significance as in **10e**), *or* $h = n$, and then **C** consists of $r + 1$ points on a line U and $r' + 1$ lines, including U, through one of these points (again with r as in **10b**, and $1 \leqq r, r' \leqq n$, but not $r = r' = n$).

This and a dual consideration leaves only the case where all ideal points have the same size, say m, and all ideal lines have the same size m'. If $m > m'$, then there is only one ideal line, of size s with $2 \leqq s \leqq n - 1$, and there are $n + 1 - s$ distinct ideal points, each of size n. This is proved similarly to **10e**. Consequently, in this case $\mathbf{S} = \mathbf{P} - \mathbf{C}$, where **C** consists of one line and $n + 1 - s$ points on it. The case $m < m'$ is dual to this. Note that here $\mathbf{S}^+ = \mathbf{P}$.

We are now left with the case that all ideal elements have the same size m. If there is only one ideal point $\mathfrak{U}$ and only one ideal line $\mathfrak{u}$, then either (i) every point in $\mathfrak{u}$ is incident with a line in $\mathfrak{U}$, or else (ii) no point in $\mathfrak{u}$ is incident with any line of $\mathfrak{U}$. In these cases, $\mathbf{S} = \mathbf{P} - \mathbf{C}$, with **C** a single point-line pair, non-incident in case (i) and incident in case (ii). Finally, if there is more than one ideal point, then there is also more than one ideal line, and $\mathbf{S} = \mathbf{P} - \mathbf{C}$ with **C** either a triangle (in case $m = n - 1$) or a non-degenerate subplane of order $n - m < \sqrt{n}$. In these cases, again $\mathbf{S}^+ = \mathbf{P}$.

Now we turn to the elliptic case. If **B** is a Baer subset of the projective plane **P** of order $n + 1$, then $\mathbf{P} - \mathbf{B}$ is an elliptic semi-plane of order n, but in contrast to **7, 10, 11** it is not known whether or not the converse is true. Instead, we prove the following weaker result:

13. *Let* $\mathbf{S} = (\mathfrak{p}, \mathfrak{L}, \mathrm{I})$ *be an elliptic semi-plane of order* n *which is not a projective plane. Then all parallel classes of* **S** *have the same size* $m > 1$, *and*

$$|\mathfrak{p}| = |\mathfrak{L}| = n(n + 1) + m, \tag{12}$$

so that $n(n+1) \equiv 0 \bmod m$. *There are only the following possibilities:*

(**a**) $m = n+1$; *then* $\mathbf{S} = \mathbf{P} - \mathbf{B}(p, L)$, *where* $\mathbf{P}$ *is a projective plane of order* $n+1$, *and* $p \mathrm{I} L$.[1])

(**b**) $m = n$; *then* $\mathbf{S} = \mathbf{P} - \mathbf{B}(p, L)$, *where* $\mathbf{P}$ *is as in case* (**a**), *and* $p \not\mathrm{I} L$.[1])

(**c**) $m \leqq n+1-\sqrt{n+1}$; *in this case the substructure of* $\mathbf{S}^+$ *formed by the ideal elements of* $\mathbf{S}$ *is a projective design with parameters*

$$(13) \quad v = b = [n(n+1)+m]m^{-1}, \qquad k = r = n(n+1-m)m^{-1}, \qquad \lambda = (n-m)(n+1-m)m^{-1}.$$

In particular, $\mathbf{S} = \mathbf{P} - \mathbf{B}$, *with* $\mathbf{P}$ *a projective plane of order* $n+1$ *and* $\mathbf{B}$ *a Baer subplane of* $\mathbf{P}$, *if, and only if,* $m = n+1-\sqrt{n+1}$.

Proof. For any line L, we get from (9) that $\pi(L) = |\mathfrak{L}| - n(n+1)$. On the other hand, $\pi(L) = \pi(p)$ for any point p; this follows easily from the hypothesis that $\mathbf{S}$ is elliptic. This proves (12).

If $m = n+1$, then $|\mathfrak{p}| = |\mathfrak{L}| = (n+1)^2$ by (12); hence every line class covers all points. Consideration of $\mathbf{S}^+$ and $\mathbf{S}^{++}$ then yields the claim of (**a**).

If $m = n$, then similarly $|\mathfrak{p}| = |\mathfrak{L}| = n(n+2)$, hence to every line class $\mathfrak{X}$ there is a unique point class $\mathfrak{x}$ with $\mathfrak{X}\, \mathrm{I}^+ \mathfrak{x}$. In this case, consideration of $\mathbf{S}^+$ and $\mathbf{S}^{++}$ shows that we have case (**b**).

Finally, suppose $m < n$. If $\mathfrak{X}$ is any ideal point, then there are

$$|\mathfrak{p}| - m(n+1) = n(n+1-m)$$

points not incident with any line in $\mathfrak{X}$. These fall into r distinct parallel classes, where r is as in (13), and by definition of $\mathbf{S}^+$, these are precisely the ideal lines incident with $\mathfrak{X}$. A dual argument shows that each ideal line is incident with $r = k$ ideal points. Hence the ideal elements form a tactical configuration with equally many points and blocks, $b = v$ being given in (13).

Now let $\mathfrak{X}$ and $\mathfrak{Y}$ be two distinct ideal points. The number of points on no line of $\mathfrak{X} \cup \mathfrak{Y}$ is then

$$|\mathfrak{p}| - m^2 - 2m(n+1-m) = m\lambda,$$

with λ as in (13). It follows from the definition of $\mathbf{S}^+$ that $\mathfrak{X}$ and $\mathfrak{Y}$ are joined by exactly λ ideal lines, and we have now proved that the ideal elements of $\mathbf{S}$ form a projective design with parameters (13), in $\mathbf{S}^+$. The inequality $m \leqq n+1-\sqrt{n+1}$ is equivalent to the obvious

[1]) The special Baer subsets $\mathbf{B}(p, L)$ were defined in 3.1; they consist of the point p and the line L, all lines through p, and all points on L. Case (**a**) is that dealt with by Cronheim 1965, p. 2, and case (**b**) that of Lüneburg 1964c, p. 441. Compare here also footnote [3]) on p. 185

$\lambda \geqq 1$, equality holding if and only if $\lambda = 1$, i.e. if the ideal elements form a projective plane.[1]) Conceivably, $m = n + 1 - \sqrt{n+1}$ must always hold in case (**c**), but this remains an open problem.

We conclude with an application of **13c**, and give the proof of result **7.1.15** promised in Section 7.1. Suppose that **S** is a partial design of class number 2, with parameters

(14) $\lambda_1 = 0,\ \lambda_2 = 1,\ v = b = q(q^3 - 1),\ r = k = q^2$, and $p^1_{22} = n_2$.

It was pointed out in 7.1 that (14) implies $p^1_{12} = 0$, so that **S** is divisible. This means that by $[x, y] = 0$ an equivalence relation is given among the points. Hence if $p \not\mathrm{I}\, L$, there is at most one $q \,\mathrm{I}\, L$ with $[p, q] = 0$. Thus at least $r - 1$ of the $r = k$ lines through p meet L, whence at most one $M \,\mathrm{I}\, p$ satisfies $[L, M] = 0$. This shows that **S** satisfies condition (2), and since (1) and (3) are obvious, **S** is an elliptic semi-plane. The order of **S** is $n = k - 1 = q^2 - 1$, and the common size of the parallel classes is given by (12):

$$m = v - n(n+1) = q^2 - q = n + 1 - \sqrt{n+1}.$$

From **13c** it now follows that $\mathbf{S} = \mathbf{P} - \mathbf{B}$, with **B** a Baer subplane of the projective plane **P** of order q^2, as we intended to show.

---

[1]) This part of **13** implies that a finite projective plane **P** with a Baer subplane **B** is uniquely determined by the substructure $\mathbf{S} = \mathbf{P} - \mathbf{B}$. Hence it is not possible to construct new finite planes by changing incidences within a Baer subplane only.

Bibliography

ABBE, E.

1962 Geometry in certain simple groups. Tohoku Math. J. **14**, 64-72.

ABRHAM, J.; DRIML, M.

1956 Über ein Problem der Kodentheorie. Časopis Pěst. Mat. **81**, 69—76.

ADHIKARY, B.

1965 On the properties and construction of balanced block designs with variable replications. Calcutta Statist. Assoc. Bull. **14**, 36—64.

AGRAVAL, H.

1963 On the dual of BIB designs. Calcutta Statist. Assoc. Bull. **12**, 104—105.

1964 On the bounds of the numbers of common treatments between blocks of semi-regular group divisible designs. J. Amer. Statist. Assoc. **59**, 867—871.

1965 A note on incomplete block designs. Calcutta Statist. Assoc. Bull. **14**, 80—83.

1966 Some generalizations of distinct representatives with applications to statistical designs. Ann. Math. Statist. **37**, 525—528.

ALANEN, J. D.; KNUTH, D. R.

1961 A table of minimum functions for generating Galois fields of $GF(p^n)$. Sankhya **23**, 128.

ALBERT, A. A.

1952 On nonassociative division algebras. Trans. Amer. Math. Soc. **72**, 296—309.

1953 Rational normal matrices satisfying the incidence equation. Proc. Amer. Math. Soc. **4**, 554—559.

1958 Finite noncommutative division algebras. Proc. Amer. Math. Soc. **9**, 928—932.

1960 Finite division algebras and finite planes. Proc. Symp. Appl. Math. **10**, 53—70.

1961a Generalized twisted fields. Pacif. J. Math. **11**, 1—8.

1961b The finite planes of Ostrom. Lecture Notes, Univ. of Chicago.

1961c Isotopy for generalized twisted fields. An. Acad. Brasil. Ci. **33**, 265—275.

1962 Knuth's construction. Mimeographed note, Univ. of Chicago.

1963 On the collineation groups associated with twisted fields. Calcutta Math. Soc. Golden Jubilee Commemoration volume (1958/59), part II, pp. 485—497.

ANDRÉ, J.

1954a Über nicht-Desarguessche Ebenen mit transitiver Translationsgruppe. Math. Zeitschr. **60**, 156—186.

1954b Über Perspektivitäten in endlichen projektiven Ebenen. Arch. Math. **6**, 29—32.

1955 Projektive Ebenen über Fastkörpern. Math. Zeitschr. **62**, 137—160.

1958a Über Perspektivitäten in endlichen affinen Ebenen. Arch. Math. **9**, 228—235.

1958b Affine Ebenen mit genügend vielen Translationen. Math. Nachr. **19**, 203—210.
1961a Über Parallelstrukturen. I. Grundbegriffe. Math. Zeitschr. **76**, 85—102.
1961b Über Parallelstrukturen. II. Translationsstrukturen. Math. Zeitschr. **76**, 155—163.
1961c Über Parallelstrukturen. III. Zentrale *t*-Strukturen. Math. Zeitschr. **76**, 240—256.
1961d Über Parallelstrukturen. IV. *T*-Strukturen. Math. Zeitschr. **76**, 311—333.
1962 Über verallgemeinerte Moulton-Ebenen. Arch. Math. **13**, 290—301.
1963a Über eine Beziehung zwischen Zentrum und Kern endlicher Fastkörper. Arch. Math. **14**, 145—146.
1963b Bemerkung zu meiner Arbeit "Über verallgemeinerte Moulton-Ebenen". Arch. Math. **14**, 359—360.
1964 Über projektive Ebenen vom Lenz-Barlotti-Typ III2. Math. Zeitschr. **84**, 316—328.

ARCHBOLD, J. W.
1955 Projective geometry over an algebra. Mathematika **2**, 105—115.
1960a Incidence matrices. Mathematika **7**, 41—49.
1960b A combinatorial problem of T. G. Room. Mathematika **7**, 50—55.
1966 Permutation mappings in finite projective planes. Mathematika **13**, 45—48.

ARCHBOLD, J. W.; JOHNSON, N. L.
1956 A method of constructing partially balanced incomplete block designs. Ann. Math. Statist. **27**, 633—641.

ARCHBOLD, J. W.; SMITH, C. A. B.
1962 An extension of a theorem of König on graphs. Mathematika **9**, 9—10.

ARF, C.
1941 Untersuchungen über quadratische Formen in Körpern der Charakteristik 2. J. r. angew. Math. **183**, 148—167.

ARTIN, E.
1940 Coordinates in affine geometry. Rep. Math. Colloq. (Univ. of Notre Dame) (2) **2**, 15—20.
1955a The orders of the linear groups. Comm. Pure Appl. Math. **8**, 355—365.
1955b The orders of the classical simple groups. Comm. Pure Appl. Math. **8**, 455—472.
1957 Geometric algebra. Interscience, New York and London.

ARTZY, R.
1956 Self-dual configurations and their Levi graphs. Proc. Amer. Math. Soc. **7**, 299—303.
1960 Relations between loop identities. Proc. Amer. Math. Soc. **11**, 847—851.
1963 Net motions and loops. Arch. Math. **14**, 95—101.
1964 Net motions and planar ternary rings. Arch. Math. **15**, 371—377.

ASSMUS, E. F.; MATTSON, H. F.
1966a Disjoint Steiner systems associated with the Mathieu groups. Bull. Amer. Math. Soc. **72**, 843—845.
1966b Perfect codes and the Mathieu groups. Arch. Math. **17**, 121—135.

ATIQULLAH, M.
1958a Some new solutions of the symmetrical balanced incomplete block design with $k = 9$ and $\lambda = 2$. Bull. Calcutta Math. Soc. **50**, 23—28.
1958b On configurations and non-isomorphism of some incomplete block designs. Sankhya **20**, 227—248.

Bachmann, F.

1959 Aufbau der Geometrie aus dem Spiegelungsbegriff. Berlin/Göttingen/Heidelberg: Springer.

Baer, R.

1939 Nets and groups. I. Trans. Amer. Math. Soc. **46**, 110—141.

1940 Nets and groups. II. Trans. Amer. Math. Soc. **47**, 435—439.

1942 Homogeneity of projective planes. Amer. J. Math. **64**, 137—152.

1944 The fundamental theorems of elementary geometry. An axiomatic analysis. Trans. Amer. Math. Soc. **56**, 94—129.

1945 Null systems in projective space. Bull. Amer. Math. Soc. **51**, 903—906.

1946a Polarities in finite projective planes. Bull. Amer. Math. Soc. **52**, 77—93.

1946b Projectivities with fixed points on every line of the plane. Bull. Amer. Math. Soc. **52**, 273—286.

1947 Projectivities of finite projective planes. Amer. J. Math. **69**, 653—684.

1948 The infinity of generalized hyperbolic planes. Courant Anniversary Volume, Interscience, New York, pp. 21—27.

1950 Free mobility and orthogonality. Trans. Amer. Math. Soc. **68**, 439—460.

1951 The group of motions of a two-dimensional elliptic geometry. Compositio Math. **9**, 241—288.

1952 Linear algebra and projective geometry. Academic Press, New York.

1961a Partitionen endlicher Gruppen. Math. Z. **75**, 333—372.

1961b Einfache Partitionen nicht-einfacher Gruppen. Math. Z. **77**, 1—37.

1962 Hjelmslevsche Geometrie. Proc. Colloq. Found. of Geometry Utrecht 1959; Pergamon 1962, pp. 1—4.

1963 Partitionen abelscher Gruppen. Arch. Math. **14**, 73—83.

Ball, R. W.

1948 Dualities of finite projective planes. Duke Math. J. **15**, 929—940.

Banerjee, K. S.

1949 On the construction of Hadamard matrices. Science and Culture **14**, 434—435.

Barlotti, A.

1955 Un'estensione del teorema di Segre-Kustaanheimo. Boll. Un. Mat. Ital. **10**, 498—506.

1956a Un'osservazione sulle k-calotte degli spazi lineari finiti di dimensione tre. Boll. Un. Mat. Ital. **11**, 248—252.

1956b Sui $\{k; n\}$-archi di un piano lineare finito. Boll. Un. Mat. Ital. **11**, 553—556.

1957a Una limitazione superiore per il numero di punti appartenenti a una k-calotta $C(k, 0)$ di uno spazio lineare finito. Boll. Un. Mat. Ital. **12**, 67—70.

1957b Le possibili configurazioni del sistema delle coppie punto-retta (A, a) per cui un piano grafico risulta (A, a)-transitivo. Boll. Un. Mat. Ital. **12**, 212—226.

1959 La determinazione del gruppo delle proiettività di una retta in sè in alcuni particolari piani grafici finiti non desarguesiani. Boll. Un. Mat. Ital. **14**, 543—547.

1962a Una costruzione di una classe di spazi affini generalizzati. Boll. Un. Mat. Ital. **17**, 182—187.

1962b Un'osservazione sulle proprietà che caratterizzano un piano grafico finito. Boll. Un. Mat. Ital. **17**, 394—398.

1964a Il gruppo delle proiettività di una retta in sè in un particolare piano non desarguesiano di ordine sedici. Matematiche **19**, 63—69.

1964b Alcuni risultati nello studio degli spazi affini generalizzati di Sperner. Rendic. Sem. Mat. Padova **35**, 18—46.

1946c Sul gruppo delle proiettività di una retta in se nei piani liberi e nei piani aperti. Rendic. Sem. Mat. Padova **34**, 135—159.
1965 Some topics in finite geometrical structures. Lecture Notes, Chapel Hill, N C.
1966a Un'osservazione intorno ad un teorema di B. Segre sui q-archi. Matematiche **21**, 23—29.
1966b Un nuovo procedimento per la costruzione di spazi affini generalizzati di Sperner. Matematiche **21**, 302—312.
1966c Una caratterizzazione grafica delle ipersuperficie hermitiane non singolari in uno spazio lineare finito di ordine quattro. Matematiche **21**, 387—395.
1967 Sulle 2-curve nei piani grafici. Rendic. Sem. Mat. Padova **37**, 91—97.

Barra, J.-R.
1963 A propos d'un théorème de R. C. Bose. C. R. Acad. Sci. Paris **256**, 5502—5504.

Barra, J.-R.; Guérin, R.
1963a Extension des carrés gréco-latins cycliques. Publ. Inst. Statist. Univ. Paris **12**, 67—82.
1963b Utilisation pratique de la méthode de Yamamoto pour la construction systematique des carrés gréco-latins. Publ. Inst. Statist. Univ. Paris **12**, 131—136.

Bartolozzi, F.
1960 Il gruppo proiettivo di un piano sopra un quasicorpo distributivo finito. Atti Accad. Sci. Lett. Palermo **19**, 115—125.
1965 Su una classe di quasicorpi (sinistri) finiti. Rendic. Mat. **24**, 165—173.
1966 Sopra una classe di piani finiti (R, r)-transitivi. Atti Accad. Naz. Lincei Rendic. **39**, 245—248.

Baumert, L. D.
1966 Hadamard matrices of orders 116 and 232. Bull. Amer. Math. Soc. **72**, 237.

Baumert, L.; Golomb, S. W.; Hall, M.
1962 Discovery of an Hadamard matrix of order 92. Bull. Amer. Math. Soc. **68**, 237—238.

Baumert, L. D.; Hall, M.
1965a A new construction for Hadamard matrices. Bull. Amer. Math. Soc. **71**, 169—170.
1965b Hadamard matrices of the Williamson type. Math. Comp. **19**, 442—447.

Bays, S.
1931 Sur les systèmes cycliques des triples de Steiner différents pour N premier (ou puissance de nombre premier) de la forme $6n + 1$. I—VI. Comm. Math. Helv. **3**, 22—41, 122—147, 307—325.
1932 Sur les systèmes cycliques des triples de Steiner différents pour N premier de la forme $6n + 1$. Comm. Math. Helv. **4**, 183—194.

Bays, S.; Belhôte, G.
1933 Sur les systèmes cycliques des triples de Steiner différents pour N premier de la forme $6n + 1$. Comm. Math. Helv. **6**, 28—46.

Bays, S.; de Weck, E.
1935 Sur les systèmes des quadruples. Comm. Math. Helv. **7**, 222—241.

Beaumont, R. A.; Peterson, R. P.
1955 Set transitive permutation groups. Canad. J. Math. **7**, 35—42.

BELOUSOV, B. D.
1960 On the structure of distributive quasigroups. Mat. Sbornik **50**, 267—298.

BENDER, H.
1968 Endliche zweifach transitive Permutationsgruppen, deren Involutionen keine Fixpunkte haben. Math. Z. **104**, 175—204.

BENNETON, G.
1944 Note sur la configuration de Kummer. Bull. Sci. Math. **68**, 190—192.
1945a Sur les configurations harmoniques. C. R. Acad. Sci. Paris **220**, 548—550.
1945b Sur les configurations de Kummer et de Klein. C. R. Acad. Sci. Paris **220**, 640—642.

BENSON, C. T.
1966a A partial geometry $(q^3 + 1, q^2 + 1, 1)$ and corresponding PBIB design. Proc. Amer. Math. Soc. **17**, 747—749.
1966b Minimal regular graphs of girths eight and twelve. Canad. J. Math. **18**, 1091—1094.

BENZ, W.
1958a Axiomatischer Aufbau der Kreisgeometrie auf Grund von Doppelverhältnissen. J. reine angew. Math. **199**, 56—90.
1958b Zur Theorie der Möbiusebenen. I. Math. Ann. **134**, 237—247.
1958c Beziehungen zwischen Orthogonalitäts- und Anordnungseigenschaften in Kreisebenen. Math. Ann. **134**, 385—402.
1958d $(8_3, 6_4)$-Konfigurationen in Laguerre-, Möbius- und weiteren Geometrien. Math. Z. **70**, 283—296.
1960a Über Möbiusebenen. Ein Bericht. Jahresber. Deutsche Math. Ver. **63**, 1—27.
1960b Über Winkel- und Transitivitätseigenschaften in Kreisebenen. I. J. reine angew. Math. **205**, 48—74.
1961 Über Winkel- und Transitivitätseigenschaften in Kreisebenen. II. J. reine angew. Math. **207**, 1—15.
1962a Zur Möbiusgeometrie über einem Körperpaar. Arch. Math. **13**, 136—146.
1962b Über eine Verallgemeinerung des Satzes von Miquel. Publ. Math. Debrecen **9**, 227—230.
1963a Süßsche Gruppen in affinen Ebenen mit Nachbarelementen und allgemeinen Strukturen. Abh. Hamburg **26**, 83—101.
1963b Fährten in der Laguerre-Geometrie. Math. Ann. **150**, 66—78.
1964a Das von Staudtsche Theorem in der Laguerregeometrie. J. reine angew. Math. **214/215**, 53—60.
1964b Pseudo-Ovale und Laguerre-Ebenen. Abh. Hamburg **27**, 80—84.
1965a Zykelverwandtschaften als Berührungstransformationen. J. reine angew. Math. **220**, 103—108.
1965b Laguerre-Geometrie über einem lokalen Ring. Math. Z. **87**, 137—145.

BENZ, W.; MÄURER, H.
1964 Über die Grundlagen der Laguerre-Geometrie. Ein Bericht. Jahresber. DMV **67**, 14—42.

BERGE, C.
1958 Théorie des graphes et ses applications. Dunod, Paris.

BERCOV, R. D.
1965 The double transitivity of a class of permutation groups. Canad. J. Math. **17**, 480—493.

BERMAN, G.
1952 Finite projective geometries. Canad. J. Math. **4**, 302—313.
1953 Finite projective plane geometries and difference sets. Trans. Amer. Math. Soc. **74**, 492—499.
1955 A three parameter family of partially balanced incomplete block designs with two associate classes. Proc. Amer. Math. Soc. **6**, 490—493.

BERZ, E.
1962 Kegelschnitte in desarguesschen Ebenen. Math. Z. **78**, 55—85.

BHAGWANDAS
1965 A note on balanced incomplete block designs. J. Ind. Statist. Assoc. **3**, 41—45.

BHATTACHARYA, K. N.
1943 A note on two-fold triple systems. Sankhya **6**, 313—314.
1944a On a new symmetrical balanced incomplete block design. Bull. Calcutta Math. Soc. **36**, 91—96.
1944b A new balanced incomplete block design. Science and Culture **9**, 508.
1946 A new solution in symmetrical balanced incomplete block designs ($v = b = 31$, $r = k = 10$, $\lambda = 3$). Sankhya **7**, 423—424.
1950 Problems in partially balanced incomplete block designs. Calcutta Statist. Assoc. Bull. **2**, 177—182.

BIRKHOFF, G.
1935 Combinatorial relations in projective geometries. Ann. Math. **36**, 743—748.
1948 Lattice Theory, 2nd ed. Amer. Math. Soc. Colloq. Publ., vol. 25.

BIRKHOFF, G.; v. NEUMANN, J.
1936 The logic of quantum mechanics. Ann. of Math. **37**, 823—843.

BLOCK, R. E.
1965 Transitive groups of collineations on certain designs, Pacif. J. Math. **15**, 13—19.
1967 On the orbits of collineation groups. Math. Z. **96**, 33—49.

BONARDI, M. T.
1964 Intorno a certe superficie cubiche dello spazio di Galois. Atti Accad. Naz. Lincei **37**, 396—400.

BOSE, R. C.
1938 On the application of the properties of Galois fields to the problem of the construction of hyper Graeco-Latin squares. Sankhya **3**, 323—338.
1939 On the construction of balanced incomplete block designs. Ann. Eugenics **9**, 353—399.
1942a A note on the resolvability of balanced incomplete block designs. Sankhya **6**, 105—110.
1942b An affine analogue of Singer's theorem. J. Ind. Math. Soc. **6**, 1—15.
1942c On some new series of balanced incomplete block designs. Bull. Calcutta Math. Soc. **34**, 17—31.
1942d A note on two series of balanced incomplete block designs. Bull. Calcutta Math. Soc. **34**, 129—130.
1942e A note on two combinatorial problems having applications in the theory of design of experiments. Science and Culture **8**, 192—193.
1945a Minimum functions in Galois fields. Proc. Nat. Acad. Sci. USA **14**, 191.
1945b On the roots of a well known congruence. Proc. Nat. Acad. Sci. USA **14**, 193.
1947a Mathematical theory of the symmetrical factorial design. Sankhya **8**, 107—166.

1947b On a resolvable series of balanced incomplete block designs. Sankhya **8**, 249—256.

1949 A note on Fisher's inequality for balanced incomplete block designs. Ann. Math. Statist. **20**, 619—620.

1951 Partially balanced incomplete block designs with two associate classes involving only two replications. Calcutta Statist. Assoc. Bull. **3**, 120—125.

1952 A note on Nair's condition for partially balanced incomplete block designs with $k > r$. Calcutta Statist. Assoc. Bull. **4**, 123—126.

1963a On the application of finite projective geometry for deriving a certain series of balanced Kirkman arrangements. Calcutta Math. Soc. Golden Jubilee Commem. vol. (1958/59), part II, pp. 341—354.

1963b Strongly regular graphs, partial geometries, and partially balanced designs. Pacif. J. Math. **13**, 389—419.

1963c Combinatorial properties of partially balanced designs and association schemes. Sankhya **25**, 109—136.

Bose, R. C.; Bush, K. A.

1952 Orthogonal arrays of strength two and three. Ann. Math. Statist. **23**, 508—524.

Bose, R. C.; Chakravarti, I. M.

1965 Hermitian varieties in a finite projective space $PG(N, q^2)$. Lecture notes, Chapel Hill, N. C.

Bose, R. C.; Chowla, S.

1945 On the construction of affine difference sets. Bull. Calcutta Math. Soc. **37**, 107—112.

Bose, R. C.; Clatworthy, W. H.

1955 Some classes of partially balanced designs. Ann. Math. Statist. **26**, 212—232.

Bose, R. C.; Clatworthy, W. H.; Shrikhande, S. S.

1954 Tables of partially balanced designs with two associate classes. North Carolina Agricultural Experiment Station Techn. Bull. **107**, Raleigh, N. C.

Bose, R. C.; Connor, W. S.

1952 Combinatorial properties of group divisible incomplete block designs. Ann. Math. Statist. **23**, 367—383.

Bose, R. C.; Mesner, D.

1959 On linear associative algebras corresponding to association schemes of partially balanced designs. Ann. Math. Statist. **30**, 21—38.

Bose, R. C.; Nair, K. R.

1939 Partially balanced incomplete block designs. Sankhya **4**, 337—372.

1941 On complete sets of Latin squares. Sankhya **5**, 361—382.

1962 Resolvable incomplete block designs with two replications. Sankhya **24**, 9—24.

Bose, R. C.; Shimamoto, T.

1952 Classification and analysis of partially balanced incomplete block designs with two associate classes. J. Amer. Statist. Assoc. **47**, 151—184.

Bose, R. C.; Shrikhande, S. S.

1959 On the falsity of Euler's conjecture about the non-existence of two orthogonal Latin squares of order $4t + 2$. Proc. Nat. Acad. Sci. USA **45**, 734—737.

1960a On the construction of sets of mutually orthogonal Latin squares and the falsity of a conjecture of Euler. Trans. Amer. Math. Soc. **95**, 191—209.

1960b On the composition of balanced incomplete block designs. Canad. J. Math. **12**, 177—188.

BOSE, R. C.; SHRIKHANDE, S. S.; BHATTACHARYA, K. N.
1953 On the construction of group divisible incomplete block designs. Ann. Math. Statist. **24**, 167—195.

BOSE, R. C.; SHRIKHANDE, S. S.; PARKER, E. T.
1960 Further results on the construction of mutually orthogonal Latin squares and the falsity of Euler's conjecture. Canad. J. Math. **12**, 189—203.

BOSE, R. C.; SRIVASTA, J. N.
1964 On a bound useful in the theory of factorial designs and error correcting codes. Ann. Math. Statist. **35**, 408—414.

BRANDIS, A.
1964 Ein gruppentheoretischer Beweis für die Kommutativität endlicher Divisionsringe. Abh. Hamburg **26**, 234—236.

BRAUER, A.
1952 Limits for the characteristic roots of a matrix. IV. Applications to stochastic matrices. Duke Math. J. **19**, 75—91.
1953 On a new class of Hadamard determinants. Math. Z. **58**, 219—225.

BRAUER, R.
1941 On the connections between the ordinary and the modular characters of groups of finite order. Ann. Math. **42**, 926—935.
1949 On a theorem of H. Cartan. Bull. Amer. Math. Soc. **55**, 619—620.
1965 On finite desarguesian planes I. Math. Z. **90**, 117—123.
1966 On finite desarguesian planes II. Math. Z. **91**, 124—151.

BRUALDI, R. A.
1965 A note on multipliers of difference sets. J. Res. Nat. Bur. Standards **69**B, 87—89.

BRUCK, R. H.
1951a Finite nets. I. Numerical invariants. Canad. J. Math. **3**, 94—107.
1951b Loops with transitive automorphism groups. Pacif. J. Math. **1**, 481—483.
1955 Difference sets in a finite group. Trans. Amer. Math. Soc. **78**, 464—481.
1958 A survey of binary systems. Berlin/Göttingen/Heidelberg: Springer.
1960 Quadratic extensions of cyclic planes. Proc. Symp. Appl. Math. **10**, 15—44.
1963a Finite nets. II. Uniqueness and imbedding. Pacif. J. Math. **13**, 421—457.
1963b Existence problems for classes of finite projective planes. Lecture notes, Canad. Math. Congress, Saskatoon.
1963c What is a loop? Studies in modern algebra. The Mathematical Association of America, pp. 59—99.

BRUCK, R. H.; BOSE, R. C.
1964 The construction of translation planes from projective spaces. J. Algebra **1**, 85—102.
1966 Linear representations of projective planes in projective spaces. J. Algebra **4**, 117—172.

BRUCK, R. H.; KLEINFELD, E.
1951 The structure of alternative division rings. Proc. Amer. Math. Soc. **2**, 878—890.

BRUCK, R. H.; RYSER, H. J.
1949 The nonexistence of certain finite projective planes. Canad. J. Math. **1**, 88—93.

DE BRUIJN, N. G.; ERDÖS, P.
1948 On a combinatorial problem. Indag. Math. **10**, 421—423.

BUEKENHOUT, F.
1963 Sur la structure linéaire du plan de translation non-arguésien d'ordre neuf. Acad. Roy. Belg. Bull. Cl. Sci. **49**, 1206—1213.
1964 Les plans d'André finis de dimension quatre sur le noyau. Acad. Roy. Belg. Bull. Cl. Sci. **50**, 446—457.
1966a Etude intrinsèque des ovales. Rendic. Mat. **25**, 1—61.
1966b Plans projectifs à ovoïdes pascaliens. Arch. Math. **17**, 89—93.
1966c Ovales et ovales projectifs. Atti Accad. Naz. Lincei Rendic. **40**, 46—49.

BUGGENHOUT, J. VAN
1965 Caractérisation de la droite affine basée sur la relation de symétrie. Acad. Roy. Belg. Bull. Cl. Sci. **51**, 476—484.

BURAU, W.
1963 Über die zur Kummerkonfiguration analogen Schemata von 16 Punkten und 16 Blöcken und ihre Gruppen. Abh. Hamburg **26**, 129—144.

BURMESTER, M. V. D.
1962 On the commutative non-associative division algebras of even order of L. E. Dickson. Rendic. Mat. **21**, 143—166.
1964a Sulla solubilità del gruppo delle collineazioni di un piano sopra un quasicorpo distributivo di ordine non-quadrato. Atti Accad. Naz. Lincei **36**, 141—144.
1964b On the non-unicity of translation planes over division algebras. Arch. Math. **15**, 364—370.

BURMESTER, M. V. D.; HUGHES, D. R.
1965 On the solvability of autotopism groups. Arch. Math. **16**, 178—183.

BURNSIDE, W.
1911 Theory of groups of finite order. 2nd ed. Cambridge. Reprint Dover, New York 1955.

BUSH, K. A.
1952a A generalization of a theorem due to McNeish. Ann. Math. Statist. **23**, 293—295.
1952b Orthogonal arrays of index unity. Ann. Math. Statist. **23**, 426—434.

BUSSEY, W. H.
1906 Tables of Galois fields. Bull. Amer. Math. Soc. **12**, 22—38.
1910 Tables of Galois fields. Bull. Amer. Math. Soc. **16**, 188—206.

BUTSON, A. T.
1962 Generalized Hadamard matrices. Proc. Amer. Math. Soc. **13**, 894—898.
1963 Relations among generalized Hadamard matrices, relative difference sets, and maximal length linear recurring sequences. Canad. J. Math. **15**, 42—48.

BÝDZOVSKÝ, B.
1954 Über zwei neue ebene Konfigurationen $(12_4, 16_3)$. Czechoslov. Math. J. **4**, 193—218.

CAMPBELL, A. D.
1932 Apolarity in the Galois fields of order 2^n. Bull. Amer. Math. Soc. **38**, 52—56.
1933 Plane quartic curves in the Galois fields of order 2^n. Tohoku Math. J. **37**, 88—93.

CARLITZ, L.
1960 A theorem on permutations in a finite field. Proc. Amer. Math. Soc. **11**, 456—459.

CARMICHAEL, R. D.
1931a Tactical configurations of rank two. Amer. J. Math. **53**, 217—240.
1931b Algebras of certain doubly transitive groups. Amer. J. Math. **53**, 631—644.
1932 Note on triple systems. Bull. Amer. Math. Soc. **38**, 695—696.
1937 Introduction to the theory of groups of finite order. Boston; Reprint Dover, New York 1956.

CHAKRABARTI, M. C.
1950 A note on balanced incomplete block designs. Bull. Calcutta Math. Soc. **42**, 14—16.

CHANG, Li-Chien
1959 The uniqueness and non-uniqueness of the triangular association schemes. Science Record **3**, 604—613.
1960 Association schemes of partially balanced designs with parameters $v = 28$, $n_1 = 12$, $n_2 = 15$ and $p_{11}^2 = 4$. Science Record **4**, 12—18.

CHANG, Li-Chien; LIU, Chang-Wen; LIU, Wau-Ru.
1965 Incomplete block designs with triangular parameters for which $k \leqq 10$ and $r \leqq 10$. Sci. Sinica **14**, 329—338.

CHANG, Li-Chien; LIU, Wau-Ru.
1964 Incomplete block designs with square parameters for which $k \leqq 10$ and $r \leqq 10$. Sci. Sinica **13**, 1493—1495.

CHOWLA, S.
1944a A property of biquadratic residues. Proc. Nat. Acad. Sci. India, Sect. A **14**, 45—46.
1944b Contributions to the theory of the construction of balanced incomplete block designs. Proc. Lahore Philos. Soc. **6**, 10—12, 17—23.
1945a On difference-sets. J. Indian Math. Soc. **9**, 28—31.
1945b A contribution to the theory of the construction of balanced incomplete block designs. Proc. Lahore Philos. Soc. **7**, 3 pp.
1949 On difference sets. Proc. Nat. Acad. Sci. USA **35**, 92—94.

CHOWLA, S.; ERDÖS, P.; STRAUS, E. G.
1960 On the maximal number of pairwise orthogonal Latin squares of a given order. Canad. J. Math. **11**, 204—208.

CHOWLA, S.; JONES, B. W.
1959 A note on perfect difference sets. Norske Vid. Selsk. Forh. Trondheim **32**, 81—83.

CHOWLA, S.; RYSER, H. J.
1950 Combinatorial problems. Canad. J. Math. **2**, 93—99.

CHOWLA, S.; SINGH, D.
1945 A perfect difference set of order 18. Proc. Lahore Philos. Soc. **7**, 52.

CLATWORTHY, W. H.
1954 A geometrical configuration which is a partially balanced design. Proc. Amer. Math. Soc. **5**, 47.
1955 Partially balanced incomplete block designs with two associate classes and two treatments per block. J. Res. Nat. Bur. Standards **54**, 177—190

1956 Contributions on partially balanced incomplete block designs with two associate classes. Nat. Bur. Standards, Appl. Math. Ser. No. **47**, Washington, D. C.

Coble, A. B.

1936 Collineation groups in a finite space with a linear and a quadratic invariant. Amer. J. Math. **58**, 15—34.

Cochran, W.; Cox, G. M.

1957 Experimental designs. 2nd ed., Wiley, New York.

Cofman, J.

1964 The validity of certain configuration-theorems in the Hall planes of order p^{2n}. Rendic. Mat. **23**, 22—27.

1965 Homologies of finite projective planes. Arch. Math. **16**, 476—479.

1966a On a characterization of finite desarguesian projective planes. Arch. Math. **17**, 200—205.

1966b On the nonexistence of finite projective planes of Lenz-Barlotti type I_6. J. Algebra **4**, 64—70.

1967a On a conjecture of Hughes. Proc. Cambridge Philos. Soc. **63**, 647—652.

1967b Double transitivity in finite affine planes. I. Math. Z. **101**, 335—342.

1967c Double transitivity in finite affine and projective planes. Proc. Proj. Geometry Conference, Univ. of Illinois, Chicago; pp. 16—19.

1968 Transitivity on triangles in finite projective planes. Proc. London Math. Soc., to appear.

Cohn, J. H. E.

1965 Hadamard matrices and some generalizations. Amer. Math. Monthly **72**, 515—518.

Cole, F. N.

1922 Kirkman parades. Bull. Amer. Math. Soc. **28**, 435—437.

di Comite, C.

1962 Sui *k*-archi deducibili da cubiche piane. Atti Accad. Naz. Lincei Rendic. **33**, 429—435.

1963 Sui *k*-archi contenuti in cubiche piane. Atti Accad. Naz. Lincei Rendic. **35**, 274—278.

Connor, W. S.

1952a On the structure of balanced incomplete block designs. Ann. Math. Statist. **23**, 57—71. Correction to this: Ann. Math. Statist. **24** (1953) 135.

1952b Some relations among the blocks of symmetrical group divisible designs. Ann. Math. Statist. **23**, 602—609.

1958 The uniqueness of the triangular association scheme. Ann. Math. Statist. **29**, 262—266.

Connor, W. S.; Clatworthy, W. H.

1954 Some theorems for partially balanced designs. Ann. Math. Statist. **25**, 100—112.

Corbas, V.

1962 Su di una classe di quasicorpi commutativi finiti e su di una congettura del Dickson. Rendic. Mat. **21**, 245—265.

1964 Omomorfismi fra piani proiettivi, I. Rendic. Mat. **23**, 316—330.

Corsi, G.

1959 Sui triangoli omologici in un piano sul quasicorpo associativo di ordine 9. Matematiche **14**, 40—66.

1964 Sui sistemi minimi di assiomi atti a definire un piano grafico finito. Rendic. Sem. Mat. Padova **34**, 160—175.

CORSTEN, L. C. A.

1960 Proper spaces related to triangular partially balanced incomplete block designs. Ann. Math. Statist. **31**, 498—501.

COSSU, A.

1960 Sulle ovali di un piano proiettivo sopra un corpo finito. Atti Accad. Naz. Lincei Rendic. **28**, 342—344.

1961 Su alcune proprietà dei $\{k; n\}$-archi di un piano proiettivo sopra un corpo finito. Rendic. Mat. **20**, 271—277.

COX, G.

1940 Enumeration and construction of balanced incomplete block designs. Ann. Math. Statist. **11**, 72—85.

COXETER, H. S. M.

1933 Regular compound polytopes in more than four dimensions. J. Math. Phys. Mass. Inst. Tech. **12**, 334—345.

1950 Self-dual configurations and regular graphs. Bull. Amer. Math. Soc. **56**, 413—455.

1956 The collineation groups of the finite affine and projective planes with four lines through each point. Abh. Hamburg **20**, 165—177.

1958 Twelve points in $PG(5, 3)$ with 95040 self-transformations. Proc. Roy. Soc. (A) **247**, 279—293.

1966 The inversive plane and hyperbolic space. Abh. Hamburg **29**, 217—242.

COXETER, H. S. M.; MOSER, W. O. J.

1957 Generators and relations for discrete groups. Berlin/Göttingen/Heidelberg: Springer.

CRAIG, R. T.

1964 Extensions of finite projective planes. I. Uniform Hjelmslev planes. Canad. J. Math. **16**, 261—266.

CRONHEIM, A.

1953 A proof of Hessenberg's theorem. Proc. Amer. Math. Soc. **4**, 219—221.

1965 T-groups and their geometry. Ill. J. Math. **9**, 1—30.

CROWE, D. W.

1959 A regular quaternion polygon. Canad. Math. Bull. **2**, 77—79.

1961 Regular polygons over $GF(3^2)$. Amer. Math. Monthly **68**, 762—765.

1964 The trigonometry of $GF(2^{2n})$ and finite hyperbolic planes. Mathematika **11**, 83—88.

1965 The construction of finite regular hyperbolic planes from inversive planes of even order. Coll. Math. **13**, 247—250.

1966 Projective and inversive models for finite hyperbolic planes. Mich. Math. J. **13**, 251—255.

CUNDY, H. M.

1952 25-point geometry. Math. Gaz. **36**, 158—166.

DADE, E. C.; GOLDBERG, K.

1959 The construction of Hadamard matrices. Michigan Math. J. **6**, 247—250.

DAI, Zong-Duo; FENG, Xu-Ning

1964a Notes on finite geometries and the construction of PBIB designs. IV. Some "Anzahl" theorems in orthogonal geometry over finite fields of characteristic not 2. Sci. Sinica **13**, 2001—2004.

1964b Notes on finite geometries and the construction of PBIB designs. V. Some "Anzahl" theorems in orthogonal geometry over finite fields of characteristic 2. Sci. Sinica **13**, 2005—2008.

DAS, M. N.

1954 On parametric relations in a balanced incomplete block design. J. Ind. Soc. Agric. Statist. **6**, 147—152.

DAVID, H. A.; WOLOCK, F. W.

1965 Cyclic designs. Ann. Math. Statist. **36**, 1526—1534.

DAVIS, E. W.

1911 A geometric picture of the fifteen schoolgirls problem. Ann. Math. **11**, 156—157.

DEMARIA, D. C.

1959 Sui piani grafici esagonali a caratteristica p. Rendic. Sem. Mat. Univ. Politecn. Torino **18**, 43—52.

DEMBOWSKI, P.

1958 Verallgemeinerungen von Transitivitätsklassen endlicher projektiver Ebenen. Math. Z. **69**, 59—89.

1959 Homomorphismen von λ-Ebenen. Arch. Math. **10**, 46—50.

1961 Kombinatorische Eigenschaften endlicher Inzidenzstrukturen. Math. Z. **75**, 256—270.

1962a Semiaffine Ebenen. Arch. Math. **13**, 120—131.

1962b Tactical decompositions of λ-spaces. Proc. Colloq. Found. of Geometry Utrecht 1959; Pergamon 1962, pp. 15—23.

1963a Scharf transitive und scharf fahnentransitive Kollineationsgruppen. Math. Ann. **149**, 217—225.

1963b Inversive planes of even order. Bull. Amer. Math. Soc. **69**, 850—854.

1964a Eine Kennzeichnung der endlichen affinen Räume. Arch. Math. **15**, 146—154.

1964b Möbiusebenen gerader Ordnung. Math. Ann. **157**, 179—205.

1964c Classes of geometric designs. Rendic. Mat. **23**, 14—21.

1965a Automorphismen endlicher Möbius-Ebenen. Math. Z. **87**, 115—136.

1965b Kombinatorik. Vorlesung Univ. Frankfurt a. M.

1965c Gruppentheoretische Kennzeichnungen der endlichen desarguesschen Ebenen. Abh. Hamburg **29**, 92—106.

1965d Sui gruppi di traslazioni d'un piano affine finito. Atti Accad. Naz. Lincei Rendic. **38**, 645—648.

1965e Die Nichtexistenz von transitiven Erweiterungen der endlichen affinen Gruppen. J. reine angew. Math. **220**, 37—44.

1966 Zur Geometrie der Suzukigruppen. Math. Z. **94**, 106—109.

1967a Berichtigung und Ergänzung zu "Eine Kennzeichnung der endlichen affinen Räume". Arch. Math. **18**, 111—112.

1967b Collineation groups containing perspectivities. Canad. J. Math. **19**, 924—937.

DEMBOWSKI, P; HUGHES, D. R.

1965 On finite inversive planes. J. London Math. Soc. **40**, 171—182.

DEMBOWSKI, P.; OSTROM, T. G.

1968 Planes of order n with collineation groups of order n^2. Math. Z. **103**, 239—258.

DEMBOWSKI, P.; PIPER, F.

1967 Quasiregular collineation groups of finite projective planes. Math. Z. **99**, 53—75.

DEMBOWSKI, P.; WAGNER, A.
1960 Some characterizations of finite projective spaces. Arch. Math. **11**, 465—469.

DICKSON, L. E.
1901 Linear groups, with an exposition of the Galois field theory. Teubner, Leipzig. Reprint Dover, New York 1958.
1905 On finite algebras. Nachr. kgl. Ges. Wiss. Göttingen, pp. 358—393.
1906 Linear algebras in which division is always uniquely possible. Trans. Amer. Math. Soc. **7**, 370—390.
1923 Algebras and their arithmetics. Univ. of Chicago Press. Chicago, Ill. Reprint Dover, New York 1960.
1935 Linear algebras with associativity not assumed. Duke Math. J. **1**, 113—125.

DIEUDONNÉ, J.
1943 Les déterminants sur un corps non commutatif. Bull. Soc. Math. France **71**, 27—45.
1955 La géométrie des groupes classiques. Berlin/Göttingen/Heidelberg: Springer.

DILWORTH, R. P.
1960 Some combinatorial problems on partially ordered sets. Proc. Sympos. Appl. Math. **10**, 85—90.

DUBREIL-JACOTIN, M. L.; LESIEUR, L.; CROISOT, R.
1953 Leçons sur la théorie des treillis, des structures algébriques ordonnées, et des treillis géometriques. Gauthier-Villars, Paris.

DULMAGE, A. L.; JOHNSON, D.; MENDELSOHN, N. S.
1959 Orthogonal Latin squares. Canad. Math. Bull. **2**, 211—216.
1961 Orthomorphisms of groups and orthogonal Latin squares. I. Canad. J. Math. **13**, 356—372.

DULMAGE, A. L.; MENDELSOHN, N. S.
1964 The exponents of incidence matrices. Duke Math. J. **31**, 575—584.

ECKENSTEIN, O.
1911 Bibliography of Kirkman's schoolgirl problem. Mess. Math. **41**, 33—36.

EDGE, W. L.
1954 Geometry in three dimensions over $GF(3)$. Proc. Roy. Soc. (A) **222**, 262—286.
1955 31-point geometry. Math. Gaz. **39**, 113—121.
1956 Conics and orthogonal projectivities in a finite plane. Canad. J. Math. **8**, 362—382.
1958a The geometry of an orthogonal group in six variables. Proc. London Math. Soc. **8**, 416—446.
1958b The partitioning of an orthogonal group in six variables. Proc. Roy. Soc. (A) **247**, 539—549.
1960 A setting for the group of bitangents. Proc. London Math. Soc. **10**, 583—603.
1963 A second note on the simple group of order 6048. Proc. Cambridge Philos. Soc. **59**, 1—9.
1965 Some implications of the geometry of the 21-point plane. Math. Z. **87**, 348—362.

EHLICH, H.
1965 Neue Hadamard-Matrizen. Arch. Math. **16**, 34—36.

ELLERS, E.; KARZEL, H.
1961 Involutorische Geometrien. Abh. Hamburg **25**, 93—104.

1963 Kennzeichnung elliptischer Gruppenräume. Abh. Hamburg **26**, 55—77.
1964 Endliche Inzidenzgruppen. Abh. Hamburg **27**, 250—264.

Elliott, J. E. H.; Butson, A. T.
1966 Relative difference sets. Ill. J. Math. **10**, 517—531.

Emch, A.
1929 Triple and multiple systems, their geometrical configurations and groups. Trans. Amer. Math. Soc. **31**, 25—42.

Erdös, P.; Rado, R.
1961 Intersection theorems for systems of finite sets. Quart. J. Math. **12**, 313—320.

Erdös, P.; Renyi, A.
1956 On some combinatorical problems. Publ. Math. Debrecen **4**, 398—405.

Esser, M.
1951 Self-dual postulates for n-dimensional geometry. Duke Math. J. **18**, 475—479.

Euler, L.
1782 Recherches sur une nouvelle espèce des quarrés magiques. Verh. Zeeuwsch. Genootsch. Wetensch. Vlissingen **9**, 85—239. See also Leonardi Euleri opera omnia, ser. I, vol. 7, pp. 291—392. Teubner, Leipzig-Berlin 1923.

Evans, T. A.; Mann, H. B.
1951 On simple difference sets. Sankhya **11**, 357—364.

Ewald, G.
1956a Axiomatischer Aufbau der Kreisgeometrie. Math. Ann. **131**, 354—371.
1956b Über den Begriff der Orthogonalität in der Kreisgeometrie. Math. Ann. **131**, 463—469.
1957a Über eine Berühreigenschaft von Kreisen. Math. Ann. **134**, 58—61.
1957b Begründung der Geometrie der ebenen Schnitte einer Semiquadrik. Arch. Math. **8**, 203—208.
1960 Beispiel einer Möbiusebene mit nichtisomorphen affinen Unterebenen. Arch. Math. **11**, 146—150.
1961a Ein Schließungssatz für Inzidenz und Orthogonalität in Möbiusebenen. Math. Ann. **142**, 1—21.
1961b Kennzeichnungen der projektiven dreidimensionalen Räume und nichtdesarguessche räumliche Strukturen über beliebigen Ternärkörpern. Math. Z. **75**, 395—418.
1961c Translationsgruppen in verallgemeinerten dreidimensionalen projektiven Räumen. Math. Z. **75**, 419—425.
1965 Erweiterungen projektiv-geometrischer Verbände. Math. Z. **87**, 101—114.
1966 Aus konvexen Kurven bestehende Möbiusebenen. Abh. Hamburg **30**, 179—187.

Fano, G.
1892 Sui postulati fondamentali della geometria proiettiva. Giornale di Matematiche **30**, 106—132.
1937 Osservazioni su alcune "geometrie finite". I., II. Atti Accad. Naz. Lincei **26**, 55—60, 129—134.

Feit, W.
1960 On a class of doubly transitive permutation groups. Ill. J. Math. **4**, 170—186.

Feit, W.; Higman, G.
1964 The nonexistence of certain generalized polygons. J. Algebra **1**, 114—131.

FEIT, W.; THOMPSON, J. G.
1962 A solvability criterion for finite groups and some consequences. Proc. Nat. Acad. Sci. USA **48**, 968—970.
1963 Solvability of groups of odd order. Pacif. J. Math. **13**, 771—1029.

FELLEGARA, G.
1962 Gli ovaloidi in uno spazio tridimensionale di Galois di ordine 8. Atti Accad. Naz. Lincei Rendic. **32**, 170—176.

FISCHER, B.
1964 Distributive Quasigruppen endlicher Ordnung. Math. Z. **83**, 267—303.

FISHER, R. A.
1940 An examination of the different possible solutions of a problem in incomplete blocks. Ann. Eugenics **10**, 52—75.
1942 New cyclic solutions to problems in incomplete blocks. Ann. Eugenics **11**, 290—299.
1949 The design of experiments. 5th ed., Edinburgh.

FISHER, R. A.; YATES, F.
1934 The 6×6 Latin squares. Proc. Cambridge Philos. Soc. **30**, 492—507.
1936 Statistical tables for biological, agricultural and medical research. Oliver and Boyd, London. 3rd ed. 1942.

FOULSER, D. A.
1962 On finite affine planes and their collineation groups. Thesis, University of Michigan, Ann Arbor.
1964a The flag-transitive collineation groups of the finite Desarguesian affine planes. Canad. J. Math. **16**, 443—472.
1964b Solvable flag transitive affine groups. Math. Z. **86**, 191—204.
1967a A generalization of André's systems. Math. Z. **100**, 380—395.
1967b A class of translation planes, $\Pi(Q_g)$. Math. Z. **101**, 95—102.

FRAME, J. S.
1938 A symmetric representation of the twenty-seven lines on a cubic surface by lines in a finite geometry. Bull. Amer. Math. Soc. **44**, 658—661.

FREEMAN, G. H.
1957 Some further methods of constructing regular group divisible incomplete block designs. Ann. Math. Statist. **28**, 479—487.

FROBENIUS, G.
1901 Über auflösbare Gruppen, IV. Berliner Sitzungsberichte 1901, 1223—1225.

FRYXELL, R. C.
1964 Sequences of planes constructed from nearfield planes of square order. Thesis, Washington State University, Pullman.

FULKERSON, D. R.
1960 Zero-one matrices with zero trace. Pacif. J. Math. **10**, 831—836.

FULKERSON, D. R.; RYSER, H. J.
1961 Widths and heights of (0, 1)-matrices. Canad. J. Math. **13**, 239—255.
1962 Multiplicities and minimal widths for (0, 1)-matrices. Canad. J. Math. **14**, 498—508.
1963 Width sequences for special classes of (0, 1)-matrices. Canad. J. Math. **15**, 371—396.

GARBE, D.; MENNICKE, J.
1964 Some remarks on the Mathieu groups. Canad. Math. Bull. **7**, 201—212.

GASSNER, B. J.
1965 Equal difference BIB designs. Proc. Amer. Math. Soc. **16**, 378—380.

GILMAN, R. E.
1931 On the Hadamard determinant theorem and orthogonal determinants. Bull. Amer. Math. Soc. **37**, 30—31.

GINGERICH, H. F.
1945 Generalized fields and Desargues configurations. Thesis, Urbana, Ill.

GLEASON, A. M.
1956 Finite Fano planes. Amer. J. Math. **78**, 797—807.

GOLDBERG, K.
1960 The incidence equation $AA^T = aA$. Amer. Math. Monthly **67**, 367.
1966 Hadamard matrices of order cube plus one. Proc. Amer. Math. Soc. **17**, 744—746.

GOLDHABER, J. K.
1960 Integral p-adic normal matrices satisfying the incidence equation. Canad. J. Math. **12**, 126—133.

GORDON, B.; MILLS, W. H.; WELCH, L. R.
1962 Some new difference sets. Canad. J. Math. **14**, 614—625.

GORENSTEIN, D.
1962 On finite groups of the form ABA. Canad. J. Math. **14**, 195—236.

GORENSTEIN, D.; HUGHES, D. R.
1961 Triply transitive groups in which only the identity fixes four letters. Ill. J. Math. **5**, 486—491.

GORENSTEIN, D.; WALTER, J. H.
1965 The characterization of finite groups with dihedral Sylow 2-subgroups. I, II, III. J. Algebra **2**, 85—151, 218—270, 354—393.

GRAVES, L. M.
1962 A finite Bolyai-Lobachevsky plane. Amer. Math. Monthly **69**, 130—132.

GRUENBERG, K. W.; WEIR, A. J.
1967 Linear geometry. Van Nostrand, Princeton/Toronto/London.

GRUNER, W.
1939 Einlagerung des regulären n-Simplexes in den n-dimensionalen Würfel. Comm. Math. Helv. **12**, 149—152.

GUÉRIN, R.
1965 Vue d'ensemble sur les plans en blocs incomplets équilibrés et partiellement équilibrés. Rev. Inst. Internat. Statist. **33**, 24—58.

GUPTA, H.
1940 A problem in combinations. Math. Student **8**, 131—132.

HABER, R. M.
1960 Term rank of (0, 1)-matrices. Rendic. Sem. Mat. Padova **30**, 24—51.
1963 Minimal term rank of a class of (0, 1)-matrices. Canad. J. Math. **15**, 188—192.

HADAMARD, J.
1893 Résolution d'une question relative aux déterminants. Bull. Sci. Math. (2) **17**, 240—246.

HAGER, A.
1943 Symmetrische Inzidenztafeln finiter Geometrien. Sitz.-Ber. Bayer. Akad. Wiss. 1943, 25—47.

HALBERSTAM, H.; LAXTON, R. R.

1963 On perfect difference sets. Quart. J. Math. **14**, 86—90.

1964 Perfect difference sets. Proc. Glasgow Math. Assoc. **6**, 177—184.

HALL, M.

1943 Projective planes. Trans. Amer. Math. Soc. **54**, 229—277.

1945 An existence theorem for Latin squares. Bull. Amer. Math. Soc. **51**, 387—388.

1947 Cyclic projective planes. Duke Math. J. **14**, 1079—1090.

1949 Correction to "Projective planes". Trans. Amer. Math. Soc. **65**, 473—474.

1952 A combinatorial problem in abelian groups. Proc. Amer. Math. Soc. **3**, 584—587.

1953 Uniqueness of the projective plane with 57 points. Proc. Amer. Math. Soc. **4**, 912—916.

1954a Projective planes and related topics. California Inst. Techn., Pasadena.

1954b Correction to "Uniqueness of the projective plane with 57 points". Proc. Amer. Math. Soc. **5**, 994—997.

1955 Finite projective planes. Amer. Math. Monthly **62**, No. 7, part II, 18—24.

1956 A survey of difference sets. Proc. Amer. Math. Soc. **7**, 975—986.

1959 The theory of groups. Macmillan, New York.

1960a Automorphisms of Steiner triple systems. IBM J. Res. Develop. **4**, 460—472.

1960b Current studies on combinatorial designs. Proc. Symp. Appl. Math. **10**, 1—14.

1962 Note on the Mathieu group M_{12}. Arch. Math. **13**, 334—340.

1965a Characters and cyclotomy. Proc. Symp. Pure Math. **8**, 31—43.

1965b Hadamard matrices of order 20. NASA Tech. Rep. No. 32-761, 41 pp.

1967a Group theory and block designs. Proc. Intern. Conference Theory of groups (Canberra 1965) pp. 115—144. Gordon & Breach, New York.

1967b Combinatorial theory. Blaisdell, Waltham, Mass.

HALL, M.; CONNOR, W. S.

1954 An imbedding theorem for balanced incomplete block designs. Canad. J. Math. **6**, 35—41.

HALL, M; KNUTH, D. E.

1965 Combinatorial analysis and computers. Amer. Math. Monthly **72**, No. 2, part II, 21—28.

HALL, M.; RYSER, H. J.

1951 Cyclic incidence matrices. Canad. J. Math. **3**, 495—502.

1954 Normal completions of incidence matrices. Amer. J. Math. **76**, 581—589.

HALL, M.; SWIFT, J. D.

1955 Determination of Steiner triple systems of order 15. Math. Tables Aids Comput. **9**, 146—155.

HALL, M.; SWIFT, J. D.; KILLGROVE, R.

1959 On projective planes of order nine. Math. Comp. **13**, 233—246.

HALL, M.; SWIFT, J. D.; WALKER, R. J.

1956 Uniqueness of the projective plane of order eight. Math. Tables Aids Comput. **10**, 186—194.

HALL, P.

1928 A note on soluble groups. J. London Math. Soc. **3**, 98—105.

HANANI, H.

1960a On quadruple systems. Canad. J. Math. **12**, 145—157.

1960b A note on Steiner triple systems. Math. Scand. **8**, 154—156.

1961 The existence and construction of balanced incomplete block designs. Ann. Math. Statist. **32**, 361—368.
1963 On some tactical configurations. Canad. J. Math. **15**, 702—722.
1964 On covering of balanced incomplete block designs. Canad. J. Math. **16**, 615—625.
1965 A balanced incomplete block design. Ann. Math. Statist. **36**, 711.

HANANI, H.; SCHONHEIM, J.
1964 On Steiner systems. Israel J. Math. **2**, 139—142.

HARDY, G. H.; WRIGHT, E. M.
1938 An introduction to the theory of numbers. Oxford. 4th ed. 1960.

HARSHBARGER, F.
1931 The geometric configuration defined by a special algebraic relation of genus four. Trans. Amer. Math. Soc. **33**, 557—578.

HARTLEY, H. O.; SHRIKHANDE, S. S.; TAYLOR, W. B.
1953 A note on incomplete block designs with row balance. Ann. Math. Statist. **24**, 123—126.

HARTLEY, R. W.
1926 Determination of the ternary collineation groups whose coefficients lie in the $GF(2^n)$. Ann. Math. **27**, 140—158.

HASSE, H.
1923 Über die Äquivalenz quadratischer Formen im Körper der rationalen Zahlen. J. reine angew. Math. **152**, 205—224.
1931 Gruppentheoretischer Beweis eines Satzes über gewisse Tripelsysteme. Norsk Mat. Tidskr. **13**, 105—107.

HAVEL, V.
1957 Über die Paare der (m, n)-Konfigurationen. Časopis Pěst Mat. **82**, 360—364.

HEFFTER, L.
1897 Über Tripelsysteme. Math. Ann. **49**, 101—112.

HENDERSON, M.
1964 Generalized finite planes. Coll. Math. **13**, 151—184.
1965 Certain finite nonprojective geometries without the axiom of parallels. Proc. Amer. Math. Soc. **16**, 115—119.

HERING, C.
1963 Eine Charakterisierung der endlichen zweidimensionalen projektiven Gruppen. Math. Z. **82**, 152—175.
1965 Eine Klassifikation der Möbius-Ebenen. Math. Z. **87**, 252—262.
1967a Eine Bemerkung über Automorphismengruppen von endlichen projektiven Ebenen und Möbiusebenen. Arch. Math. **18**, 107—110.
1967b Endliche zweifach transitive Möbiusebenen ungerader Ordnung. Arch. Math. **18**, 212—216.
1968a Zweifach transitive Permutationsgruppen, in denen zwei die maximale Anzahl von Fixpunkten von Involutionen ist. Math. Z. **104**, 150—174.
1968b Über projektive Ebenen vom Lenz-Typ III. Math. Z. **105**, 219—225.

HERTZIG, D.
1963 Rational normal matrices satisfying the incidence equation. Proc. Amer. Math. Soc. **14**, 878—879.

HEUZÉ, G.
1965 Sur les schémas d'association et les plans partiellement équilibrés à deux classes. C. R. Acad. Sci. Paris **261**, 630—631.

HIGMAN, D. G.
1962 Flag transitive collineation groups of finite projective spaces. Ill. J. Math. **6**, 434—446.
1964 Finite permutation groups of rank 3. Math. Z. **86**, 145—156.
1966 Primitive rank 3 groups with a prime subdegree. Math. Z. **91**, 70—86.
1968 On finite affine planes of rank 3. Math. Z. **104**, 147—149.

HIGMAN, D. G.; MCLAUGHLIN, J. E.
1961 Geometric ABA-groups. Ill. J. Math. **5**, 382—397.
1965 Rank 3 subgroups of finite symplectic and unitary groups. J. reine angew. Math. **218**, 174—189.

HIGMAN, D. G.; SIMS, C. W.
1967 A simple group of order 44,352,000. Preprint of forthcoming publication. Univ. of Michigan, Ann Arbor, Mich.

HIGMAN, G.
1969 On the simple group of D. G. Higman and C. W. Sims. Ill. J. Math., to appear.

HILBERT, D.
1899 Grundlagen der Geometrie. Teubner, Leipzig und Berlin. 9. Auflage, Stuttgart 1962, herausgeg. von P. Bernays.

HINKELMANN, K.
1964 Extended group divisible partially balanced incomplete block designs. Ann. Math. Statist. **35**, 681—695.

HOFFMAN, A. J.
1951a On the foundations of inversion geometry. Trans. Amer. Math. Soc. **71**, 218—242.
1951b Chains in the projective line. Duke Math. J. **18**, 827—830.
1952 Cyclic affine planes. Canad. J. Math. **4**, 295—301.
1960a On the uniqueness of the triangular association scheme. Ann. Math. Statist. **31**, 492—497.
1960b On the exceptional case in a characterization of the arcs of a complete graph. IBM J. Res. Develop. **4**, 487—496.
1960c Some recent applications of the theory of linear inequalities to extremal combinatorial analysis. Proc. Symp. Appl. Math. **10**, 113—127.
1963 On the duals of symmetric partially-balanced incomplete block designs. Ann. Math. Statist. **34**, 528—531.
1965 On the line graph of a projective plane. Proc. Amer. Math. Soc. **16**, 297—302.

HOFFMAN, A. J.; NEWMAN, M.; STRAUS, E. G.; TAUSSKY, O.
1956 On the number of absolute points of a correlation. Pacif. J. Math. **6**, 83—96.

HOFFMAN, A. J.; RAY-CHAUDHURI, D. K.
1965a On the line graph of a finite affine plane. Canad. J. Math. **17**, 687—694.
1965b On the line graph of a symmetric balanced incomplete block design. Trans. Amer. Math. Soc. **116**, 238—252.

HOFFMAN, A. J.; RICHARDSON, M.
1961 Block design games. Canad. J. Math. **13**, 110—128.

HOFFMAN, A. J.; SINGLETON, R. R.
1960 On Moore graphs with diameters 2 and 3. IBM J. Res. Develop. **4**, 497—504.

HSU, N. C.
1962 Note on certain combinatorial designs. Proc. Amer. Math. Soc. **13**, 682—684.

Hughes, D. R.

1955a A note on difference sets. Proc. Amer. Math. Soc. **6**, 689—692.

1955b Planar division neo-rings. Trans. Amer. Math. Soc. **80**, 502—527.

1955c Additive and multiplicative loops of planar ternary rings. Proc. Amer. Math. Soc. **6**, 973—980.

1956 Partial difference sets. Amer. J. Math. **78**, 650—674.

1957a Regular collineation groups. Proc. Amer. Math. Soc. **8**, 165—168.

1957b A class of non-Desarguesian projective planes. Canad. J. Math. **9**, 378—388.

1957c A note on some partially transitive projective planes. Proc. Amer. Math. Soc. **8**, 978—981.

1957d Collineations and generalized incidence matrices. Trans. Amer. Math. Soc. **86**, 284—296.

1957e Generalized incidence matrices over group algebras. Ill. J. Math. **1**, 545—551.

1959a Collineation groups of non-desarguesian planes. I. The Hall Veblen-Wedderburn systems. Amer. J. Math. **81**, 921—938.

1959b Review of some results in collineation groups. Proc. Symp. Pure Math. 1 (1959) 42—55.

1960a Collineation groups of non-desarguesian planes. II. Some seminuclear division algebras. Amer. J. Math. **82**, 113—119.

1960b On homomorphisms of projective planes. Proc. Symp. Appl. Math. **10**, 45—52.

1962 Combinatorial analysis. t-designs and permutation groups. Proc. Symp. Pure Math. **6**, 39—41.

1964a On k-homogeneous groups. Arch. Math. **15**, 401—403.

1964b Sottopiani non-Desarguesiani di piani finiti. Atti Accad. Naz. Lincei Rendic. **36**, 315—318.

1965a On t-designs and groups. Amer. J. Math. **87**, 761—778.

1965b Extensions of designs and groups: Projective, symplectic and certain affine groups. Math. Z. **89**, 199—205.

Hughes, D. R.; Kleinfeld, E.

1960 Seminuclear extensions of Galois fields. Amer. J. Math. **82**, 389—392.

Hughes, D. R.; Thompson, J. G.

1959 The H_p-problem and the structure of H_p-groups. Pacif. J. Math. **9**, 1097—1102.

Huppert, B.

1955 Primitive, auflösbare Permutationsgruppen. Arch. Math. **6**, 303—310.

1957a Lineare auflösbare Gruppen. Math. Z. **67**, 479—518.

1957b Zweifach transitive, auflösbare Permutationsgruppen. Math. Z. **68**, 126—150.

1962 Scharf dreifach transitive Permutationsgruppen. Arch. Math. **13**, 61—72.

Hussain, Q. M.

1945a On the totality of the solutions for the symmetrical incomplete block designs: $\lambda = 2$, $k = 5$ or 6. Sankhya **7**, 204—208.

1945b Symmetrical incomplete block designs with $\lambda = 2$, $k = 8$ or 9. Bull. Calcutta Math. Soc. **37**, 115—123.

1946 Impossibility of the symmetrical incomplete block design with $\lambda = 2$, $k = 7$. Sankhya **7**, 317—322.

1948a Structure of some incomplete block designs. Sankhya **8**, 381—383.

1948b Alternative proof of the impossibility of the symmetrical design with $\lambda = 2$, $k = 7$. Sankhya **8**, 384.

Ishaq, M.

1950 A note on Hadamard matrices. Ganita **1**, 13—15.

Ito, N.
1960 Über die Gruppen $PSL_n(q)$, die eine Untergruppe von Primzahlindex enthalten. Acta Sci. Math. Szeged **21**, 206—217.
1962 On a class of doubly transitive permutation groups. Ill. J. Math. **6**, 341—352.
1963 Transitive permutation groups of degree $p = 2q + 1$, p and q being prime numbers. Bull. Amer. Math. Soc. **69**, 165—192.
1964 Transitive permutation groups of degree $p = 2q + 1$, p and q being prime numbers. II. Trans. Amer. Math. Soc. **113**, 454—487.
1965 Transitive permutation groups of degree $p = 2q + 1$, p and q being prime numbers. III. Trans. Amer. Math. Soc. **116**, 151—166.
1967a On a class of doubly, but not triply transitive permutation groups. Arch. Math. **18**, 564—570.
1967b On permutation groups of prime degree p which contain (at least) two classes of conjugate subgroups of index p. Rendic. Sem. Mat. Padova **38**, 287—292.

James, A. T.
1957 The relationship algebra of an experimental design. Ann. Math. Statist. **28**, 993—1002.

Järnefelt, G.
1951 Reflections on a finite approximation to Euclidean geometry. Physical and astronomical prospects. Ann. Acad. Sci. Fenn., No. 96.

Järnefelt, G.; Kustaanheimo, P.
1949 An observation on finite geometries. 11. Skand. Math. Kongress Trondheim, pp. 166—182.

Järnefelt, G.; Qvist, B.
1955 Die Isomorphie eines elementargeometrischen und eines Galois-Gitterpunktemodells. Ann. Acad. Sci. Fenn., No. 201.

John, P. W. M.
1964 Balanced designs with unequal numbers of replicates. Ann. Math. Statist. **35**, 897—899.

Johnsen, E. C.
1964 The inverse multiplier for abelian group difference sets. Canad. J. Math. **16**, 787—796.
1965 Matrix rational completions satisfying generalized incidence equations. Canad. J. Math. **17**, 1—12.
1966a Integral solutions to the incidence equation for finite projective plane cases of orders $n \equiv 2 \pmod 4$. Pacif. J. Math. **17**, 97—120.
1966b Skew-Hadamard abelian group difference sets. J. Algebra **4**, 388—402.

Jones, B. W.
1950 The arithmetic theory of quadratic forms. Wiley, New York.

Jónsson, B.
1953 On the representation of lattices. Math. Scand. **1**, 193—206.
1954 Modular lattices and Desargues' theorem. Math. Scand. **2**, 395—314.

Jónsson, W.
1963 Transitivität und Homogenität projektiver Ebenen. Math. Z. **80**, 269—292.
1965 (C, γ, μ)-homogeneity of projective planes. Canad. J. Math. **17**, 331—334.
1966 Doubly-transitive groups, nearfields and geometry. Arch. Math. **17**, 83—88.

Jordan, C.
1871 Théorèmes sur les groupes primitifs. J. Math. Pures Appl. **16**, 383—408.

Jung, W.; Melchior, E.
1936 Symmetrische Geradenkomplexe. Ein Beitrag zur Theorie der Konfigurationen. Deutsche Math. **1**, 239—255.

Kantor, W. M.
1968a 4-homogeneous groups. Math. Z. **103**, 67—68.
1968b Automorphism groups of designs. Math. Z., to appear.
1969a Characterizations of finite projective and affine spaces. Canad. J. Math. to appear.
1969b 2-transitive symmetric designs. Trans. Amer. Math. Soc. to appear.

Karzel, H.
1960 Verallgemeinerte elliptische Geometrien und ihre Gruppenräume. Abh. Hamburg **24**, 167—188.
1962 Kommutative Inzidenzgruppen. Arch. Math. **13**, 535—538.
1964 Ebene Inzidenzgruppen. Arch. Math. **15**, 10—17.
1965a Normale Fastkörper mit kommutativer Inzidenzgruppe. Abh. Hamburg **28**, 124—132.
1965b Projektive Räume mit einer kommutativen transitiven Kollineationsgruppe. Math. Z. **87**, 74—77.
1965c Bericht über projektive Inzidenzgruppen. Jahresb. Deutsch. Math. Ver. **67**, 58—92.
1965d Zweiseitige Inzidenzgruppen. Abh. Hamburg **29**, 247—256.
1965e Unendliche Dicksonsche Fastkörper. Arch. Math. **16**, 247—256.

Keedwell, A. D.
1963 On the order of projective planes with characteristic. Rendic. Mat. **22**, 498—530.
1964 A class of configurations associated with projective planes with characteristic. Arch. Math. **15**, 470—480.
1965 A search for projective planes of a special type with the aid of a digital computer. Math. Comp. **19**, 317—322.

Kegel, O. H.
1961a Produkte nilpotenter Gruppen. Arch. Math. **12**, 90—93.
1961b Die Nilpotenz der H_p-Gruppen. Math. Z. **75**, 373—376.
1961c Nicht-einfache Partitionen endlicher Gruppen. Arch. Math. **12**, 170—175.
1961d Aufzählung der Partitionen endlicher Gruppen mit trivialer Fittingscher Untergruppe. Arch. Math. **12**, 409—412.

Kegel, O. H.; Lüneburg, H.
1963 Über die kleine Reidemeisterbedingung. II. Arch. Math. **14**, 7—10.

Kegel, O. H.; Wall, G. E.
1961 Zur Struktur endlicher Gruppen mit nicht-trivialer Partition. Arch. Math. **12**, 255—261.

Keiser, V. H.
1966 Finite affine planes with collineation groups primitive on the points. Math. Z. **92**, 288—294.
1967 Finite projective planes with non-solvable transitive collineation groups. Amer. Math. Monthly **74**, 556—559.

Kerawala, S. M.
1946 Note on symmetrical incomplete block designs: $\lambda = 2$, $k = 6$ or 7. Bull. Calcutta Math. Soc. **38**, 190—192.
1953 Symmetrical incomplete block designs with $\lambda = 2$. Scient. Pakistan **1**, 1—24.

KILLGROVE, R. B.
1964 Completions of quadrangles in projective planes. Canad. J. Math. **16**, 63—76.
1965 Completions of quadrangles in projective planes. II. Canad. J. Math. **17**, 155—165.

KIRKMAN, T. P.
1847 On a problem in combinations. Cambridge and Dublin Math. J. **2**, 191—204.
1850 Query. Lady's and Gentleman's diary (1850), p. 48.

KISHEN, K.; RAO, R. C.
1952 An examination of various inequality relations among parameters of the balanced incomplete block designs. J. Ind. Soc. Agric. Statist. **4**, 137—144.

KLEINFELD, E.
1951 Alternative division rings of characteristic 2. Proc. Nat. Acad. Sci. USA **37**, 818—820.
1959 Finite Hjelmslev planes. Ill. J. Math. **3**, 403—407.
1960 Techniques for enumerating Veblen-Wedderburn systems. J. Assoc. Comput. Mach. **7**, 330—337.

KLINGENBERG, W.
1952 Beziehungen zwischen einigen affinen Schließungssätzen. Abh. Hamburg **18**, 120—143.
1954 Projektive und affine Ebenen mit Nachbarelementen. Math. Z. **60**, 384—406.
1955a Desarguessche Ebenen mit Nachbarelementen. Abh. Hamburg **20**, 97—111.
1955b Beweis des Desarguesschen Satzes aus der Reidemeisterfigur und verwandte Sätze. Abh. Hamburg **19**, 158—175.
1956 Projektive Geometrien mit Homomorphismus. Math. Ann. **132**, 180—200.
1957 Affine Ebenen mit Orthogonalität. Arch. Math. **8**, 199—202.

KLUG, L.
1932 Desmische Vierecke. Math. naturw. Ber. Ungarn **38**, 1—52.
1934 Erweiterungen der desarguesschen Konfiguration. Mat. természett. Értes. **51**, 77—133.

KNUTH, D. E.
1965a A class of projective planes. Trans. Amer. Math. Soc. **115**, 541—549.
1965b Finite semifields and projective planes. J. Algebra **2**, 182—217.

KOMMERELL, K.
1941 Die Pascalsche Konfiguration 9_3. Deutsche Math. **6**, 16—32.
1949 Mehrfach ausgeartete Desarguessche Konfigurationen. Math. Z. **52**, 472—482.

KÖNIG, D.
1936 Theorie der endlichen und unendlichen Graphen. Akad. Verl. Ges. Leipzig.

KONTOROWITSCH, P.
1943 Sur les groupes à base de partition. Mat. Sbornik **12**, 56—70.

KUO, P. T.
1948 Projective correspondences in the finite projective geometry $PG(3, 2)$. Acad. Sinica Science Rec. **2**, 171—178.

KUROSAKI, T.
1941 Über die mit einer Kollineation vertauschbaren Kollineationen. Proc. Imp. Acad. Tokyo **17**, 24—28.

KUSTAANHEIMO, P.
1950 A note on a finite approximation of the Euclidean plane geometry. Soc. Sci. Fenn. Comm. **15**, No. 19, 11 pp.

1952 On the fundamental prime of a finite world. Ann. Acad. Sci. Fenn. **129**.
1957a On the relation of congruence in finite geometries. Rendic. Mat. **16**, 286—291.
1957b On the relation of order in finite geometries. Rend. Mat. **16**, 292—296.

Kustaanheimo, P; Qvist, B.
1954 Finite geometries and their application. Nordisk Mat. Tidskr. **2**, 137—155.

Kusumoto, K.
1965 A necessary condition for the existence of regular and symmetrical PBIB designs of T_3 type. Ann. Inst. Statist. Math. **17**, 149—165.

Lafon, M.
1957a Construction de blocs incomplets partiellement équilibrés à $s+1$ classes associées. C. R. Acad. Sci. Paris **244**, 1714—1717.
1957b Blocs incomplets partiellement équilibrés à deux classes associées avec quatre répétitions. C. R. Acad. Sci. Paris **244**, 1875—1877.

Lauffer, R.
1954a Die nichtkonstruierbare Konfiguration (10_3). Math. Nachr. **11**, 303—304.
1954b Über die Struktur der Konfigurationen (10_3). Math. Nachr. **12**, 1—8.

Lebesgue, V. A.
1850 Sur l'impossibilité en nombres entiers de l'équation $x^m = y^2 + 1$. Nouv. Ann. **9**, 178—181.

Lehmer, E.
1953 On residue difference sets. Canad. J. Math. **5**, 425—432.
1954 On cyclotomic numbers of order sixteen. Canad. J. Math. **6**, 449—454.
1955 Period equation applied to difference sets. Proc. Amer. Math. Soc. **6**, 433—442.

Leisenring, K. B.
1954 A theorem in projective n-space equivalent to commutativity. Mich. Math. J. **2**, 35—40.

Lenz, H.
1953 Beispiel einer projektiven Ebene, in der einige, aber nicht alle Vierecke kollineare Diagonalpunkte haben. Arch. Math. **4**, 327—330.
1954a Kleiner desarguesscher Satz und Dualität in projektiven Ebenen. Jahresber. Deutsche Math. Ver. **57**, 20—31.
1954b Zur Begründung der analytischen Geometrie. Sitz.-Ber. Bayer. Akad. Wiss. 1954, 17—72.
1961 Grundlagen der Elementarmathematik. Deutscher Verlag d. Wiss., Berlin.
1962 Quadratische Formen und Kollineationsgruppen. Arch. Math. **13**, 110—119.
1965 Vorlesungen über projektive Geometrie. Akad. Verl. Ges. Geest & Portig, Leipzig.

LeVeque, W. J.
1956 Topics in number theory, 2 volumes. Addison-Wesley; Reading, London.

Levi, F. W.
1929 Geometrische Konfigurationen. Hirzel, Leipzig.
1942 Finite geometrical systems. University of Calcutta.

Levi, H.
1965 Plane geometries in terms of projections. Proc. Amer. Math. Soc. **16**, 503—511.

Liebmann, H.
1934 Synthetische Geometrie. Teubner; Leipzig, Berlin.

LINGENBERG, R.
1962 Über Gruppen projektiver Kollineationen, welche eine perspektive Dualität invariant lassen. Arch. Math. **13**, 385—400.
1964 Über die Gruppe einer projektiven Dualität. Math. Z. **83**, 367—380.
1966 Metrische Geometrie der Ebene und S-Gruppen. Jahresber. Deutsche Mathematiker-Vereinig. **69**, 9—50.

LIU, Chang-Wen
1963 A method of constructing certain symmetrical partially balanced designs. Sci. Sinica **12**, 1935—1937.

LIU, Chang-Wen; CHANG, Li-Chien.
1964 Some PBIB(2)-designs induced by association schemes. Sci. Sinica **13**, 840—841.

LIU, Wau-Ru; CHANG, Li-Chien.
1964 Group divisible incomplete block designs with parameters $v \leqq 10$ and $r \leqq 10$. Sci. Sinica **13**, 839—840.

LIVINGSTONE, D.; WAGNER, A.
1965 Transitivity of finite permutation groups on unordered sets. Math. Z. **90**, 393—403.

LJAMZIN, A. I.
1963 Ein Beispiel eines Paars orthogonaler lateinischer Quadrate der Ordnung 10. Usp. Mat. Nauk **18**, 173—174.

LJUNGGREN, W.
1942 Einige Bemerkungen über die Darstellung ganzer Zahlen durch binäre kubische Formen mit positiver Diskriminante. Acta Math. **75**, 1—21.

LOMBARDO-RADICE, L.
1953a Su alcuni modelli di geometrie proiettive piane finite. Atti 4. Congr. Un. Mat. Ital., Taormina 1951. Roma 1953, pp. 370—373.
1953b Una nuova costruzione dei piani grafici desarguesiani finiti. Ricerche Mat. **2**, 47—57.
1953c Piani grafici finiti a coordinate di Veblen-Wedderburn. Ricerche Mat. **2**, 266—273.
1954 L'inversione come dualità nei piani su sistemi cartesiani. Ricerche Mat. **3**, 31—34.
1955 Sul rango dei piani grafici finiti a caratteristica 3. Boll. Un. Mat. Ital. **10**, 172—177.
1956 Sul problema dei k-archi completi in $S_{2,q}$ ($q = p^t$, p primo dispari). Boll. Un. Mat. Ital. **11**, 178—181.
1957 Questioni algebrico-geometriche ai teoremi "$p = 0$". Convegno italofrancese di algebra astratta. Padova 1956, p. 37—47.
1959 Piani grafici finiti non desarguesiani. Denaro, Palermo.
1962 La decomposizione tattica di un piano grafico finito associata a un k-arco. Ann. Mat. Pura Appl. **60**, 37—48.

LÜNEBURG, H.
1960 Über die beiden kleinen Sätze von Pappos. Arch. Math. **11**, 339—341.
1961a Zentrale Automorphismen von λ-Räumen. Arch. Math. **12**, 134—145.
1961b Über die kleine Reidemeisterbedingung. Arch. Math. **12**, 382—384.
1962 Affine Hjelmslev-Ebenen mit transitiver Translationsgruppe. Math. Z. **79**, 260—288.
1963 Steinersche Tripelsysteme mit fahnentransitiver Kollineationsgruppe. Math. Ann. **149**, 261—270.

1964a Endliche projektive Ebenen vom Lenz-Barlotti-Typ I-6. Abh. Hamburg **27**, 75—79.

1964b Finite Möbius planes admitting a Zassenhaus group as group of automorphisms. Ill. J. Math. **8**, 586—592.

1964c Charakterisierungen der endlichen desarguesschen projektiven Ebenen. Math. Z. **85**, 419—450.

1965a Ein neuer Beweis eines Hauptsatzes der projektiven Geometrie. Math. Z. **87**, 32—36.

1965b Fahnenhomogene Quadrupelsysteme. Math. Z. **89**, 82—90.

1965c Über projektive Ebenen, in denen jede Fahne von einer nichttrivialen Elation invariant gelassen wird. Abh. Hamburg **29**, 37—76.

1965d Zur Frage der Existenz von endlichen projektiven Ebenen vom Lenz-Barlotti-Typ III-2. J. reine angew. Math. **220**, 63—67.

1965e Die Suzukigruppen und ihre Geometrien. Lecture Notes in Mathematics. Berlin/Heidelberg/New-York: Springer 1965, 111 pp.

1966a Scharf fahnentransitive endliche affine Räume. Arch. Math. **17**, 78—82.

1966b Some remarks concerning the Ree groups of type (G_2). J. Algebra **3**, 256—259.

1966c On Möbius-Planes of even order. Math. Z. **92**, 187—193.

1966d Über die Struktursätze der projektiven Geometrie. Arch. Math. **17**, 206—209.

1967a Gruppentheoretische Methoden in der Geometrie. Ein Bericht. Jahresber. Deutsche Math. Ver. **70**, 16—51.

1967b An axiomatic treatment of ratios in an affine plane. Arch. Math. **18**, 444—448.

1967c Kreishomogene endliche Möbius-Ebenen. Math. Z. **101**, 68—70.

Lunelli, L.; Sce, M.

1958 k-archi completi nei piani proiettivi desarguesiani di rango 8 e 16. Centro Calcol. Numerici, Politecnico Milano. 15 pp.

1964 Considerazioni aritmetiche e risultati sperimentali sui $(K; n)_q$-archi. Ist. Lombardo Accad. Sci. Lett. Rendic. **98**, 3—52.

Machtinger, L. A.

1965 Multiple transitivity of primitive permutation groups. Proc. Amer. Math. Soc. **16**, 168—172.

MacInnes, C. R.

1907 Finite planes with less than eight points on a line. Amer. Math. Monthly **14**, 171—174.

MacMahon, P. A.

1916 Combinatory Analysis. 2 vol.'s, Cambridge Univ. Press. Reprint Chelsea, New York, 1960.

MacNeish, H. F.

1922 Euler squares. Ann. Math. **23**, 221—227.

Magari, R.

1958 Le configurazioni parziali chiuse contenute nel piano, P, sul quasicorpo associativo di ordine 9. Boll. Un. Mat. Ital. **13**, 128—140.

Maier, K.

1939 Die Desarguessche Konfiguration. Deutsche Math. **4**, 591—641.

Maisano, F.

1960 Sulle ovali steineriane nel piano sopra il corpo commutativo di ordine 27. Matematiche (Catania) **15**, 108—120.

1964 Sulle collineazioni lineari del piano sopra il quasicorpo commutativo di ordine 27. Atti Accad. Sci. Lett. Arti Palermo (Parte I) **23**, 299—320.

MAJINDAR, K. N.
1962 On the parameters and intersection of blocks of balanced incomplete block designs. Ann. Math. Statist. **33**, 1200—1205.
1963 On incomplete and balanced incomplete block designs. Proc. Amer. Math. Soc. **14**, 223—224.
1966 On integer matrices and incidence matrices of certain combinatorial configurations. I. Square matrices. II. Rectangular matrices. III. Rectangular matrices. Canad. J. Math. **18**, 1—5, 6—8, 9—17.

MAJUMDAR, K. N.
1953 On some theorems in combinatorics relating to incomplete block designs. Ann. Math. Statist. **24**, 377—389.
1954 On combinatorial arrangements. Proc. Amer. Math. Soc. **5**, 662—664.

MAKOWSKI, A.
1963 Remarks on a paper of Tallini. Acta Arithm. **8**, 469—470.

MAMMANA, C.
1959 Sui quasicorpi di ordine p^2. Matematiche (Catania) **14**, 10—30.
1960 Sui quasicorpi distributivi finiti. Matematiche (Catania) **15**, 121—140.

MANN, H. B.
1942 The construction of orthogonal Latin squares. Ann. Math. Statist. **13**, 418—423.
1943 On the construction of sets of orthogonal Latin squares. Ann. Math. Statist. **14**, 401—414.
1949 Analysis and design of experiments. Analysis of variance and analysis of variance designs. Dover, New York.
1950 On orthogonal Latin squares. Bull. Amer. Math. Soc. **50**, 249—257.
1952 Some theorems on difference sets. Canad. J. Math. **4**, 222—226.
1964 Balanced incomplete block designs and abelian difference sets. Ill. J. Math. **8**, 252—261.
1965a Difference sets in elementary abelian groups. Ill. J. Math. **9**, 212—219.
1965b Addition theorems: the addition theorems of group theory and number theory. Interscience, New York.
1967 Recent advances in difference sets. Amer. Math. Monthly **74**, 229—235.

MANN, H. B.; RYSER, H. J.
1953 Systems of distinct representatives. Amer. Math. Monthly **60**, 397—401.

MARCUS, M.; MINC, H.
1963 Disjoint pairs of sets and incidence matrices. Ill. J. Math. **7**, 137—147.

MARTIN, G. E.
1965 On Mammana division rings. Matematiche (Catania) **20**, 1—6.
1967a On arcs in a finite projective plane. Canad. J. Math. **19**, 376—393.
1967b Projective planes and isotopic ternary rings. Amer. Math. Monthly **74**, 1185—1195.

MASUYAMA, M.
1961 Calculus of blocks and a class of partially balanced incomplete block designs. Rep. Statist. Appl. Res. Un. Japan Sci. Engrs. **8**, 59—69.
1964a Construction of PBIB designs by fractional development. Rep. Statist. Appl. Res. Un. Japan Sci. Engrs. **11**, 47—54.

1964b Linear graphs of PBIB designs. Rep. Statist. Appl. Res. Un. Japan Sci. Engrs. **11**, 147—151.

McFarland, R. L.

1965 A generalization of a result of Newman on multipliers of difference sets. J. Res. Nat. Bur. Standards **69B**, 319—322.

McFarland, R. L.; Mann, H. B.

1965 On multipliers of difference sets. Canad. J. Math. **17**, 541—542.

Melchior, E.

1937 Untersuchungen über ein Problem aus der Theorie der Konfigurationen. Schr. Math. Sem. u. Inst. Angew. Math. Univ. Berlin **3**, 181—206.

Mendelsohn, N. S.

1956 Non-Desarguesian projective plane geometries which satisfy the harmonic point axiom. Canad. J. Math. **8**, 532—562.

Menon, P. K.

1960 Difference sets in abelian groups. Proc. Amer. Math. Soc. **11**, 368—376.

1962a Certain Hadamard designs. Proc. Amer. Math. Soc. **13**, 524—531.

1962b On difference sets whose parameters satisfy a certain relation. Proc. Amer. Math. Soc. **13**, 739—745.

1964 On certain sums connected with Galois fields and their applications to difference sets. Math. Ann. **154**, 341—364.

Mesner, D. M.

1964 Traces of a class of (0, 1)-matrices. Canad. J. Math. **16**, 82—93.

1965 A note on the parameters of PBIB association schemes. Ann. Math. Statist. **36**, 331—336.

1967 Sets of disjoint lines in $PG(3, q)$. Canad. J. Math. **19**, 273—280.

Metelka, V.

1955a Über ebene Konfigurationen $(12_4, 16_3)$. Časopis Pěst. Mat. **80**, 133—145.

1955b Über gewisse ebene Konfigurationen $(12_4, 16_3)$, welche mindestens einen D-Punkt enthalten. Časopis Pěst. Mat. **80**, 146—151.

1957 Über ebene Konfigurationen $(12_4, 16_3)$, die mindestens einen D-Punkt enthalten. Časopis Pěst. Mat. **82**, 385—439.

Minkowski, H.

1890 Über die Bedingungen, unter welchen zwei quadratische Formen mit rationalen Koeffizienten ineinander rational transformiert werden können. J. reine angew. Math. **106**, 5—26. See also Ges. Abh. I, Teubner, Berlin and Leipzig 1911, 219—239.

Mitchell, H. H.

1911 Determination of the ordinary and modular ternary linear groups. Trans. Amer. Math. Soc. **12**, 207—242.

1913 Determination of the finite quaternary linear groups. Trans. Amer. Math. Soc. **14**, 123—142.

Moore, E. H.

1893 Concerning triple systems. Math. Ann. **43**, 271—285.

1896 Tactical memoranda. Amer. J. Math. **18**, 264—303.

1899 Concerning the general equation of the seventh and eighth degrees. Math. Ann. **51**, 417—444.

Morgan, D. L.; Ostrom, T. G.

1964 Coordinate systems of some semi-translation planes. Trans. Amer. Math. Soc. **111**, 19—32.

MOUETTE, L.
1957 Recherches sur la théorie des triades. Mathesis **66**, 283—287.

MOUFANG, R.
1931 Zur Struktur der projektiven Geometrie der Ebene. Math. Ann. **105**, 536—601.
1932a Die Schnittpunktsätze des projektiven speziellen Fünfecksnetzes in ihrer Abhängigkeit voneinander (Das A-Netz). Math. Ann. **106**, 755—795.
1932b Ein Satz über die Schnittpunktsätze des allgemeinen Fünfecksnetzes [Das (A, B)-Netz]. Math. Ann. **107**, 124—139.
1933 Alternativkörper und der Satz vom vollständigen Vierseit. Abh. Hamburg **9**, 207—222.
1935 Zur Struktur von Alternativkörpern. Math. Ann. **110**, 416—430.

MOULTON, F. R.
1902 A simple non-desarguesian plane geometry. Trans. Amer. Math. Soc. **3**, 192—195.

MULDER, P.
1917 Kirkman-Systemen. Dissertation Groningen. Leiden 1917.

MURTY, V. N.
1964 On the block structure of PBIB designs with two associate classes. Sankhya **26**, 381—382.

NAGAO, H.
1966 On multiply transitive groups I. Nagoya Math. J. **27**, 15—19.

NAGELL, T.
1921 Des équations indéterminées $x^2 + x + 1 = y^n$ et $x^2 + x + 1 = 3y^n$. Norsk matematisk forenings skrifter, ser. I, No. 2.
1951 Introduction to number theory. Almqvist & Wiksells, Uppsala.

NAIR, C. R.
1962 On the methods of block section and block intersection applied to certain PBIB designs. Calcutta Statist. Assoc. Bull. **11**, 49—54.
1964 The impossibility of certain PBIB designs. Calcutta Statist. Assoc. Bull. **13**, 87—88.

NAIR, K. R.
1943 Certain inequality relationships among the combinatorial parameters of incomplete block designs. Sankhya **6**, 255—259.
1950 Partially balanced incomplete block designs involving only two replications. Calcutta Statist. Assoc. Bull. **3**, 83—86.
1951a Some two-replicate PBIB designs. Calcutta Statist. Assoc. Bull. **3**, 174—176.
1951b Some three-replicate PBIB designs. Calcutta Statist. Assoc. Bull. **4**, 39—42.
1953 A note on group divisible incomplete block designs. Calcutta Statist. Assoc. Bull. **5**, 30—35.

NAIR, K. R.; RAO, C. R.
1942 A note on partially balanced incomplete block designs. Science and Culture **7**, 568—569.

NANDI, H. K.
1945 On the relations between certain types of tactical configurations. Bull. Calcutta Math. Soc. **37**, 115—123.
1946a Enumeration of non-isomorphic solutions of balanced incomplete block designs Sankhya **7**, 305—312.

1946b A further note on non-isomorphic solutions of incomplete block designs. Sankhya **7**, 313—316.

NAUMANN, H.

1954 Stufen der Begründung der ebenen affinen Geometrie. Math. Z. **60**, 120—141.

NETTO, E.

1893 Zur Theorie der Tripelsysteme. Math. Ann. **42**, 143—152.

1927 Lehrbuch der Combinatorik. 2nd edition Teubner, Leipzig. Reprint Chelsea, New York.

NEUMANN, B. H.

1940 On the commutativity of addition. J. London Math. Soc. **15**, 203—208.

NEUMANN, H.

1955 On some finite non-desarguesian planes. Arch. Math. **6**, 36—40.

NEWMAN, M.

1963 Multipliers of difference sets. Canad. J. Math. **15**, 121—124.

NIKOLAI, P. J.

1960 Permanents of incidence matrices. Math. Comp. **14**, 262—266.

NORMAN, C. W.

1965 Note on the fixed point configuration of collineations. Math. Z. **89**, 91—93.

1968 A characterization of the Mathieu group M_{11}. Math. Z., to appear.

NORTON, H. W.

1939 The 7×7 squares. Ann. Eugenics **9**, 269—307.

NOVÁK, J.

1959 Anwendung der Kombinatorik auf das Studium ebener Konfigurationen $(12_4, 16_3)$. Časopis Pěst. Mat. **84**, 257—282.

1963 Contributions to the theory of combinations. Časopis Pěst. Mat. **88**, 129—141.

OEHMKE, R. H.; SANDLER, R.

1963 The collineation groups of division ring planes I. Jordan algebras. Bull. Amer. Math. Soc. **69**, 791—793.

1964 The collineation groups of division ring planes. I. Jordan division algebras. J. reine angew. Math. **216**, 67—87.

1965 The collineation groups of division ring planes. II. Jordan division rings. Pacif. J. Math. **15**, 259—265.

OGAWA, J.

1959 A necessary condition for existence of regular and symmetrical experimental designs of triangular type, with partially balanced incomplete blocks. Ann. Math. Statist. **30**, 1063—1071.

1960 On a unified method of deriving necessary conditions for existence of symmetrically balanced incomplete block designs of certain types. Bull. Instit. Internat. Statist. **38**, 43—57.

O'KEEFE, E. S.

1961 Verification of a conjecture of Th. Skolem. Math. Scand. **9**, 80—82.

OKUNO, C.; OKUNO, T.

1961 On the construction of a class of partially balanced incomplete block designs by calculus of blocks. Rep. Statist. Appl. Res. Un. Japan Sci. Engrs. **8**, 113—139.

ORE, O.

1962 Theory of graphs. American Mathematical Society, Providence, R. I.

D'ORGEVAL, B.

1960 Sur certains $(k, 3)$-arcs en géométrie de Galois. Bull. Sci. Acad. Roy. Belgique **44**, 597—603.

OSBORN, J. M.

1961a New loops from old geometries. Amer. Math. Monthly **68**, 103—107.

1961b Vector loops. Ill. J. Math. **5**, 565—584.

OSTROM, T. G.

1953 Concerning difference sets. Canad. J. Math. **5**, 421—424.

1955 Ovals, dualities, and Desargues's theorem. Canad. J. Math. **7**, 417—431.

1956 Double transitivity in finite projective planes. Canad. J. Math. **8**, 563—567.

1957 Transitivities in projective planes. Canad. J. Math. **9**, 389—399.

1958a Dual transitivity in finite projective planes. Proc. Amer. Math. Soc. **9**, 55—56.

1958b Correction to "Transitivities in projective planes". Canad. J. Math. **10**, 507—512.

1959 Separation, betweenness, and congruence in planes over nonordered fields. Arch. Math. **10**, 200—205.

1960 Translation planes and configurations in desarguesian planes. Arch. Math. **11**, 457—464.

1961a Planar half-loops. Arch. Math. **12**, 151—158.

1961b Homomorphisms of finite planar half-loops. Arch. Math. **12**, 462—469.

1962a A class of non-desarguesian affine planes. Trans. Amer. Math. Soc. **104**, 483—487.

1962b Ovals and finite Bolyai-Lobachevsky planes. Amer. Math. Monthly **69**, 899—901.

1963 Planar loops as direct sums. Arch. Math. **14**, 369—372.

1964a Semi-translation planes. Trans. Amer. Math. Soc. **111**, 1—18.

1964b Finite planes with a single (p, L)-transitivity. Arch. Math. **15**, 378—384.

1964c Nets with critical deficiency. Pacif. J. Math. **14**, 1381—1387.

1965a Collineation groups of semi-translation planes. Pacif. J. Math. **15**, 273—279.

1965b Derivable nets. Canad. Math. Bull. **8**, 601—613.

1965c A characterization of the Hughes planes. Canad. J. Math. **17**, 916—922.

1966a The dual Lüneburg planes. Math. Z. **92**, 201—209.

1966b Replaceable nets, net collineations, and net extensions. Canad. J. Math. **18**, 666—672.

1966c Correction to: "Finite planes with a single (p, L)-transitivity. Arch. Math. **17**, 480.

1967 Closure with respect to subplanes. Math. Z. **97**, 57—65.

1968 Vector spaces and construction of finite projective planes. Arch. Math. **19**, 1—25

OSTROM, T. G.; SHERK, F. A.

1964 Finite projective planes with affine subplanes. Canad. Math. Bull. **7**, 549—559.

OSTROM, T. G.; WAGNER, A.

1959 On projective and affine planes with transitive collineation groups. Math. Z. **71**, 186—199.

OTT, E. R.

1937 Finite projective geometries, $PG(k, p^n)$. Amer. Math. Monthly **44**, 86—92.

PAIGE, L. J.

1949 Neofields. Duke Math. J. **16**, 39—60.

1951 Complete mappings of finite groups. Pacif. J. Math. **1**, 111—116.

PAIGE, L. J.; HALL, M.
1955 Complete mappings of finite groups. Pacif. J. Math. **5**, 541—549.

PAIGE, L. J.; WEXLER, C.
1953 A canonical form for incidence matrices of finite projective planes and their associated Latin squares. Portug. Math. **12**, 105—112

PALEY, R. E. A. C.
1933 On orthogonal matrices. J. Math. Phys. Mass. Inst. Tech. **12**, 311—320.

PALL, G.
1945 The arithmetical invariants of quadratic forms. Bull. Amer. Math. Soc. **51**, 185—197.

PANELLA, G.
1955 Caratterizzazione delle quadriche di uno spazio (tridimensionale) lineare sopra un corpo finito. Boll. Un. Mat. Ital. **10**, 507—513.
1958 Un insieme di piani di traslazione isomorfi. Convegno Intern. Reticole e Geometrie proiettive. Palermo 1957, pp. 109—119.
1959 Isomorfismo tra piani di traslazione di Marshall Hall. Ann. Mat. Pur. Appl. **47**, 169—180.
1960a Una nuova classe di quasicorpi. Atti Accad. Naz. Lincei Rendic. **28**, 44—49.
1960b Le collineazioni nei piani di Marshall Hall. Riv. Mat. Univ. Parma **1**, 171—184.
1964 Osservazioni sulla costruzione dei piani di Hughes. Rendic. Mat. **23**, 331—350.
1965 Una classe di sistemi cartesiani. Atti Accad. Naz. Lincei Rendic. **38**, 480—485.

PARKER, E. T.
1957 On collineations of symmetric designs. Proc. Amer. Math. Soc. **8**, 350—351.
1959a Construction of some sets of mutually orthogonal Latin squares. Proc. Amer. Math. Soc. **10**, 946—949.
1959b Orthogonal Latin squares. Proc. Nat. Acad. Sci. USA **45**, 859—862.
1959c On quadruply transitive groups. Pacif. J. Math. **9**, 829—836.
1962a Nonextendibility conditions on mutually orthogonal Latin squares. Proc. Amer. Math. Soc. **13**, 219—221.
1962b On orthogonal Latin squares. Proc. Symp. Pure Math. **6**, 43—46.
1963a Computer investigations of orthogonal Latin squares of order 10. Proc. Symp. Appl. Math. **15**, 73—81.
1963b Remarks on balanced incomplete block designs. Proc. Amer. Math. Soc. **14**, 729—730.
1965 Essential denominators in normal completions. Proc. Symp. Pure Math. **8**, 89—94.

PARKER, E. T.; KILLGROVE, R. B.
1964 A note on projective planes of order nine. Math. Comp. **18**, 506—508.

PASSMAN, D. S.
1967 Permutation groups, Lecture Notes, Yale Univ., Hartford, Conn.

PAYNE, S. E.
1967 A consequence of the non-existence of certain generalized polygons. Duke Math. J. **34**, 555—560.
1969 Symmetric representations of nondegenerate generalized polygons. Proc. Amer. Math. Soc., to appear.

PELTESOHN, R.
1938 Eine Lösung der beiden Heffterschen Differenzenprobleme. Compos. Math. **6**, 251—257.

PERASSI, R.
1950 Sulla geometria elementare a tre dimensioni in un corpo finito. Atti Accad. Sci. Torino, Class. Sci. Fis. Mat. Nat. **84**, 189—207.

PETERSEN, J.
1902 Les 36 officiers. Annuaire des Math., Paris 1901—1902, pp. 413—427.

PICKERT, G.
1952a Der Satz vom vollständigen Viereck bei kollinearen Diagonalpunkten. Math. Z. **56**, 131—133.
1952b Nichtkommutative cartesische Gruppen. Arch. Math. **3**, 335—342.
1955 Projektive Ebenen. Berlin/Göttingen/Heidelberg: Springer.
1956a Eine nichtdesarguessche Ebene mit einem Körper als Koordinatenbereich. Publ. Math. Debrecen **4**, 157—160.
1956b Projektive Ebenen über Neokörpern. Wiss. Z. Friedr.-Schiller-Univ. Jena **5**, 131—135.
1959a Gemeinsame Kennzeichnung zweier projektiver Ebenen der Ordnung 9. Abh. Hamburg **23**, 69—74.
1959b Eine Kennzeichnung desarguesscher Ebenen. Math. Z. **71**, 99—108.
1959c Der Satz von Pappos mit Festelementen. Arch. Math. **10**, 56—61.
1959d Bemerkungen über die projektive Gruppe einer Moufang-Ebene. Ill. J. Math. **3**, 169—173.
1963 Lectures on projective planes. Canad. Math. Congress, Saskatoon.
1964 Geometrische Kennzeichnung einer Klasse endlicher Moulton-Ebenen. J. reine angew. Math. **214/215**, 405—411.
1965a Koordinatenbereiche von Semitranslationsebenen. Math. Z. **89**, 62—70.
1965b Zur affinen Einführung der Hughes-Ebenen. Math. Z. **89**, 199—205.
1965c Algebraische Methoden in der Theorie der projektiven Ebenen. Séminaire Dubreil-Pisot, 18e anné, no. 21, 14 pp.
1967 Die cartesischen Gruppen der Ostrom-Rosati-Ebenen. Abh. Hamburg **30**, 106—117.

PIERCE, W. A.
1953 The impossibility of Fano's configuration in a projective plane with eight points per line. Proc. Amer. Math. Soc. **4**, 908—912.
1961a Binary systems for finite planes. Riv. Mat. Univ. Parma **2**, 115—132.
1961b Moulton planes. Canad. J. Math. **13**, 427—436.
1964a Collineations of affine Moulton planes. Canad. J. Math. **16**, 46—62.
1964b Collineations of projective Moulton planes. Canad. J. Math. **16**, 637—656.

PIPER, F. C.
1963 Elations of finite projective planes. Math. Z. **82**, 247—258.
1965 Collineation groups containing elations. I. Math. Z. **89**, 181—191.
1966a Collineation groups containing elations. II. Math. Z. **92**, 281—287.
1966b On elations of finite projective spaces of odd order. J. London Math. Soc. **41**, 641—648.
1967 Collineation groups containing homologies. J. Algebra **6**, 256—269.
1968a The orbit structure of collineation groups of finite projective planes. Math. Z. **103**, 318—332.
1968b On elations of finite projective spaces of even order. J. London Math. Soc. **43**, 459—464.

PLACKETT, R. L.
1947 Cyclic intrablock subgroups and allied designs. Sankhya **8**, 275—276.

PLACKETT, R. L.; BURMAN, J. P.
1945 The design of optimum multi-factorial experiments. Biometrika **33**, 305—325.

PLESS, V.
1965 The number of isotropic subspaces in a finite geometry. Atti Accad. Naz. Lincei Rendic. **39**, 418—421.

PRIMROSE, E. J. F.
1951 Quadrics in finite geometries. Proc. Cambridge Philos. Soc. **47**, 299—304.
1952 Resolvable balanced incomplete block designs. Sankhya **12**, 137—140.

PUHAREV, N. K.
1963 Some properties of finite Lobačevsky planes. Perm. Gos. Univ. Učen. Zap. Mat., no. 103, 61—63.
1965 On A_n^k-algebras and finite regular planes. Sibirsk Mat. Ž. **6**, 892—899.

QUATTROCCHI, P.
1966 Un metodo per la costruzione di certe strutture che generalizzano la nozione di piano grafico. Matematiche (Catania) **21**, 377—386.

QVIST, B.
1952 Some remarks concerning curves of the second degree in a finite plane. Ann. Acad. Sci. Fenn., no. 134, pp. 1—27.

RHAGAVARAO, D.
1960 On the block structure of certain PBIB designs with two associate classes having triangular and L_2 association schemes. Ann. Math. Statist. **31**, 787—791.
1962a Symmetrical unequal block arrangements with 2 unequal block sizes. Ann. Math. Statist. **33**, 620—633.
1962b On balanced unequal block designs. Biometrika **49**, 561—562.
1963 A note on the block structure of BIB designs. Calcutta Statist. Assoc. Bull. **12**, 60—62.

RAGHAVARAO, D.; CHANDRASEKHARARAO, K.
1964 Cubic designs. Ann. Math. Statist. **35**, 389—397.

RAMAKHRISHNAN, C. S.
1956 On the dual of a PBIB design, and a new class of designs with two replications. Sankhya **17**, 133—142.

RAMAMURTI, B.
1933 Desargues configurations admitting a collineation group. J. London Math. Soc. **8**, 34—39.

RANKIN, R. A.
1964 Difference sets. Acta Arithm. **9**, 161—168.

RAO, M. B.
1965 A note on incomplete block designs with $b = v$. Ann. Math. Statist. **36**, 1877.

RAO, R. C.
1944 Extensions of the difference theorems of Singer and Bose. Science and Culture **10**, 57.
1945 Finite geometries and certain derived results in the theory of numbers. Proc. Nat. Inst. Sci. India **11**, 136—149.

1946 Difference sets and combinatorial arrangements derivable from finite geometries. Proc. Nat. Inst. Sci. India **12**, 123—135.
1949 On a class of arrangements. Proc. Edinburgh Math. Soc. **8**, 119—125.
1951 A simplified approach to factorial experiments and the punched card technique in the construction and analysis of designs. Bull. Inst. Internat. Statist. **23**, part II, 1—28.
1956 A general class of quasifactorial and related designs. Sankhya **17**, 165—174.
1961 A study of BIB designs with replications 11 to 15. Sankhya **23**, 117—127.

Rao, V. A.
1958 A note on balanced designs. Ann. Math. Statist. **29**, 290—294.

Rashevskij, P. K.
1940 Sur une géométrie projective avec de nouveaux axiomes de configuration. Mat. Sbornik **8**, 183—204.

Ray-Chaudhuri, D. K.
1962a Some results on quadrics in finite projective geometry based on Galois fields. Canad. J. Math. **14**, 129—138.
1962b Application of the geometry of quadrics for constructing PBIB designs. Ann. Math. Statist. **33**, 1175—1186.
1965 Some configurations in finite projective spaces and partially balanced incomplete block designs. Canad. J. Math. **17**, 114—123.

Ree, R.
1961a A family of simple groups associated with the simple Lie algebra of type (F_4). Amer. J. Math. **83**, 401—420.
1961b A family of simple groups associated with the simple Lie algebra of type (G_2). Amer. J. Math. **83**, 432—462.
1964 Sur une famille de groupes de permutations doublement transitifs. Canad. J. Math. **16**, 797—820.

Reidemeister, K.
1929 Topologische Fragen der Differentialgeometrie. V. Gewebe und Gruppen. Math. Z. **29**, 427—435.

Reiman, S.
1963 Su una proprietà dei piani grafici finiti. Atti Accad. Naz. Lincei Rendic. **35**, 279—281.

Reiss, M.
1859 Über eine Steinersche combinatorische Aufgabe, welche im 45sten Bande dieses Journals, Seite 181, gestellt worden ist. J. reine angew. Math. **56**, 326—344.

Repphun, K.
1965 Geometrische Eigenschaften vollständiger Orthomorphismensysteme von Gruppen. Math. Z. **89**, 206—212.

Richardson, M.
1956 On finite projective games. Proc. Amer. Math. Soc. **7**, 458—465.

Rickart, C. E.
1940 The Pascal configuration in a finite projective plane. Amer. Math. Monthly **47**, 89—96.

Rigby, J. F.
1965 Affine subplanes of finite projective planes. Canad. J. Math. **17**, 977—1014.

RINGEL, G.
1957 Über Geraden in allgemeiner Lage. Elem. Math. **12**, 75—82.
1959 Färbungsprobleme auf Flächen und Graphen. Dtsch. Verl. d. Wiss., Berlin.

RIORDAN, J.
1958 An introduction to combinatorial analysis. Wiley, New York.

RODRIQUEZ, G.
1959 Sui quasicorpi distributivi finiti. Atti Accad. Naz. Lincei **26**, 458—465.
1964 Le sub-collineazioni nei piani di traslazione sopra i quasicorpi associativi. Matematiche (Catania) **19**, 11—18.

ROSATI, L. A.
1957a L'equazione delle 27 rette delle superficie cubica generale in un corpo finito. I. Boll. Un. Mat. Ital. **12**, 612—626.
1957b Piani proiettivi desarguesiani non ciclici. Boll. Un. Mat. Ital. **12**, 230—240.
1957c Sui piani desarguesiani affini "non-ciclici". Atti Accad. Naz. Lincei **22**, 443—449.
1958a L'equazione delle 27 rette delle superficie cubica generale in un corpo finito. II. Boll. Un. Mat. Ital. **13**, 84—89.
1958b I gruppi di collineazioni dei piani di Hughes. Boll. Un. Mat. Ital. **13**, 505—513.
1960a Unicità e autodualità dei piani di Hughes. Rend. Sem. Mat. Univ. Padova **30**, 316—327.
1960b Su una generalizzazione dei piani di Hughes. Atti Accad. Naz. Lincei Rendic. **29**, 303—308.
1961 Il gruppo delle collineazioni dei piani di Hughes infiniti. Rivista Mat. **2**, 19—24.
1963 Su una nuova classe di piani grafici. Atti Accad. Naz. Lincei Rendic. **35**, 282—284.
1964 Su una nuova classe di piani grafici. Ric. Mat. **13**, 39—55.

ROTH, R.
1964a Collineation groups of finite projective planes. Math. Z. **83**, 409—421.
1964b Flag transitive planes of even order. Proc. Amer. Math. Soc. **15**, 485—490.
1965 Correction to a paper of R. H. Bruck. Trans. Amer. Math. Soc. **119**, 454—456.

ROUSE BALL, W. W.
1962 Mathematical recreations and essays. 11th edition, revised by H. S. M. Coxeter. Macmillan, New York.

ROY, J.; LAHA, R. G.
1956a Classification and analysis of linked block designs. Sankhya **17**, 115—132.
1956b Two associate partially balanced designs involving three replications. Sankhya **17**, 175—184.
1957 On partially balanced linked block designs. Ann. Math. Statist. **28**, 488—493.

ROY, P. M.
1952 A note on the resolvability of balanced incomplete block designs. Calcutta statist. Assoc. Bull. **4**, 130—132.
1953a A note on the relation between BIB and PBIB designs. Science and Culture **19**, 40—41.
1953b A note on the unreduced balanced incomplete block designs. Sankhya **13**, 11—16.
1954a On the method of inversion in the construction of partially balanced incomplete block designs from the corresponding b. i. b. designs. Sankhya **14**, 39—52.

1954b Inversion of incomplete block designs. Bull. Calcutta Math. Soc. **46**, 47—58.
1954c On the relation between b. i. b. and p. b. i. b. designs. J. Ind. Soc. Agric. Statist. **6**, 30—47.

RUSSO, G.
1961 Una classe di decomposizioni tattiche di un piano di Galois. Riv. Mat. Univ. Parma **2**, 183—190.

RYSER, H. J.
1950 A note on a combinatorial problem. Proc. Amer. Math. Soc. **1**, 422—424.
1951 A combinatorial theorem with an application to Latin rectangles. Proc. Amer. Math. Soc. **2**, 550—552.
1952 Matrices with integer elements in combinatorial investigations. Amer. J. Math. **74**, 769—773.
1955 Geometries and incidence matrices. Amer. Math. Monthly **62**, no. 7, part II, pp. 25—31.
1956 Maximal determinants in combinatorial investigations. Canad. J. Math. **8**, 245—249.
1957 Combinatorial properties of matrices of zeros and ones. Canad. J. Math. **9**, 371—377.
1958a The term rank of a matrix. Canad. J. Math. **10**, 57—65.
1958b Inequalities of compound and induced matrices with applications to combinatorial analysis. Ill. J. Math. **2**, 240—253.
1960a Matrices of zeros and ones. Bull. Amer. Math. Soc. **66**, 442—464.
1960b Traces of matrices of zeros and ones. Canad. J. Math. **12**, 463—476.
1960c Compound and induced matrices in combinatorial analysis. Proc. Sympos. Appl. Math. **10**, 149—168.
1963 Combinatorial mathematics. Wiley, New York.

SADE, A.
1948a Enumération des carrés latins du 6me ordre. Marseille.
1948b Enumération des carrés latins. Application au 7me ordre. Marseille.
1951 An omission in Norton's list of 7×7 squares. Ann. Math. Statist. **22**, 306—307.
1958 Groupoïdes orthogonaux. Publ. Math. Debrecen **5**, 229—240.

SALKIND, C. T.
1965 Factorization of $a^{2n} + a^n + 1$. Math. Mag. **38**, 163.

SANDLER, R.
1962a A note on some new finite division ring planes. Trans. Amer. Math. Soc. **104**, 528—531.
1962b Autotopism groups of some finite non-associative algebras. Amer. J. Math. **84**, 239—264.
1962c The collineation groups of some finite projective planes. Portugal. Math. **21**, 189—199.
1964 Some theorems on the automorphism groups of planar ternary rings. Proc. Amer. Math. Soc. **15**, 984—987.

SAVUR, S. R.
1939 A note on the arrangement of incomplete blocks, when $k = 3$ and $\lambda = 1$. Ann. Eugenics **9**, 45—49.

SAXENA, P. N.
1950 A simplified method of enumerating Latin squares by MacMahon's differential operators. J. Ind. Soc. Agric. Statist. **2**, 161—188.

1951 A simplified method of enumerating Latin squares by MacMahon's differential operators. II. J. Ind. Soc. Agric. Statist. **3**, 24—79.

SCAFATI, M.

1958 Sui 6-archi completi di un piano lineare $S_{2,8}$. Convegno Intern. Reticole e Geometrie proiettive. Palermo 1957, pp. 128—132.

SCARPIS, U.

1898 Sui determinanti di valore massimo. Rendic. R. Istit. Lombardo Sci. Lett. **31**, 1441—1446.

SCE, M.

1958 Sui K_q-archi di indice h. Convegno Intern. Reticole e Geometrie proiettive. Palermo 1957, pp. 133—135.

1960 Preliminari ad una teoria aritmetico-gruppale dei k-archi. Rendic. Mat. **19**, 241—291.

SCHELLEKENS, G. L.

1962 On a hexagonic structure. I, II. Indag. Math. **24**, 201—217, 218—234.

SCHLEIERMACHER, A.

1967 Bemerkungen zum Fundamentalsatz der projektiven Geometrie. Math. Z. **99**, 299—304.

SCHÖNHEIM, J.

1964 On coverings. Pacif. J. Math. **14**, 1405—1411.

SCHULZ, R. H.

1967 Über Blockpläne mit transitiver Dilatationsgruppe. Math. Z. **98**, 60—82.

SCHUR, I.

1904 Über die Darstellung der endlichen Gruppen durch gebrochene lineare Substitutionen. J. reine angew. Math. **127**, 20—50.

1907 Untersuchungen über die Darstellungen der endlichen Gruppen durch gebrochene lineare Substitutionen. J. reine angew. Math. **132**, 85—137.

1908 Neuer Beweis eines Satzes von W. Burnside. Jahresb. Deutsche Math. Ver. **17**, 171—176.

SCHÜTZENBERGER, M. P.

1949 A non-existence theorem for an infinite family of symmetrical block designs. Ann. Eugenics **14**, 286—287.

1951 An extension problem in the theory of incomplete block designs. J. Roy. Statist. Soc. (Ser. B) **13**, 120—125.

SCHWAN, W.

1919 Streckenrechnung und Gruppentheorie. Math. Z. **3**, 11—28.

SEGRE, B.

1954 Sulle ovali nei piani lineari finiti. Atti Accad. Naz. Lincei Rendic. **17**, 141—142.

1955a Ovals in a finite projective plane. Canad. J. Math. **7**, 414—416.

1955b Curve razionali normali e k-archi negli spazi finiti. Ann. Mat. Pura Appl. **39**, 357—379.

1957 Sui k-archi nei piani finiti di caratteristica due. Rev. Math. Pures Appl. **2**, 289—300.

1958 Sulle geometrie proiettive finite. Convegno Intern. Reticole e Geometrie proiettive. Palermo 1957, pp. 46—61.

1959a Le geometrie di Galois. Ann. Mat. Pura Appl. **48**, 1—96.

1959b Le geometrie di Galois. Archi ed ovali; calotte ed ovaloidi. Conferenze Sem. Mat. Univ. Bari **43—44**, 32 pp.

1959c On complete caps and ovaloids in three-dimensional Galois spaces of characteristic two. Acta Arithm. **5**, 315—332.

1959d Intorno alla geometria di certi spazi aventi un numero finito di punti. Archimede **9**, 1—15.

1960a On Galois geometries. Proc. Intern. Congr. Math. Edinburgh 1958, pp. 488—499. Cambridge Univ. Press.

1960b Gli spazi grafici. Rend. Sem. Mat. Fis. Milano **30**, 223—241.

1961a Lectures on modern geometry. Cremonese, Roma.

1961b Alcune questioni su insieme finiti di punti in geometria algebrica. Atti conv. Intern. Geometria Algebrica Torino, pp. 15—33.

1962a Ovali e curve σ nei piani di Galois di caratteristica due. Atti Accad. Naz. Lincei Rendic. **32**, 785—790.

1962b Geometry and algebra in Galois spaces. Abh. Hamburg **25**, 129—132.

1964a Arithmetische Eigenschaften von Galois-Räumen. I. Math. Ann. **154**, 195—256.

1964b Teoria di Galois, fibrazioni proiettive e geometrie non desarguesiane. Ann. Mat. Pura Appl. **64**, 1—76.

1965a Forme e geometrie hermitiane, con particolare riguardo al caso finito. Ann. Mat. Pura Appl. **70**, 1—202.

1965b Istituzioni di geometria superiore. Lecture Notes, 3 vol.'s, Univ. di Roma.

SEIDEN, E.

1950 A theorem in finite projective geometry and an application to statistics. Proc. Amer. Math. Soc. **1**, 282—286.

1961 On a geometrical method of construction of partially balanced designs with two associate classes. Ann. Math. Statist. **32**, 1177—1180.

1963a A supplement to Parker's "Remarks on balanced incomplete block designs". Proc. Amer. Math. Soc. **14**, 729—730.

1963b A method of construction of resolvable BIBD. Sankhya **25**, 393—394.

1963c On necessary conditions for the existence of some symmetrical and unsymmetrical triangular PBIB designs and BIB designs. Ann. Math. Statist. **34**, 348—351.

SHAH, B. V.

1959 On a generalization of the Kronecker product designs. Ann. Math. Statist. **30**, 48—54.

SHAH, S. M.

1963 On the upper bound for the number of blocks in balanced incomplete block designs having a given number of treatments common with a given block. J. Ind. Statist. Assoc. **1**, 219—220.

1964 An upper bound for the number of disjoint blocks in certain PBIB designs. Ann. Math. Statist. **35**, 398—407.

1965 Bounds for the number of common treatments between any two blocks of certain PBIB designs. Ann. Math. Statist. **36**, 337—342.

SHAUB, H. C.; SCHOONMAKER, H. E.

1931 The Hessian configuration and its relation to the group of order 216. Amer. Math. Monthly **38**, 388—393.

SHRIKHANDE, S. S.

1950 The impossibility of certain symmetrical balanced incomplete block designs. Ann. Math. Statist. **21**, 106—111.

1951a On the nonexistence of certain difference sets for incomplete group designs. Sankhya **11**, 183—184.

1951b On the nonexistence of affine resolvable balanced incomplete block designs. Sankhya **11**, 185—186.

1951c The impossibility of certain affine resolvable balanced incomplete block designs. Ann. Math. Statist. **22**, 609.

1952 On the dual of some balanced incomplete block designs. Biometrics **8**, 66—72.

1953a Cyclic solutions of symmetrical group divisible designs. Calcutta Statist. Assoc. Bull. **5**, 36—39.

1953b The non-existence of certain affine resolvable balanced incomplete block designs. Canad. J. Math. **5**, 413—420.

1954 Affine resolvable balanced incomplete block designs and non-singular group divisible designs. Calcutta Statist. Ass. Bull. **5**, 139—141.

1959a On a characterization of the triangular association scheme. Ann. Math. Statist. **30**, 39—47.

1959b The uniqueness of the L_2 association scheme. Ann. Math. Statist. **30**, 781—798.

1962 On a two-parameter family of balanced incomplete block designs. Sankhya **24**, 33—40.

1963a A note on finite Euclidean plane over $GF(2^n)$. J. Ind. Statist. Assoc. **1**, 48—49.

1963b Some recent developments on mutually orthogonal Latin squares. Math. Student **31**, 167—177.

1964 Generalized Hadamard matrices and orthogonal arrays of strength two. Canad. J. Math. **16**, 736—740.

1965 On a class of partially balanced incomplete block designs. Ann. Math. Statist. **36**, 1807—1814.

SHRIKHANDE, S. S.; BHAGAVANDAS

1965 Duals of incomplete block designs. J. Ind. Statist. Assoc. **3**, 30—37.

SHRIKHANDE, S. S.; RAGHAVARAO, D.

1964 A note on the nonexistence of symmetric balanced incomplete block designs. Sankhya **26**, 91—92.

SHRIKHANDE, S. S.; SINGH, N. K.

1962 On a method of constructing symmetrical balanced incomplete block designs. Sankhya **24**, 25—32.

1963 A note on balanced incomplete block designs. J. Indian. Statist. Assoc. **1**, 97—101.

SILVERMAN, R.

1960 A metrization for power sets with applications to combinatorial analysis. Canad. J. Math. **12**, 158—176.

SIMS, C. C.

1967 Graphs and finite permutation groups. Math. Z. **95**, 76—86.

SINGER, J.

1938 A theorem in finite projective geometry and some applications to number theory. Trans. Amer. Math. Soc. **43**, 377—385.

1942 A pair of generators for the simple group $LF(3, 3)$. Amer. Math. Monthly **49**, 668—670.

1960 A class of groups associated with Latin squares. Amer. Math. Monthly **67**, 235—240.

SINGH, N. K.; PANDEY, K. N.

1964 Impossibility of some partially balanced incomplete block designs. J. Sci. Res. Banaras Hindu Univ. **14**, No. 2, 180—188.

SINGH, N. K.; SHUKLA, G. C.
1963 The non-existence of some partially balanced incomplete block designs with three associate classes. J. Indian Statist. Assoc. **1**, 71—77.

SINGH, N. K.; SINGH, K. N.
1964 The non-existence of some partially balanced incomplete block designs with three associate classes. Sankhya **26**, 239—250.

SINGLETON, R. R.
1966 On minimal graphs of maximal even girth. J. Combin. Theory **1**, 306—332.

SKOLEM, T.
1931 Über einige besondere Tripelsysteme mit Anwendung auf die Reproduktion gewisser Quadratsummen bei Multiplikation. Norsk Mat. Tidskr. **13**, 41—51.
1958 Some remarks on the triple systems of Steiner. Math. Scand. **6**, 273—280.

SKORNYAKOV, L. A.
1949 Natural domains of Veblen-Wedderburn projective planes. Isv. Akad. Nauk SSSR **13**, 447—472. Amer. Math. Soc. Transl. No. 58 (1951).
1950 Alternative fields. Ukrain. Math. J. **2**, 70—85.
1951a Right-alternative fields. Isv. Akad. Nauk SSSR **15**, 177—184.
1951b Projective planes. Uspehi Mat. Nauk **6**, 112—154. Amer. Math. Soc. Transl. No. 99 (1953).
1957 Homomorphisms of projective planes and T-homomorphisms of ternaries. Mat. Sbornik **43**, 285—294.

SPENCER, J. C. D.
1960 On the Lenz-Barlotti classification of projective planes. Quart. J. Math. **11**, 241—257.

SPERNER, E.
1960 Affine Räume mit schwacher Inzidenz und zugehörige algebraische Strukturen. J. reine angew. Math. **204**, 205—215.

SPROTT, D. A.
1954 A note on balanced incomplete block designs. Canad. J. Math. **6**, 341—346.
1955a Balanced incomplete block designs and tactical configurations. Ann. Math. Statist. **26**, 752—758.
1955b Some series of partially balanced incomplete block designs. Canad. J. Math. **7**, 369—381.
1956 Some series of balanced incomplete block designs. Sankhya **17**, 185—192.
1959 A series of symmetrical group divisible incomplete block designs. Ann. Math. Statist. **30**, 249—251.
1962 Listing of BIB designs from $r = 16$ to $r = 20$. Sankhya **24**, 203—204.
1964 Generalizations arising from a family of difference sets. J. Ind. Soc. Agric. Statist. **2**, 197—209.

STANTON, R. G.
1957 A note on BIBDs. Ann. Math. Statist. **28**, 1054—1055.

STANTON, R. G.; SPROTT, D. A.
1958 A family of difference sets. Canad. J. Math. **10**, 73—77.
1964 Block intersections in balanced incomplete block designs. Canad. Math. Bull. **7**, 539—548.

v. STAUDT, G. K. C.
1856 Beiträge zur Geometrie der Lage. Vol. I., Nürnberg.

Steck, M.

1936 Über finite Geometrien und ihren Zusammenhang mit der Axiomatik der projektiven Geometrie. Deutsche Math. **1**, 578—588.

1938 Grundlegung einer Theorie der reellen Inzidenzabbildungen in endlichen Geometrien. I. Eine geometrische Deutung der zyklischen Gruppen. J. reine angew. Math. **179**, 37—64.

Steiner, J.

1853 Combinatorische Aufgabe. J. reine angew. Math. **45**, 181—182.

Stern, E.

1939 General formulas for the number of magic squares belonging to certain classes. Amer. Math. Monthly **46**, 555—581.

1941 Über eine zahlentheoretische Methode zur Bildung und Anzahlbestimmung neuartiger lateinischer Quadrate. Bull. Sci. École Polytechn. Timisoara **10**, 101—131.

Stettler, R.

1950 Über endliche Geometrien. Ann. Acad. Sci. Fenn. No. 72, 45 pp.

Strickler, W.

1955 Über die endlichen klein-desarguesschen Zahlsysteme. Dissertation, Zürich, 33 pp.

Straus, E. G.; Taussky, O.

1956 Remarks on the preceding paper. Algebraic equations satisfied by roots of natural numbers. Pacif. J. Math. **6**, 97—98.

Suzuki, M.

1951 A characterization of the simple groups $LF(2, p)$. J. Fac. Sci. Univ. Tokyo **6**, 259—293.

1959a On characterizations of linear groups, I. Trans. Amer. Math. Soc. **92**, 191—204.

1959b On characterizations of linear groups, II. Trans. Amer. Math. Soc. **92**, 205—219.

1960a A new type of simple groups of finite order. Proc. Nat. Acad. Sci. USA **46**, 868—870.

1960b Investigations on finite groups. Proc. Nat. Acad. Sci. USA **46**, 1611—1614.

1961a On a finite group with a partition. Arch. Math. **12**, 241—254.

1961b Finite groups with nilpotent centralizers. Trans. Amer. Math. Soc. **99**, 425—470.

1962a On a class of doubly transitive groups. Ann. Math. **75**, 105—145.

1962b On generalized (ZT)-groups. Arch. Math. **13**, 199—202.

1962c On characterizations of linear groups III. Nagoya Math. J. **21**, 159—183.

1963a A class of doubly transitive permutation groups. Proc. Internat. Congr. Math. 1962 Stockholm, pp. 285—287.

1963b Two characteristic properties of (ZT)-groups. Osaka Math. J. **15**, 143—150.

1964a On a class of doubly transitive groups: II. Ann. Math. **79**, 514—589.

1964b Finite groups of even order in which Sylow 2-groups are independent. Ann. Math. **80**, 58—77.

1965a A characterization of the 3-dimensional projective unitary group over a finite field of odd characteristic. J. Algebra **2**, 1—14.

1965b Finite groups in which the centralizer of any element of order 2 is 2-closed. Ann. Math. **82**, 191—212.

Swift, J. D.

1964 Chains and graphs of Ostrom planes. Pacif. J. Math. **14**, 353—362.

1965 Existence and construction of non-desarguesian projective planes. Semin. 1962/63 Anal. Alg. Geom. Topol., Vol. 1, Ist. Naz. Alta Mat. Roma, pp. 153—163.

SYLVESTER, J. J.
1867 Thoughts on inverse orthogonal matrices etc. Phil. Mag. (4) **34**, 461—475. See also Coll. Math. Papers, vol. II, pp. 615—628, Cambridge Univ. Press 1908.

SZAMKOŁOWICZ, L.
1962 On the problem of existence of finite regular planes. Colloq. Math. **9**, 245—250.
1963 Remarks on finite regular planes. Colloq. Math. **10**, 31—37.
1964 Sur une classification des triplets de Steiner. Atti Accad. Naz. Lincei Rendic. **36**, 125—128.

TAKEUCHI, K.
1962 A table of difference sets generating balanced incomplete block designs. Rev. Inst. Internat. Statist. **30**, 361—366.
1963 On the construction of a series of BIB designs. Rep. Statist. Appl. Res. Un. Japan Sci. Engrs. **10**, 226.

TALLINI, G.
1956a Sulle k-calotte di uno spazio lineare finito. Ann. Mat. Pura Appl. **42**, 119—164.
1956b Sulle k-calotte degli spazi lineari finiti. I, II. Atti Accad. Naz. Lincei Rendic. **20**, 311—317, 442—446.
1957a Sui q-archi di un piano lineare finito di caratteristica $p = 2$. Atti Accad. Naz. Lincei Rendic. **23**, 242—245.
1957b Caratterizzazione grafica delle quadriche ellitiche negli spazi finiti. Rendic. Mat. **16**, 328—351.
1958 Una proprietà grafica caratteristica della superficie di Veronese negli spazi finiti. I, II. Atti Accad. Naz Lincei Rendic. **24**, 19—23, 135—138.
1959 Caratterizzazione grafica di certe superficie cubiche di $S_{3,q}$. Atti Accad. Naz. Lincei Rendic. **26**, 484—489, 644—648.
1960 Le geometrie di Galois e le loro applicazioni alla statistica e alla teoria dell'informazione. Rendic. Mat. **19**, 379—400.
1961a On caps of kind s in a Galois r-dimensional space. Acta Arithm. **7**, 19—28.
1961b Sulle ipersuperficie irreducibili d'ordine minimo che contengono tutti i punti di uno spazio di Galois $S_{r,q}$. Rendic. Mat. **20**, 431—479.
1964 Calotte complete di $S_{4,q}$ contenenti due quadriche ellittiche quali sezioni iperpiane. Rendic. Mat. **23**, 108—123.

TALLINI SCAFATI, M.
1964 Archi completi in un $S_{2,q}$, con q pari. Atti Accad. Naz. Lincei Rendic. **37**, 48—51.

TARRY, G.
1900 Le problème des 36 officiers. C. R. Assoc. Franç. Avanc. Sci. nat. **1**, 122—123.
1901 Le problème des 36 officiers. C. R. Assoc. Franç. Avanc. Sci. nat. **2**, 170—203.

TAUSSKY, O.
1960 Matrices of rational integers. Bull. Amer. Math. Soc. **66**, 327—345.
1961 Commutators of unitary matrices which commute with one factor. J. Math. Mech. **10**, 175—178.

THOMPSON, H. R.
1956 On a new class of partially balanced incomplete block designs. Calcutta Statist. Assoc. Bull. **6**, 193—195.

THOMPSON, J. G.

1960a Normal p-complements for finite groups. Math. Z. **72**, 332—354.

1960b A special class of non-solvable groups. Math. Z. **72**, 458—462.

1964 Normal p-complements for finite groups. J. Algebra **1**, 43—46.

THOMPSON, W. A.

1958 A note on P.B.I.B. design matrices. Ann. Math. Statist. **29**, 919—922.

TITS, J.

1952 Généralisations des groupes projectifs basées sur leurs propriétés de transitivité. Mem. Acad. Roy. Belg. **27** (2) 115 pp.

1956 Sur les analogues algébriques des groupes semi-simples complexes. Colloque d'algèbre supérieure, Bruxelles 1956, p. 261—289.

1959 Sur la trialité et certains groupes qui s'en déduisent. Publ. Math. I.H.E.S. Paris **2**, 14—60.

1960 Les groupes simples de Suzuki et de Ree. Séminaire Bourbaki, 3e année, No. 210.

1962a Ovoïdes à translations. Rendic. Mat. **21**, 37—59.

1962b Ovoïdes et groupes de Suzuki. Arch. Math. **13**, 187—198.

1962c Géométries polyédriques et groupes simples. 2e Reunion Math. d'expression latine (Firenze-Bologna 1961). Cremonese, Roma, pp. 66—88.

1963 Groupes simples et géométries associées. Proc. Intern. Congress Math. 1962 Stockholm, pp. 197—221.

1964a Géométries polyédriques finies. Rendic. Mat. **23**, 156—165.

1964b Sur les systèmes de Steiner associés aux trois "grands" groupes de Mathieu. Rendic. Mat. **23**, 166—184.

1966 Une propriété caractéristique des ovoïdes associés aux groupes de Suzuki. Arch. Math. **17**, 136—153.

1967 Le groupe de Janko d'ordre 604, 800. Preprint of forthcoming publication, Univ. Bonn.

TODD, J. A.

1933 A combinatorial problem. J. Math. Phys. Mass. Inst. Tech. **12**, 321—333.

1959 On representations of the Mathieu groups as collineation groups. J. London Math. Soc. **34**, 406—416.

TREHAN, A. M.

1963 On the bounds of the common treatments between blocks of balanced incomplete block designs. J. Indian Statist. Assoc. **1**, 102—103.

TSUZUKU, T.

1966 On primitive extensions of rank 3 of symmetric groups. Nagoya Math. J. **27**, 171—177.

TURYN, R. J.

1964 The multiplier theorem for difference sets. Canad. J. Math. **16**, 386—388.

1965a Character sums and difference sets. Pacif. J. Math. **15**, 319—346.

1965b The nonexistence of seven difference sets. Ill. J. Math. **9**, 590—594.

VARTAK, M. N.

1955 On an application of Kronecker product of matrices to statistical designs. Ann. Math. Statist. **26**, 420—438.

1959 The non-existence of certain PBIB designs. Ann. Math. Statist **30**, 1051—1062.

VEBLEN, O.; BUSSEY, N. J.

1906 Finite projective geometries. Trans. Amer. Math. Soc. **7**, 241—259.

VEBLEN, O.; WEDDERBURN, J. H. M.
1907 Non-Desarguesian and non-Pascalian geometries. Trans. Amer. Math. Soc. **8**, 379—388.

VEBLEN, O.; YOUNG, J. W.
1916 Projective geometry, 2 vol.'s. Ginn & Co., Boston.

VIJAYARAGHAVAN, T.; CHOWLA, S.
1945 Short proofs of theorems of Bose and Singer. Proc. Nat. Acad. Sci. India Sect. A, **15**, 194.

VRIES, J. DE
1936 Konfigurationen von Punkten und Kreisen. Akad. Wetensch. Amsterdam, Proc. **39**, 486—488.

VAN DER WAERDEN, B. L.; SMID, L. J.
1935 Eine Axiomatik der Kreisgeometrie und der Laguerre-Geometrie. Math. Ann. **110**, 753—776.

WAGNER, A.
1956 On finite non-desarguesian planes generated by 4 points. Arch. Math. **7**, 23—27.
1958 On projective planes transitive on quadrangles. J. London Math. Soc. **33**, 25—33.
1959 On perspectivities of finite projective planes. Math. Z. **71**, 113—123.
1961 On collineation groups of finite projective spaces. I. Math. Z. **76**, 411—426.
1962a On involutions of projective planes. Arch. Math. **13**, 529—534.
1962b Perspectivities and the little projective group. Proc. Colloq. Found. of Geometry Utrecht 1959; Pergamon 1962, pp. 199—208.
1964 A theorem on doubly transitive permutation groups. Math. Z. **85**, 451—453.
1965 On finite affine line transitive planes. Math. Z. **87**, 1—11.
1966 Normal subgroups of triply-transitive permutation groups of odd degree. Math. Z. **94**, 219—222.

WALKER, A. G.
1947 Finite projective geometry. Edinburgh Math. Notes **36**, 12—17.

WALKER, R. J.
1963 Determination of division algebras with 32 elements. Proc. Symp. Appl. Math. **15**, 83—85.

WAN, Cheh-Hsian
1964 Notes on finite geometries and the constructions of PBIB designs. I. Some "Anzahl" theorems in symplectic geometry over finite fields. II. Some PBIB designs with two associate classes based on the symplectic geometry over finite fields. Sci. Sinica **13**, 515—516, 516—517.
1965 Notes on finite geometries . . . VI. Some association schemes and PBIB designs based on finite geometries. Sci. Sinica **14**, 1872—1876.

WAN, Cheh-Hsian; YANG, Ben-Fu
1964 Notes on finite geometries and the construction of PBIB designs. III. Some "Anzahl" theorems in unitary geometry over finite fields and their applications. Sci. Sinica **13**, 1006—1007.

WANG, Yuan
1964 A note on the maximal number of pairwise orthogonal Latin squares of a given order. Sci. Sinica **13**, 841—843.

WATSON, G. L.
1960 Integral quadratic forms. Cambridge Univ. Press.

WEDDERBURN, J. H. M.
1905 A theorem on finite algebras. Trans. Amer. Math. Soc. **6**, 349—352.

WEISS, E. A.
1931 Über ein Bild der R_4-Konfiguration $(10_6, 15_4)$ im Linienraum. J. reine angew. Math. **164**, 256—258.
1936 Die geschichtliche Entwicklung der Lehre von der Geraden-Kugel-Transformation. I. Deutsche Math. **1**, 23—37.

WESSON, J. R.
1955 Finite plane projective geometries. Amer. Math. Monthly **62**, No. 7, part II, pp. 32—40.
1957 On Veblen-Wedderburn systems. Amer. Math. Monthly **64**, 631—635.
1966 The construction of projective planes from generalized ternary rings. Amer. Math. Monthly **73**, 36—40.

WHITE, A. S.; COLE, F. N.; CUMMINGS, L. D.
1925 Complete classification of triad systems on fifteen elements. Memoirs Nat. Acad. Sci. **14**, 2nd memoir, pp. 1—89.

WHITEMAN, A. L.
1957 The cyclotomic numbers of order sixteen. Trans. Amer. Math. Soc. **86**, 401—413.
1960a The cyclotomic numbers of order ten. Proc. Amer. Math. Soc. **10**, 95—111.
1960b The cyclotomic numbers of order twelve. Acta Arithm. **6**, 53—76.
1962 A family of difference sets. Ill. J. Math. **6**, 107—121.

WIELANDT, H.
1951 Über das Produkt paarweise vertauschbarer nilpotenter Gruppen. Math. Z. **55**, 1—7.
1954 Zum Satz von Sylow. Math. Z. **60**, 407—408.
1956 Primitive Permutationsgruppen vom Grad 2p. Math. Z. **63**, 478—485.
1958 Über Produkte von nilpotenten Gruppen. Ill. J. Math. **2**, 611—618.
1959a Ein Beweis für die Existenz der Sylowgruppen. Arch. Math. **10**, 401—402.
1959b Zum Satz von Sylow. II. Math. Z. **71**, 461—462.
1960 Über den Transitivitätsgrad von Permutationsgruppen. Math. Z. **74**, 297—298.
1962 Gedanken für eine allgemeine Theorie der Permutationsgruppen. Rendic. Sem. Math. Univ. Politecn. Torino **21**, 31—39.
1964 Finite permutation groups. Academic Press, New York and London.
1967 Endliche k-homogene Permutationsgruppen. Math. Z. **101**, 142.

WIELANDT, H.; HUPPERT, B.
1958 Normalteiler mehrfach transitiver Permutationsgruppen. Arch. Math. **9**, 18—26.

WILLIAMSON, J.
1944 Hadamard's determinant theorem and the sum of four squares. Duke Math. J. **11**, 65—81.
1947 Note on Hadamard's determinant theorem. Bull. Amer. Math. Soc. **53**, 608—613.

WINTERNITZ, A.
1940 Zur Begründung der projektiven Geometrie: Einführung idealer Elemente unabhängig von der Anordnung. Ann. Math. **41**, 365—390.

WITT, E.

1931 Über die Kommutativität endlicher Schiefkörper. Abh. Hamburg **8**, 413.

1937 Theorie der quadratischen Formen in beliebigen Körpern. J. reine angew. Math. **176**, 31—44.

1938a Die 5fach transitiven Gruppen von Mathieu. Abh. Hamburg **12**, 256—264.

1938b Über Steinersche Systeme. Abh. Hamburg **12**, 265—275.

WOLF, J. A.

1967 Spaces of constant curvature. McGraw-Hill, New York.

WYLER, O.

1953 Incidence geometry. Duke Math. J. **20**, 601—610.

YAMAMOTO, K.

1954 Euler squares and incomplete Euler squares of even degrees. Mem. Fac. Sci. Kyushyu Univ. **8**, 161—180.

1961 Generation principles of Latin squares. Bull. Inst. Intern. Statist. **38**, 73—76.

1963 Decomposition fields of difference sets. Pacif. J. Math. **13**, 337—352.

1965 On an orthogonal basis of the eigenspaces associated with partially balanced incomplete block designs of a Latin square type association scheme. Mem. Fac. Sci. Kyushyu Univ. **19**, 99—104.

YANG, Ben-Fu

1965 Studies in finite geometries and the construction of incomplete block designs. VII. An association scheme with many associate classes constructed from maximal completely isotropic subspaces in symplectic geometry over finite fields. VIII. An association scheme with many associate classes constructed from maximal completely isotropic subspaces in unitary geometry over finite fields. Acta Math. Sinica **15**, 812—825, 826—841. Also in Chinese Math. **7**, p. 547—576.

YANG, Chung-Tao

1947 Certain chains in a finite projective geometry. Acad. Sinica Sci. Rec. **2**, 44—46.

1948a Certain chains in a finite projective geometry. Duke Math. J. **15**, 37—47.

1948b Projective collineations in a finite projective plane. Acad. Sinica Sci. Rec. **2**, 157—164.

YAQUB, J. C. D. S.

1961a On projective planes of class III. Arch. Math. **12**, 146—150.

1961b The existence of projective planes of class I 3. Arch. Math. **12**, 374—381.

1966a On two theorems of LÜNEBURG. Arch. Math. **17**, 485—488.

1966b On projective planes of Lenz-Barlotti class I 6. Math. Z. **95**, 60—70.

1967a The nonexistence of finite projective planes of Lenz-Barlotti class III 2. Arch. Math. **18**, 308—312.

1967b The Lenz-Barlotti classification. Proc. Proj. Geometry Conference, Univ. of Illinois, Chicago, pp. 129—160.

1968 On the group of projectivities on a line in a finite projective plane. Math. Z. **104**, 247—248.

YATES, F.

1933 The formation of Latin squares for use in field experiments. Empire J. experim. agric. **1**, 235—244.

1936 Incomplete randomized blocks. Ann. Eugenics **7**, 121—140.

YOUDEN, W. J.

1951 Linked blocks: a new class of incomplete block designs. Biometrics **7**, 124.

ZACHARIAS, M.

1941 Untersuchungen über ebene Konfigurationen (12_4, 16_3). Deutsche Math. **6**, 147—170.

1948 Eine neue ebene Konfiguration (12_4, 16_3). Math. Nachr. **1**, 332—336.

1949 Neue Wege zur Hesseschen Konfiguration. Math. Nachr. **2**, 163—170.

1951a Streifzüge im Reich der Konfigurationen: Eine Reyesche Konfiguration (15_3), Stern- und Kettenkonfigurationen. Math. Nachr. **5**, 329—345.

1951b Die ebenen Konfigurationen (10_3). Math. Nachr. **6**, 129—144.

1952 Konstruktionen der ebenen Konfigurationen (12_4, 16_3). Math. Nachr. **8**, 1—6.

1953 Über eine mit der Bydzovskyschen Konfiguration (12_4, 16_3) verbundene Hessesche Konfiguration. Math. Nachr. **10**, 187—196.

ZADDACH, A.

1956 Über Anti-Fano-Ebenen. Math. Z. **65**, 353—388.

1957 Bemerkungen über spezielle Anti-Fano-Ebenen. Arch. Math. **7**, 425—429.

ZAIDI, N. H.

1963 Symmetrical balanced incomplete block designs with $\lambda = 2$ and $k = 9$. Bull. Calcutta Math. Soc. **55**, 163—167.

ZAPPA, G.

1953 Sui piani grafici transitivi e quasi-transitivi. Ric. Mat. **2**, 274—287.

1954a Sui piani grafici finiti h-l-transitivi. Boll. Un. Mat. Ital. **9**, 16—24.

1954b Sulle omologie dei piani h-l-transitivi e dei piani su quasicorpi. Ric. Mat. **3**, 35—39.

1957 Sui gruppi di collineazioni dei piani di Hughes. Boll. Un. Mat. Ital. **12**, 507—516.

1958 Piani affini finiti con traslazioni. Ric. Mat. **7**, 241—253.

1960 Piani grafici a caratteristica 3. Ann. Mat. Pura Appl. **49**, 157—166.

1964a Sui piani quasi di traslazioni secondo Lingenberg. Rendic. Mat. **23**, 124—127.

1964b Sugli spazi generali quasi di traslazione. Matematiche (Catania) **19**, 127—143.

ZASSENHAUS, H.

1935a Über endliche Fastkörper. Abh. Hamburg **11**, 187—220.

1935b Kennzeichnung endlicher linearer Gruppen als Permutationsgruppen. Abh. Hamburg **11**, 17—40.

1935c Über transitive Erweiterungen gewisser Gruppen aus Automorphismen endlicher mehrdimensionaler Geometrien. Math. Ann. **111**, 748—759.

1952 A group-theoretic proof of a theorem of Maclagan Wedderburn. Proc. Glasgow Math. Assoc. **1**, 53—63.

ZELEN, M.

1954 A note on partially balanced designs. Ann. Math. Statist. **25**, 599—602.

ZEMMER, J. L.

1961 On a class of doubly transitive groups. Proc. Amer. Math. Soc. **12**, 644—650.

1964 Nearfields, planar and non-planar. Math. Student. **32**, 145—150.

ZORN, M.

1931 Theorie der alternativen Ringe. Abh. Hamburg **8**, 123—147.

1933 Alternativkörper und quadratische Systeme. Abh. Hamburg **9**, 395—402.

Dictionary

Terminology of this book	Other terms frequently used[1])
Absolute (elements of a polarity)	Self-conjugate (BACHMANN 1959, LENZ 1965)
Affine design	Affine resolvable balanced incomplete block design (BOSE 1942a, SHRIKHANDE 1951b)
Bundle (in inversive plane)	Elliptic pencil (COXETER 1966)
Cartesian group	Cartesian number system (BAER 1942)
Central collineation	Perspective collineation (VEBLEN & YOUNG 1916)
Collineation	Projectivity (BAER 1952)
Concircular	Concyclic (BENZ 1960a, COXETER 1966)
Design	Balanced incomplete block design (BOSE 1939); BIB design
Divisible (partial design)	Group divisible (BOSE & SHIMAMOTO 1952)
Egglike (inversive plane)	Ovoidal (DEMBOWSKI 1964b)
Exceptional nearfield	Dickson nearfield of order 9 (KARZEL 1965c)
Flock (in inversive plane)	Hyperbolic pencil (COXETER 1966)
Hall multiplier	Numerical multiplier
Hjelmslev plane	Ebene mit Nachbarelementen (KLINGENBERG 1954)
Inversive plane	Möbiusebene im engeren Sinne (BENZ 1960a); Möbius-plane (LÜNEBURG 1964b); conformal plane.
Irregular (nearfield)	Exceptional
Knot (of oval in plane of even order)	Nucleus (SEGRE 1961a)
Linear ternary field	Linear planar ternary ring (HUGHES 1955c); Ternärkörper mit 1. Zerlegbarkeitsbedingung (PICKERT 1955)

[1]) The references given in this column are convenient samples; they are not meant to designate the historical origin of the terms in question.

Nest (of subspaces in a projective geometry)	Flag (HIGMAN 1962)
Norm residue symbol	Hilbert symbol (JONES 1950)
Oval (in plane of order n)	$(n+1)$-arc (SEGRE 1961a)
Ovoid (in $\mathscr{P}(3, q)$, $q > 2$)	Ovaloid (SEGRE 1959c)[1]
Pappus (Theorem of)	Pascal (HILBERT 1899)
Parallelism, design with	Resolvable balanced incomplete block design (BOSE 1942a)
Partial design	Partially balanced incomplete block design (BOSE & NAIR 1939); PBIB design
Partial plane	Inzidenzstruktur (PICKERT 1955)
Pencil	Tangent bundle (Berührbüschel, BENZ 1960a); parabolic pencil (COXETER 1966)
Projective design	Symmetric balanced incomplete block design (BOSE 1939); symmetric BIB design; (v, k, λ)-configuration (RYSER 1963); λ-plane (BRUCK 1955); λ-space (DEMBOWSKI & WAGNER 1960)
Quasifield	Veblen-Wedderburn system (HALL 1943, 1959); VW-system
Regular nearfield	Dickson nearfield (KARZEL 1965c)
Regular (permutation group)	Semiregular (WIELANDT 1964)
Semifield	Distributive quasifield (PICKERT 1955); non-associative division ring (HALL 1959)
Semi-translation plane	Normal semi-translation plane (PICKERT 1965a)
Sesquilinear	Semi-bilinear (BAER 1952)
Spread	Fibration (SEGRE 1964b); congruence
Symplectic polarity	Null system (BAER 1952)
Ternary field	Ternary ring (HALL 1943, 1959); planar ternary ring (HUGHES 1955c)

[1] Segre's definition of "ovaloid" differs essentially from that of "ovoid", but it can be shown that in $\mathscr{P}(3, q)$, $q > 2$, both terms are equivalent.

Special notations

m	number of blocks in parallel class	71
(M)	Theorem of Miquel	255
$\mathbf{M}(q)$	miquelian inversive plane of order q	104, 258
(N)	Nondegeneracy condition	6
$n = r - \lambda$	order of design	59
$N(n, q)$	regular nearfield of q^n elements, with centre $GF(q)$	34
$N(t) = \prod_{\alpha \in \mathsf{A}} t^\alpha$	norm of t with respect to automorphism group A	232
$o(\alpha)$	order of group element (permutation, automorphism, collineation) α	80, 172
$\mathfrak{O}(p, C)$	set of circles through p orthogonal to C	265
$O_n(q)$, $O_n(q, \varepsilon)$	subgroup of $GL_n(q)$ leaving nondegenerate quadratic form (of index $(n - 1 + \varepsilon)/2$ if n is even) invariant	46
$\mathscr{P}(d, q)$	d-dimensional projective geometry over $GF(q)$	28
$\mathbf{P}_{s,t}(d, q)$	partial design of s- and t-subspaces of $\mathscr{P}(d, q)$	291
$\mathbf{P}_t(d, q) = \mathbf{P}_{0,t}(d, q)$	design of points and t-subspaces of $\mathscr{P}(d, q)$	28
$\mathbf{P}(q) = \mathbf{P}_1(2, q)$	desarguesian projective plane of order q	29
$\mathscr{P}(\mathbf{I})$	set of pencils of inversive plane $\mathbf{I}$	253
PH	collineation group of $\mathscr{P}(d, q)$ induced by subgroup H of $GL_{d+1}(q)$	31, 46
r	number of blocks per point in tactical configuration	5
(R, m), ($\overline{\text{R}}$, n)	regularity conditions	5
$\mathbf{R}(q)$	unital determined by group of Ree 1961b	105
SH	intersection of $SL_n(q)$ with subgroup H of $GL_n(q)$	31, 54
$SL_n(q)$	group of all unimodular automorphisms of $V(n, q)$	31
$\mathbf{S}(q)$	inversive plane connected with $Sz(q)$	104, 275
$Sz(q)$	group of Suzuki 1960a	52
$Sp_n(q)$	subgroup of $GL_n(q)$ leaving antisymmetric form (symplectic polarity) invariant	46
$\mathbf{T}(\Gamma)$	Lenz-Barlotti class of collineation group Γ of projective plane	123
$\mathscr{T}(\Gamma)$	Hering class of automorphism group Γ of inversive plane	261
$\mathfrak{T}(o, e, u, v)$	ternary field determined by quadrangle o, e, u, v in projective plane	128
$\mathfrak{T}(p, C)$	set of circles through p tangent to C	264
$T(p^d, p^m, c)$	quasifield of Albert 1958 ("Twisted field")	242
$\mathbf{T}(q)$	Steiner triple system of Netto 1893	98
$\mathbf{U}(s)$	unital determined by $PSU_3(s^2)$	104
v	number of points of finite incidence structure	3
$V(r, q)$	vector space of rank r over $GF(q)$	28
$\mathbf{W}(q)$	tactical configuration defined by symplectic polarity of $\mathscr{P}(3, q)$	51
$Z(m)$	ring of integers mod m	223

Index

For items with more than one reference, numbers in italics refer to definitions

36. Putnam: Commutation Properties of Hilbert Space Operators and Related Topics. DM 28,—; US $ 7.00
37. Neumann: Varieties of Groups. DM 46,—; US $ 11.50
38. Boas: Integrability Theorems for Trigonometric Transforms. DM 18,—; US $ 4.50
39. Sz.-Nagy: Spektraldarstellung linearcr Transformationen des Hilbertschen Raumes. DM 18,—; US $ 4.50
40. Seligman: Modular Lie Algebras. DM 39,—; US $ 9.75
41. Deuring: Algebren. DM 24,—; US $ 6.00
42. Schütte: Vollständige Systeme modaler und intuitionistischer Logik. DM 24,—; US $ 6.00
43. Smullyan: First-Order Logic. DM 36,—; US $ 9.00
44. Dembowski: Finite Geometries. DM 68; US $ 17.00
45. Linnik: Ergodic Properties of Algebraic Fields. DM 44,—; US $ 11.00
46. Krull: Idealtheorie. DM 28,—; US $ 7.00

Livingstone-Wagner th p.92
Kantor's th p.98
PP10 see p177

721/24/68 — III/18/203